Other books by Larry E. Arnold:

The Parapsychological Impact of the Accident at
Three Mile Island

The Reiki Handbook

THE MYSTERIOUS FIRES OF ABLAZE!

SPONTANEOUS HUMAN COMBUSTION

Larry E. Arnold

M. Evans and Company, Inc.
New York

Text copyright © 1995 by Larry E. Arnold

All photographs and illustrations copyright © 1995 by Larry E. Arnold unless noted otherwise.

All rights reserved. No part of this book may be reproduced or transmitted in any form or by any means without the written permission of the publisher.

M. Evans and Company, Inc.
216 East 49th Street
New York, New York 10017

Selections from *AND THE SUN IS UP: Kundalini Rises in the West* by W. Thomas Wolfe are copyright © 1978 by W. Thomas Wolfe and are used with permission of the publisher, Sun Publishing Co. All rights reserved.

Selections from *LOST IN THE FIRE* by David Clewell are copyright © 1993 by David Clewell and are used by permission of the publisher, Garlic Press. All rights reserved.

Library of Congress Cataloging-in Publication Data
Arnold, Larry E.
    Ablaze! : the mysterious fires of spontaneous human combustion / Larry E. Arnold.
      p. cm.
    ISBN 0-87131-789-3 (cl)
    1. Combustion, Spontaneous human. I. Title.
RA1085.A76 1995
001.9—dc20   95-34593
                                                               CIP

Design by Charles A. de Kay

Typeset by Classic Type, Inc.

Manufactured in the United States of America

9 8 7 6 5 4 3 2 1

Dedicated
to
the quest for discovery
understanding
and truth
wherever it is to be found
wherever it may lead.

CONTENTS

Credits	ix
Preface	xv
Introduction	xvii
1. The Baffling Burning of Dr. Bentley	1
2. The Burning Question: What the Blazin's Going On Here?	13
3. "By the Visitation of God" Did They Die	29
4. Firewater and Demon Flames	38
5. The Verdict Is In?	53
6. Mary Reeser: The Remarkable "Cinder Woman"	73
7. Thermogenesis: The (Quantum) Physics of Fire Within	92
8. The Parade of Personal Fire Marches On	107
9. Burning Down Not Up: "I Thought I'd Just Drop In..."	137
10. SuperHot Humans: Stoking Up the Body's Furnace	152
11. Kundalini, Chi, and Chakras: The Serpent-Fire Within	165
12. Driving the Fire: Human Auto-Cremation	180
13. Eyewitnesses: Seeing Is Believing?	195

14. SHC Fables and Select Myths	213
15. Surviving Self-Incineration: The Angel Who Beat Hell Fire!	227
16. "Light My Fire"... Beating Back These Innermost Flames	237
17. Beamed Up and Burned Out: Close Encounters of the Combustible Kind	263
18. Death-Fire: Searing Suicides and Combusting Corpses	283
19. Art Imitating Life: The Literary Side of SHC	298
20. On the Road to Incredible Incinerations	317
21. Revelations from the Shadowed Past	331
22. Earthfire! The Cartography of Combustion	342
23. Triangles of Fire: Putting Fire-Leynes to the Test	362
24. It's Only a Matter of Time	378
25. The Mott Case: All the Symptoms... and New Puzzles	393
26. More Clues to Human Combustion Conundrums	412
27. Phoenix Fire: Wrapped in the Flames of Ascension	458
Epilogue	469
Notes	471

CREDITS

No book is an island, especially when it is science nonfiction. To effect research that spans the globe, enthusiasm alone is insufficient. Assistance of others is prerequisite. Vital.

Probably a thousand people have contributed in various ways to the twenty-year gestation of *ABLAZE!* To everyone cited in the text, I am grateful for your help; whether in agreement or disagreement, corresponding or speaking with you has been among the rewards of writing this book.

Those who provided leads, facilitated field investigations, offered guidance, expertise or special assistance include Bev Adams; Gene Alexander; James R. Allen, California State Fire Marshal; Dr. Michael Anteski; Dale Auer and the Pennsylvania Cremation Society; John Barracato, Deputy Chief Fire Marshal for New York City; James B. Beal; Boyce Beattey; Dr. Robert O. Becker; Marybeth Beechen; Dr. William E. Belanger; Police Chief Howard Bougherty, Lower Saucon Township, Pennsylvania; Ruth Brown; Tolly and Peggy Burkan; Linda Bush; Ann Cartwright; Robin Champagne; Charla Cohen; Loren Coleman; Amy Collins; Trevor James Constable; William R. Corliss and *The Sourcebook Project*; Vernon (Komar) Craig; Marlene Csandl; Bobbi Curry; Jerry Deardorff; John DeSalvo; Jim Dixon and retired chief Raymond Church, Pensacola Fire Department; Sharon Dogherty; Dr. Irving Feller, Director of Burn Center, University of Michigan; David Fideler; Pascal L. Foucault; Warren Freiberg; Mike Frizzell and The Enigma Project; Frances Garcia; Lorena Gardenshire; Congressman George W. Gekas; John C. Grenoble; Fire Chief Jerry Grimmett of Gilbert, West Virginia; Art Gustin; Deputy Fire Marshal Robert H. Hall of the West Virginia State Fire Marshal's Office; Jo Hamby; Rev. J. Audrey Havice; Professor Gerald Hawkins; Dolores Higgey; Betsey Hooker; Gene Hoy; Young Stephen James; Evelyn Johnson; Wilbur Johnson, Iowa State Fire Marshal; Peter and Barbara Jones; Bill Knell; Rev. Thomas Allen Long; Matthew Lotride; Gary Mangiacopra; Ron Mangravite; Ray Manners; Maryland State Fire Marshal's Office; Sandra

Matuschka; Hannah Mazzaratti; Jack Merz; Mickey McWilliams; James E. Meskill, Philadelphia Deputy Fire Marshal; Ernestine Mitchner; Ken Mouk; Elizabeth and Les Nachman; Dr. Ogden Nash; Carl Natale; Debbie Newman; Joe Nickell; Elizabeth K. Norris; John A. Norris, editor-in-chief, *American Journal of Law & Medicine*; Roxan O'Brien; LaVerne Olsen; Cheryl Paserman; Jerry Pearson; the Pennsylvania State Police Fire Marshal's Office, including former Commissioner Charles Henry, and retired Cpls. Jack Lotwick and William Sweet; Fred Putsche, Jr.; James T. Randall; Diane B. Rayner; John Farrow Reno; Frank Rossi of *The Philadelphia Inquirer*; Professor Beverly Rubik, director of Temple University's Center for Frontier Sciences; Dr. Lee S. Sannella; Helen Schreiber; John F. Schuessler; Professor William R. and Abigail Shearer; Dr. Eleanor Shields; Shirlee G. Snyder; Marilyn Spanvill; Charlotte Speece; Dr. Claus Speth and Ptl. William Wright of Gloucester County, NJ; Brad Steiger; Jim Stoddard; Greg Thayer; Sal Trapani; Frank Tribbe; Howard (Buzz) Triebold; Dr. Douglas Ubelaker; John Wakefield; Lorraine Warren; William Waters of *The Arizona Daily Star*; Novella Watson-Lee, Nevada State Fire Marshal Division; Dan Weinberg; H. Newton Wells; Yonnie Wells; Dr. C. Samuel West; John W. White; Suzanne B. Williams; R. Martin Wolf; Marian Wurster; V. Yuls; Joe Zarzynski.

I appreciate your input and the friendships that have resulted.

My research has crossed national boundaries, as have contributions to it.

In Australia: Chief Librarian D. R. Andrews, *The Australian*; Walter Geerlings; and Joan Parker, *Sydney Morning Herald* Information Services. In Brazil: Alberto Aguas. In Canada: Shelagh Kendal; Dr. Douglass Mills; D. A. Rhydwen, chief librarian, *The Globe and Mail*; H. S. Turner; Dwight Whalen; and X. In Denmark: Dr. Mogens Thomsen, Department of Plastic Surgery and Burns Unit, Hvidovre Hospital, Copenhagen. In France: Michel Granger. In Spain: Instituo Geografico y Catastral of Madrid; Ignatio Darnaude. In Sweden: Dr. Björn Nordenström.

In the United Kingdom: Stuart Blades; G. M. Brooks, editor, *The West Sussex Gazette*; Dr. James Malcolm Cameron, Secretary-General of The British Academy of Forensic Sciences; Peter Christie; Miss J. Crowther and R. G. Roberts of Leisure Services, Humberside County Council; Russell Danby; William Deedes, editor, *The (London) Daily Telegraph*; Louth Fire Chief Alfred Espin; D. Foreman, senior librarian, *The (Newcastle) Journal*; the Rev. Donald Galloway; Dr. David Gee; Assistant Editor Peter Goodman, *The (Sheffield) Star*; S. S. Kind, The Forensic Science Society, North Yorkshire; Her Majesty's Coroner Dr. Gavin Thurston; R. A. Wardale, editor, *Southern Evening Echo*; J. A. Whithead, editor of *Hull Daily Mail*; Her Majesty's Coroner Dr. Richard Whittington.

CREDITS xi

Special acknowledgment goes to *Fortean Times* and its contributors, especially intrepid editor and friend, Robert J. M. Rickard. Also to Phyllis Benjamin, Al Rosenzweig, and Richard Leshuk of the International Fortean Organization (INFO); Robert Warth and the Society for the Investigation of the Unexplained (SITU); Scott Colborn and the Fortean Research Center of Lincoln, Nebraska; and the Foundation of Truth in Atlanta, Georgia.

Without the willingness of fire officials, police and legal officers, physicians and other professionals, plus a very select group of humans who agreed to speak about the pivotal roles they played in "impossible" events, my research would have been significantly diminished. As would be history. Their open, generous cooperation has been invaluable.

Deserving recognition of merit in this category are Jack Angel; Peter Calhoun; John Heymer; retired Tpr. Fred Klages, Pennsylvania State Police Fire Marshal Division; Wilton Marion Krogman for generously recalling the experiences of his distinguished career, especially of Mary Reeser; Fire Marshal Robert C. Meslin; Joseph A. Nuzum; Marilyn O.; Lincolnshire Fire Brigade Commander John Shenton; London Fire Brigade Commander John Stacey; Dr. Edward J. Sullivan, Commander U.S. Navy retired.

In the case of John Bentley: Potter County Historical Society curator Robert Curran, Deputy Coroner John Dec, The Potter Enterprise editor William Fish, Donald E. Gosnell, Richard C. Lindhome, Carl D. McCloud, and Paul C. Toombs.

In the case of Elijah Currin: Fire Chief James E. Anderson, Deputy Fire Chief Bob Norman, Deputy Sheriff Max Fields, Coroner Darr Williamson of Cassville, Missouri.

In the case of Anthony Gianninoto: Fire Marshal Ward and Deputy Fire Marshal Dave Herring, and Maryland Deputy State Fire Marshal Bob Thomas.

In the case of George Mott: staff at Adirondack-Burlington Crematorium, Ticonderoga Fire Department Capt. Newton Brown, Michael Connery, Kimberly Ferguson, Coroner Jack Harland, Mrs. Thomas Labonte, New York State Trooper Richard Lavalley, Kendall Mott, Essex County Emergency Preparedness Director Robert Purdy, Essex County Medical Examiner Dr. C. Frances Varga, and *especially* Anthony T. Morette.

In the Beatrice Oczki case: Fire Marshal Vincent Calcagno, Fire Chief Terrence Droogan, and her son Frank Oczki.

For recollections about the world-famous passing of family member Mary Reeser: Jan Bowen, Mrs. Edward L. Jones, and Dr. Richard Reeser Jr.; also Walter G. Tipton.

Without librarians and the treasures they safeguard, this book would be considerably shorter; research made infinitely more difficult; and history less rich.

I am pleased to mention Dorothy W. Bowers, chief reference librarian, Dickinson College; Claire Brown, reference librarian at Reddick's Library, for clippings of the original coverage of Mrs. Patrick Rooney in *The Ottawa Republican* and *Ottawa Free Trader Journal*; Thomas D. Burney, Assistant to the Chief of the Rare Book and Special Collections Division of the Library of Congress; Messrs. Cousins and Goossmas of the Colindale Branch of the British Museum, for favorable (and welcomed) rule-bending; Eleanor M. Gehres and Lynn Spenst of the Denver Public Library; the staffers who loaded (and unloaded) cartloads of obscure dust-covered books for me at the Library of the College of Physicians, Philadelphia, and at the Library of Medicine at the National Institutes of Health; Evelyn Stephenson, historian and archivist for Peabody College.

The broadcast media assisted in many ways, including Alan Landsburg Productions and *That's Incredible!*; Hana Gartner, Alex Powell, and Gordon Stewart of CBC-TV's *the fifth estate*; Nancy Nickel and Toni Stevens of CFTO-TV's *Lifetime*; Alex Williamson and KWY-TV's *People Are Talking*; WGAL-TV's *Live!*; WLS-TV's *A.M. Chicago*; Stephanie Burke and WJZ-TV's *People Are Talking*; Gregory Fein and Fox-TV's *Sightings*; Meredith Robertson and *The Other Side*; Marissa Melatti, formerly with WHTM-TV; and Mike Pintek of KDKA radio.

To Vincent Gaddis for special contributions and for setting a standard with his own book about mysterious fires; to Michael Harrison for same; and to Ron Dobbins for a voluminous supply of news clips and citations, thanks.

Anne Hurley, Ray Manners, Anna Inez Matus, Teresa Miskell, Norman and Debbie Mitchell, Jerry Pearson, Ginger Sewell, William and Abigail Shearer, Barbara Meister Vitale, Lynn Volpe, and John W. White critiqued the manuscript, offering improvements you will never know about but are benefiting from.

Thanks also to David Clewell, award-winning poet, for permission to quote his verse and prose from *Lost in the Fire* (Garlic Press, 606 Rosewood, St. Louis, MO 63122).

Sharon Jarvis earns special recognition. She went far beyond the customary role of agent in bringing this book to market—especially after one publisher sat for almost four years on the manuscript-under-contract because it was "too good to publish." (No, I'm not making that up!) I honor your perseverance.

No adjective should limit the gratitude reserved for Sandra K. Nevius. Her contributions to every aspect of this book, plus support and encouragement that eased frustrations and cheered the discoveries made together, have been delightfully indispensable.

To those overlooked in these acknowledgments, sincerest apologies. Notes get lost or misplaced (as anyone who has seen my office will quickly understand!) during two decades. To those who have requested anonymity, plus those unknown and unknowables that have contributed to this book's fruition, I thank you all.

Last, I acknowledge you, the reader. Without your interest now, the others' contributions to this book would be in vain.

When it's a storage shed, garage, nightmare of an attic
or one more derelict warehouse lit up in the night
and authorities can't find any neighborhood kid
with his pocket full of matches and a bad idea,
we might have to settle for the textbook explanation:
rags and gasoline, the invariably poor ventilation, heat
gradually building to the delicate point of ignition and
suddenly it's a plausible last resort: spontaneous combustion.

The smallest leap of faith should tell us it can happen
this way too: people burn, it's humanly possible
and for no good reason experts can agree on, burn
in minutes from the inside out with something so hot
it takes people who burn for a living, professionals
at their incinerators, hours of patient stoking
to achieve that degree of obliteration.

...In the coroner's office they've always been fond of death
by misadventure, hinting at some vague, misguided bravery.
In centuries of monasteries it's a matter of pious record
where, being eternal optimists, they call it The Fire
From Heaven. When it struck, it struck completely down.

> —David Clewell, "A Starter Kit,"
> *Lost in the Fire* (1993)

Preface

The word the fireman used was meshuga—known by those with even a smattering of Yiddish to mean "crazy." He was looking right at me when he said it. "You must be meshuga!" he declared when told I was researching a book about spontaneous human combustion. Perhaps he thought he could dampen my enthusiasm, like water quells flame.

My quest to understand this long-rumored enigma was not deterred, however. (Though this would certainly not be the only time my intellectual competence was questioned.) As the nineteenth-century American poet laureate Ralph Waldo Emerson observed, "Nothing great was ever achieved without enthusiasm."

Whether what follows is *meshuga* or great or something else, is your judgment to make. But the enthusiasm I have invested in this book has amassed upstart evidence that has been pushed aside for far too long, yet stubbornly won't go away. I find the evidence provocative, challenging, and, well, downright fascinating. I trust that you will, too.

I realize that what you are about to read can be terribly, even frightfully disconcerting. There is no way around this. Beyond the shock and sensationalism, lies revelation and new marvels.

I do ask that you not (until at least having finished this book) leap to judgment as once did the famous physician Jean Bouillaud, member of the prestigious French Academy of Sciences who, upon hearing the phonograph first demonstrated, grabbed the exhibitor by the throat and shook him violently while screaming "You wretch! How dare you try and deceive us with the ridiculous tricks of a ventriloquist!"

Like life itself, the subject of *Ablaze!* will at times be contradictory. Confusing. Credibility may be stretched many times. But as strength comes from stretching muscles, learning comes from stretching the mind with ideas that exercise it. Usually, one must be accustomed to incredible happenings before one can embrace them. Therefore, what follows will—to some extent—be repetitive; not only to earn acceptance through reiteration but because of the surprising number of cases themselves, many chronicled here for the first time.

Where the facts lead your thinking is, of course, for you to allow. You may embrace all the evidence presented. You might dismiss only some of the suppositions I propose. You could damn all the conjecture offered. What you cannot do, I submit, is ignore the facts of the cases which are the foundation for this adventure.

The best thinkers, I believe, come with an open mind unprejudiced, ready to consider—if not immediately embrace—any new idea supportable by insight and force of evidence.

"Sit down before fact as a little child," instructed the philosopher-scientist Dr. Thomas Henry Huxley; "be prepared to give up every preconceived notion, follow humbly wherever and to whatever abysses nature leads, or you shall learn nothing."

The topic at hand demands no less of you.

Introduction

"The universe is full of magical things, patiently waiting for our wits to grow sharper," Eden Phillpotts wryly wrote in *A Shadow Passes*. You are about to embark on an adventure where many magical things abound. A headlong journey into impossibilities.

Impossibilities imposed upon nature...like those established by the great scientist Lavoisier, when in 1772 he stood before the august French Academy of Sciences and declared that "stones cannot fall from the sky" because no stones are there; by the eminent physicist Rutherford, who said "moonshine" to the idea that the indivisible atom could be split; by no less an authority than Albert Einstein, who announced in 1932 that "There is not the slightest indication that nuclear energy will ever be obtainable."

Impossibilities placed upon the body...like the declarations that man cannot survive traveling faster than a horse can run; that the four-minute mile is unattainable; that climbing Mt. Everest without oxygen is as suicidal as is breathing water.

Impossibilities misplaced upon intellect...like the U.S. Office of Patents being superfluous because its commissioner in 1889, Charles Duell, said "Everything that can be invented has been invented," like the selling of computers is a fast track to the poor house because Thomas Watson (before he became chairman of the board of IBM) said "I think there is a world market for about five computers."

Science has not solved the universe. Learned men are not omniscient. Experts have been proven wrong. History is full of notable but dubious misjudgments based on innocence; based on unwise rejection of facts; based on a refusal born of haughty reputation to acknowledge evidence; based on the quagmires of scientism that suffocate innovative ideas and stifle paradigm breakthroughs.

Thousands of trains and planes, hundreds of thousands of patents and computers, millions of meteorites, and countless fissioned atoms later, these presumptuous pronouncements from bygone days no longer define the impossible. Not even the magical. In retrospect, we collectively laugh at the myopia of earlier generations.

Yet each generation defines a consensus for reality beyond which certain "magical things" may not, cannot exist. Designated phenomena are disallowed. People—especially scientists—believe that nothing special is happening when, in fact, something quite extraordinary is happening; sometimes before their very eyes. It is not uncommon, especially when one explores anomalies and occurrences deemed paranormal, to see defended *status quo* opinions that are untenable, even palpably absurd in the face of overwhelming evidence to the contrary.

This book is about something that, after many generations, *still* remains magical and, more so, controversial: spontaneous combustion of human beings.

"We have no evidence of the human body suddenly bursting into flame. The whole thing is a fairy tale," averred philosophy professor Paul Kurtz in 1986, when he was (and remains) chairman of the Committee for the Scientific Investigation of Claims of the Paranormal (CSICOP).

"I don't know of any verified cases where a *human being* has burst into flames without heat from an outside source," answered Marilyn vos Savant (possessor of the world's highest IQ according to the Guinness Hall of Fame) to a question about spontaneous combustion in *Parade* (September 30, 1990).

Are they right? Once great thinkers think all is neatly knowable in the predictable universe, then one can anticipate the arrival of an unaddressed upstart observation, an incompatible fact, a datum which shouldn't be and which stubbornly won't go away...demanding that a higher plateau of knowledge be established, with new horizons to ponder.

Spontaneous human combustion offers many upstart incidents, replete with conundrums screaming to anyone with ears to hear—if not always with eyes to see—that we live in an uncommon, magical world.

As the examples aforementioned show, certain phenomena were not allowed by the orthodoxy of their day. Substantiating proof got condemned. For a while. What changed, sometimes, was strength of the evidence itself. Other times, definitions of the possible expanded to make inroads into the dark regions of impossibility.

This book examines the varied viewpoints expounded on the world stage about spontaneous human combustion. It attempts to determine which have merit and which do not. It spans the full range of human psychological reaction to an event decreed impossible; to an event that, if *not* impossible, is by its nature extraordinarily distressing. And fearsome.

This is unavoidable.

So too is the challenge to reigning scientific dogma which this enigma presents. Hostility and censorship battle the magical; the mysterious; the phenomenal. Today, attacks on evidence have become part of the evolution of knowledge to solve the mysteries of the universe. Virulent, even irrational attacks can regrettably be expected when predominating convictions face being discredited. Nowhere is this more evident than in spontaneous human combustion—a concept wholly irreconcilable within the consensus currently permitted for human experience.

Which brings us to the heart of *Ablaze!* The cases themselves. Upon them all discussion, conjecture, and, ultimately, conclusions must be based.

Some cases have been acquired through exclusive interviews. Others were culled from obscure documents forgotten in history's attic, to resurface here for the first time since their initial recording. In the pages to come, there will be something new for every reader.

As the parade of cases passes by, your attentive patience and active participation will be called upon to grasp the breadth and complexity of this subject. As is always the case in good detective work and good science, facts must first be *retrieved*; then *examined*; finally *understood* against what is currently known (and not known) by science, by medicine, and, perhaps, by yourself regarding the forces that govern our bodies and our world.

Because so much error plagues the history of this subject, wherever possible and practical I have attempted to find "primary sources" for documenting it. A primary source is, ideally, an eyewitness to an event. Even then, as with two eyewitnesses to the same car accident, there can be disagreement as to detail, even to cause. The more controversial an event, the more one can anticipate variations in detail. Conversely: the nearer one *can* get to the original evidence, the more reliable will be the account, free of time's diluting effects on memory and a need to second-guess or to alter evidence to defend one's credibility

Likewise, scholarship mandates that all quotations in *Ablaze!* represent actual written or spoken testimony. I have contrived no conversation, nor invented dialogue, for dramatic effect. I prefer utilizing (as space permits) the *original* words of those who personally confronted the impossible, to allow you to feel the vitality in what iconoclast Charles Fort aptly called "the damned."

Along with the importance of including original sources, is the challenge of how to present what has become a voluminous amount of material. (I could easily write separate books about each of a dozen cases, there is so much information.) After first anchoring the subject with a modern case, I have chosen to introduce the evidence chronologically—because this allows insights into the psychology that has underlain the phenomenon for hundreds of years.

Evolution of the response to the problematic can prove no less fascinating than the phenomenon itself. Later, cases will be categorized by the phenomenon's numerous subsets.

At times, the testimony that follows might seem incredulous, as is the very premise itself that humans can burn from within. But things that are unlikely may nevertheless be true. Sir Oliver Lodge's axiom from 1924 is no less valid today: "Our knowledge of nature is not so extensive that we are able to say beforehand what is possible and what is not possible in a novel reign of enquiry."

Are you ready to embark upon an excursion into nature acting at its most bizarre? To experience awe in the face of the inexplicable? To explore one of the most incredible enigmas you may ever have to ponder?

Larry E. Arnold
June 30, 1995

"This fire consumes our minds, let's bid adieu,
Plumb Hell or Heaven, what's the difference?
Plumb the Unknown, to find out something new."
—Charles Baudelaire

"It is characteristic of nature that science
can predict fully the motion of a planet or a comet
but not as yet the burning of a match."
—Philip Morrison,
Scientific American, March 1975

1

The Baffling Burning of Dr. Bentley

"That was the oddest thing you ever seen!"

—Deputy Coroner John Dec
on the Bentley case

There was horror in the morning out on Main Street in Coudersport on December 5, 1966.

Dr. Bentley's burnt up!

The thought rang like a fire alarm gong inside Donald Gosnell's brain as he bolted from the bathroom of the elderly physician. Gosnell felt chilled to the bone, not by northern Pennsylvania's record-cold weather of this December morning, but by the shock of the horrific discovery he had just made inside the old doctor's apartment.

He fled the house, ran down Main Street, and burst into the office of the North Penn Gas Company where he worked as a meter reader. Panic, fear, abject terror overwhelmed him. What, his coworkers wondered, could have happened to so rattle his accustomed composure.

Acting as though he had just seen a ghost, Gosnell gasped for breath, desperate to give them an answer. He had none. Fright stifled thought, unnerved his mind. He could but cry out four haunting words: "*Dr. Bentley's burnt up!*"

It was the understatement of the year.

What Gosnell saw that Monday morning was incomprehensibly impossible. Everyone who entered Dr. Bentley's apartment shook their heads in amazement. Some declared the scene "Impossible!" despite what their own eyes beheld.

Coudersport had lost a physician... the world had gained a mystery.

AN OLD-FASHIONED DOCTOR

Dr. John Irving Bentley lived on the first floor of a two-story, white clapboard house at 403 North Main Street in Coudersport. To say the townspeople held him in high esteem would be an understatement. They loved him!

Born of British parents in Granville, Vermont, young John was blessed with a keen intelligence. First tutored by his mother, the precocious lad at age twenty-one earned his M.D. from the University of Pennsylvania Medical College. Young Dr. Bentley interned at a charity hospital for a year, then established the first of several medical offices that kept him in Pennsylvania for the rest of his life.

In the midst of the Roaring '20s he moved to Coudersport, where in 1945 the Medical Society of Pennsylvania honored him with an award for "50 years of faithful and loyal service"—a commitment he continued until his retirement in 1953. The county newspaper, *The Potter Enterprise*, said that Bentley's greatest satisfaction was "to have helped more than 2,000 babies into the world. And, when the time came, to have helped an old person out of it as painlessly as possible."

In 1956, Dr. Bentley fell over a rocking chair and broke his hip, partially paralyzing his left leg. Many of the townsfolk now returned his years of kindness by stopping in to visit. These were usually spirited encounters! Though Bentley once imbibed liquor in moderation, the only spirit he now indulged in was conversation. "He was a great arguer over anything, anytime!" said Florence Fourness, his housekeeper for thirty years. "His mind was good."

After his third wife died in 1959, the octogenarian widower lived alone. He hobbled slowly through his efficiency apartment with the aid of a four-legged aluminum walker. Otherwise, he was almost totally dependent on others. "He was not able to dress himself," recalled Mary Nicholson, who volunteered to be his personal nurse. "He was able to remove his robe"—which, along with a flannel undergarment that doubled as a nightgown, was his standard attire—"but that was all." Sometimes she even had to help him put on his pair of soft leather slippers.

Unlike earlier times when medical emergencies made the general practitioner's life unpredictable, Bentley's retirement years were characterized by an unchanging daily routine. Life in his cozy apartment centered around an overstuffed chair in the middle of the fifteen-foot-square living/sleeping quarters; the adjoining bathroom, twelve feet away; and the walker that enabled him to traverse that distance. He would sit in the chair from 10:00 A.M. until two o'clock the next morning, said Mrs. Fourness, forever trying to keep an old pipe lit while he entertained his daily guests. At 2:00 A.M., he would turn off the gas heater and get the eight hours of sleep he had always recommended to his patients.

Sometime during the day he'd eat his only meal: a mug of coffee and a bowl of shredded wheat. Doubts over his diet aside, his nurse insisted Dr. Bentley remained "alert and exhibited no signs of senility." He still "loved to visit with people," and kept track of every child and former patient in town.

Nevertheless, she and others in Coudersport became increasingly concerned about Dr. Bentley. It was now 1966, and his ninety-two-year-old body no longer kept pace with his agile mind. "It was very hard for him to get about," confided Mrs. Nicholson. Both she and Mrs. Fourness worried about the clumsiness with which he handled the wooden matches, kept in each pocket of his day robe, to light the pipe. His robes were dotted with "spark holes," as his nurse called them, from fumbled matches and spilled tobacco embers. But the old doctor enjoyed smoking, and no amount of fretting by his friends would deter him. All the fretting in the world, however, could not have deterred the unimaginable destiny that was to occur.

Sunday morning, December 4, Mrs. Nicholson dropped by to see that everything was in order. The heater comfortably warmed the apartment; the tobacco pipe was in its holder on the small wooden table next to his chair; the walker stood on the other side, ready at any time for its next twelve-foot journey. A spare robe hung over the rim of the bathtub, placed there earlier by his housekeeper. And, as always, Dr. Bentley entertained Mrs. Nicholson with memories of years gone by and with questions about the days to come. He seemed as alert as ever.

Around nine o'clock that evening Mary Nicholson, along with her husband who had come by later in the day, bid good evening to their longtime friend and braved the frigid night air to return home. No chilling nightmare could rival what was about to happen as the town slumbered that night, for they were the last people to see... *literally* see... John Irving Bentley.

THE METER MAN'S NIGHTMARE

Five inches of newly fallen snow crunched under each step Don Gosnell took as he walked down the tree-lined sidewalk of North Main Street, gas meter book in hand. Towering behind him above the whitened branches, the autumn-yellow clock tower of the Potter County Hall showed 9:05 A.M. as he stepped onto the porch of number 403 this bitter cold December 5.

"Gas man!" Gosnell yelled, as he opened the door and strode into the dark corridor that led past Dr. Bentley's apartment on the left to the basement stairway. As was customary, he paused, awaiting the physician's usual reply.

Silence.

Mildly perplexed, Gosnell headed for the basement to read the gas meter. Probably ol' doc was up visiting with friends last night and now lay asleep, the meterman mused. He was right on one point.

Then, as he grasped the doorknob on the door to the dungeonlike basement, Gosnell noticed something odd. A light-blue smoke with an unusual smell filled the corridor. "It was a strange, sickening sweet odor that I wasn't familiar with," he said, "like that of a new heating system—an oil film burning, somewhat sweet."

Now more perplexed, he cautiously descended the narrow plank staircase, doubling up his five-foot-eleven-inch body to clear the beam above the last riser. In the haze-dimmed glow of his flashlight, he recorded the amount of gas used. Turning to leave, he glimpsed a pile of ashes in the far corner of the rock-walled subterranean chamber.

Gosnell strode over to examine the unexpected cone of ash, fourteen inches in diameter. Being a volunteer fireman, he instinctively scattered the five-inch-high mound with his boot and found nothing aflame. "I looked up overhead and saw this hole," he said. "There was no fire whatsoever...embers only around the hole." The cherry-red glow of smoldering floorboards revealed that three nine-inch-thick oak beams had been burned from above, one almost completely through.

Gosnell pushed his fingers into one beam's charred wood; it was cold to the touch. "They had a fire," he concluded to himself, thinking it must have been a minor one since no fire alarm sounded during the night. "Well, they got doc out of here," he silently decided, "but why didn't they put out the embers?"

He turned away to leave the basement.

Upstairs now, Gosnell pondered anew the origin of the honey-sweet smell he had never before encountered and why embers from a small blaze had been left to smolder. He decided to look in Dr. Bentley's apartment and make sure everything was alright. To this day he wishes he hadn't.

A wisp of light-blue smoke with the same weirdly sweet scent issued from the apartment as he opened the semi-invalid's door. Not seeing the physician, Gosnell entered and peered around the doorway leading into the bathroom.

"A hole about two-and-a-half feet wide and no longer than four feet had burned through the bathroom floor, exposing pipes leading to the lavatory and running across the ceiling of the basement" is how *The Potter Enterprise* (December 7, 1966) described the scene now before Gosnell.

But there was more. Beside the charred cavity was, to quote Gosnell, a "browned leg from the knee down. I thought it was a mannequin. I had to bend over and look really close to tell the difference. I realized it was a human leg. I didn't look further!"

DR. BENTLEY'S LEG...ACY AND MORE

"I can still remember what I saw as plain as can be," Gosnell confessed nine years later. Having to recall his extraordinary encounter with the bizarre still visibly disturbed him, I noticed. "And after I realized what I saw, they couldn't pull me back in there with a D-8 [a bulldozer]! What bothered me most was what I *didn't* find."

What wasn't found was Dr. Bentley—at least in recognizable human form.

The fire department was called as soon as Gosnell blurted out his discovery. Fire Chief John E. Pekarski's men responded quickly. Gosnell ran back to the physician's house to forewarn his colleagues. He didn't want anyone else to walk in unprepared to see that brown stump of a left leg. Gosnell himself resolutely refused to re-enter the building.

"I was told not to go in," recalled Fred Sallade, probably the first fireman to arrive after the alarm sounded. He did anyway. His laconic report: "The body burnt up, made a hole through the floor. And *that* was all!"

As others arrived, Gosnell explained that water hoses weren't needed since there was no fire to extinguish. Hoses were connected anyway.

Each newcomer refused to believe Gosnell's incredible tale that Dr. Bentley—or any human being—could have simply disintegrated. But each man who entered the house realized Gosnell was telling the truth, however incredible it sounded.

Richard C. Lindhome, a Coudersport undertaker whose Victorian-style funeral home was a few blocks away, soon arrived to investigate the macabre scene by what he called his "normal procedures." Normal procedures weren't adequate this day. He quickly called the coroner's office.

Meanwhile the editor of *The Potter Enterprise*, William Fish, stopped by to collect material for what would become his newspaper's lead story in the issue due out in two days. Fortunately he not only brought a camera, but used it. Then firemen sprayed water on the room to make *sure* the fire was out.

John Dec, the deputy coroner for Potter County, thought at first he was being called to just another fire death, in which the body is covered with flayed blackened skin and a noxious, acrid stench pervades the room. He was as perplexed as Gosnell had been by the scene before his eyes now. "All I found was a knee joint, which was atop a post in the basement," he stated to me, along with the lower leg and the now-scattered ashes. "There was *no evidence of flame or nothing*—just a light-blue smoke. And a sweet odor, like perfume."

Neither Dec nor Gosnell, in his hasty retreat from the death chamber, recognized any other anatomical parts belonging to Dr. Bentley. However: local historian Robert R. Lyman, Sr. later mentioned in his book, *Amazing Indeed* (1973), "the gruesome sight of...Dr. Bentley's head."

This apparent contradiction was resolved in 1980. During the filming of a segment about this incident for ABC-TV's *That's Incredible!* I asked Mr. Lindhome whether any portion of the physician's head had been found among the ashen rubble. "Oh yes! It was on the pipes," he replied. "I can't recall if part of the jawbone was there or not. The skull was on the water pipe and the leg was on the floor. Outside of that, you wouldn't have known it was a human being." Simply, the incineration was so thorough it masked any easy identification of his head.

Carl 'Chick' McCloud, another involved witness, agreed with this astonishing situation. "I helped the undertaker pick up the remains and we put it [the skull] in a rubber bag and carried it to the car," he reminisced. "I would say there wasn't much of him remaining."

"We didn't stay in there too long!" McCloud would later add, though his laughter that followed failed to mask the memory of horror recalled. "On entering the bathroom, we smelled no body odor, no smoke. And the house, as old as it was, should've caught fire—if it *was* a fire."

That there had been a fire seemed obvious. After all, an entire human body had been reduced almost totally to ashes; an ember-rimmed hole through the linoleum and floor boards marked the place where the victim *literally* collapsed; three oak beams on sixteen-inch centers below were burned, one nearly through its nine-inch thickness.

Yet at the epicenter of the burned hole, where Dr. Bentley's body had fallen and the greatest heat generated, his aluminum walker lay tipped against the bathtub. The walker's frame was intact. As were, miraculously, the *unmelted* rubber caps on two of its legs!

"I don't see where it's possible in such an intense heat that they weren't burned completely off," Lindhome ruminated about the rubber caps in 1980. "But they aren't! This is something I can't figure out."

And only inches away—directly above the hole's perimeter—the claw-legged bathtub's white enamel paint, though blackened, had not even blistered! The portion of his spare robe draped over the outside of the tub had burned away, allowing the other half to fall into the tub. The highly combustible tar-backed linoleum floor was barely singed beyond the ominous hole, a fact so baffling to Lindhome that he said he wouldn't have believed it had he not *seen* it for himself.

"It was mystifying!" exclaimed fireman Fred Sallade.

"So bewildering!" confided Don Gosnell.

"That was the oddest thing you ever seen!" chimed in Deputy Coroner Dec, whose wife added: "It's funny how one can burn up so completely, and yet not burn the house down."

COUDERSPORT'S VERY CURIOUS COMBUSTION

What fantastic flaming fate befell John Irving Bentley? A number of ideas were offered at the time, but the mystery deepened.

Speculation briefly turned to foul play. That required a motive, a suspect, and evidence of an accelerant capable of reducing a body to ashes—something a bathrobe alone cannot achieve (despite Dec's pronouncement that clothing burns).

"No one wanted to kill him," exclaimed his nurse, horrified by the suggestion of murder. "I never spoke to anyone who harbored ill will against him." And with money found lying in the apartment, it seemed illogical to search for a transient who tried to cover up a robbery by torching his victim.

An investigation by Dec and county coroner Dr. Herman C. Mosch found no support for foul play. "I'm satisfied it was accidental," Dec informed the newspaper. Mosch agreed. "There is no evidence of any outside influence," Mosch reaffirmed nine years later. "No suicide. It was accidental."

A contention that the ashes belonged to someone other than Dr. Bentley was never seriously considered. "Well," Dec quipped sardonically, "if he's living *you* produce the body! Where is he? Besides, he was an old man anyway."

The rational solution was that the old man, an inveterate pipe smoker, accidentally set his robe—and subsequently himself—on fire.

"As he dozed off, the pipe must have flipped over and lighted his clothes," John Dec suggested, "because there are no traces of where he walked across the room from the bedroom to the bathroom. And that's where he reached for water, where he slumped—that's why there's a hole in the floor. So naturally..."

Dec paused, as if the consequence was foreordained. Then, for emphasis, he added: "Clothing *will* burn you up."

The Potter Enterprise proposed the same scenario though with less conviction, suggesting that Dr. Bentley "may have dozed off in his bedroom chair." But the keen-eyed editor had noticed one interesting fact when examining the area around the doctor's chair. "His pipe was carefully placed in a stand beside it," Bill Fish wrote without equivocation.

It seemed unreasonable, therefore, to have the doctor—waking to find his robe ablaze—reach into his blazing lap to retrieve the pipe and neatly place it in its holder before trying to douse the life-threatening flames. The pipe's placement was the first hole—beside the one in Bentley's bathroom floor—in John Dec's reconstruction of the fatal fire.

The editor, seeing this dilemma, suggested an alternative: that "a live ash from his pipe had fallen on his night clothes, or perhaps it was the spark from a barn-burner type of match he used."

Florence Fourness was among those espousing this theory. "I can tell you what happened to him. Do you want to know?" she asked. "I think that ol' pipe got to going, dropped out of his hand and fell on his robe, and lit the matches in his pocket. He must have really got on fire quick! He then collapsed in the bathroom and couldn't get back up."

If not the pipe-in-the-lap theory *per se*, which common sense ruled out, is it plausible the physician's unsteady hand jarred a tobacco ember from the pipe as he was returning it to the table, and while he slept the unnoticed ember smoldered its way into one of the matchboxes in his robe's pockets? Or could he have struck a match before reaching for the pipe, and fatally dropped the miniature torch on his lap?

Yet if the doctor's garments became thus inflamed, why did he not remove the robe in the room where he sat rather than hobble with his walker to a source of water to squelch an intensifying fire? His friends all professed to the physician's mental alertness. Surely his education made him aware (1) that the *longer* one is enveloped in flames, the *less* chance there is for survival; and (2) that movement through air *fans* a fire's furor.

Not only were these reconstructed "actions" of Dr. Bentley incompatible with his medical training, but his nurse shed further doubt on the physician's alleged dash for life. It took Bentley "*a full five minutes*" with his walker to traverse the distance from his chair to the toilet, Mary Nicholson told me, not realizing the impact of her statement. Even Mrs. Fourness, who advocated the pipe-in-the-lap theory, acknowledged that "he was *so* slow."

Had the aging gentleman indeed ignited in his favorite chair, as nearly everyone surmised, the ravenous flames would have at least asphyxiated (if not first burned to death) his infirm body *long* before he could have reached the apartment's only source of water fully *five tottering minutes away!*

There were other puzzling problems, too.

Dec, for instance, marveled at the airtightness of the fire scene. "It's hard to believe!" he conceded. "The house was tight; didn't get no oxygen. So how in the world? How could you burn with the house closed like that, no air inside? So it must have consumed whatever air was in there; just smotherin'..."

But a smoldering fire does not reduce a corpse to ashes overnight.

"The thing that we remarked about time and again was why that linoleum didn't ignite," pondered Lindhome, the undertaker, and his colleagues. "Because once linoleum goes, it's pretty hard to put out! And there was so *little* of the body left..."

Why, then, had this highly flammable floor covering *failed* to ignite as Dr. Bentley ostensibly hobbled across it, bits of blazing cloth falling from his robe? Why had it not burned *anywhere* beyond the immediate circle where his body collapsed?

Others took the question one step further. "Well, it's funny how the *house* didn't burn up," declared Chick McCloud.

Dr. Bentley's residence had been built around 1850–1860. "As with all these older homes, the tinders are very very dry," amplified Pastor Ken Lewis, who had carefully examined the Bentley house before purchasing it around 1970 and beautifully remodeling its interior. "And the *amazing* thing would be that when this fire started, *somehow it didn't continue to burn the house down!*"

I asked him about the condition of the house when he purchased it, for it had stood vacant since Bentley's demise. "All the windows, all the downstairs walls had a film," Lewis replied, characterizing it as oily soot. "The ceiling and walls were covered with a film as a result of the fire."

Pastor Lewis escorted me to the basement, where he again pondered the flames that had curiously cut a six-square-foot hole above the three burned beams. "As a builder and a carpenter," he said, rubbing his fingers over the charcoal surface of the deeply gouged center beam, "I am baffled how this area could burn away and yet the rest of the house not burn. That, to me, is the phenomenon that has to be explained. To my mind, the *big mystery* is how a man could have been completely consumed in this size of a fire and the house not go along with it!"

Furthermore, to make a final mockery of common sense and reason, there was *absolutely no fire damage to the bathroom ceiling directly above Bentley's cremated corpse.*

Pennsylvania State Fire Marshal Fred Klages taught me in his arson investigation course that in an accidental fire "there'll be three times the damage above the point of origin than on the floor." Instead, the heat of Dr. Bentley's flames was directed *downward*—clearly right through the floor.

And what of that hole? Fire science textbooks say that it takes one hour and twenty minutes for a blowtorch to burn one inch into a redwood beam. To burn nearly through the nine-inch beam under Bentley—an oak beam is harder, more dense than redwood—would require more than ten continuous hours under the jet of a blowtorch. Almost the entire time Dr. Bentley was alone. Yet the flammable linoleum didn't ignite...nor were the aluminum struts (with a 1200°F melting point) of the walker even deformed, nor even its legs' rubber tips.

How?

FIGURING OUT THE FANTASTIC FLAMES

By now, one might think that all Coudersport would be trying to figure out the answers to these mysteries in their midst. They weren't.

"Well," Gosnell recollected, "people were shocked. 'This is something that just doesn't happen.' So they never pursued it in the least that I could see." People avoided him, he noticed. He was, by virtue of his shocking discovery, associated with this hometown tragedy, and people wanted no reminder of the tragedy in their midst. Gosnell himself would be all too happy to forget that December morning also, if only he could.

"The town, including a reporter," wrote local historian Lyman, "failed to recognize it as anything unusual. They reasoned that the doctor's clothing had caught on fire and that had caused his death."

The coroner's certificate of death for Dr. Bentley gave the townsfolk little reason to question the physician's fate, though it raised a number of questions itself. This document as filed in Pennsylvania (or any other state) is to provide an accurate record and evaluation of a person's physical transition. Dr. Bentley's death certificate stated the following:

> IMMEDIATE CAUSE OF DEATH: "asphyxiation and 90% burning of body."
> INTERVAL BETWEEN ONSET AND DEATH: "1/2 hr."
> ESTIMATED TIME OF DEATH: "5 a.m. EST."
> MODE OF ARRIVAL: "burial."
> WAS AN AUTOPSY PERFORMED?: "Yes."

Could asphyxiation be the cause of death? No trachea existed to be examined for inhaled carbon deposits, and half a leg doesn't give (if you'll pardon the pun) much evidence to stand on.

The "burning of body"—more precisely the *disintegration* of body—contributed to his death, certainly. But what caused the voracious fire, which ultimately killed its target, in the first place?

And what of the *extent* of burning? "A knee and one part of the leg was left," Dec emphatically stressed nine years later as I sat with him at his kitchen table. "I wrote down 90 percent [missing] on the coroner's report, but I believe it was more like 98 percent.... As a matter of fact, I'd call it 99 percent." Facts were beginning to change, and moving even farther from the norm.

No one I spoke to could offer a reason for the assigned time of death. It had to occur within the twelve hours prior to 9:05 on the morning of December 5, but "5 A.M." was nothing more than a groundless guess.

The alleged duration for destruction was glaringly conspicuous. "It takes 3 to 6 hrs. for [a] body to smolder in fire," Dec wrote in the Certificate's notes, basing that estimate on the favored burning-clothes theory. On the other hand, lungs do not require thirty minutes of exposure to flame before a person will fatally choke. Besides, the predicted one-half-hour interval is too brief for combustion of any *household* type to produce such complete physiological disintegration.

The mention of an autopsy did give hope of finding clues to the true cause of death. The hope was short-lived. Dec expressed incredulity at our request for the autopsy report. *"There's nothing to autopsy!"* he almost shouted. "Ashes, yes! But how do you do that?" However (and to whatever extent) an autopsy was done, *The Potter Enterprise* reported the coroner's conclusion "that Dr. Bentley had not suffered a skull injury."

The remains of Dr. Bentley were buried, said the death certificate. "I don't know how they can bury anything like that—one leg!" Dec sarcastically remarked. But buried his sparse remains were: a granite headstone in the West Hill Cemetery at Galeton, Pennsylvania, marks the site of Bentley's interment.

Ironically, the box labeled "CREMATION" on the death certificate was not checked. Yet only in a crematorium could one begin to incinerate a body to the degree of what was shoveled from Bentley's home.

One should treat cautiously declarations made on official documents. As Gosnell confided: "This whole thing was handled pretty much as if it didn't happen. It took someone with a special interest to get it going."

A Hint about What Happened

So what *did* happen to Dr. John Irving Bentley?

More than a dozen years after his horrific discovery, Gosnell was still being called a liar when he'd describe the fire scene. "They just don't understand it," he said woefully to me. "Yes, I'd like to see it proved in documentation where it can't be disputed!"

Gosnell never had reason to change his testimony about that memorable December morning. Others involved in the original investigation, however, began to question their own earlier conclusions.

Dec, for example, encountered another case of human incineration: a Volkswagen crash in which flaming gasoline was spewed on its three youthful occupants. The heat of the ensuing inferno was so intense, Dec remembered, that no attempt could be made to rescue the trapped lads; their burns so severe that no one, save the boys' mothers, could identify the charred bodies. *Yet each victim's severely burned body was still easily recognizable as human—rib cage, arms, neck, teeth, and skull intact and easily distinguishable.* How, puzzled Dec now, had Dr. Bentley undergone a far more dramatic destruction under less catastrophic conditions?

"I just thought the fella caught fire and died," confessed Lindhome about the days following December 5. "But the more I thought about it," he confided fourteen years later, "the fact there was nothing but a hole in the floor—you couldn't do that if you soaked the body in gasoline!" If it really mattered anyway, the coroner told me no gasoline had been found.

Also, what produced the sweet-smelling ethereal haze in lieu of the anticipated dense black smoke and noxious odor associated with burned flesh? And how did the rubber caps on the walker's legs and the bathtub's paint escape a blaze so searing it cut through oak beams? And why didn't this blaze consume the rest of the structure's dry timbers or even singe the ceiling overhead, let alone highly flammable linoleum?

To the few who knew all the facts and thought about them, it had become a *most* unusual fire. It defied logic, reason, and belief, this baffling burning of Dr. Bentley.

"In thirty years," remarked the undertaker, "I've seen a lot of car fires, house fires, and so on. *Nothing*"—he paused to chuckle—"was quite like this. The body was consumed by such an intense heat that it couldn't have been done intentionally *even* if gasoline or anything like that had been used. The fact that the body burned the linoleum in the form of the body and dropped down through the floor into the cellar, and didn't catch the linoleum on fire, is hard to understand. It was like a *ball of fire* burned right through the floor!"

Whatever happened, no one could dispute that the aged, frail, thin body of John Irving Bentley, in its last moments of life, had somehow lost all resistance to fire.

Nobody paid any attention at the time to a Pennsylvania state fire marshal who, disconcertingly, quietly muttered three simple words: "*spontaneous human combustion.*"

2

THE BURNING QUESTION: WHAT THE BLAZIN'S GOING ON HERE?

> It is essential to find out whether a phenomenon does in fact happen, even if we know not how it comes about or why it occurs, regardless whether tradition or authority are opposed.
>
> —Lester S. King, M.D.
> *The Growth of Medical Thought*

Everyone agrees that under just the right conditions damp hay, piled lawn clippings, rags soaked with linseed oil, coal dust, and certain metal powders can spontaneously ignite—that is, inflame due to chemical action, fermentation, or oxidation without a spark or open fire. But the human body?

Spontaneous human combustion (SHC) is the phenomenon and process whereby a person burns in the absence of a known, conventional external source of fire. Classic SHC culminates in nearly total reduction of the person to ashes while surrounding combustibles escape largely unburned.

Talk of SHC rankles the Newtonian-Cartesian scientific community today. It is a tired superstition; a medieval relic of mistaken reasoning; an impossible fate deserving derision, diatribe, and debunking. Today, anyone who dares suggest that a human body can burn of its own accord risks a fusillade of scientific scorn.

The concept is grisly, to be sure. Gruesome. Unthinkable.

"I don't know anyone in our unit who has seen anything like normal SHC," Los Angeles County fire investigator Dick Moon told me in 1995. He proposed another acronym: SWAG—a "scientific wild-assed guess!" We both laughed. Likewise, Washington State's King County Medical Examiner, Dr. Donald Reay, dismissed SHC when telling *The Seattle Times* (April 1, 1990) that he'd never seen it in his seventeen-year career: "there's no scientific evidence or literature to support that this kind of spontaneous combustion occurs."

How could it be otherwise? Fire kills about five thousand people in the United States each year—about 80 percent happen at home. The vast majority of these fire fatalities occur needlessly, say the experts at the National Fire Prevention Association. Each October, on the anniversary of the great Chicago fire of 1871, the NFPA sponsors National Fire Prevention Week. People are taught that the majority of one- and two-fatality fires are residential; that careless use of smoking materials, mostly cigarettes, is the leading cause (32 percent) of one- and two-fatality fires in the home; that about two-thirds of fire deaths happen between 8 P.M. and 8 A.M.; that most fire deaths involve furniture or bedding as fuel load; and that many of these deaths occur when the fire had a large head start—"over 40 minutes before discovery for 38 percent of the deaths," according the Dr. Louis Derry in *The 1983 Fire Almanac*.

If something defined as SHC does indeed happen, there is ample opportunity, given all these deaths annually and all the educational efforts by the NFPA (and firefighting organizations worldwide), for people to have heard about it. For firemen to have seen it. Yet nowhere in NFPA's comprehensive *Almanac*—the 1983 edition has 792 pages—is there mention, statistically or historically, of persons burning to ash amid unburned surroundings.

S. A. Smith and F. S. Fiddes, two renowned toxicologists, address the burning question of self-igniting people this way in their classic 1955 book, *Forensic Medicine: A Textbook for Students and Practitioners*:

> Spontaneous combustion of the human body cannot occur, and no good purpose can be served by discussing it.

Their unequivocal statement succinctly sums up the consensus about SHC. Theirs is the reigning belief among agencies that investigate human trauma: fire and police departments, biophysics, the medico-legal specialties of forensics and pathology among others. Extremely rare is the professional in any of these disciplines who has even heard of the concept of a human fireball; rarer still to find one who won't unhesitatingly scoff at the idea.

You will not find SHC listed in *Cumulative Medicus Index*, the bible of the medical profession. You will not find SHC on Medine (a computerized database of general and forensic medicine articles); the Library of Congress's Congressional Research Service searched Medine for SHC in 1988...and got no "hits."

Burn injury classification does not grant status to SHC. Burns of the first-degree involve reddening of skin; second-degree produces blistering, though hair follicles remain intact; third-degree causes loss of skin, scarring, and nerve damage; and fourth-degree yields charring of underlying skin and muscles. Surely though, Dr. Bentley's far exceeds the criteria for a fourth-degree burn!

Disbelief in SHC is easy to understand. "That a body composed four-fifths of water should take fire from a spark or flame, and continue to burn till only ashes are left," remarked E. Muirhead Little in a letter to *The British Medical Journal* (1905), "is surely incredible enough without the added wonder of spontaneity..."

> Wonder rather than doubt is the root of knowledge.
> —Abraham Joshua Heschel

Further, to burn a body supposedly requires *prodigious* amounts of heat and fuel. Even then it is no easy task, as countless examples throughout history can attest.

Medieval France, for example, refined burning at the stake to an art. There, a baker's boy in Caen was found guilty of petty theft. For his crime—probably stealing a loaf of bread—two large carts of bundled branches were required to fuel his fiery execution. According to Paris and Fonblanque's account in *Medical Jurisprudence* (1823), parts of the boy still remained after ten hours on his judgment pyre.

Nationality was no factor either in the difficult task of burning up a person. Citing an instance from England, Paris and Fonblanque said "the extreme incombustibility of the human body was exemplified in the case of a Mrs. King, who having been murdered by a Foreigner, was afterwards burnt by him; but in the execution of this plan he was engaged for several weeks, and after all did not succeed in its completion."

This murderer could have benefitted from the expertise of Dr. Strong, a nineteenth-century American physician who professed "considerable experience burning up or getting rid of human remains by fire." Yet it was a chore even Dr. Strong sweated through in more ways than one. He wrote in *A Treatise on Medical Jurisprudence* (1855):

> Once in particular I had a pirate given me by the United States' Marshal, for dissection; and it being warm weather, I wanted to get rid of the flesh, and only preserve the bones. He was a muscular stout man, and I began upon it one night, with a wood fire, in a large old-fashioned fire-place. I built a rousing fire and sat up all night, piling on the wood and the flesh, and had not got it consumed by morning. I was afraid of a visit from the police; and by eleven o'clock they gave me a call, to know what made such a smell in the street. I finished it up, somehow, that afternoon; but I look upon it as no small operation to burn up a body.

Further evidence of the near-imperviousness of the human body to fire is found in the East Indian ritual of suttee (widow burning), conducted as recently as September 4, 1987, at Deorala near Jaipur. A considerable fuel load, burning for hours, was required to cremate the widow—as hundreds of years of history had proven it would. And the heinous legacy left by Nazi crematoria at Auschwitz-Berkenau attests a millionfold plus to the challenges of burning up a human body; as do persons burned (but not cremated) by napalm in later wars.

As Gonzales et al. summarized in *Legal Medicine, Pathology and Toxicology* (1954), "The body is not usually completely consumed by the fire unless the conflagration is intense or prolonged, or unless the body is subjected to professional cremation."

While modern cremation technology indeed makes that task easier, it also amplifies the complications inherent in the burning of a human being.

John J. Grenoble, noted lecturer on cremation and former operator of the East Harrisburg Cemetery crematorium, explains what happens to a body inside a retort oven. "As the fire jets heat up, hair on the corpse quickly burns with a blue flame, the head snaps back, and the body tries to actually sit up in its casket. Bones begin to explode at the relatively low temperature of 700–800 degrees Fahrenheit, and the chest cavity might open. Rapidly soaring heat reddens the body, its skin turning black then splitting; the flesh carbonizes, then is oxidized.

> Late morning, October 7, 1983, a tank truck laden with 9,000 gallons of gasoline careened across Philadelphia's Schuylkill Expressway, crashed into a hapless motorist, and exploded in a monstrous fireball so hot the concrete roadway buckled.
>
> Nevertheless: so resistant to destruction by fire is the human body that the Medical Examiner's office could identify the victims by comparing the teeth in their bodies—not destroyed in this highway crematorium—with dental records.

"The corpse is typically subjected to 2200 degrees for ninety minutes. Muscles and organs gradually burn from the bones, which continue crackling in innumerable small explosions. The retort's temperature is then lowered to around 1700 degrees and the bone fragments cooked for another hour to an hour-and-a-half," says Grenoble. A normal cremation consumes 40 to 45 gallons of fuel oil, he added.

(For comparison, the worst house fires generate temperatures rarely exceeding 1700°F at the ceiling where the heat rises and usually not above 1200°F—about *half* the incredible heat a crematorium generates. At 1800°F, rock melts and flows like hot syrup from the bowels of the earth at Hawaii's Kilauea volcano.)

Therefore, under conditions considered scientifically ideal for combustion, in which an accelerant (oil or gas) is always available and the draft is optimized, up to *four hours* could be needed to maximize the destruction of a body by fire.

Yet what comes out of the retort oven is *not* the proverbial ashes and dust, Grenoble emphasizes, but fragments of the skull, vertebrae, and limb bones up to six inches long. Fragments readily identifiable as human.

Even more torrid than crematoria are thermite fires. These industrial blazes are *so* hot that steel I-beams melt, yet bodies pulled out of these holocausts can be identified. "If bodies can survive a thermite fire—the hottest fire—and remain identifiable as human," marveled Fire Marshal Klages as I handed him the photograph of Dr. Bentley, "then this [SHC] is something *very* different!"

The dilemma of SHC is immediately obvious: *there seems no way in which burning human tissue can alone generate the colossal heat necessary for nearly total human disintegration.* And, if such elevated temperatures were encountered *externally*, the effects certainly would not be confined to the body while leaving adjacent, more readily combustible objects untouched.

This is why it is so difficult to understand the ease with which certain people have been, it is said, reduced to piles of ashes...and *precisely* as Dr. Bentley was!

CAN SHC *REALLY* BE REAL?

But SHC is impossible, is it not?

The idea of SHC has long been dismissed by the rationalism of modern science as a myth. Otherwise, there is the distressing ramification that your body can turn into the enemy. Something from the inside, attacks the outside.

"It is evident that the weight of authority is not in favor of human spontaneous combustion. No one of any position or authority has ever seen a case," declared Charles M. Tidy in the second volume of his *Legal Medicine* (1882). Twelve years later in their 1894 classic, *Medical Jurisprudence, Forensic Medicine and Toxicology*, authors Witthaus and Becker mention SHC "for the sole purpose of stating that no trustworthy evidence of the possibility of any such condition or result exists."

Carrying this rejection into the twentieth century is Dr. Gavin Thurston, who developed a historical interest in SHC while serving as Her Majesty's Coroner for Inner West London, England. As editor of *The Medical-Legal Journal*, he acknowledged in 1961 that while "extensive combustion of the human body can occur with minimal damage to surrounding objects" he insisted it was always a case of *preternatural combustibility* (PC). "The combustion is not spontaneous," he wrote, "but started by an external source of heat."

In 1976, for Michael Harrison's book *Fire from Heaven*,[1] Dr. Thurston again rejected SHC.

That made unequivocally clear, all one has to do, then, is find the "external fire" that superheated Dr. Bentley beyond the capability of a crematorium.

As we have seen, the scenario that the ninety-two-year-old doctor accidentally ignited his robe in his living/bedroom, then hobbled with his walker for five minutes to reach a pitcher of water in the bathroom, succumbs to scrutiny on several points. The evidence better supports a scenario not considered by others at the time: that Dr. Bentley became enveloped by fire *in situ* while standing in his bathroom.[2] Yet no heater—gas or electric—was in the bathroom, according to Pastor Lewis and the heater in his living/bedroom had not malfunctioned.

> *Le Revue de l'Hypnotisme* (Paris, June 1890) chronicles the successful experiments of Russian Dr. Rybalkin at the Hôpital Marie, St. Petersburg, to produce blisters and burns on patients solely by suggestion.
>
> If this is not spontaneous human combustion to Mssrs. Witthaus and Becker, what then is it?
>
> Numerous cases of burns, blisters, rashes, and otherwise inflamed tissues produced under hypnosis have been well-documented since then... and still denial of SHC persists.

So where is the "external fire" demanded by Thurston that is compatible with the facts surrounding the hole in Dr. Bentley's bathroom? Even invoking spontaneous combustion of the matches alleged to be in his pocket fails to resolve the thoroughness of his burning.

Perhaps there is merit to the concept of spontaneous human combustion, after all. But how much merit?

UNRAVELING HISTORY'S TAPESTRY OF SHC

The evidence of self-igniting people threads a meandering, often twisted trail through the last five hundred years of human experience, and may extend right to the very edge of time where the mists of recorded memory become lost in prehistory. Its exploration has engaged some of the world's best medical minds in bitter controversy and promises to be among the more provocative and profound human adventures yet undertaken.

In the pursuit of knowledge, one must be skeptical, yes; but skepticism mustn't be so strong that there's wholesale denial of the situation.

Renaissance chronicler Thomas H. Bartholini describes in his extraordinarily rare 1654 booklet, *Historiarum Anatomicarum Rariorum*, a risk faced by some medieval unfortunates: the predicament of wine imbibers igniting internally. I present Bartholini's own words, translated probably for the first time ever from the original archaic Latin, as they appeared in his *Historiarum*, Centuries I and II, chapter 70, titled "Flames from the Belly":

> To stir up a fire in the belly from excessive inhaling of wine spirits [that is, drinking too much wine] is foreign neither to reason nor to experience. Doctors from Lyon, at the confluence of the Saone and the Rhone rivers, studying the corpse of a certain woman to determine the cause of death, report than an enormous flame, filling the entire abdominal cavity all around, burst forth but was quickly extinguished. An account has been written for Renatus Moraum, the most glorious of Parisian scholars, from which great man I have this report. The cause of the flame was believed [to be] wine often swallowed by that woman, or a rather hot Theriacus★ water, which sick persons take under doctors' orders.
>
> A knight, Polonus, during the time of good queen Bona Sforza, having consumed two ladles of strong wine, vomited a flame and was thereupon totally consumed, according to a report from his parents, Mrs. and Mrs. Eberhard Vorstius. The older son of the aged father, living in Leiden, narrated an account to me. A deposition to the Academic Senate of Copenhagen mentions a similar case.

First: an anonymous Frenchwoman in the vicinity of Lyon, dead of fire issuing from her gut sometime before 1654. Second: the human fire-spewer named Polonus Vorstius, "consumed" by hellacious halitosis sometime during 1468–1476 when Queen Sforza reigned. (Though others have mentioned the incident, until now no one has identified the victim by name since Bartholini published his obscure report.)

As Bartholini indicates, Copenhagen's Academic Senate already had on file another early incident of someone who belched fire Polonus-like and was consumed.

Two cases of drinkers vomiting flame, with no mention of an external ignition source. Both linked to consumption of alcohol. (In the case of Polonus, Bartholini would later specifically cite "brandy wine" as the culprit.)

Apparently, however, this bizarre fate was not all that uncommon in the middle of the second millennium. It must have become sufficiently rampant by the sixteenth century so that learned men on the Council of the University of Copenhagen actually convened to decide whether it was possible to drink oneself to death *without* flames gushing from the throat to char lips and body!

★ An antidote for venomous bites.

I believe I have identified this incident the Council discussed more than 300 years ago. It was the death of Peter Petraeus of Houff, who allegedly "was able to be consumed by flames from drinking wine, when no flames from the jaws were present." They said the sketchy information provided to them neither proved nor disproved "that flames were produced from his jaws, either splitting or consuming the lips." Neither did they know his age, his temperament, his drinking habits, his health. "Therefore we were unable to determine nothing with certainty in this case," the Council stated in the University's *Actum Kjøbenhaffn aff Consistorio* (December 14, 1635). Bartholini repeated their noncommittal verdict in his 1662 *Cista Medica Hafniensis*—which translates as *A Copenhagen Medical Chest*.

Preacher John Hilliard expressed no equivocation in his 1613 pamphlet, *Fire from Heauen* [sic], whose rarity is rivaled only by its exceptional subject. He told of a frightful burning that began on June 26, 1613, in Christchurch on the English Channel. The burning was inside John Hittchell. And inside him it continued, wrote the pamphleteer in the spelling of his day, until Hittchell "was consumed to ashes, and no fire seene, lying there-in smoking and smothering three dayes and three nights, not be quenched by water, nor the help of man's hand." It was, the preacher confessed, a strange thing "the like of which was neuer seene nor heard of." It is supposed lightning was the cause, but of this no proof is given.

And in 1644, the abdomen of an anonymous woman of Lyon spit forth fire, attested witnesses and physicians in the time of Moraum.*

Meanwhile, word about the plight debated among Copenhagen's learned must not have reached two German noblemen in Courland,† where one night circa 1653 they engaged each other and a third acquaintance in a drinking bout. It proved fatal. Sturmius' *German Ephemerides* (1670) recorded that in the midst of quaffing strong liquor "two of them died...scorched and suffocated by a flame forcing itself from the stomach." Incidentally, Sturmius believed that such searing stomachaches were prevalent only "in the northmost countries." He thus became the first to suggest a geographic or/and ethnic aspect to these enigmatic inflamings.

Next, in 1662 an "arme frau"—a pauper woman—died in a localized blaze that reduced the poor woman and nothing else to ashes in her Paris home.

---

* In *Ancient Journal de Médecine* (1786), a Dr. René Moreau cites the 1644 case of a Lyon woman's anomalous burning. As Moraum is probably Dr. Moreau latinized, we can date Bartholin's otherwise vague incident with reasonable confidence.

† Cf. Rolli (1745) and Knott (1905), where the spelling is Curland.

In *Collection Academique* (Actes de Copenhague) for 1671–1672, Professor Mathiaeus Jacobaeus described the death of a woman of the lower class who for her last three years had consumed liquor "to the point of not taking any other nourishment." She sat down one evening on a straw chair to sleep, and during the night "was consumed along with her chair by an internal fire" to an extent that "only her skull and extreme joints of her fingers were found; all the rest was reduced to cinders."

Clearly, Jacobaeus' case predates 1672. I strongly suspect he refers to the 1662 Parisian pauper★ mentioned previously, for in the following year Bartholini (Jacobaeus' father-in-law) republished the case in his own *Acta Medica et Philosophica Hafniensia (1673)*. Thus the standard claim by others who have written about SHC: this unfortunate Parisian's fate earned her postmortem distinction as the earliest death definable as classic spontaneous human combustion to be recorded in a medical document.

I submit instead that Polonus Vorstius deserves the honor, if you will, of being the *first* medically documented death by SHC—predating the Parisian pauper by nineteen years when Bartholini recorded his death notice in *Historiarum Anatomicarum Rariorum*, itself a rare history about medical events.

MADAME MILLET: THE INNKEEPER'S WIFE IGNITES

A half century would elapse before this anomalous fiery fate reentered recorded history.

This time the woman's name is known, and two physicians witnessed her strange demise.

On Whit Monday evening of 1725, Jean Millet went to bed with his wife around eight o'clock. The following day would be a very busy one for Millet, the innkeeper of Le Lion d'Or in Rheims, and his shrewish and drunken wife, Madame Nicole Millet. Monsieur Millet was neither disappointed nor concerned when his wife, unable to sleep, arose and went to the kitchen where he thought she would again be warming herself externally by the fireplace and internally with firewater.

Around two o'clock in the morning, a strange odor from the kitchen pervaded the hotel. Millet sprang from his bed and shouted the dreaded alarm that all lodgers fear: *Fire!* Among those awakened was Claude Nicolas LeCat

---

★ This case is cited by Henry Bohanser's *Le Nouveau phosphore enflammé* and assigned to 1671 or 1672; by *Gazette salutaire* (1767); by Pierquin's *Reflexions* (1829); by Devergie (1852); by several other early compilations. The *London Mirror* and Lair (1812) are among many references which erroneously (and impossibly) assign it to 1692. Lair compounds his error by duplicating it as a second case (cf. 162 and 165). Such is the confusion in sorting out some of these early episodes.

(1700–1768), an apprentice surgeon who quickly followed his nose and his curiosity. The "infectious odor," as LeCat described it, led him into the kitchen. There, what everyone found burning was not the innkeeper's inn but the innkeeper's wife!

This unexpected discovery profoundly affected LeCat. Here are his own words describing the scene, translated directly from his rare *Memoire postume sure les incendie spontanes de l'economie animale*:

> This woman was found consumed on the 20th of February, 1725, at the distance of a foot and a half from the hearth in her kitchen. A part of the head only, with a portion of the lower extremities and a few of the vertebrae, had escaped combustion. A foot and a half of the flooring under the body had been consumed, but a kneading-trough and a powdering-tub, which were very near the body, had sustained no injury. M. Chretien, a surgeon, examined the remains of the body with every judicial formality.

There was little—if any—smoke in the kitchen to hinder viewing this macabre scene. LeCat clearly saw that the drunken woman's body had almost wholly disintegrated, *including part of her skull*, without burning the adjacent wooden furniture. As to the cause of Mme. Millet's death, LeCat is equally clear: it was, he unhesitatingly avowed, death "by the visitation of God"—*spontaneous human combustion*.

> SHC is never discussed medically...
> —Deloris Ament, in her article
> "SHC: it does fire the imagination,"
> *Seattle Sunday Times* (April 1, 1990)

THE BURNING OF THE COUNTESS BANDI

This aura of early medical respectability for SHC failed to convince many later medico-legal experts, however. "Most of the cases," C. M. Tidy complained in his 1882 book, *Legal Medicine*, "are marred by a lack of precision, or by an evident credulity on the part of those who relate them." That was true too often, unfortunately.

Dr. LeCat, who went on to distinguish himself in medicine as "the most renowned lithotomist in France" and never, so far as can be determined, revoked his conclusion about Mme. Millet's death, should not be accused of "evident credulity"—a polite euphemism for today's charge of gross gullibility. Nor was this case devoid of precision. A French court had sufficient evidence to exonerate Jean Millet of fatal spousal abuse and to issue a verdict in favor of spontaneous human combustion.

Tidy's complaint about a lack of precision in the evidence for SHC does, however, apply to the curious fate of Duke Antonio Ferdinando, a "teetotaler" from the area of Guastalla, Italy. His death by the administration of alcohol had been predicted for April 19, 1729, but the duke did not succumb that day to overindulgence. Instead, according to *Ripley's Believe It or Not* (16th Series), when later his body was being rubbed with alcohol, he caught fire and burned to death. Believe it or not...

Rev. Giuseppe Bianchini, a prebendary of the Catholic Church in Verona, Italy, believed. Believed in something comparably strange. In a letter to Ottolini and in his trail-blazing *Dissertazione epistolare istorico-filosofica*, Bianchini detailed at length the unnatural death of Countess Cornelia Zangari and Bandi, which happened on March 14, 1731.

At age sixty-two, Cornelia Bandi lived comfortably and in good health at Cesena, a city near the Adriatic Sea in east-central Italy. On this March evening while dining, she experienced a "dull and heavy" drowsiness and retired to her bedchamber, accompanied by her maidservant. After three hours of conversation and prayer, the Countess fell asleep. Her handmaid quietly drew the bed curtains and closed the door behind her.

The next morning the maidservant entered the bedchamber with a cheery greeting and undoubtedly—like Don Gosnell more than two centuries later—forever afterward wished she had not. According to Bianchini, four feet from the Countess's bed lay "a heap of ashes, two legs untouched from the foot to the knee, with their stockings on; between them was the lady's head, whose brains, half of the back part of the skull, and the whole chin, were burned to ashes, among which were found three fingers, blackened; all the rest was ashes, which had this particular quality, that they left in the hand, when taken up, a greasy and stinking moisture."

Soot hung in the air, and clung to an empty oil lamp upon the floor. Candlesticks upon a table stood upright, their wicks drooping and devoid of tallow. Her bedsheets and blankets were—apart from a film of "moist and ash-colored soot"—undamaged. Everywhere, soot "penetrated into the chest of drawers even to foul the linens" and "hung on the walls, moveables, and utensils." Of the room above the Countess's bedchamber, Bianchini wrote "that from the lower part of the windows trickled down a greasy, loathsome, yellowish liquor, and thereabout they smelt a stink without knowing of what."

What befell this hapless countess on a fine Italian night? "It is impossible that, by any accident, the lamp could have caused such a conflagration," remarked Royal Society member Paul Rolli in translating Bianchini's account for *The Philosophical Transactions of the Royal Society of London* (1745). "There is no room to suppose any supernatural cause," he also concluded of the conflagration that confined itself singularly to the countess. Rolli believed the "likeliest cause" was a flash of lightning—which can pack 100 million volts of electricity and heat air in its path above 60,000°F. But no thunderstorms had rumbled over Cesena that night; nor are lightning bolts known to ash their human targets.

Having assessed *all* the evidence, Bianchini concluded—insightfully—that Countess Bandi "burnt to ashes while standing, as her skull was fallen perpendicular between her legs; and that the back-part of her head had been damaged more than the fore-part was." Perhaps she had arisen to regurgitate her dinner which she complained had made her dizzy... and instead threw up flames that then devoured her entire body.

No one called the Countess an imbiber of spirituous liquors, a factor so prevalent in these early human infernos that it finally began to be noticed. Indeed, she never drank alcohol. Liquor was implicated nevertheless. The Marquis Scipio Maffei, who verified Bianchini's account, reported how townspeople whispered that the Countess "was accustomed to bathe all her body in camphorated spirit of wine." Asserted Maffei in *Journal de Medecine* (vol. 26): "The use of this drug is one of the causes of this phenomenon."

A FORGOTTEN SELF-SEARING RESURFACES

Maffei thought he had solved the odd burnings of Countess Bandi and others before her, though he failed to explain how a film of alcohol on one's skin could so incinerate the bather!

His "solution" would fail to resolve another case in 1731, similar to Countess Bandi.

Spanish surgeon Dr. Pierquin wrote about a curious case he read in the Swiss *Journal d'Yverdon* (December 1731).[3] It is a case as different from its forerunners as steam is to ice; yet it is the same. Here, for the first time, this forgotten case has been retrieved from its obscure source and translated into English.

On November 10, 1731, a fierce but localized fire had ravaged a sixty-eight-year-old female in Vitry, a remote village in Haute-Saône, France. The woman, who lived alone, fell and struck her head on the hearth, which had "a little fire in it," said the *Journal*. News of her death spread like wildfire the next day when neighbors discovered that "her body, though reduced to ashes, appeared at first to be quite whole. One surprisingly noticed the folds of her dress, upon which her right arm and fingers were like firebrands.[4] On touching them one found no more than ashes; all the bones, even the largest, were consumed."

The woman's head, which lay directly upon the hearth and hence closest to the heat of any flames present therein, was "where the fire stopped"—leaving the tongue intact! "Regarding the legs," Pierquin said, "the fire had lost its activity towards the middle of the right leg, and had not in any way burned the stocking." There is no mention of alcohol.

Despite the widespread interest of the local townspeople, the episode escaped the notice of every physician until Pierquin. And of everyone since, *until now*! Perhaps no one else wanted to risk being associated with such an outlandish tale of a lady burned to dust between her chin and her right knee; burned to dust where not in contact with any proven external flames; burned to dust *inside her unburned clothing*.

The mind-boggling fate of the Vitry Frenchwoman is *not* unique, however.

IPSWICH'S *AMAZING* GRACE

The Pett family, if anyone had asked them, would probably have scoffed at such a mind-boggling fate for anyone...until they awoke one morning in 1744 to find themselves reduced in number by one. Precisely, the reduction was in Mrs. Pett. Literally.

Grace Pett, about sixty years old, was said to be fond of gin, no doubt for the relief it offered her impoverished years as the wife of a poor fisherman in Ipswich, Sussex, England. Her low-class life would have been lost to history, swallowed up among the unremarkable lives of the masses, were it not for the remarkable way in which she gave it up.

"On Tuesday Morning a Woman in St. Clement's Parish was found burnt to Death in her own House," announced the Saturday edition of *The Ipswich Journal* (April 14, 1744). The newspaper alluded to "several extraordinary circumstances," but didn't elaborate. It would not be the last time an editor found a newsworthy story about SHC too hot to handle.

Fortunately the prestigious *Philosophical Transactions of the Royal Society of Great Britain* (1744–1745) retrieved Grace Pett[5] from obscurity and gave her historical immortality.

"On the night of the 9th of April, 1744," said the *Transactions*, "she got up from bed as usual. Her daughter, who slept with her, did not perceive she was absent till next morning when she awoke, soon after which she put on her clothes, and going down to the kitchen, found her mother stretched out on the right side, with her head near the grate; the body extended on the hearth, with the legs on the floor, which was of deal [pinewood], having the appearance of a log of wood, consumed by a fire without apparent flame.

"On beholding this spectacle," continued the *Transactions*, "the girl ran in great haste and poured over her mother's body some water contained in two large vessels in order to extinguish the fire; while the foetid odour and smoke which exhaled from the body almost suffocated some of the neighbours who had hastened to the girl's assistance. The trunk was in some measure incinerated, and resembled a heap of coals covered with white ashes. The head, the arms, the legs, and the thighs, had also participated in the burning."

One old woman who "had drunk a large quantity of spirituous liquor" had burned up in a way that Royal Society scholars deemed worthy of mention. "There was no fire in the grate," said the Royal Society to those who alleged she stumbled and fell into the fireplace's roaring flames. However, near the consumed body lay "the clothes of a child and a paper screen, which had sustained no injury from the fire."

The Royal Society met this dilemma with silence, and ended its report.[6]

Yet the *Transactions'* editor excluded many fascinating facts about the lady who had become the toast of Ipswich.

A Mr. R. Love attended the coroner's inquest "on ye old woman that was burnt," and his letter[7] to the Royal Society stated the fire occurred between 10 P.M. and 6:00 A.M., and left behind very little of its victim [spelling and punctuation is the original]:

> Ye feet and lower part of ye Leggs were not burnt...ye Stockings on yet parts remaining, not Singed; very little of yet parts of ye Leggs yt were burnt lay on ye wood floor the rest of ye body on a Brick hearth in ye Chimney; part of the Head not burnt a Body of ffire in her breast in ye morning found, which was Quenched with water, her bone chiefly calcined and ye whole so farr reduced to ashes as to be put in ye Coffin with a Shovel...

Love added that, contrary to popular reports, Mrs. Pett was "not in liquor nor addicted to drink Gin."

Everyone did agree on this point: no surrounding combustibles fueled the fierce flames which carbonized Mrs. Pett except for her cotton gown. So isolated was this fire that Rev. Notcutt of Ipswich and a Mr. Gibbons (who interviewed the victim's daughter and two boarders) told the Society that even the pinewood flooring beneath the victim "was neither singed nor discolour'd." They added that a greasy film "as not to be scoured out" covered the hearth where she burned.

The parallels to Countess Bandi's burning are astounding, with one difference. Mrs. Pett was *seen* aflame.

A Mr. B. informed the obscure *Ipswich Magazine* (1799) "that some ship-carpenters going to work about five o'clock in the morning, saw a great light in the room, broke into it, and found her in a blaze: they then procured some bowls of water from a pump which was near the door, in the street; on throwing it on the body it made the same hssing [*sic*] as if thrown on red-hot iron."

Maybe the hissing blaze they extinguished had been burning since eight o'clock the previous evening. Or maybe it had begun *just as they walked by*.

What fearsome force befell this poor fishmonger's wife on April 10, 1744, that transformed her overnight (or quicker) into "a great light" and so atomized her body that it had to be shoveled up like spent embers?

The coroner's jury didn't know. Fellows of the Royal Society could not explain, either.

THE EVIDENCE MOUNTS

At a different unscorched house the next year, 1745, an unnamed woman was discovered totally consumed by fire "except a part of the head and limbs." She, according to LeCat, who (you'll recall) twenty years earlier personally investigated the strange combustion of Madame Millet, was likewise found on a hearth—no mention of any fire being within it. She, too, said the *Journal of the American Temperance Union* (1837), was "habitually drunken."

In the small French town of Plerquér, on or about February 7, 1749,[8] Madame de Boiseon turned herself into a body of fire within a few minutes and died. The consuming fire came like a devil incognito, and departed like a thief in the night. No lambent flicker was seen playing upon her alcohol-steeped body like the graceful flames that dance over Christmas pudding. Combustion gently began somewhere in her grease-fat—the chair in which she burned was itself only slightly scorched—and worked its way to her extremities like fire progresses through smoldering cotton. Except, here, the outcome of transition by unholy flame was inaminate potash. Fire left behind little but a bad reputation for French priest Boinneau to ponder as he committed his thoughts about this eighty-year-old woman's fate to paper, a letter which LeCat referenced in his *Memoire*. Was it by the "act of God" or less godly means that she died?

> The leading proponent of SHC... is Larry E. Arnold... Elsewhere SHC has little support today even among maverick scientists. If it is a real phenomenon, a convincingly documented case of it has yet to be recorded.
> —Jerome Clark, *Unexplained!* (1993)

And sometime just prior to February 22, 1749, a woman of Dinan in northwestern France burned outside the city's gate "under circumstances very similar" to those of Madame de Boiseon. So reported Boinneau[9] in a letter to LeCat. If this case is distinct from de Boiseon, as the nineteenth-century SHC scholars Pierre-Aime Lair and Marie Devergie claimed, the similarities and proximity are striking.

As the eighteenth-century neared its midpoint, men of medicine faced a growing body of evidence for which they had no cure. And no explanation. A mystery worthy to challenge any Sherlock Holmes in medicine.

That evidence—observed by Drs. Chretien and LeCat, described by Bartholini and Bianchini and Dr. Pierquin, and reported in testimony before a coroner's inquest and also to the august Royal Society—suggested a disquieting possibility: that people, especially females, could indeed spontaneously combust. But how?

3

"By the Visitation of God" Did They Die

> In Natural as well as civil history there are facts presented to the meditation of the observer, which, though confirmed by the most convincing testimony, seem on the first view to be destitute of probability. Of this kind is that of people consumed by coming into contact with common fire, and of their bodies being reduced to ashes.
>
> —Pierre-Aime Lair
> *On the Combustion of the Human Body* (1808)

"By the visitation of God" did Nicole Millet leave this world, declared LeCat of her passing in 1725.

As the second half of the 1700s unfolded, the parade of thoroughly—well, *almost* thoroughly—incinerated people continued to increase in number. Divine intervention and alcohol became its marching theme.

On New Year's Eve 1770, Hannah Bradshaw, already intoxicated by holiday festivities and alcohol, crumbled to dust in her New York City apartment. A special feature of her demise will be discussed later.

The widow Mary Clues[1] lived two years and two months beyond Hannah Bradshaw before meeting her own personal holocaust in Coventry, England. During much of this period she daily drank "at least half a pint of rum or aniseed water." Not surprisingly, her health gradually declined; then sometime during the early morning of March 1, 1773, her health problems abruptly ended. At 5:30 A.M., smoke issued from the widow's window. Alarmed neighbors on Gosford Street speedily broke open her locked door and extinguished the small flame they saw with equal haste.

On the floor between her bed and the fireplace (a distance of only three feet) lay Mrs. Clues, calcined with "whitish efflorescence" upon the one leg and thigh which remained of her fifty-year-old body. So observed the eminent surgeon Dr. Wilmer, when he visited the scene two hours later. In the *Annual Register for 1773*, he said "nothing except the body exhibited any strong traces of fire"—*not even her nearby feather bed and clothes had burned!*

Standing before the incomprehensible—which had happened in no more than six hours—Dr. Wilmer perhaps wondered whether alcohol or the visitation of God had dealt the final hand in the macabre sight now before his eyes.

(In 1829, Dr. Thomas Newell would have something to say about alcohol and the death of Mary Clues of Coventry, after he himself encountered an even more complete human incineration. Quoting Dr. Newell from the *Midland Medical and Surgical Reporter* for May 1829: "In the Coventry case, as well as in some others, it is true, the subjects had been in the habit of drinking a great deal of spirit; but if this circumstance is capable of rendering the body liable to spontaneous combustion, instead of its being so very rare an occurrence, we should have instances of it every day.")

Three years after Mary Clues, in 1776, an even more macabre sight struck down a man of God. The priest Gio Maria Bertholi entered his cell in Fivizzano, Italy, to read his prayerbook before retiring one evening. A few minutes later screams and flame interrupted the monastery's solitude. Both came from the priest. The surgeon Battaglia rushed in and found Bertholi sprawled on the floor, surrounded by a self-limiting flame, which receded when approached until finally vanishing.

> Some things... arrive on their own mysterious hour, on their own terms and not yours, to be seized or relinquished forever.
> —Gail Godwin

The priest explained to Battaglia that "a sharp blow" had struck his right hand and, at the same time, a pale bluish flame attacked his shirt, which was immediately reduced to ashes (except for its cuffs). His sackcloth garment beneath was said by *London Medical and Surgical Journal* (November 25, 1837) not to be affected. Although not a hair on his head was burnt, his cap was entirely consumed. Bertholi insisted there had been no fire in the room *until* that moment.

Bertholi's skin subsequently detached from the flesh and hung from his body in shreds, and the "nails of his left hand separated" from his fingertips. He stoically lived for four agonizing days, complaining of "burning thirst" only. If the priest ever got an explanation for his self-limiting blaze, it came in the afterlife.

In February 1779, Marie-Anne Jauffret of Aix-en-Provence, France, unknowingly mimicked Mary Clues when she likewise suddenly turned to ash. Corpulent and addicted to alcohol, Madame Jauffret burned up but one chair and herself. All that remained identifiable as human was her skull, one hand, and one foot. The surgeon Roccas testified that she accomplished this in no more than eight hours in a room having *no fire* in the hearth, as reported by Muraire to the *Journal de Médecine* (1783).

France lost many of its citizens to the disruption of revolution in the eighth decade of the eighteenth century. A few more were lost to the disruption caused by fire within.

In 1780 or 1781, a woman in Caen, in her sixties, who drank *and* bathed in alcohol, died in a very personal revolution that began within her own body. In an era when people were (quite literally) losing their heads, this Frenchwoman left behind *only* her head and two feet in a room with no damage but a greasy oil covering everything around her otherwise vanished body. Said Louis Valentin in his report to Foderé: "she was a very little pile of cinders."

Caen, a city just inland from the English Channel in northwest France, hosted *another* "divine visitation" in 1782. Mademoiselle Thuars, an "exceedingly corpulent" sexagenarian was likewise addicted to spirituous liquors. On June 3, Miss Thuars burned up even more thoroughly than had her predecessor in Caen. Only her feet and seven bones of her body remained identifiable, said Merille, a surgeon who examined the scene and reported his findings to the king's officers and *Journal de Médecine* (1783). Bones, he wrote, which easily "became dust by the least pressure." Her right foot "was found entire," severed by flame at the ankle; "the left was more burnt."

Merille also noted that the combustion had to have occurred "in less than seven hours" in a room that, as you might anticipate by now, exhibited minimal fire damage. The chair on which Thuars had sat, he observed, was "absolutely untouched." Even twigs in the fireplace were but "little burnt."

Speaking of the fireplace, so often blamed in these inflammatory predicaments, Merille noted that although a fragment of Thuars' head rested against one of the andirons, it was her anatomy *farthest* from the grate that (excepting her feet) burned to "nothing but a mass of ashes."

Fat and firewater, if we may so phrase it, was establishing itself as a potently incendiary (and fatal) combination.

Sometime prior to 1796, the fire settled briefly upon Capel, England. Local artist John Constable appended this notation to one of his sketches: "Curious circumstances happened in this cottage a few years since, a poor woman being burnt entirely to ashes." Regrettably the rest of his notation is unclear, so we'll never know if this pauper in Capel tried to drown the pain of poverty in alcohol. Intriguingly, Capel is not far from Ipswich where, you will recall, the drunkard Grace Pett in 1744 was "so farr [*sic*] reduced to ashes as to be put in ye Coffin with a Shovel..."

Back across the Channel again, I find a case of possible interest from 1799 in the *Revue Médicale*. A male, probably French, succumbed to "incomplete" combustion as the result of abusing alcohol, specifically brandy. Did the *Revue* mean incomplete in the sense that only a foot or a hand or a head remained amid a pile of ashes on an unburned wooden chair, or that the imbiber died from first- or second-degree burns that left his body largely intact? I do not know. I do know the *Journal of the American Temperance Union* (March 1837) used this incident in its crusade that employed SHC as a deterrent against the evils of alcohol.

The glow of the fire that cremates people lingered in France. On December 10, 1799, it visited Pont Neuf, a southeastern French town. There Madame Bias, wife of the police inspector, inflamed to an extent that her "whole trunk was a mass of carbon" resembling a burned-out log. Her "sex was no longer distinguishable" and only one foot escaped "with its natural color," said Neveux, the region's officer of health. In reporting his firsthand observations to *Recueil Periodique de la Société de Médecine de Paris* (1800), Neveux added that neighbors had spoken with Madame Bias less than two hours before the fire he likened to spontaneous combustion had ravaged her.

Pierquin listed another episode with a French connection in the late 1700s. A man serving as a cook for the 106th French regiment burned to death within fifteen minutes when the regiment was stationed at Ferrara, a town to the northeast of Bologna, Italy. Whether the victim carelessly caught himself afire over a cook stove and died unremarkably *or* died rapidly by cooking himself from the inside-out, Pierquin doesn't say. I only know that Pierquin felt this incident warranted inclusion in his catalogue of SHC cases for "Recherches sur la combustibilite humaine pathologique" in *Journal des Progres des Sciences et Institutions Médicales en Europe* (vol. 17). And I pass it along in that context.

SHC Invades the Nineteenth Century

Of the twenty-three examples of seemingly inexplicable human incineration that history recorded by the year 1800, at least nineteen cases involved alcohol. As the Age of Reason and rationalism took firm root, the idea blossomed that alcohol and humans could be a *thoroughly* inflammable combination—one especially hazardous to women.

America's Bay Colony, March 16, 1802. A Massachusetts grandmother died with a difference. "The body of an elderly woman evaporated and disappeared from some internal and unknown cause, in the duration of about one hour and a half," reported the *Philosophical Magazine for 1802–1803*. It added that "there was a sort of greasy soot and ashes... and an unusual smell in the room."

Unfortunately missing was any explanation about how it happened—except to label it another instance of SHC and bury the case in obscure medical documents such as *Journal général de médec* (vol. 35) and *The Medical Repository* (1802). The latter listed the grandmother's death as "Rapid disorganization of the human body."

Another French woman, age twenty-eight, became a medical footnote after a blaze consumed her body so completely that her corpse fell apart when it was moved, circa 1804. In describing the incident for *Leroux's Journal de Médecine*, Dr. Prouteau claimed the fire migrated from the "interior to the exterior" of her body; "all outside causes were foreign to this case." Dr. Prouteau left room for no interpretation *other* than SHC.

Another little-known publication, *De la médecine légale*, alluded to the death by SHC-style incineration of a Parisian woman, age sixty-eight, on either Tuesday, Christmas Day, 1804, or the next day. The episode, initially reported by Vigné, was saved from almost certain obscurity by Foderé, whose own compilation of problematic human combustion is overlooked by nearly all medical historians today.

Paris again; January 1805. Madame Boyer, age sixty-eight, died in a way that gets her obituary memorialized in two references devoted to spontaneous human combustion.

Back to Britain. Lord Ferres is hanged after testimony against him by his wife, Lady Campbell, but not before he curses her with a death "even more painful" than his own. In 1807, according to A. J. C. Hare's book *In My Solitary Life* (1953), the widow Lady Frederick Campbell burns up in the manor house's tower... "from spontaneous combustion it was said." History? Or folklore? "Nothing was found of her but her thumb, she was so completely consumed..."

Southeast of Paris by 135 miles: the town of Saulieu in the province of Côte d'Or. There, in 1808, sixty-year-old Madame Laire ignited oddly and became another victim of SHC, according to Beck's *Elements of Medical Jurisprudence* (1851).

Remember Caen, France? SHC returned there for the *third* time with a case reported by Dr. Bouffet, the author of *Essay on Intermittent Fevers*. The victim, another woman of Caen's lower class, died at the Place Villars sometime in 1808 or earlier. As a consequence of—or despite—her addiction to liquor, the fire spared only her extremities as it rampaged throughout her incinerated body while failing to ignite other combustibles.

Back across the Channel, again, this time to Coote Hill in County Cavan, Ireland. Mrs. Stout retired one evening in 1808 and during the night arose for an unknown reason. The next morning she lay "burned to a cinder on the floor." Smoke was still rising from her charred body when her husband discovered he was a widower. The Irishwoman was an inveterate intoxicator, to use James Apjohn's description of the deceased; but that didn't explain why her corpse "crumbled" to dust when touched, nor how her "chemise and nightcap escaped uninjured."

Little wonder that Apjohn declared Mrs. Stout had "died by the visitation of God" in a fire which totally dehydrated her body yet *failed to scorch the nightgown and cap she wore*.

Ireland, circa 1808. In the city of Dublin lived another imbiber, Mrs. Anne Nelis, about forty-five years old. Her torso, found upright in a chair, was "burned to a cinder" yet her extremities *and clothing* were untouched by fire, said Apjohn. As was the chair, *except* where in contact with Mrs. Nelis's torso—and then it was merely charred. Her head was scorched; her hair *unsinged*. There was no damage beyond the victim. Finally, said Apjohn, though there had been a candle in her chamber, it had earlier been removed. Finding any external source of fire that could have ignited her proved as illusive as a candle flame's shadow.

Ireland, 1808, again. County Down, now part of Northern Ireland. Another Irishwoman, Mrs. A.B., also about sixty years old and a voracious drinker of alcohol, did battle with the banshee of death and lost. Smoke with an unspecified odor but sufficiently uncharacteristic to remark on, alerted her family to the tragic predicament in their midst. Rev. Ferguson investigated and said that "no flame" was seen issuing from the woman; yet, said he, it was her "body burning with internal fire" that left a lasting impression on all who viewed her powdery remains. Note that Rev. Ferguson spoke of an *internal* fire that burned *without flame*—a very curious way for a priest or anyone else in the early nineteenth century to describe combustion!

> Spontaneous human combustion...
> this most gruesome of mysteries.
> —Arthur C. Clarke, author of *2001*;
> in "Arthur C. Clarke's Mysterious
> Universe" (January 26, 1995)

If you think this parade of incinerated bodies across the (mostly) European landscape has by now become tedious and unimaginatively repetitive, be alerted. According to Rev. Ferguson, this unnamed Irishwoman gave the phenomenon a new and horrifying twist: she burned while *lying in the same bed with her daughter, who was unaffected*. Even the bedsheets, like all other combustibles in the room, suffered no damage whatsoever.

What a lasting impression that true-life nightmare must have made on this Irish family! Particularly on the daughter who would have rolled over in bed to discover the very localized cremation of her mother just inches away. Not at all the way to start a new day!

Ferguson added one more factor to this very curious episode: the woman's body, when lifted to be placed into its coffin, crumbled to pieces in the lifters' hands.

LAIR: "ON THE COMBUSTION OF THE HUMAN BODY"

If national pride at being top-rated is ever an issue with the French, they were by this time winning hands down the distinction of hosting the greatest number of citizens lost to SHC-style immolations. Of the twenty-eight cases of nearly total human incineration chronologically recounted to this point, sixteen involved the French and twenty-two of them involved prolonged use of alcohol... a percentage that should make imbibers anywhere (particularly in France, I suspect) sit up and take notice.

Frenchman Pierre-Aime Lair was one scholar who, in 1808, took note of the propensity of intemperate individuals to inflame beyond reasonableness. "I confess that at first they appeared to me worthy of very little credit," he said of these accounts, "but they are presented to the public as true by men whose veracity seems unquestionable."

Several respected men of medicine had already written about spontaneous human combustion, notably Jonas Dupont's *De Incendiis Corporis Humani Spontaneis* (1763). Dupont tried to give a scientific explanation in his thesis, saying that chemical processes in a person or frictional heat of particles rubbing together in the bloodstream could so raise body temperature that only a small external cause—he named alcohol—would ignite runaway combustion. Others who said they had seen its aftermath called it, simply and less scientifically, an anathema.

The constancy of their reports (explanations aside) inspired Lair to collect twenty examples of the phenomenon. These formed the basis for his treatise titled *Essai sur les combustion humaines, produites par un long abus des liqueurs spiritueuses* (1800), later published as "On the Combustion of the Human Body, produced by the long and immoderate use of Spirituous Liquors" in the *Journal de Physique* (1808). It is a landmark in the study of SHC.

Lair identified eight characteristics which he ascribed to SHC:

1. The persons who experienced the effects of this combustion had for a long time made an immoderate use of spirituous liquors.
2. The combustion took place only in women.
3. These women were far advanced in life.
4. Their bodies did not take fire spontaneously but were burnt by accident.

5. The extremities, such as the feet and the hands, were generally spared by the fire.
6. Sometimes water, instead of extinguishing the flames that proceeded from the parts on fire, gave them more activity.
7. The fire did very little damage and often spared the combustible objects that were in contact with the human body at the moment when it was burning.
8. The combustion of these bodies left a residue of fat and fetid ashes, and an unctuous, stinking, and very penetrating soot.

These "general observations"—to use Lair's words—became the criteria on which nearly all future writers about SHC have based their arguments.

What Lair excluded from his list of factors associated with SHC was the "visitation by God." He believed the cause lay inside the victims themselves, not in an outside deity. He named alcohol as the principal agent.

"Such excess, in regard to the use of spirituous liquors," he wrote, "must have had a powerful action on the bodies of the persons to whom I allude. All their fluids and solids must have experienced its fatal influence [and] is absorbed into every part of their bodies." The body eliminated the liquid portion of spirituous liquors, he contended, but stored "the alcoholic" inside fat until the unsuspecting imbiber became a human tinderbox filled with ignitable fuel.

These walking, drunken tinderboxes who so often left behind their legs amid a mound of amorphous ashes were, he declared, always women! "I will not pretend to assert that men are not liable to combustion in the same manner," Lair conceded, "but I have never yet been able to find one well certified instance of such an event."

The reasons he gave for this (false) contention are sure to offend Women's Liberation advocates: "The female body is in general more delicate than that of the other sex. The system of their solids is more relaxed; their fibres [sic] are more fragile and of a weaker structure, and therefore their texture more easily hurt. Their mode of life also contributes to increase the weakness of their organization." It was the sedentary, melancholy homemaking lifestyle of eighteenth-century women that, he advocated, made them "more subject than men to become corpulent." Also, by implication, more readily burnable.

Hence women—especially if poor and elderly—who preferred sitting with a glass of brandy instead of working, are apt to overcharge their bodies with hydrogen and thus, he contended, "during the moment of intoxication, should experience the effects of combustion."

But as his fourth point reveals, Lair could not bring himself to call this combustion spontaneous. He acquiesced to "accidental"—ironic in that he provided no external fire source as the trigger for said accidental burning.

Lair's contribution to medical science, and science generally, is that he compiled facts about mysterious human inflamings and attempted to explain their cause in a way more quantified than "visitations by God." And he hoped his SHC research would benefit his fellowman.

"I shall consider myself happy if this picture of the fatal effects of intoxication makes an impression on those afflicted with this vice," Lair ended his essay, "and particularly on women, who most frequently become the victims of it. Perhaps the frightful details of so horrid an evil as that of combustion will reclaim drunkards from this horrid practice."

Time would show his comments about women to be chauvinistic. But the characteristics of SHC and PC that he identified, particularly the association of alcohol and human fireballs, would become reinforced with time.

4

FIREWATER AND DEMON FLAMES

> Yet, admitting that the phenomenon of preternatural inflammability is opposed to the laws of combustion as far as we know, we should not reject as untrustworthy of belief, the many curious and authentic facts on record. They may be true, however incorrectly accounted for.
>
> —Dr. W. H. Watkins, on human combustibility,
> *The New Orleans Journal of Medicine* (1870)

Without argument, the firewater of demon rum and ardent spirits can inflame human nature. But can "spirituous liquors" be inflammatorily lethal to the *body* as well? Lair thought so. So did Orfila, one of the most famous figures in forensic medicine, who declared that SHC occurs "chiefly in aged corpulent females, and especially in persons long addicted to the use of spirituous liquors."

Lair's monograph, remarkable even by today's standards for its lucidity and depth, did not however bring an end to the subject it studied. The path of human fire continued, blazing new trails across Europe.

In the low mountains some 100 miles west-southwest of Paris, seventy-two-year-old Kaufmannswittwe Laurent of Alençon inflamed in classic SHC style on June 3, 1809, according to Hergt's *Annalen der Staatsarznerkunde* (1837).

In Waertelfeld, Germany, lived a forty-eight-year-old man named Ignatius Meyer—"a very intemperate man." Dr. Scherf of Detmold said Meyer died on January 17, 1811, from the combustible combination of fire and alcohol...a combustion that *failed to affect the bedclothes he wore at the time!* So affirmed the victim's nephew, who witnessed the bizarre fire.

As if to vindicate Lair's belief that SHC singled out intemperate females, Mme. François-Marie Rousseau of Morigny died by death "accidental"—that is, not murdered, according to two doctors—on December 22, 1812. Because her body had been totally burnt *within no more than four hours*, they first suspected she had been murdered. But they could find nothing amid the black soot and charred chair in the otherwise undamaged room to confirm their suspicion of foul play.

The *Philosophical Magazine* (1813) told of an early nineteenth-century English gentleman, drunk on tincture of valerian and gum guaiacum, who rolled out of his bed whereupon fire attacked "his saturated body, and reduced it to a cinder, without materially injuring the bed furniture." Served him right for abusing his body with that concoction!

Meanwhile, life continued to get too hot for more residents of France. A French kitchenmaid of Drôme, a region in southeast France some eighty miles north of Marseille, was found burned in her room sometime in 1816. No point of fire existed outside herself. She was in good health and never drank, reported Pierquin in *Reflexions theoriques*. Nevertheless, the fire left only her arms and head intact.

In one remarkable instance in Nevers, southeast of Paris, life became *doubly* hot. Ninety-year-old Madame P– and her servant lived together, drank alcohol excessively together, slept together, and died together in bed by fire with "no explanation" on January 12, 1820.[1] A physician investigated and reported that all that was left of the old woman was the lower part of one leg still wearing a shoe. Of the servant, only her skull was not utterly destroyed as she lay beside her employer. This time the bed, including its sheets and curtains, were burned. But no damage occurred to the walls and ceiling except for a covering of "humid" vapors—and *there was no fire in the room's fireplace*.

> Every fire is a puzzle, and you've got to figure it out. You learn something new every day. You never have two identical fires.
> —Arson investigator, LAFD; CBS-TV's "Hard Evidence," 48 Hours (May 13, 1992).

What stood out about this case is that two people died in a fire some members of medical science said could not happen even singularly. Charpentier himself admitted to having "no explanation" for this double conundrum—a case of *multiple* SHC?

A heavy drinker named Thomasse Goret, fifty-seven, burned to ashes on New Year's Eve, 1820, amid *no other fire* in her home at No. 85 Arpens Road, Rouen, France. Her husband, who admitted regularly giving his wife four francs to drink at a neighborhood tavern and who was accustomed to seeing her quite drunk, was altogether unprepared to end the year a widower.

Though Mme. Goret's face was intact, little else of her was, reported Dr. Hellis to the *Journal General de Médicine, de Chirurgie, de Pharmacie* (1826). "I had observed this myself," Dr. Hellis affirmed, and he had smelled the "empyreumatique excessivement fétide" in her smoke-filled bedroom. Her fireplace, he stated, was "without fire." Furnishings very near Mme. Goret likewise showed no fire damage. In this woman, he concluded, are found the reported characteristics in like cases of spontaneous combustion.

A fifty-five-year-old alcoholic woman in Marseille, Hacquin de Bar-de-Duc, left as her contribution to the SHC parade one of her feet and half a leg, some vertebrae, unidentified parts of her head, and a few interior organs. She was, said the *Journal de Paris* (April 11, 1822), "a novel example of the phenomenon called spontaneous human combustion."

January 1825; Hamburg, Germany. Margaret Heins burned up in a fashion that emulated the priest Bertholi almost fifty years earlier, according to Hecker's *Annalen*.

December 1825; Rouen, France. Madame Soret, age fifty-seven, failed to see her next birthday because of her SHC-style death. Dr. Hellis likened the incident to the ghastly demise of Mme. Goret as he chronicled this new case for *Medico-Chirurgical Review*.

Mrs. Lappiter's Lamentable Demise

A few years prior to 1829, a very odd death occurred in Cheltenham, England. Dr. Thomas Newell, a surgeon to the king and resident of this city some thirty miles west-northwest of Oxford, personally examined the victim's body prior to documenting his detailed findings in a paper he titled "Observations upon Spontaneous Combustion of the Human Body." It appeared in the *Midland Medical and Surgical Reporter* (May 1829).

Jane Lappiter,* a widow in her sixties, was last seen alive shortly after ten o'clock one evening when her live-in companion, Mrs. Roper, put her to bed. At half-past two the next morning, a loud noise and smell of something burning awakened Mrs. Roper. She jumped from her own bed, flung open the window, and yelled the alarm.

A Mr. Overbury responded. He burst into Jane's upstairs apartment and observed "a slight lurid appearance, near the hearth, but not bright enough to enable him to judge, exactly, from what it proceeded." In a minute or two, due to the influx of air through the opened door, a small blaze broke out where Overbury had seen the faint glow. He and his wife, now present, procured three or four pails of water and doused the flames.

---

* The victim's name is also spelled Lappitter in Dr. Newell's paper; apparently a typesetter's error, as it appears only once this way.

There, in the blackened room filled with smoke "which had a disagreeable empyreumatic smell," one of the many gathering bystanders stooped to pick up a foot covered with a shoe and stocking at the end of a strip of ashes on the stone hearth. It was the left foot of Jane Lappiter.

The poor woman discovered, her room was quickly vacated. Within hours, the Cheltenham coroner hastily convened an inquest jury—which swiftly returned a verdict of "accidentally burned"—and what little remained of Mrs. Lappiter was placed in a coffin for interment. But not before Dr. Newell made his own observations at the scene of this remarkable calamity.

"The most striking and important peculiarity in this case," Newell wrote, "was the manner in which the combustion spread, or extended itself. The burned parts were separated from the skin and flesh that were uninjured, by an accurately defined line, on one side of which, utter destruction had taken place, while immediately adjoining, there was not the smallest injury done, the skin being sound and natural." He likened the burning of this human being to a piece of ignited charcoal "which spreads itself in the same way, by a perfectly defined line, reducing in its progress, to ashes, the portions it comes in contact with."

Newell devoted two pages to describing Mrs. Lappiter's anatomical destruction, and absence of surrounding fire damage. At the time, it was history's most detailed account of such a fire fatality. Suffice it to say here that her thorax, abdomen, and pelvis "were reduced to ashes, not a vestige of them" remaining. The muscles of her chest, back, and abdomen "were consumed, leaving the ribs on the right side, and the back bone, without any covering whatever." The ribs themselves had burned away to within inches of the vertebrae; her left leg bones were transformed to ashes, "to within about three inches of the ancle [sic] joint." In striking contrast, her left foot—the one picked up—and its shoe showed not "the least alteration, but were in a natural state." Her right foot, leg, and thigh "were uninjured, and preserved a natural appearance."

At the other end of her body, the fire "took an oblique direction, across the face towards the right side of the head, in such a way, as to destroy the lower portion of the nose, and the right eye," Newell noted. Yet her left eye and upper nose "had escaped, and were uninjured." The crown of her head, its hair, and the cotton cap she wore, were all undamaged. Likewise, her right hand and arm up to the shoulder had escaped intact.

To Newell, it appeared a scythe of searing heat had swept diagonally across Mrs. Lappiter's body, incinerating her flesh and bone from above the ankle to across her face; burning fiercely her left side, but leaving much of her right side (including those portions of the woolen gown and two petticoats she wore) unscathed as she lay sprawled on the floor.

A chair standing near the site of her body "had no mark of fire," stated Newell. Nor did it appear to him there had been a fire in the fireplace the night of the incident. Jane was a pipe smoker, Newell noted. He found her pipe "empty" and "standing upright upon the iron of the [fireplace] grate."

In his meticulous inspection, he saw no evidence to connect her burning to either pipe or fireplace. Rather, Dr. Newell concluded "it will appear more probable that the combustion originated in the body itself, the causes of which, we are at present unacquainted with."

In other words, Mrs. Lappiter died by fire within.

A fire that had been intensely hot, if but momentary, because Newell determined the loud noise that had alerted Mrs. Roper to her housemate's demise had been caused by heat splitting the hearthstone across its center, where Jane lay burning.

> What is the peculiar condition under which spontaneous combustion takes place, we have at present no means of knowing; but that some important change must be induced, there can be no doubt; for if the body, in its ordinary state, was capable of taking it on, it would be a frequent occurrence, instead of its being so extremely rare.
>
> —Dr. Thomas Newell

"In the case of spontaneous combustion," the good doctor observed, "the heat must either be more intense than in accidental burning, or some change must have been induced in the body, rendering it more susceptible of the action of fire."

Newell, noting also that persons found in similar circumstances were said to be besotted, pointed out that this woman "had scarcely ever tasted spirits in her life time. She was a person of the most temperate habits, her drink consisting mostly of tea, with now and then, though seldom, a glass of beer." Thus, Mrs. Lappiter is among the earliest victims of this odd fate to be freed from the stigma of abusing alcohol.

Adding to the noteworthiness of her extraordinary death, she was "of very spare habit of body"... that is, thin, not corpulent.

THE COMBUSTIBLE CONUNDRUMS CONTINUE

Faced with all these tales of human beings victimized by a nearly all-consuming fire, Julia de Fontanelle (1790–1842) decided in 1828 to apply the scientific method to this continuing conundrum. He took thin pieces of meat, and saturated them with alcohol to simulate the body tissue of chronic alcoholics. (His experiment, by design, precluded applicability to Jane Lappiter.) He then set fire to his samples.

The result? The meat only roasted, he reported in the *Revue Medécin Française Etrang.* (1828). He poured more alcohol onto his samples and set them alight again. After the fire had been extinguished a second time, the meat was still only roasted on the surface. His experiment failed to prove his hypothesis that spirituous liquor was the fuel that, lit by an outside flame, transformed people into ashes.

To his credit, he tried. Unfortunately his work was (as often happens in science) forgotten. A more famous scientist would get accolades for doing the same experiment two decades later.

Undeterred, human combustion roared on.

November 15, 1829; Lexington, Kentucky. In an obscure incident carefully documented by investigating physician Dr. Charles Short in *The Transylvania Journal* (1830), an unnamed Kentucky woman, age sixty-five to seventy, died by burning up in her home. Yes, she drank; no, she was not corpulent. Her body, found in the fireplace, was first "mistaken for a piece of wood." Her left arm and leg simply did not exist; what did remain weighed "not more than 30 pounds" he concluded. To those who would argue that she fell into a roaring fire and lay there unconscious until consumed, Dr. Short would want you to know this: *a potato he found in the fireplace, alongside the woman's remains, was* not *baked!*

The topic is, after all, *localized* combustion, is it not?

The parade of burning bodies continued for the next twenty years. Physicians, town mayors, and other dependable witnesses were often on hand to testify to its passing. Reports of spontaneous human combustion showed up throughout Europe in medical journals, too.

On Christmas Day, 1829, alcoholic Maria Jeanne Antoinette Bally, age fifty or fifty-one, burned away to charred bone—except for her stockings and legs—in a chair in her otherwise fire-free seven-by-ten-foot room in Paris, France.

On August 20, 1831, gendarmes flung Charles François Francoy, age forty-one, onto a Paris dung heap to sober him up. Instead, he burned up. Although Ogston included Francoy in his compilation of SHC, this Frenchman probably slow-roasted upon the heat of decaying organic waste and died from pseudo-preternatural combustion of a distasteful but perfectly acceptable fashion.

Devonald reported a case from 1833 of a person enigmatically burnt in London. No details were provided.

Around the same time, a Frenchman dining in London presented a scene no less extraordinary than if he had been praising nineteenth-century British cooking. His beard and chest erupted in flames while he ate! Onlookers doused him with pitchers of water, but his "face, neck and chest were horribly charred." No mention of whether he had brandy with his last supper, or whether, *if* imbibing, he had spilled alcohol-rich brandy into his beard and then reached for a pipe and... or how other diners' appetites had been affected.

Canada's *St. Catharines Journal* (May 12, 1836) reported that an old drunken bachelor in Hamilton, Ontario, lay down opposite his fireplace "when a burning stick falling out, communicated with his clothes, and no assistance being at hand, he was burned to death." Maybe, the bachelor died a death actually due to SHC but blamed on a fireplace.

An encore *double bill* of inexplicable fire played in the Commune de Surville, France, on September 6, 1836. Monsieur and Madame Bernard Larivière, ages seventy-three and sixty-five respectively, both known for their consumption of alcohol, were destroyed by fire. Dr. Joly visited their room, which was covered in gray soot. It reeked of a "strong empyreumatic smell," he remarked. In front of the hearth, beside a table strewn with glasses and bottles of brandy, he found "the cinders of an extinguished fire" arrayed in an X on the floor, depicting the pattern in which one victim had burned atop the other. Mr. and Mrs. Larivière now consisted of two pairs of legs burnt off an inch above the knees, their black wool stockings and cloth slippers intact; a few pieces of whitish bone and vertebrae; and the rest "a black, shapeless, carbonaceous mass... calculated not to exceed four pounds." All this, said Joly, accomplished by a fire which lasted "less than 14 hours!"[2]

Imagine two adults together weighing less than four pounds—surely you would say this fire had almost totally *consumed the couple!* Would you not?

Dr. Joly rummaged through the melted fat of what had been their torsos and located livers and lungs shrunken to half size; their brains,

> Few things are harder to put up with than the annoyance of a good example.
> —Mark Twain

"about the size of a hen's egg." Unfortunately, his report in *Journal des Connaissances Médico-chirurgicales* (September 1836) neglects to mention whether their skulls also shrank. (This point is important, not for its gruesomeness but because it could have had important bearing on cases that would occur more than a century later.)

The fire was markedly localized, if Dr. Joly's report titled "Remarkable Case of Spontaneous Combustion" is to be believed. A chair and its straw cushion beside the victims was but partially burned. A table at their feet was unburnt. A wooden shoe, unmarked, rested next to Mr. Larivière. A few inches above the bodies was a broom "made of rush, which was scarcely singed on one side, and some matches, the sulphurous end of which projected beyond a sabot which contained them." The matches had not ignited. There was no fire in the fireplace.

The *British Foreign and Medical Review* (1837) reprinted Joly's report, and noted that it emulated other cases of spontaneous combustion. With one difference. "It offers also an example of two individuals placed in such identical physiological conditions," said the *Review* editors, "that the combustion affected them both in the same degree and in the same parts."

Sometime prior to October 1836, fire struck down another citizen of France addicted to brandy. This time it was a seventy-four-year-old obese woman of Aunay in Avalon. Overnight she became, said the *Medico-Chirurgical Review* (October 1836), "a heap of something burnt to cinders, at the end of which was a head, a neck, the upper part of a body, and one arm. At the other end were some of the lower parts, and one leg still retaining a very clean shoe and stocking." The mayor of Avalon beheld "a blue flame which played along the surface of a long train of grease (or serous liquor)"—fat which had exuded from the combusting flesh? He concluded that she died by "that which is called spontaneous combustion of the human frame" after she blew on some embers.

That blowing on an ember would beget a body-ashing backfire defines a new hazard for mankind! Not spontaneous combustion perhaps, but preternatural to be sure. Maybe, on the other hand, his conclusion as to cause was in error. The ember was innocent, for human combustion without external fire does seem to have precedent.

Theresa Lemaitre, a sixty-year-old alcoholic in Paris, mimicked her French predecessors in the way she burned on December 15, 1836. Her body looked "more like a mummy" than a recently deceased woman and it crumbled to dust when touched, reported Dr. Patrix in *La Lancette Française* (December 1836). He found no outside source anywhere for the fire.

Mrs. Bonjours, a forty-three-year-old Belgian widow and heavy drinker, had a not-so-good day at Nieppe on March 30, 1837. According to Hergt in his monograph on SHC, she "died of Selfcombustion" and "only a few charred remnants of her body [were] found."

The fire that burns some alcoholics with a vengeance appeared next inside Abdallah Ben-Ali, an Algiers Moor aged forty-five to fifty. One day in October 1839, onlookers watched him become enveloped by a blue flame. Dr. Bubbe-Liévin stated in the *Journal des Connoissances Médicales* that when the flame vanished, so too had three-quarters of Ben-Ali's body.

Later in 1839, a New York City woman died four hours after being discovered engulfed in flames. Neighbors who had seen her only one hour before were alerted to her plight by "volumes of smoke" gushing from her body. The fire was extinguished "with great difficulty," said *St. Catharines Journal* (December 19, 1839). Investigators deduced she had died "by that well attested but very rare occurrence, spontaneous combustion." No mention was made whether she was an alcoholic.

Two years later, 1841, saw publication of a thesis by Benjamin Frank, titled *De Combustione Spontanea Humani Corporis. Commentatio historica physiologica et medico-forensis de sententia gratiosi medicorum ordinis in Certamine Literario civium Academiae Georgiae Augustae.* (They don't title books today the way they used to, eh?) Frank's monograph is almost as thick as its title is long. In it, he catalogued forty cases culled from the medical literature, cases which appeared to him (or to those who originally reported them) to be spontaneous combustion of the human body. For many cases he gave multiple citations—the mark of a scholar—many of which are impossibly obscure today. His own treatise is no less difficult to find. I have held in my hands perhaps the only original edition available in the United States. It is the finest example of SHC-related literature I have uncovered up to 1841, and some of the cases presented here are courtesy of Frank's scholarship.

Several more cases happened in the early 1840s; cases that, naturally, Frank did not chronicle. Too bad. Details of these episodes are sketchy, at best. In 1841 or earlier, an English woman, age forty, was found one morning still burning. With trepidation (I suspect), Dr. F. S. Reynolds informed the Manchester Pathological Society that her femora had carbonized and knee-joints had split open *inside her unharmed stockings!*

A German(?) male suffered facial burns that he claimed resulted from vomiting the alcoholic contents of his stomach, recorded Henke's *Zeitsch. E. H.* (1842). An incentive for reducing (if not to give up entirely) one's imbibing, for sure.

Henke also mentioned a fat, drunken German(?) female, seventy, who—like Countess Bandi—bathed in alcohol and whose "body completely carbonized," leaving behind only her hand and feet. Henke cited a third incident: a sixty-year-old German(?) beggar who tried to sleep off his alcoholic stupor overnight in a tavern but was found the next morning "reduced to ashes" amid an unspecified odor and soot. His lower legs and shoes lay on the floor, untouched by fire, and his bread pouch was undamaged where it lay across the ashes of his chest.

On January 6, 1847, Monsieur Charbonnier was discovered surrounded by a whitish flame that receded when approached by one of the witnesses to it. The story produced a brief media sensation—and a trial for murder. After all, surely he must be a victim of foul play! The seventy-one-year-old Charbonnier was neither corpulent nor a drunkard, but that could not save him from death by SHC, declared French physician Dr. Masson, who recommended to the *Gazette Médicale* that the incident "be referred to the class of spontaneous combustion." The court agreed. The defendant was acquitted of murder, and the verdict in favor of SHC quickly vanished into the obscurity of jurisprudence.

As the decade came to a close, renown Paris alcoholic Xavier C– made a bet at a tavern near the Barrière de l'Etoile that he could eat a lighted candle. Scarcely had he brought the candle to his lips when he uttered a slight cry and fell helpless to the floor. He had lost his bet. The *Gazette des Tribunaux* (February 25, 1850)[3] explained why: "A bluish flame was seen to flicker about his lips, and on an attempt being made to offer him assistance the bystanders were horrorstruck to find that he was burning internally. At the end of half an hour his head and the upper part of his chest were reduced to charcoal.... A handful of dust on the spot where the victim lay is all that remains."

Two doctors declared that "Xavier had fallen a victim to spontaneous combustion." It was a determination no one in the tavern would bet against.

COUNTESS VON GÖRLITZ AND BARON VON LIEBIG: THE ROYAL DOWNFALL OF SHC?

These individual apocalyptic inflamings were being reported too often and causing too much controversy within the medical-scientific community to be ignored. *Was* or *was not* this tantalizing phenomenon happening? If SHC truly occurred, it meant a large gap existed in knowledge about the human body. If the events resulted from misrepresented observations or uninformed conclusions, then a lot of men of science needed to smarten up and the mystery proven illusionary once and for all.

With men of medicine and no less than two courts of law affirming the occurrence of spontaneous combustion in humans by the middle of the nineteenth century, why was this phenomenon so little known (and vigorously denied) nearly 150 years later?

> Today hardly anybody believes in this strange phenomenon...
> —Mogens Thomsen, M.D., Ph.D.;
> "Combustio humana spontanea Fup og faktum," *Særtryk fra Dansk medicinhistorisk årbog* (1978)

The answer resides in the conjunction of two esteemed professions of the period: the legal and the scientific. And in their shared role in one milestone event.

The event began unfolding on June 13, 1847, in the house of Count von Görlitz of Germany. That afternoon, the count had left to attend a party, leaving his countess at home with a manservant named Stauff. When Count von Görlitz returned in the evening, he could not find his wife. By eleven o'clock, he had tradesmen break down the locked door to his wife's bedroom, where she often spent hours in solitude. There they found great disorder: the countess's writing desk was afire and a stinking smell pervaded the bedchamber. The greatest disorder, however, was evident in the Countess von Görlitz herself. Her head was badly charred, though her mouth was wide open and her tongue distended. Deep burns lacerated most of her torso, and the joints of her arms were exposed. Her lower extremities were untouched.

Dr. J. A. Graff, the medical director of nearby Darmstadt, was summoned early the following morning. Graff found no signs of violence to the body. And despite the fire having failed to significantly destroy the body, Graff suggested that the countess had succumbed not to murder or a dropped candle but to so-called spontaneous human combustion.

Debate raged immediately as to whether this royal's death was due to SHC or she had been the victim of a crime. A second doctor judiciously rejected Graff's assessment, while a third concurred with Graff. The body was removed from its coffin, and reexamination revealed a long crack in the skull, which could have resulted from either a concussion or the effects of fire. To answer such questions, the Supreme Court convened to hear evidence amid frenzied speculation among the populace and media. It was the O. J. Simpson trial of its day.

Called to the stand were two specialists in chemistry, professors T. L. W. Bischoff (1792–1870) and Baron Justus von Liebig (1803–1873).

Bischoff gave an account[*] of experiments he had undertaken when he burned a corpse at the Anatomical Institute. He was unable to produce any crack in the skull like the countess herself exhibited.

But the big gun of this legal Dream Team was the Baron. With conviction and aptitude, von Liebig applied his considerable scientific reputation to the challenge of SHC and to putting an end to the howling debate that stormed among his colleagues.

Von Liebig was a luminary in the nineteenth century scientific community. He helped establish the science of organic chemistry, the study of compounds that contain carbon (which includes the human body). He devised techniques to improve methods of organic analysis and increased knowledge about the preparation of foods through his book *Chemistry of Food*. Born in Darmstadt, Germany, he later directed one of the world's most renown chemical laboratories at the University of Giessen. His reputation was already well-established by 1850.

When von Liebig spoke, science listened.

Perhaps because von Liebig had connections with both Darmstadt and Paris— each city now associated with the subject of SHC—it was natural for him to

---

[*] Bischoff, T. L. W. (1850): "Zusätze und Bemerkungen zu der Abhandlung des Herrn Med.-Direktor Dr. Graff zu Darmstadt uber die Todesart der Gräfin v. Görlitz, mit besonderem Hinblicke auf die Lehre von der Selbstverbrennung," Z. Stattsarzneik, vol. 60.

focus his considerable intellect and chemical expertise on the controversy. In 1850, von Liebig spoke to the Court about spontaneous human combustion, testimony that was published later that year in his treatise *Zur Beurtheilung der Selbstverbrennungen des menschlichen Körpers*. He reviewed the cases known to him and discovered that nearly all of his fifty cases of alleged SHC shared a mere six (remember, Lair had earlier identified eight) characteristics:

1. SHC occurred in winter.
2. SHC victims were alcoholics and drunk at the time of death.
3. SHC occurred in rooms heated by open fireplaces or pans containing glowing charcoal.
4. SHC never occurred in the presence of witnesses.
5. SHC was never observed by any physician, and no physician ever tried to determine conditions preceding the fatal fire.
6. SHC occurred over an unknown length of time, no one ever having determined the combustion's elapsed time.

Already you know that *each* point von Liebig established as criteria for the episodes labeled SHC can be tossed out. Just as von Liebig chose to toss out three cases listed by Frank in which there was no external fire in the victims' vicinities because, said von Liebig, these three cases were "totally unworthy of belief."

Von Liebig did not stop there. Being a laboratory scientist, he performed quantitative experiments on human tissue to see *if* what he considered ideal circumstances *might* burn the human body so thoroughly to ashes. For him, the ideal circumstances had to involve flesh wholly saturated by spirituous liquors. As any organic chemist must have known, alcohol will burn, often with a bluish flame. Tissue absorbs alcohol. All victims of alleged SHC were notorious drunkards. Weren't they? It was a relationship that seemed as tight as enamel on teeth.

A relationship that, nevertheless, von Liebig doubted. After all, he pointed out, the *corpses* of drunkards had never been seen to spontaneously combust, so why should living imbibers?

Nonetheless, the Baron tested the hypothesis (as had his colleague Bischoff, and Fontanelle, before him). His experiment was definitive; the result negative. Tissue did not become ash. Hypothesis disproved.

Consequently, on March 11, 1850, the servant Johan Stauff was accused of robbery, fire-raising, and murder. Found guilty by the Supreme Court, he was sentenced to life imprisonment.

Years later, Stauff confessed that he had, in fact, entered the bedchamber of Countess von Görlitz, who caught him in the act of purloining jewelry and money from her writing desk. A struggle ensued, and he viciously struck and then strangled the countess. To conceal the crime, he heaped combustibles atop her body and set the pile alight. He was guilty of designing the countess' funeral pyre, one which proved ineffective to hide his evildoing.

For Stauff, the court's verdict was correct. But was its verdict right for SHC?

PUTTING THE BARON'S VERDICT ON APPEAL

What Bischoff and von Liebig showed to the world was this: flesh saturated with alcohol for a great length of time is not catastrophically combustible. When it is ignited, the alcohol burns off and the flames die. Combustion stops, having produced at best only the most superficial charring.

An illustrative example, if unpleasant conceptually, comes from *A Handbook of the Practice of Forensic Medicine* (1861): a five-month fetus "which had lain in spirits as an anatomical preparation for an unknown length of time, and whose tissues must therefore have been even more saturated with combustible matter than those of the most inveterate toper could ever become, was so exposed to the most intense flame of a small chemical glass and metal-smelting furnace, that the apex of the cone of flame was applied to the body. In a few minutes the skin commenced to burn, the flame was at once removed, and the body—instantaneously ceased to burn." The experimenter repeated this procedure at least ten times with the same result: merely patches of burnt skin, "by no means a 'spontaneous combustion' of the perfectly alcoholized body."

I have done the same experiment—but with a cut of fatty beef marinated for nine months in a blend of gin, vodka, whiskey, and beer. What happened? Just to ignite the meat after its removal from solution required *several* lighted matches; after the alcohol flared off and the pretty blue flames ceased, lo and behold! the beef *wasn't the least bit incinerated*. As anyone who cooks on a grill already knows: the most well-done marbled steak doesn't dissolve into ashes over the charcoal! Thank goodness.

Von Liebig and Bischoff had determined that alcohol-saturated bodies do *not* burn themselves up, even in the presence of external flame. Therefore, Baron von Liebig starkly *decreed* that SHC must disappear from the framework of science.

Not everyone agreed. In 1852, Marie Guillaume Devergie published a more comprehensive study of SHC than had von Liebig. In his "De la Combustion Humaine Spontanée," Devergie sharply disagreed with von Liebig's conclusion. He challenged the interpretation of the latter's experiments, and boldly asserted that molecular changes may be occurring in the living body by which it then becomes combustible from the absorption of alcohol; or, once inside a person, from alcohol's conversion to more flammable compounds.

Dr. Marc, another physician who took an interest in SHC, suggested that flammable gases—even phosphorated hydrogen (which under certain circumstances will ignite in contact with air)—may be generated in the living body, hence giving rise to true spontaneous combustion in humans. J. H. Kopp had suggested the same theory way back in 1811, in *Ausführliche Darstellung*.

Devergie, Marc, Kopp, and of course Fontanelle did not have Baron von Liebig's reputation, however, and their protestations fell largely on the ears of the deaf scientific community. Von Liebig contended that the concept of, and belief in, SHC had arisen at a time when mankind entertained entirely false views about combustion and its causes. In their uninformed ignorance about combustion, they readily assigned marvelous and miraculous traits to mysterious fires...especially fires that *seemed* to originate inside people. Unsophisticated people are, after all, prone to ascribe to the marvelous what they don't understand.

But it was now 1850! Science had explained combustion, von Liebig pointed out, thanks largely to the research of Davy fifty years earlier. Jurisprudence and science had just joined forces in a famous trial, irreversibly replacing a romantic philosophy about nature with modern logical, experimental natural science. The old-fashioned marvel of SHC was an antique medical delusion that had attained a dubious sort of credence; now it must be discarded as a relic of naive misunderstanding.

Is this what von Liebig's experiments actually demonstrated? No. They proved something quite different from what he concluded.

Let's imagine that the Baron found a Lamborghini automobile in his laboratory, with a note instructing him to put fuel in the tank and turn the ignition key. Von Liebig pours a blend of whiskey and brandy in the tank, knowing each will burn and is therefore a proper fuel. He turns the key, and goes nowhere. See, he exclaims, the car doesn't move! It is worthless as transportation; give me a horse! His conclusion is wrong, of course, because his reasoning was in error: he chose the wrong fuel. By pouring in gasoline instead, von Liebig would be motoring wide-eyed down carriage paths at speeds he never could have imagined!

What von Liebig had actually demonstrated was that *one* type of combustion—the oxidizing of alcohol—could not cremate flesh. That von Liebig could not imagine SHC without the exothermic burning of alcohol is a limitation of his reasoning. That he failed to demonstrate SHC is the result of his choosing the wrong fuel—and inadequately comprehending the full range of effects that alcohol can produce inside the human body once it is ingested.

What von Liebig did accomplish was to dismiss one aspect of one alleged cause behind the deaths of people who inflamed in a fashion certainly out of the ordinary. His experimental work did not negate SHC *per se*.

Science did not see it this way, however. Liebig had the last word. His pronouncement would stand, even though it was not reasoned scientifically. SHC would henceforth be a historical curiosity, forever laid to rest as a relic of an Age of Silliness. As a popular 1905 encyclopedia, *The Student's Reference Work*, remarked about spontaneous [human] combustion: "Liebig says a dead body filled with alcohol may burn of itself, but not a living one, in which the blood is circulating." At last, SHC was dead and buried.

> It is not a burning question. It is not even a question. Not a single textbook on forensic medicine written this century considers the phenomenon a possible cause of death.
> —science writer Martin Gardner, on SHC; "A book for burning," *Nature* (Nov. 26, 1992)

Despite von Liebig's pronouncement, not everyone discounted the evidence at hand.

In 1850, German physician von Gorup-Besanez wrote to *Carl Christian Schmidt's Jahrbücher der in und ausländischen gesammten medicin* that, while he agreed the Countess Görlitz had not died by *selbstverbrennung*, nonetheless "the possibility of a selfcombustion appears not to be refuted by Liebig's writing, even in regard to the chemical cause of it."

In 1854, Dr. Joel Shew proclaimed that "long-continued intemperance" causing combustion "easily excited or spontaneous, is abundantly proved." He called this "rare" phenomenon *Catacausis Ebriosa*.

In 1857, Dr. R. T. Trall recognized the same ailment, calling it *alcoholic diathesis* and saying it is "too well authenticated to be longer doubted." And somewhat belatedly, Archie Stockwell reviewed the evidence and found sufficient cause to announce in *The Therapeutic Gazette* (1889) a new malady caused by drinking too much alcohol: "Catacausis Ebriosus (Spontaneous Combustion)."

Clearly the final verdict was not yet in. And von Liebig notwithstanding, people continued to burn up with impossible thoroughness.

5

THE VERdict Is IN?

> Approach each new problem not with a view of finding what you hope will be there, but to get the truth, the realities that must be grappled with. You may not like what you find.... But do not deceive yourself as to what you do find to be the facts of the situation.
>
> —Bernard Baruch

Verdicts stand and verdicts fall on the weight of evidence and the wisdom of those who judge the evidence. In legal parlance, it is the principle of "Judgment contrary to the evidence"—which is to say, when all evidence points away from the judgment, the judgment cannot stand.

That principle was invoked in 1725 when Jean Millet stood accused of killing his wife by fire, a crime for which he was convicted *until* LeCat pointed out that the evidence spoke not to murder but to SHC. The verdict against Monsieur Millet was overturned on appeal, and jurisprudence admitted internal fire as a credible defense. This determination was subsequently upheld when a defendant accused in Charbonnier's death was exonerated as the verdict of guilt again fell to SHC.

This legal precedent was later overturned, not by the courts, but in the laboratory of Baron von Liebig, who exemplified the risk inherent in what Oswald Spangler's *The Decline of the West* portrayed as rationalism's tendency of "always looking for *the* solution to *the* question. It was never seen that many questioners implies many answers..."

Von Liebig imagined only *one* solution to SHC and, finding it wanting, determined myopically that no further proof against SHC was needed. The case for SHC was slammed shut tighter than a fire door.

Never mind that humans continued to incinerate themselves thoroughly, and raise *the* question von Liebig and his disciples now ignored. Indeed, human beings were burning up so often in Western Europe by this time that the greatest writer of the time felt compelled to explore the phenomenon.

SHC AND THE DICKENSIAN MINDSET

Charles Dickens (1812–1870) needs no introduction beyond his name. With a keen eye for absurdity, humor, mystery, and tender pity, he championed the poor and exposed the villainous policies of heartless industrialists. With an equally keen mind, he pursued what Dickensian scholar Ann Y. Wilkinson called an obsession with all aspects of fire and combustion.

In his Faraday series on science for *Household Words* (September 1850), Dickens explains the means of generating bodily heat and asks if people shouldn't be liable to inflame occasionally? "It is said," comes the disquieting answer, "that spontaneous combustion does sometimes happen; particularly in great spirit drinkers."

Followers of Dickens could not have guessed what he planned for them next. If *Household Words* was the set-up, *Bleak House* would be the sting.

Bleak House was first published as a magazine serial during 1852–1853. It received accolades for its attack on the insufferable lawsuits in English chancery courts. Today, however, it is probably best known for the way its sinister protagonist Chancellor Krook got his comeuppance in Chapter 33.

As London's clocks toll midnight, Tony and Mr. Guppy arrive at Krook's small apartment. Although Tony had visited Krook only moments earlier, the room is now silent and dim with "a dark greasy coating on the walls and ceiling." Evil haunts the place. Where Krook was seated, now some "crumbled black thing is upon the floor." The two lads look closer:

> [S]omething on the ground, before the fire and between the two chairs. What is it? Hold up the light.
>
> Here is a small burnt patch of flooring; here is the tinder from a little bundle of burnt paper, but not so light as usual, seeming to be steeped in something; and here is—is it the cinder of a small charred and broken log of wood sprinkled with white ashes, or is it coal? O Horror, he IS here! and this from which we run away, striking out the light and overturning one another into the street, is all that represents him.
>
> Help, help, help! come into this house for Heaven's sake!

Plenty will come in, but none can help. The Lord Chancellor of that Court, true to his title in his last act, has died the death of all Lord Chancellors in all Courts, and of all authorities in all places under all names soever, where false pretences are made, and where injustice is done. Call the death by any name Your Highness will, attribute it to whom you will, it is the same death eternally—inborn, inbred, engendered in the corrupted humours of the vicious body itself, and that only—Spontaneous Combustion, and none other of all the deaths than can be died.

While most of his contemporaries were striving to bury SHC, Dickens grasped its inherent literary power. As J. Hillis Miller would comment a hundred years later in *Twentieth Century Interpretations of Bleak House* (1958): "Krook is transformed into the basic elements of the world of the novel, fog and mud. The heavy odor in the air, as if bad pork chops were frying, and the 'thick yellow liquor' which forms on the window sill as Krook burns into the circumambient atmosphere, are particularly horrible versions of these elements."

The evil which Krook personified is removed, cleansed, banished; society is symbolically and literally purged by a magical fire that utterly destroys the wickedness and nothing else. By imposing moral implications on SHC, Dickens used Krook's death to convey the horrible punishment commensurate with the great sin that the champion of the downtrodden so abhorred: indifference to moral wrongs. Krook leaves not a trace of himself behind in which his evil can live on. All grossness is  sublimated, refined by the chemical magic of SHC into symbolic art; into matter purified.

Through SHC, Dickens's apocalyptic symbology succeeded. To some Victorian minds, though, it succeeded in exceeding literary propriety.

Dickens ignited a literary firestorm among his critics, notably the rationalist George Henry Lewes. Writing to *The Leader*, Lewes told the world he "objected to the episode of Krook's death by spontaneous combustion as overstepping the limits of fiction and giving currency to a vulgar error." (I didn't realize fiction needed limits to be fiction!)

In the first bound printing of *Bleak House* (1853), Dickens defended himself. "I have no need to observe that I do not wilfully or negligently mislead my readers," Dickens wrote about SHC in his preface, "and that before I wrote that description I took pains to investigate the subject." Using Countess Bandi (misspelled by him as "Baudi Cesenate") as the pattern for Krook, he said he also relied on "the recorded opinions and experiences of distinguished medical professors, French, English, and Scotch" who supported SHC. Concluded Dickens: "I shall not abandon the facts until there shall have been a considerable Spontaneous Combustion of the testimony on which human occurrences are usually received."

Lewes persisted. He wrote haughtily to Dickens, "I believe you will not find one eminent organic chemist who credits Spontaneous Combustion."

Dickens would not capitulate. Two years before his death, in the 1868 edition of *Bleak House*, he again countered Lewes. Not only was there evidence of SHC past, Dickens insisted, there is SHC present. "Another case, very closely described by a dentist," he stated, "occurs at the town of Columbus, in the United States of America, quite recently. The subject was a German who kept a liquor shop, and was an inveterate drunkard."

Nonetheless, Lewes continued to criticize. So vitriolic was his attack against Dickens in the *Fortnight Review* (February 1872) that a Mr. Forster responded by calling Lewes and his article "odious by intolerable assumptions of an indulgent superiority." Ah, how the Victorians could turn a phrase.

Towards a Vindication of the Literary Champion of SHC

As von Liebig had triumphed in eliminating SHC from scientific consideration, Lewes was victorious in the literary ring: his intellectual fisticuffs scored a TKO on SHC despite its championing by the foremost English literatus. SHC was trounced once more. However, Lewes could have spent his time more productively (and in the long run less embarrassingly) had he read his country's newspapers and researched medical journals instead of bashing Mr. Dickens.

Shortly before November 16, 1867; Downpatrick, England. Hugh McMullen is asphyxiated by a blaze that killed his mother, Mary McMullen, in an adjacent room.

> What is research but a blind date with knowledge.
> —anonymous

Reported *The Downpatrick Recorder*, only hands, breast, both feet, and lower legs comprised the "few fragments of what had been his mother." Dr. Newport White noted minimal damage in the house, and said he was "strongly inclined" to rule in favor of SHC regarding Mrs. McMullen. But officially, and inoffensively, he recorded simply "Burnt to death."

In early 1868 or earlier, a drunken German-American burned up in Columbus, said a Dr. Watkins, who investigated. The body was found in a fireplace, true; but Dr. Watkins felt the extent of injury was too complete to be attributed to it. The body was charred; its genitals totally destroyed; its muscles "dissolved." Yet the old man's silk scarf and shirt were unburned. Dr. Watkins was right in looking beyond the fireplace for an explanation! He said the octogenarian died in Arkansas or Mississippi or Texas.

Of these three states, only Texas has a large town named Columbus. Thus, I believe, I have both pinpointed the location of this otherwise overlooked incident of SHC... and identified the basis for the "recent" case of the liquor shop proprietor who Dickens cited in his second edition of *Bleak House*.

March 14, 1869; Aberdeen, Scotland. Mrs. Warrack (or Ross) drank one too many Scotches or did something else that caused her thorough combustion. Within *no more than sixty minutes* a fire blackened the bones of this sixty-six-year-old Scot; made a "black mass" of her intestines; roasted her right thigh like beef; totally burnt her muscles and skin; and made a "shapeless cinder" of her right foot and left leg and foot. Otherwise, she was greasy charcoal, said Dr. Ogston, who viewed the scene. Her clothing burnt off, but there was no other damage beyond the stairstep and railing where she had collapsed. Ogston could not even tell if she had been climbing or descending the stairs, so little remained of her.

Around August 1, 1869, another intemperate Frenchwoman in Paris, thirty-seven, defied science and Mr. Lewes. The police sent for Dr. Bertholle, who filed a clinical and detailed report: left arm totally consumed, right hand a cinder, no trace of internal organs in the thorax, abdominal organs unrecognizable. Bedclothes, mattresses, curtains, all other things in the small room remained isolated from the raging furnace within her. Neighbors had heard no cry of alarm or pain. Her unique passing earned her a footnote in history.

December 1869; Alexandria, Virginia. A. B. Flowers inflamed in a fashion found noteworthy of mention by *Scientific American* (1870).

November 12, 1871; New York City. Mary Sullivan, eighty-five, was found burned to death amid little fire damage to her furnishings. An invalid, she *may* have fallen on a candle, said *The New York Times* (November 13, 1871).

The Combustible Widow Eliza Collier

As I scanned through volume after obscure, dusty volume of *The Philadelphia Medical Times* at the College of Physicians, I suddenly felt like the prospector who chances upon a ten-pound nugget lying hidden in the gravel of a river bed. *Pay dirt!* The Eureka! moment.

Gold of a different sort was retrieved that morning from the muck of history forgotten. Here, I reintroduce to the world the century-lost story of Mrs. Eliza Collier.

On the morning of December 1, 1872, Dr. J. M. Collier conducted an inquest at the weathered-pine home of the late widow Collier (apparently no relation) on the outskirts of Orion, Georgia. There he saw, about nine feet from the fireplace, the astonishingly burned and split-asunder body of the aged woman. Dr. Collier found it sufficiently amazing to describe with anatomical precision the state of her cadaver to the *Times* (April 12, 1873). Suffice it to say that her ribs lay upon the earthen floor; her heart had shrunken to half-size; her left lung looked "like the upper leather of an old shoe which had been burned." Her head was so destroyed that her own daughter could not identify it. The only sign of flesh, he wrote, was on the face and left hand.

The evening before, a servant said, Mrs. Collier had been in her usual health. Now Dr. Collier had to use a hoe to gather up the charred pieces of the woman, saying "I could have put it all into a small candle-box."

He noticed several bruises on her head but found them insufficient to have caused death; nor could he find any blood stains. He attributed death solely to fire.

Dr. Collier looked diligently for the source of that fire. A burned shawl lay nearby, but no ash trail led either to it or the widow or the fireplace. No other combustibles in the fire-prone house had burned, not even a candle on the mantle. He checked the fireplace: "no evidence of any unusual fire having been in the fireplace, but just the contrary... the few remaining chunks were not scattered," which, in fact, indicated the fireplace had not been lit that night. Cobwebs showed no new deposit of soot, proving the burnt body had generated but little smoke.

Dr. Collier had previously seen bodies taken from fires that ravaged whole buildings. This case was unlike those, however. This scene, he conceded, "so far exceeds anything which I have seen [that it] brings me...to believe that it is a case of what may truly be called spontaneous (?) combustion."

Dr. Collier inserted a question mark because, he explained, "I do not believe in any such; spontaneous is not the word; but there must have been some predisposing cause for such burning, and I trust you will be able to throw light on it." Neither the editor nor readers of the *Times* could do so.

Dr. Collier himself admitted his meticulous observation discerned no external cause! He further admitted that what he observed duplicated dozens of similar inflamings chronicled by Devergie, except that Mrs. Collier differed in being "the merest shadow of any living person I ever saw" rather than obese.

It was the diagnosis, not the symptoms, that Dr. Collier would not accept. He needed to heed the words of Dickens, who in 1852 had anticipated such baseless rejection of SHC when he wrote "Call the death by any name..."

WHAT *DO* WE CALL THESE DEATHS?

In 1876, the Reverend Adams of the Stockcross parish in Newbury, England, was staying in a New York City hotel where, it was reported, the clergyman one day gave up his holy ghost to the flames of spontaneous human combustion.

During the first week of September, 1883, death visited Mrs. Murphy of Norwood, Australia. Corpulent, eccentric, and living alone, the widow was found "covered with burns...and half-consumed," reported the *Adelaide Observer* (September 15, 1883). Confoundingly, "the body was clothed in a nightdress which was uninjured, and one hand showed no marks of fire except a scorched thumb." The inquest hearing must have been a wonder to behold, confronted as it was with such befuddling facts. Its jury did its duty, and announced its findings: (1) lack of criminal intent, and (2) "deceased came to her death by burning, but that there was not sufficient evidence to show how it happened."

How might von Liebig have designed an experiment to duplicate the circumstances which Mrs. Murphy presented?

On January 24, 1884, the clothing upon a Mrs. Sheller suddenly took fire and burned her. Dr. Carter dressed the burns. Then the bandages themselves incongruously inflamed. In reporting this curiosity to the *American Medical Journal of St. Louis*, he blamed "spontaneous combustion" due to an escape of "caloric"★ from her burns, which then ignited the oil in the dressings he applied to those burns. Or not.

The next instance in the chronology of curious combustion brings us back to Britain, and fifteen years forward in time. The date is January 5, 1899, at the home of John Henry Kirby, and his estranged wife, Sara Ann Kirby. Though the couple lived in two dwellings distant by a mile in Sowerby Bridge, west-southwest of Leeds in West Yorkshire, they—or their domiciles (both

---

★*Old Physics*—a hypothetical fluid in matter which determined its thermal state.

physically apart)—may have been linked in ways unimagined then, and not understood now.

At eleven o'clock on that Thursday morning, Mrs. Kirby went outside her Hargreaves Terrace home to fetch water at a well, some twenty yards away. She returned within two minutes, to hear wretched screaming from four-year-old daughter Amy Kirby. Amy was ablaze. A neighbor who rushed in would later say that flames a yard high flared from little Amy's head. Mrs. Kirby would tell the inquest jury, "If she had had paraffin oil [kerosene] thrown over her she would not have burned faster." Somehow, the fire in Amy died out, and an ambulance was called for.

Overcome by grief, Mrs. Kirby trudged down Fall Lane to tell her husband about the tragedy which had befallen their daughter. Horror soon compounded distress. Halfway across the valley to Mr. Kirby's home at 45 Wakefield Road, she encountered a messenger who was racing to find her. Amy's sibling—five-year-old Alice Kirby—had herself shortly before been found ablaze! Grandmother Susan Kirby, who was babysitting Alice, said she had left her granddaughter in bed and, after a brief visit to a neighbor, returned to find the little girl "enveloped in flames and almost burned to death"—*at exactly eleven o'clock.*

Amy and Alice arrived at the Royal Halifax Infirmary, transported by the same horse-drawn ambulance. There—in a playful twist of fortean lexicon—a Dr. Wellburn attended to his young burn patients as best he could. For naught. Alice died at three o'clock in the afternoon; her sister, just before midnight.

No spent matches, no charred paper, no embers from a fireplace were found at either fire scene, the inquest jury was told. The coroner, according to contemporary accounts that were summarized in an article appearing in the *Halifax Evening Courier* (April 13, 1985), spoke in terms of "strange" and "remarkable" and "shocking coincidence"—then dismissed the Kirby siblings' deaths as an accident. Admitted one juror, though: "We have no evidence to show how the fire occurred in either case."

> Coincidence is such an easy word for those things that are inexplicable.

Jenny Randles and Peter Hough, to whom credit is given for confirming the details of this astounding incident (which brought profound distress to the family), acknowledge in their book *Spontaneous Human Combustion* (1992) that crucial information is lost to history. Nothing is known about how localized these twin blazes were, for instance; or anything about the burn injuries themselves. "But in all fairness," these journalists ask, "just what are the chances of two sisters, living apart but separated by just a mile across a valley and whose homes were in sight of one another, catching fire at exactly the same time and with the same lack of evidence of a rational cause?"

BLAZING A TRAIL INTO THE TWENTIETH CENTURY

Despite these many cases, people at the turn of the present century still believed Lewes had prevailed in the match fought over SHC, as a letter by E. Muirhead Little to *The British Medical Journal* (August 26, 1905) attests. "If an intelligence so active and acute as that of Dickens was imposed upon by the pseudo-scientific evidence in favor of this marvel," said Little about SHC, "it is small wonder that a belief in it dies hard among the wonder loving and superstitious people who form the majority of the public."

Pity the lowly common man who knows not how ignorant he collectively is.

But wait! Why are those who believe in spontaneous human combustion not dissuaded by all the evidence to the contrary? Because all the evidence to the contrary is not entirely dissuasive.

Immediately following Little's denunciation of poor gullible Dickens and other SHC advocates, the BMJ printed a letter from Dr. E. G. Archer, member of the Royal College of Surgeons. In delightfully ironic juxtaposition, Dr. Archer described a case he investigated in 1903–1904, which would offer little comfort to Mr. Little.

Since Dr. Archer's report has been overlooked by history, I quote at length his testimony about a very intemperate elderly Englishwoman who lived alone in a small house and was last seen alive in the evening reading a magazine by candle light:

> The following morning early a policeman passing noticed smoke issuing from the closed shutters of the sitting-room window, and the house was broken into. The upper part of the walls and the ceiling of the room were much scorched, but the furniture in the room was intact. No trace of the occupant was found at first, but a small heap of black debris was noticed on the floor in front of the chair, which was an iron one, and the chir's [*sic*] cover of which was destroyed. I was sent for, and found the small heap to consist of the broken calcined bones of a human body. They were lying in a small pyramid, on top of which lay the skull. All the bones were completely bleached and brittle, every particle of soft tissue had been consumed. A table covering with a baize cloth within 3 ft. of the remains was not even scorched.

Even if the woman's clothing took fire by candle, reasoned Dr. Archer, "how is one to account for the absolute cremation of a body in the midst of a sitting room filled with furniture? I may say the remains were seen with me by a brother practitioner, and we were both agreed that several features of this case were beyond comprehension."

Dr. Archer was not called to the inquest—I cannot imagine why!—but he did learn from the jury foreman that death was ruled "spontaneous combustion."

> When you have eliminated the impossible, whatever remains, however improbable, must be the truth.
> —Sherlock Holmes,
> *The Sign of the Four*

Two physicians confronted the real-life remains of a drunken woman reduced to a skull atop a pyramid of human ashes in a scene grimly resembling the demise of the fictional Mr. Krook, and they could not explain it.

Did life imitate art before the eyes of Dr. Archer and his colleague? Was Dickens vindicated at last? Had Lewes finally been handed his comeuppance?

Through the summer and fall of 1903, the Carter household of Rockbeare in Devon, England, was plagued with poltergeist-type* activity. In particular, Elizabeth Carter, nineteen, was plagued by accidents and deaths that focused around her; on her own body "the Devil's mark" appeared, said John Mackin in Reading's *Evening Post* (April 12, 1975). Just after Christmas a piercing scream brought Mr. Carter running into his daughter's room, where intense heat repelled him. Then worse. The discovery of Elizabeth; "her body was charred to a cinder."

December 16, 1904; Stirlingshire, Scotland. Mrs. Thomas Cochrane, fifty-one, "burned beyond recognition" in her Falkirk apartment. Of the woman, only "ashes remained"—though surrounding pillows were but slightly scorched and there was no fire in the room, declared next day's *Falkirk Herald*.

February 2, 1905; north-central London. Mrs. Maria Hall was jolted awake in her chair in front of the fireplace by fire. A fire in herself! To the consternation of any debunker eager to blame a leaping ember for her fatal combustion, *Lloyd's Weekly News* (February 5, 1905) politely pointed out that coroner Dr. Wynn Westcott stated the thirty-seven-year-old woman of the lower class had been sitting *facing* the fire—yet it was her *back* that had burned. Perhaps an ember flew out of the fireplace, circled the room, then cleverly and invisibly tunneled its way through the back of her chair before (finally) burning into her back.

February 4, 1905; Homerton, England. George Horn, 1 year old, suffered slight burns on his face and right arm, declared *Lloyd's Weekly News* (February

---

* Poltergeist—German for "noisy spirit"—encompasses psychokinetic phenomena, such as movement and manipulation of objects without direct physical contact hy buman agents.

5, 1905), when "fire originated through some linen which was airing catching alight." I suppose if an ember can navigate a complex trajectory to burn a middle-aged woman, so too can linen hung out to dry spontaneously singe a toddler. Lloyd's headlined its frontpage story, "Mystery Burns."

Still later that month, on February 26, Mr. and Mrs. Kiley of Butlock Heath, near Southampton, England, were found in their burning home. On the floor lay Mr. Kiley, dead of burns; in a nearby chair sat his wife, "badly charred, but recognizable." Evidence indicated both had died long before the house-fire began, and although an overturned lamp was found there, examination revealed it did not cause any of the fires. *The Hampshire Advertiser* (March 4, 1905) gave the jury's verdict: "Accidental death, but by what means, they were unable to determine."

Around eleven o'clock the night of January 27, 1907, neighbors noticed a "bright light" inside the Houck home in Pittsburgh, Pennsylvania. Mr. Houck arrived home at 1:00 A.M., and found himself locked out when his wife didn't respond to his knocks. He broke in through a basement window and, said *The Pittsburgh Post* (January 28, 1907), he "searched room after room but was unable to find his wife." Finally he discerned a faint odor of burning rags, and traced it to a bedroom. There, said *The Pittsburgh Press* (same date), the "cremated body" of Kate Houck, fifty-two, "was found in a sitting position against the wall." A physician said she had been dead two hours, which coincided with the burst of light. Police faced a mystery, said *The Post*, because of a peculiar circumstance: nothing but her body and a waste paper basket had burned up.

Near Dinapore, India, around May 13, 1907, two constables happened upon a woman living at Manner whose body was smoking. They searched her home for fire damage, reported the *Madras Mail* (May 13, 1907), and found none—not even her clothing was damaged. They carried the smoldering corpse in its unscorched raiment to the District Magistrate. The paperwork on this case must have been particularly troublesome to fill out.

On December 12, 1912, at Topsfield, Massachusetts, fifty-year-old farmhand James Welch was "found with the bones of his head burned to ashes and his clothes still smouldering, in the dining room of his home on the Dodd estate. The man had fallen with his head in the big fireplace." The hot embers needed to kindle his ignition were not to be mentioned, or found, by the reporter for the *Niagara Falls Gazette* (December 21, 1912).

On the morning of December 23, 1916, proprietor Thomas W. Morphey found his housekeeper, Lillian Green, badly burned body on the floor of the Lake Denmark Hotel located seven miles from Dover, New Jersey. The fire did not extend beyond the floorboards, which were only slightly charred. She could not explain anything to anyone about her predicament and succumbed soon afterward. There had been a blaze at a neighboring hotel overnight, but a "massive search" turned up nothing suspicious to link it to the fatally burned lady.

At 2:40 A.M. on April 6, 1919, firemen in Dartford, England, arrived at Hawley Manor because the house was ablaze... problematically ablaze. *The Dartford Chronicle* (April 11, 1919) said that the fire had rampaged throughout the house until it reached the door of one room, and there it stopped. In this room there was no fire, but therein lay J. Temple Thurston dead and scorched. Upon his clothes were no trace of scorch marks. The inquest pondered how his thighs and lower legs had burned red; pondered that he may have been dead hours before the fire outside his room began; pondered that evidence indicated he may have been unaware of his predicament, hence made no effort to escape or call for help.

Circa 1921 in the northern lumber camps around Calumet, Minnesota, Matti Kivimaki and another lumberjack were drinking wood alcohol with immediately fatal results. According to Rita Jones Lewis who heard the story from her mother, "it went through their systems so fast it just caught fire. They burned up, but not the cabin." They were found "all charred." Could the two loggers have been another prototype for *double* SHC?

Anomalous observations accumulate.

For four days in 1922, from March 9 through 12, the threat of SHC terrorized twenty-three-year-old Mrs. Ona Smith. Two mattresses, a wall calendar, a shawl worn by the invalid, and other items

> Anomalies are what science is about.
> —Ralph Markson,
> astmospheric scientist;
> Massachusetts Institute of Technology

already had ignited and burned to ashes as "blue flames burst from the air" in her room in Alva, Oklahoma, said *The New York Times* (March 14, 1922). Bedside watchers were kept busy protecting her and her possessions, while striving to keep one another from inflaming amid these frightful sparks—"their origin a mystery." The persecuting flames eventually returned from whence they came and Ona escaped physical injury, though her psychological trauma surely endured long afterward.

During the summer of 1922, sixty-eight-year-old Euphemia Johnson of Sydenham (London), England, was "consumed so by fire that on the floor of her room there was only a pile of calcined bones." The moderately plump widow had made herself a cup of tea, sat down to drink it, then fell to the floor afire (one presumes). Fort detailed this incident in *Lo!* and raised many questions about how the fire did what it did "if in an ordinary sense it was a fire..."

Shortly before January 24, 1930, Mrs. Stanley Lake passed away at Kingston, New York. The coroner investigated, said the *New York Sun* (January 24,

1930): "Although her body was severely burned, her clothing was not even scorched." No other damage reported.

Sometime in 1931, Father Edward W. Lynch was found dead lying crosswise upon his bed at St. Mary's College in Winona, Wisconsin. Said Brad Steiger in his 1973 newspaper column, "The corpse was charred to a crisp, but there was not a trace of fire anywhere else in the room. Not even the bed itself bore the slightest trace of flames." Only one other item had burned: the priest's prayer book.

During July 1931, at the Simpson home in Brighton Beach, Brooklyn, New York, unusual events including spontaneous fires centered around thirteen-year-old Elinor Simpson. One fire centered itself *under* her, as her bed ignited of its own accord, said Julie Murray's account in *Beyond Reality* (February 1980). What if Elinor had been the *cause* of these fires? I have dozens of cases where furniture appears to have self-destructed in fire of its own making. And what if Elinor had been lying *in* bed at the time?

The *New York Sun* (February 2, 1932)—Bladenboro, North Carolina: "Fires, which apparently spring from nowhere, consuming the household effects of C. H. Williamson, here, have placed this community in a state of excitement, and continue to burn. Saturday [January 30th] a window shade and curtain burned in the Williamson home. Since then fire has burst out in five rooms. Five window shades, bed coverings, tablecloths, and other effects have suddenly burst into flames, under the noses of the watchers. Williamson's daughter stood in the middle of the floor, with no fire near. Suddenly her dress ignited. That was too much, and household goods were removed from the house."

Enough fires of this type have occurred to fill a book. What makes this one of special interest is (1) the flames were blue, as at Ona Smith's house a decade earlier; (2) items taken outside did not catch fire, yet when taken back inside they did; (3) a young lady *narrowly escaped SHC* when the garment she wore ignited itself paranormally; and (4) the son of Charles H. Williamson personally confirmed to me that all these events

> Rathbone Oliver read to the American Association of the History of Medicine a paper in which he declared: "Perhaps, there were once spontaneous combustions, but now they have disappeared."... Absence of fireplaces and oil-lights reduces the chance a drunkard sotted with bad alcohol will come in contact with an open flame.
>
> He ends with a comforting conclusion: "We may, therefore, feel more or less assured that if we restrict ourselves to really good Scotch and Rye and if we avoid open fires and sit on the radiators, we shall not, in all probability, combust spontaneously."
> —J. Rathbone Oliver, "Spontaneous Combustion— a literary curiosity," *Bulletin of History and Medicine* (1936)

happened as reported in 1932. What was said then can still be said today: to explain the self-igniting household there is "no scientific theory."

May 28, 1938; Ipswich, again. This time nine Ipswich men were burned or otherwise injured by a "meteorite"—so the front page of *Reynold's News* for May 29 announced. We have scores of episodes of fires from heaven; always strange, occasionally fatal.

It was the start of a terribly inflammatory period in Great Britain, as Eric Frank Russell set out to document:

During September 11 through 19, 1938, a building in Leeds was plagued with fifteen inexplicable blazes in nine days, reported the *Daily Telegraph & Morning Post*. No report of any associated SHC.

September 19, 1938; South Woodford, Essex. Mrs. Janet Hadrill was "preparing breakfast when her dress caught fire," reported *The Daily Telegraph* (September 20, 1938), "and she died soon after admission to hospital." Perhaps she was careless; perhaps her dress ignited; perhaps *she* ignited. The information necessary to answer these possibilities is lost to history, buried along with Mrs. Hadrill.

September 19, 1938; Firth Park, Sheffield. Widowed Amelia Ridge, seventy-eight, burned at home on the floor near her fireplace; when found, her clothes still smoldered. Said *The Daily Telegraph* the next day: "She is believed to have fallen on the fire." Russell thought otherwise, that she might have fallen into the Unknown.

September 19, 1938; Carlisle. Mrs. Ellen Wright, seventy, was discovered "lying dead in the wreckage" of fire in her house. It was believed, said *The Daily Telegraph* (September 20, 1938), that she became "trapped while in bed." Russell said SHC might have entrapped her there.

September 19, 1938; the Aston area of Birmingham. Mrs. Sarah Pegler, age seventy-eight and a widow, was found burned and "lying dead on the floor." Russell suggested SHC, though it could be a common house fire.[1]

Around October 30 or 31, 1938; Newcastle-upon-Tyne. An unidentified woman blazed on Elswick Road. Based on a *Reynolds News* story, Russell posits SHC.

November 6, 1938; Summer Street, Stroud. Mrs. Sarah Butcher and her son Fred succumb to flames so intense, would-be rescuers are driven back, says Russell in quoting *Reynolds News* (November 6, 1938). A double tragedy? Or tragedy doubly compounded by mystery fire?

December 26, 1938; Croydon (London). Mrs. Florence Hill, age forty-five or fifty, had been ill until her bed inflamed this day. Then she died of "burns and shock" en route to Croydon General Hospital. No witnesses. Her death was attributed to SHC at home. I have accounts of bedding itself spontaneously combusting. Perhaps Mrs. Hill fell victim, if not to SHC, to spontaneous *bed combustion*.

December 26 or 27, 1938; Brixton. Mrs. Agnes Flight neither smoked nor was near a fire; nevertheless, she was the victim of fire. *The Daily Telegraph* (December 27, 1938) said her clothing caught fire naturally—whatever *that* means!

December 26 or 27, 1938; Downham, Kent. Mrs. Louise Gorringe is cited by Russell as a case of SHC. I cannot find a site in Kent for Downham, only in Essex and Cambridgeshire. I have been unable to find anything more to add about this incident.

Around December 28, 1938; Warrington. Harriet Lawless is said by Russell to have burned in her home under circumstances devoid of witnesses and explanation.

January 7, 1939; Makin Street, Liverpool. Mrs. N. Edwards was seen burning; rescuers were repelled by the heat of her combustion. There was no fire in the house at the time; she did not smoke tobacco. Said Russell: "Burnt to death by her own clothes, which caught fire nobody knows how."

August 3, 1941; location unknown. Lois Irene Chapman is found "sitting dead on a burning davenport...little damage to surroundings." *Doubt*, and later *Pursuit* (April 1975), put it down to SHC.

Around January 9 through 12, 1943; Deer Isle, Maine. On this island in Penobscot Bay south of Bangor, Allen M. Small, eighty-two, was found badly burned upon the carpeting of his Eggemoggin Reach Road home. His remains were found on January 14, though the Ellsworth *American* (January 20, 1943) said he had been "dead several days." The carpet beneath him was charred, but no other fire damage was to be seen. I visited Ellsworth in the 1970s to seek more information, but the trail of evidence had long ago gone cold.

February 1, 1943; Lancaster, New York. Invalid Arthur Baugard, thirty-nine, burned beyond recognition in his room, which exhibited no damage; nor was there a fire elsewhere in the dwelling. "Sheriff's office launched investigation" said *Doubt* (June 1943). But today no one I spoke to in Lancaster remembers what (if any) solution was agreed to.

November 19, 1943; Aldingbourne, England. Mrs. Madge Knight, forty-three and "perfectly happy" in her marriage even when not drinking five to six whiskeys a day, burned up as she lay in bed early that morning. Nothing else did. Doctors treated the "very extensive burns" on her torso, astonished at the absence of burnt flesh odor in the house. Her injuries, one said, resembled a corrosive liquid burn for which police could find no source. That not a hair on her head was scorched only added to the mystery. Described as "very head-strong" generally and "excited" the night before, Mrs. Knight had not been intoxicated, friends said. She died December 6, unable to explain what happened. Nor could the inquest. Coroner N. H. Hignett, said the *West Sussex Gazette* (December 23, 1943), left a verdict "open as to how the burns were sustained."

Around November 28, 1943; St. John's Terrace, Plumstead, England. Old Jimmy Evans lived alone in his apartment until a neighbor found the hallway filled with smoke. He yelled to Evans who enigmatically replied, "the flames were dying down." Evans said no more. Perhaps he couldn't. Checking on Evans a short time later, the neighbor found him burnt to death on his bed. The fire had been *under* the bed, said author Harry Price in *Poltergeist Over England*. Coroner W. R. H. Heddy called it "a remarkable case."

Late evening, March 27, 1946; Utica, New York. Mrs. Theodore Gifford, seventy, died alone from a fire that affected only herself and the kitchen chair she sat on. The kitchen was otherwise intact. Reported *The New York Times* two days later: "cause of the fire has not been determined."

October 17, 1947; Liverpool, England. A ten-year-old Liverpool lad is discovered aflame amid a setting free of fire damage, said the *Liverpool Echo* that day.

June 12, 1948; Butterworth, England. An unidentified woman from Butterworth in England died when her clothes "inexplicably" caught fire. Said Russell, who said he was relying on *Reynolds News*, "There was nothing to account for it."

December 14, 1949; Manchester, New Hampshire. Mrs. Ellen King Coutres, fifty-three, incinerated in her wood-frame home inside clothing found to be still smoldering. Said *Doubt*: "woman must have been a human torch"—yet there was no surrounding damage to the tinderbox house in which she burned up.

January 13, 1950; Fulham, England. Mrs. Annie Coleshill, sixty-six, died of intense burning.

January 13, 1950; Brighton, England. Mrs. Mary Forge, ninety-four, died in a blaze alleged to be unusual.

Greater London Council long ago had destroyed its files on these two fires "in accordance with home office instructions," a record office spokesman told me. A double dead end.

January 13, 1950; Orpington (London), England. Spinster Kate Ellen Kelly, eighty-three, died in a fire that *Doubt*'s editor suggested might be SHC. *The Orpington, St. Mary Cray and Kentish Times* (January 20, 1950) said she cried out to her niece, "Come quick; I have upset the candle and caught fire to my skirt." The niece ran in, found her aunt *facing* the table with the tipped taper, and quickly doused flames jetting from the *back* of her clothing. A pathologist pronounced "extensive burns on the body." Probably a candle *is* at fault; perhaps the candle is meaningless coincidence. We speculate that maybe her rheumatoid arthritis prevented her from turning to see the unsuspected truth of her predicament, which she blamed on the only fire source she knew to be at hand—when instead the short-lived fire (and the considerable injury it caused) had originated *inside* her back!

January 19, 1950; Indianapolis, Indiana. Another spinster, Mary Louise Spencer, fifty-eight years old and partially crippled, burned up while sitting on a chair in her kitchen. When found, remnants of her clothes still smoldered though most had burned away. Problematic was the fact police "could find no other evidence of fire in the house" even though she, a nonsmoker, had been seen alive *only a few minutes earlier*, said the *Warsaw Times-Union* (January 20, 1950).

In 1950 (or before); Bloomington, Illinois. Aura Troyer, a fifty-nine-year-old bank janitor, was sweeping the basement steps one night, as he had done so many nights before, before going home. One night his routine changed. He suddenly collapsed to the floor, almost all his clothing burned away. He survived, but briefly, said *Doubt* (1950). Said he, it happened "all of a sudden." No more could he think to say; no more did he have to say. *"It happened all of a sudden"* became his laconic epitaph. The mystery of his death lives quietly on.

In 1950; Pittsburgh, Pennsylvania. Carl Brandt, thirty-three, was found in the steel city lying on the sidewalk, nearly all of his clothing burned off. Police who investigated remained perplexed.

For one hundred years after von Liebig dismissed SHC as scientifically absurd, human beings persevered in their ability to enigmatically inflame. Belief in SHC had good reason to persist! On much less evidence have nations gone to war and subatomic physics been constructed.

Still, SHC is not allowed.

Coming Face to Face with "the Damned"

It is the history of discovery that those who challenge the unassailable consensus risk ridicule. Censorship is *nowhere more encountered* than with spontaneous human combustion. To paraphrase Charles Fort, the "damned" shall be damned; there shall be no heresy against what is agreed to.

Two noted medical scholars of the nineteenth century, J. Ayston Paris and J. S. M. Fonblanque, faced the dilemma of "the damned." The timeless lesson they learned is instructive here. Having declared emphatically in their three-volume masterwork *Medical Jurisprudence* (1823) that human combustion "has been erroneously designated as *spontaneous*" and that a body cannot be totally burned up, they personally came face-to-face with an account of a person burned totally to ash. It forced them to *reconsider* their stand against SHC:

> the phenomenon is contrary to all our preconceived views, and must therefore require more than ordinary testimony for its support, although we are ready to admit, that upon any other less miraculous subject, evidence even less powerful than that produced on the present occasion, would be deemed amply sufficient.

In other words, if SHC weren't so "miraculous" because of its rarity and inexplicability they could more readily believe in it. Even then, here was one case that made them consider the miraculous *could* be possible... that SHC *can* indeed happen!

Still, prejudices toward SHC continue to prevail... as do harangues against those who study it seriously. Notable among these antagonists is Joe Nickell.

As America's most outspoken critic of SHC today, Nickell has never met a case of SHC that he didn't like... to discard. He dogmatically states there is "no credible evidence" for SHC, and he urges proponents of SHC to "temper their understandable thirst for mystery by doing more careful research..."

Because Nickell's outspoken opposition is part of SHC lore, it is useful to understand his approach and determine whether he has earned the right to declare unassailed that SHC—along with preternatural combustibility?—is not part of the greater reality about which Bernard Baruch spoke.

Nickell, a stage magician and former investigator for America's best known detective agency (which he wouldn't name but coyly associates with President Abraham Lincoln), teaches technical writing at the University of Kentucky's Lexington campus and conducts workshops on techniques of investigation for the Committee for the Scientific Investigation of Claims of the Paranormal (CSICOP). "My background is as an investigator and," Nickell informed me personally, "if I do say so myself, I solved lots of cases. I didn't make up explanations."

In teaching the arts of writing and investigation, Nickell would seem ideally suited to research SHC... research which must go beyond armchair musings, even beyond library book stacks, before one puts one's fingers to keyboard. The pursuit of SHC, as with any apparent anomaly, requires *quality* detective work: tracking down leads, finding informants, coming up with authentic information to ponder.

Nickell the writer is a skilled wordsmith; Nickell the SHC investigator, you can evaluate for yourself.

Nickell tells me he is a skeptic, not a debunker. Debunking, he says in his book *Secrets of the Supernatural* (1988), "implies bias, suggesting—rightly or wrongly—that the results are known prior to investigation and will always be negative."

Debunking has no place in investigation, Nickell states rapturously; only facts should support a determination. Meritorious advice! As a self-described nondebunker, then, has he resolved SHC by researching its mystery out of existence?

In his article "Spontaneous Human Combustion" for *The Fire and Arson Investigator* (1984) and in his book *Exploring the Supernatural* (1988)—both co-written with John F. Fischer—Nickell liberated Madame Millet and Countess Bandi of SHC by "suggesting" a hearth fire ignited the former; a lamp, the latter. To dismiss Grace Pett, he strung together a stream of qualifiers before implicating either her pipe or a "knocked over" candle. The "supposed" cause for Mary Clues' calcined skeleton became an ember (or a candle) that brushed against her dress. The same "possibility"

> You will be tempted to accept the unreasonable and be distracted by the irrelevant.
> —Marcia Clark, prosecutor; opening arguments in the O.J. Simpson trial (January 24, 1995)

got rid of anything bothersome about Madame de Boiseon. "Possibly, while raking the coals, Mrs. Peacock [*sic*] had accidentally set her clothes afire," he guessed.

For the other twenty-five cases he tabulated, a variety of presumed external ignition sources are prefaced by "apparently...possibly [four times]...might have [twice]...may have [thrice]...actions of the apparent criminal...entirely possible that...if...the possibility that..." In other words, second-guessing. Second-guessing that is many times refuted by primary source documents, had he read them during the two years he claims to have spent investigating SHC.

For problematic cases like Mrs. A.B., Nickell simply deigns to give them credence.

Gee, that was easy. Nickell's make-sense approach to SHC: append a string of qualifiers and supposition to a few cases, ignore others, and SHC is banished. How could physicians be silly enough to sometimes see, then debate for hundreds of years, a nonexistent phenomenon so easily dismissed?

If I said Rush Limbaugh *might* be the most liberal commentator in America, would that make it so? If I said the added qualifier proved my claim, would you then believe the premise?

As for Dr. Bentley's death, Nickell concluded "he had set his clothes afire and had made his way, with the aid of his walker, to the bathroom, where he had vainly attempted to extinguish the flames." As further support that the doctor died in a quite normal fire, he contended that Bentley shed his burning robe, later "found smoldering in the bathtub."

This scenario collapses quicker than the elderly physician probably did. Evidence refutes it. Recall that Dr. Bentley's nurse said she placed a *second* robe over the edge of the bathtub; hence disrobing wasn't required. In fact, Nickell's solution mandates that Dr. Bentley did *not* disrobe: otherwise there would have been no flammable fuel load necessary to support combustion leading to disintegration—which Nickell wants to attribute to a burning robe its wearer has already removed! And why had this fire failed to blister even a square centimeter of paint on the bathtub? Nickell ignores this quandary.

Naturally, neat cases—ones that can be very firmly closed—are always preferable. However, one's desire for the universe to be simple and comprehensible does not compel nature to comply.

As the eminent forensic anthropologist Wilton Marion Krogman once wrote of the lessons his long career had taught him, "There's one rule...that I follow: *some things are improbable, but nothing is impossible.*"

In 1951, nature was preparing to put Krogman and his axiom to the ultimate test.

6

Mrs. Reeser: The Remarkable "Cinder Woman"

As I review it in memory, the short hairs of my neck bristle with a vague fear. Were I living in the Middle Ages I'd mutter something about "black magic."

—William M. Krogman, Ph.D.,
about Mary Hardy Reeser

Cemeteries are silent places, fields of miniature monoliths marking sites hallowed and revered, frequented and adorned, or merely forgotten. If stones could speak about the lives they memorialize, what wondrous tales some could tell... especially one tombstone at the Chestnut Hill Cemetery, outside Mechanicsburg in southern Pennsylvania.

The story that lies interred there began on March 8, 1884, with the birth of a little girl named Mary Hardy Purdy. She grew up in the Pennsylvania Dutch country of Lancaster County, where she married Richard Reeser, a young physician. He was quite a catch for a young woman who "never went beyond the fourth or fifth grade," said Jan Bowen, a great aunt of Mrs. Reeser.

Dr. Reeser established a general practice in Columbia, and his new bride busied herself with raising a family and participating in local social and charity events. The Reeser household became a center for gala gatherings, recalled several Columbia residents who told me they fondly remembered the lavish lawn parties the Reesers hosted in the 1930s and 1940s. They all described Mary as a kindly, gregarious person, a bit vain with a streak of rebellion in her, always interested in gossip. Several remembered that she was fond of beer. Four years after her husband's death in 1947, Mrs. Reeser moved to

St. Petersburg, Florida, to be near her son, Richard Jr., himself a physician. She would live there for a mere five weeks.

As the summer of 1951 sweltered over Florida's west coast, Mary Hardy Reeser hoped to return north to the cooler climate of south-central Pennsylvania and attend a memorial service for her husband. There were problems with the arrangements and this depressed Mary, Ms. Bowen told me. But she took the delays in stride, and awaited news that everything would work itself out for her trip back home.

On the evening of July 1, 1951, Mary's son and oldest granddaughter, Mary Carol, dropped by to visit. When they left, Dr. Reeser bent down and kissed his mother good night. Mrs. Pansy M. Carpenter, owner of the Allamanda Apartments at 1200 Cherry Street N.E. and Mary's landlady, stopped in later to visit briefly. Bidding pleasant dreams to her tenant, who had just moved in the month before, Mrs. Carpenter walked down the hall to her own apartment, leaving Mary sitting in an overstuffed chair beside a small stand stacked with reading materials. Her rayon nightgown, where exposed by her cotton housecoat, fluttered in the warm breeze entering through a nearby window. A breeze gently nurtured by two electric fans. Her legs were stretched out before her, her feet clad in satin slippers. A pack of cigarettes lay nearby. An hour earlier, Mary had taken two seconal tablets (a mild sedative) to help her sleep. In all, a most familiar scene.

The time was 9:00 P.M.

The world was (and still is) ill-prepared for what the door to Mary Reeser's apartment would reveal when it next opened. Dreams had become nightmare. It would be, said one commentator, "a scene macabre beyond words."

Revelations in Mrs. Reeser's Room

Around five o'clock the next morning, July 2, Mrs. Carpenter awakened to a sweetish smell in the air. She checked the apartment's hot water heater, which had been acting up, but it seemed fine. She shut it off anyway and went back to bed. An hour later she got up to get the morning newspaper.

At 8:07 A.M., Western Union messenger Richard Bruce arrived at Mrs. Reeser's door to deliver a telegram. He knocked. No response. He walked to the other end of the hallway to see if the landlady could help.

"Got a telegram here for Mrs. Mary Reeser," said the messenger to Mrs. Carpenter. "I knocked on her door but don't get any answer. You take it?"

Pansy Carpenter did—and expressed concern that her tenant, known as a light sleeper, had failed to answer the messenger's knocks. Carpenter's raps became thuds, then pounding, as her knuckles struck ever more sharply on

Mary's door. No sound was heard from the other side. Already worried about Mary's silence and now further concerned because she didn't hear the radio broadcasting the eight o'clock news that Mary always tuned in, Mrs. Carpenter grasped the brass doorknob in order to...

A scream of pain. Or was it fright? The doorknob was too hot to turn!

Albert Delnet and L. P. Clements, two house painters working across the street, responded to her alarm. They broke down Reeser's door. A blast of hot air rushed out, briefly forcing them to retreat. Clements eventually peeked inside and quickly advised Mrs. Carpenter to call the fire department. The landlady's concern about Mrs. Reeser was justified: beyond the superheated doorway, the scene so familiar the night before was now a mound of ashes amid a faintly disgusting stickiness.

St. Petersburg firefighters arrived almost instantly. Station 4 responded "within three to five minutes, probably four," one of its members, Nelson Aters, told me. Already on the scene was another fire company. City police soon followed. Both agencies were quick in their initial assessment: this was *no normal fire.*

"I went in, took a hand pump and did start to squirt water on the debris," reminisced Aters. "Then it was called to my attention not to do that, because it was not *just* debris but what was left of the lady!"

Dr. Reeser, notified by Mrs. Carpenter that an accident had occurred, rushed to his mother's residence. Emergency personnel quickly told him it would be best that he *not* go in. Dr. Reeser waited outside.

Inside Mary's apartment, consternation, confabulation, and chaos reigned, due to no fault of the officials. They were, simply, quite unprepared for anything the likes of which they now confronted.

"Incredible! No rhyme or reason to it.... It was quite a phenomenon!" Aters exclaimed to me, remembering what his eyes beheld that Monday morning more than four decades ago.

In the far side of the apartment, severely heat-eroded coil springs protruded from an amorphous mound of ashes, once a chair. Of the end table adjacent to the chair, there was now no trace; it too had disintegrated. Sifting through these ashes, the investigators made a fearsome discovery. They now spoke in terms of a "Circle of Death"—for therein, to their horror, they retrieved *a few pieces of calcined vertebrae, "crisp ash," something that resembled a liver, a skull "shrunk uniformly to the size of an orange" and, just beyond the ashes, a "wholly untouched left foot still wearing its slipper."*

Although smoke pervaded the apartment, reporter John Perry wrote in the *Tampa Tribune* (July 4, 1951) that "There was no odor of burning flesh." In this remarkable way did Mary Hardy Reeser leave this world.

Beyond that macabre circle—which you could trace, turning around with the end of a yardstick balanced on your hip—the room was little affected, even though investigators knew the heat from the fire had to be frightful.

Throughout the room above a line four feet above the floor, a moist black soot coated everything. Below the four-foot level, the room was *practically pristine*—save the chair and its incinerated occupant. Wall paint behind the chair was faintly browned. Plastic outlet covers above the four-foot line had melted "into comic shapes." Fuses, though, had not blown. Baseboard electrical outlets were *not* damaged—except for one into which was plugged a lamp next to Mrs. Reeser's chair, and another one directly across the room.

A wall mirror ten feet from the chair was cracked, probably by heat. On a small buffet near the mirror a lace cloth, not burned, was covered with a puddle of hardened pink wax from two melted candles. The unburned wicks were draped over the candle holders and stretched out across the top of the buffet.

Magazines and newspapers on a table within arm's length of the chair escaped intact. (Newsprint will auto-ignite at 446°F.) Mrs. Reeser's bed was three feet away, sheets turned down; it was "as spotless as if the linens had just been taken from the closet." Nor did the carpet *beyond* her incinerated chair show signs of fire damage!

Inside a kitchen cabinet, the waxpaper lining of a cereal box was crinkled by heat; the box itself unharmed. And there was this enigma, mentioned exclusively by the *Tampa Tribune*: "Inside the bathroom, drooped from a porcelain holder, was what was left of a plastic water glass. Toothbrushes, dangling just below, had not been harmed."

An electric clock plugged into one of the melted outlets had stopped at precisely 4:20.37—less than four hours before officials arrived—yet the clock began running when plugged into one of the unmelted baseboard outlets. Next to the clock sat a lamp with heavy sooting on its shade, and a radio, its upper left corner just slightly melted.

As the *Tribune* quipped: "The fire played strange tricks."

Firemen knocked away the wall behind the overstuffed (and overburned) chair in order to extinguish a feeble flame that had eaten its way into it.

It was the only spot that required their fire-fighting skills.

Magistrate Edward Silk, the acting coroner, stood before a scene that was like no other he'd ever seen. He ordered that the remains be removed. So completely did fire destroy inside its circle of death that fireman Buddy Standish and a colleague had to *shovel up* the woman and her chair, like shoveling spent ashes from a fireplace hearth.

Notwithstanding this extraordinary cleanup, coroner Silk listed "third-degree burns" as cause of death on Mrs. Reeser's Death Certificate, according to Robin Champagne at the Rhodes Funeral Home, which conducted Mary's funeral.

What happened next is fraught with more controversy and contradictions. The former is to be expected of such an astonishing fire scene; the latter are the result of opinions and prejudices that may never be settled.

AUTHORITIES AND THE PUBLIC INVESTIGATE AND SPECULATE

It looked as though Mary Hardy Reeser had dissolved in a flux of disentangled atoms.

The St. Petersburg Police Department (SPPD) began using the adjectives "weird," "unbelievable," and "fantastic" to describe the mystery they were duty-bound to solve.

Police Chief Jake R. Reichert promptly requested assistance from the Federal Bureau of Investigation. The shoveled-up remnants of the chair, carpet, and Mrs. Reeser were boxed and sent to J. Edgar Hoover's agency for microanalysis. A month later the FBI returned a five-page report on the Reeser remains.

The general conclusion of the FBI Laboratory's report—though not the specifics leading to it—I found in a copy of an official SPPD statement from 1951, on file at St. Petersburg's city library:

> An examination of all specimens sent to the FBI laboratory for analysis shows no oxidizing chemicals, petroleum hydrocarbons or other volatile fluids commonly used as accelerants or any chemical substances used to initiate or accelerate combustion.

The FBI findings did little to solve the quandary in Florida, except to categorically rule out the presence of fire-enhancing fuels.

The *St. Petersburg Times* gave the case extensive coverage, from which I learned about three more key points:

- tests revealed no evidence of any drugs that might have caused death.
- no criminal intent could be identified.
- the "large, robust" woman had weighed 170 to 175 pounds, yet after the fire her remains—the shrunken head, the whole foot, the bits of calcined vertebrae, and a minute section of tissue tentatively identified as liver—plus the chair and end table weighed *less than ten pounds!*

Mrs. Reeser's stocky five-foot-four-inch frame had undergone some type of natural, albeit never-to-be-a-fad, weight-reduction program without equal.

Without question, Mrs. Reeser's death was sensational. And the local press did not fail to publicize vigorously the predicament facing the puzzled police. It seemed everybody either wanted to know *what* befell the gregarious widow or thought they knew *how* it happened.

"This fire is a curious thing," Detective Chief Cass Burgess told the *St. Petersburg Times* (5 July 1951), "and I've been deluged by letters and phone calls offering solutions to the problems facing us."

One postcard addressed to "Cheif of Detectiffs" [*sic*] sent anonymously by an alleged eyewitness said "a ball of fire came through the open window and hit her. I seen [*sic*] it happen."

The experts had already considered lightning. Three factors ruled out this conjecture: (1) Although the area had a storm that evening, the U. S. Weather Bureau reported no "severe lightning anywhere in St. Petersburg" during the night; (2) There was no trace of lightning-stroke burning on the window sills, and no fuses had blown in Pansy Carpenter's building, and (3) Characteristics of lightning itself refuted this speculation. Julius H. Hagenguth, engineering coordinator of man-made lightning experiments for General Electric, explained to Atlanta's *Journal and Constitution* that although a lightning bolt "could have paralyzed or killed" the widow and set her clothing afire, the "stroke could not, however, have lasted long enough to completely consume her chair, leaving only the springs, a shrunken skull and a backbone. That is fantastic.... It's like a first-class Oppenheimer mystery."

An "electrical induction current, passing through the body, caused by faulty wiring" was, in turn, ruled out by the fire inspectors because the wiring proved not to be faulty. The bottled gas radiant heater near her chair was suspect, until the police proved it was turned off and its pipes gas-free. Appliances were examined: the kitchenette oven was found to be disconnected, and the refrigerator functioned normally.

An "atomic pill" was proposed. Sleeping tablets were the only known pill regularly taken by Mrs. Reeser—and the Federal Food and Drug Administration has never issued a warning that seconal poses a combustible hazard.

Was Mrs. Reeser murdered? people wanted to know. So did the police.

According to the victim's great aunt, at one point even Dr. Reeser Jr. "was a suspect, because there was money, substantial." But facts weighed against foul play. The apartment had not been looted. Mrs. Reeser was not known to have provoked murderous antagonism in anyone. No one, including the landlady, heard or saw indications of intruders breaking in or out of Mary's apartment. Common accelerants capable of concealing a crime—and absolutely necessary to have fueled the fire's ferocity—had already been ruled out. So attention turned to exotic materials like "ether, kerosene, napalm (jelly fire), thermite bombs, magnesium and phosphorus." These too were excluded: no telltale odors or trace elements were left behind, reported Silk.

Detective Lieutenant Cass Burgess ruled that Mrs. Reeser had not been murdered by her son or anyone else.

St. Petersburg Fire Chief Nesbit and Edward S. Davies, a respected arson specialist with Tampa's National Board of Underwriters, ruled out arson for the same reasons the police dismissed murder. "I never saw anything like this in the 48 years I have been investigating fire cases," commented a stumped Davies on July 3. "We do know that it was a fire... we just don't know what could have caused it."

Davies called it "an accident of the most tragic kind." Of the most *mysterious* kind, as well.

Did Mrs. Reeser commit suicide? Her behavior gave no hint that she contemplated it. A *Tribune* reporter checked labels on bottles in her medicine cabinet anyway, and found "nothing more than standard digestive aids." Concluded the acting coroner: "It is fantastic to suppose the fire itself could have been a act of suicide. Her death appears to be accidental and there is nothing to indicate it was anything but accidental."

Had Mrs. Reeser died accidentally by her own hand, dropping a lit cigarette onto the chair?

Executives of the Jacksonville firm that manufactured the chair that cradled the widow's cremated body were questioned about their product's composition. It was wood, metal springs, and cotton batting sterilized by steam—just like most domestic overstuffed chairs of the time, they said. They adamantly denied using inflammable chemicals at any stage of the chair's construction. Another manufacturer noted that an overstuffed chair's batting may contain felt, hair-pads, or foam rubber, each capable of slow smoldering but *not apt to inflame violently*.[1]

> No! No! Not the comfy chair!
> —Monty Python

Indeed, a fire in an overstuffed chair can evolve to room flashover in only about two minutes. However: here, many hours (potentially eleven) proved inadequate to ignite the apartment. The sought-for solution remained elusive.

"It was the biggest thing that had happened in years," says Jerry Blizen when I spoke to him about the story he covered admirably for the *St. Petersburg Times*. The public was loudly demanding an answer to this mystery in their midst. The authorities were frustrated by failure to find one.

Then a ray of hope; of resolution. Dr. LeMoyne Snyder, member of the American Medical Association and American Bar Association, and medico-legal director for the Michigan State Police, questioned in his book *Homicide Investigations* why a case like Mrs. Reeser should be so perplexing: "Underneath the skin is a layer of fat which may be quite thick in heavy individuals. This burns readily and the destruction of the tissue after a comparatively small fire may be great."

Mrs. Reeser was plump. So, according to Dr. Snyder, resolving Mrs. Reeser's death would require nothing more than finding the source for "a comparatively small fire." The officials, thinking they had no recourse but Snyder's human-candle contention, finally affixed blame: *a dropped cigarette.*

Gee, it was all so simple after all. How had all the crime and arson experts at the scene managed to overlook it before?

The SPPD issued their official, five-hundred-word statement on August 8, 1951, proposing this reconstruction of Mrs. Reeser's last hours:

> There is conclusive evidence that the deceased body could be consumed by fire, as in this case, if the body had become ignited.... Due to the fact Mrs. Reeser had taken a considerable amount[2] of sedatives at night... there is every possibility that Mrs. Reeser while sitting in the overstuffed chair in her apartment, could have become drowsy or fallen asleep while smoking a cigarette, thus igniting her clothes.
>
> At that time she was clad in a rayon acetate nightgown and a housecoat. The nightgown being highly inflammable, could have been ignited by a burning cigarette, causing immediate death, if the deceased was in a semi-conscious condition.
>
> Naturally, when her clothes became afire they would also set the chair afire creating intense heat which completely destroyed the chair and a nearby end table. Once the body became ignited, almost complete destruction occurred from the burning of its own fatty tissues. For once the body starts to burn, there is enough fat and inflammable substances to permit varying amounts of destructions to take place. Sometimes this destruction by burning will proceed to a degree which results in almost complete combustion of the body, as in this case.

Still, the police seemed a bit unsure of themselves. Look at all the qualifiers they imbedded into their report. The clothes-to-chair-to-fat-to-ash sequence stood on the strength of conviction born of helplessness, not evidence based on actual experimentation. And why had Mrs. Reeser so baffled them for weeks, if people who drop cigarettes onto themselves—certainly not an uncommon human habit—will readily incinerate so completely? As Vincent H. Gaddis, himself a newspaper man, would later write in his fine book, *Mysterious Fires and Lights* (1967; 1968): "The report... would probably satisfy the unthinking."

> Why didn't the room flash over?... It's a mystery. I don't know what else to call it. Maybe someday, someone will figure it out.
> —Fire Chief John Polk, St. Petersburg Fire Department, on the Reeser case; personal interchange (Jan. 27, 1995)

Indeed, the problems inherent in the SPPD report were about to be compounded. Dramatically. Enter Wilton Marion Krogman, Ph.D.

DR. KROGMAN: THE FORENSIC ASPECT

"In 1951 I ran into a case which for sheer horror and bizarreness beat anything in my experience," wrote Professor Krogman in *Pageant* magazine (October 1952).[3] He referred to the burning death of Mary Hardy Reeser. His experience *is* vast, and a brief biography will help you appreciate the significance of his comments about this case.

Krogman's skill at "bone identification" began at age eleven, and his talent was recognized by the American judicial system in 1929 when he established the first of many legal precedents during his career by showing how evidence derived solely on skeletal remnants could identify a victim's age and sex and hence win million-dollar judgements. He had proven "once and for all," he pointed out with pride, "that the forensic anthropologist had the techniques so careful and so precise that he could, in fact, go on the stand as an expert witness and testify."

In the 1950s, Krogman served as professor of Physical Anthropology at the University of Pennsylvania's Graduate School of Medicine. In the 1960s, he became internationally recognized as an authority in forensic medicine and wrote a textbook classic, *The Human Skeleton in Forensic Medicine* (1962). And of greatest pertinence here, he is a renown expert on the effects of fire on the human body.

It was no idle boast, then, that led Krogman to write about himself: "From the skeleton or its remains it is possible for me to... determine cause of death..."

Possible, that is, until mid-1951.

At that time, Krogman was visiting his children in Florida when a friend told him about Mrs. Reeser. "I learned then the pertinent details," he would write, "*becoming more and more baffled* as the story unfolded."

Professional curiosity prompted the bone detective to interrupt his vacation and apply his expertise to what he called "serious problems—problems which strain belief beyond the breaking point."

The SPPD reconstruction of Reeser's passing, already tenuous, was about to collapse.

Krogman identified seven serious problems the widow's demise posed, which he discussed in *The General Magazine and Historical Chronicle* (1964).

First: *the extent of the reduction of her body*. Even in his experiments that used powerful fans to create strong drafts to intensity combustion for twelve hours at 3000°F, he said. "At the end of that time there was scarcely a bone that was not present and completely recognizable as a human bone. It was calcined... but *it was not ashed or 'powdered'*."

Until he encountered the restricted firestorm of the Reeser case.

Second: *the localization of the tremendous heat that ashed Mrs. Reeser's bones.* "Only at 3000 degrees F., plus," he wrote, "have I seen bone fuse—or melt, so that it ran and became volatile. These are very great heats—they would sear, char, scorch, or otherwise mar or affect anything and everything within a considerable radius. What I'm driving at is this: the terrific destruction of Mrs. Reeser's body (bones included) must have been accompanied by such heat that the room, itself, should have been burned much more than it was."

Until he encountered the Reeser case.

Third: *the time available for the reduction of Mrs. Reeser's body*. He knew how *extremely difficult* it is to burn an animal or human. A continuous oxygen-providing draft is necessary to keep the fire burning. "In Mrs. Reeser's room, a draft was present from windows to the cracks under and around the door." But a draft of hurricane ferocity? The sky had been clear over western Florida that evening. Without benefit of accelerants, could the few small fans in her apartment an d its open windows create a veritable whirlwind that fanned the voracious flames (allegedly created by a dropped cigarette) engulfing her body?

And what of the time factor? At most, Mary Reeser had only eleven hours to burn herself more completely than Krogman's cremation experiments achieved at 3000°F for twelve hours—and she did so certainly without the aid of optimum conditions. Most likely the burning occurred around 4:20 A.M.[4] when the clock stopped—leaving *less than four hours* for total cremation to unfold before the painters broke down her door. This was too short a time to so completely incinerate a person, Krogman knew.

Until he encountered the Reeser case.

When entering Mrs. Reeser's death chamber, police and fire officials discovered the eight to ten pounds of residue left little of what the night before had been a contented widow. What they did find was what Krogman said he had never seen before, not even in a crematorium: *an adult human reduced almost entirely to powdered ash, including the skeleton!*

Fourth: *the absence of noxious smell.* "When human flesh burns it gives off an acrid, evil-smelling odor," confessed Krogman, "especially if burning free in a room or in the open. How 175 pounds of mortal flesh could burn with no detectable or discernible smoke or odor permeating the entire building—well, experience says differently!"

Until he encountered the Reeser case.

Fifth: *the shrunken skull.* This damnable artifact—variously described as a "baseball," "grapefruit," "orange," or "teacup" in size—perplexed the good doctor more than any other aspect of a case which had already broken his belief system. Skulls just don't naturally shrivel or symmetrically reduce in a fire.[5] "In the presence of heat sufficient to destroy soft tissues the skull would burst—explode, literally—into many pieces. I have experimented on this, using cadaver heads," Krogman affirmed, "and can but say I've never known an exception to this rule."

Until he encountered the Reeser case.★

Sixth: *the affront to logic and common sense that Mrs. Reeser left behind.* "The body is over 90 percent water—perhaps even a higher percentage in very fat people," penned Krogman. "I find it hard to believe that a human body, once ignited, will literally consume itself—burn itself out, as does a candlewick, guttering in the last residual pool of melted wax."

> Why, sometimes I've believed as many as six impossible things before breakfast.
> —The Queen, in *Alice in Wonderland*

Until he encountered the Reeser case.

Seventh: *the similarity of Mrs. Reeser's death to other strange deaths by fire.* Krogman knew about nine earlier cases of mysteriously complete natural human cremations, all cited in Dickens' *Bleak House.* He found them as problematic as the one with which he was now so personally familiar. "Here, then," he wrote, "is a monotonously consistent pattern which almost gives the lie to mere coincidence. Such things *can't* happen! But they seemed to have happened—and again and again and again!"

★ Dr. Douglas Ubelaker, curator of anthropology at the Smithsonian Institution's National Museum of Natural History and called "truly one of the finest forensic sleuths of our time," would presumably be no less perplexed today by Mrs. Reeser than was Krogman in 1951. In his fascinating 1992 book, *Bones,* Dr. Ubelaker describes in its "Burning Questions" chapter how he can reconstruct crime evidence from bone fragments of bodies blown up by dynamite or exposed to days of incineration. He explains that firing of bones *can* yield shrinkage, from 1 percent to 25 percent, depending on bone density and a 1300°F to 1650°F temperature range. Nowhere does he allude to a case resembling SHC. "Well, it sounds kinds of preposterous," he chuckled as I described Mrs. Reeser to him on March 16, 1995. "No, I guess I'd have to say I never encountered that."

And it all happened again in the Reeser case, despite these seven challenges to Krogman and to science.

"I have posed the problem to myself again and again, and I always end up rejecting it in theory—but facing it in apparent fact!" he confessed.

Given these facts, Mrs. Reeser at the very least succumbed to preternatural combustion. *If* she dropped a cigarette onto her lap. Since there is no conclusive evidence she did this, one cannot rule out the corollary to PC: spontaneous human combustion.

Nor could Krogman. He himself posed the key question about her death: "By spontaneous or 'internal' combustion—that is, by fire starting in the body tissues *without any external cause?* But this is held impossible!" So were meteorites once upon a time.

The world-renown forensic anthropologist concluded his contemporary analysis by expressing the thinking of most investigators who puzzled over Mary Reeser: "Just what *did* happen on the night of July 1–2, 1951, in St. Petersburg, Florida? We'll never know. The 'Cinder Woman'... couldn't happen, but it did."

TRYING TO REOPEN THE REESER CASE

That's where history left the fascinating last twelve hours of Mary Hardy Reeser's life in 1951. As SPPD Chief Reichert admitted, after guiding the investigation as far as his resources allowed, "This is the most unusual case I've seen during my almost twenty-five years of police work in the City of St. Petersburg.... I am not closing the door yet on the case."

Apart from periodic retelling in anthologies about the paranormal, which helped ensure she would be—in the popular mind—the best known death attributed to SHC, the case itself had stagnated.

In 1975, as my interest in SHC accelerated, I began searching for original documents and new information about this woman who once lived and now was buried near my home.

One of our visits to Mrs. Reeser's grave fortuitously coincided with a burial of another member of the Reeser family. (Ain't synchronicity grand?) In talking to family members after the funeral, I learned that Dr. Richard Reeser, Jr., still lived in St. Petersburg. Who better to ask about insights the intervening decades might have shed on his mother's passing? I wrote a letter, delicately explaining my interest in knowing more about his beloved mother. A courteous reply was soon in my hands.

"I am afraid, and I do believe, that the case of my Mother has been worked over time and time again," Dr. Reeser wrote, "and personally I have no doubts in my mind as to what happened. She was a heavy smoker; she fell asleep in an over-stuffed chair—there were fans blowing, and I think she literally was consumed by her own fat." Actually, this is the decision of the FBI who were called in and who stated that there were several other,

> She was smoking a cigarette and seemed content.
> —Dr. Richard Reeser Jr., on closing his mother's door the night of July 1

in fact, there may have been six other instances recorded.[6]

"I do not believe there was any spontaneous combustion," he added. "So, really I do not believe anything further can be gained by any re-hash of the old events which were recorded many times. But as I say, in summary, I do believe the facts are as recorded."

The facts as recorded, yes. But about the *interpretation* of those facts, I wasn't so convinced.

I wrote back. Dr. Reeser replied: "Finally, to me, this whole incident is one that I would like to forget, and I think every angle has been examined, traced and discussed for lo, these many years."

However, *I* could not forget it. His mother's death puzzled me. And new puzzlement awaited, as I was about to discover.

One document I sought was the FBI Laboratory report, which I assumed would end the contentious controversy over whether Mrs. Reeser's head had fragmented or, as published contemporaneous accounts said, had indeed shrunk "to the size of a baseball."

Regrettably I have been unable to resolve this dispute. FBI Director Clarence M. Kelley denied my Freedom of Information Act request to release the results of his FBI Laboratory analysis on Mrs. Reeser's remains. "Please be advised that the FBI did not conduct any investigations concerning the death of Mrs. Mary Hardy Reeser," the Director wrote me. He did acknowledge the FBI laboratory had examined "certain evidence" received from Chief Reichert. He indicated the SPPD had this information, and upon "written authorization" the FBI would release its documents under FOIA.[7]

In line with department policy, Chief M. E. Handley of the SPPD Records Division twice turned down my requests for the Reeser file. Later, when I walked into SPPD headquarters in 1977, I was informed that their file on Reeser had been destroyed, and no photographs existed to be studied. Yet in 1980 the Reeser case was *still open*, according to personal communication from Sue A. McDonald of the SPPD records division. She added that Mrs. Reeser "truly was and still is a baffling case," and regretted being unable to furnish any information but a photocopied department handout (which contained several errors of fact).

As Charles Fort once observed, "One measures a circle, beginning anywhere." The trouble is, if you're on the outside of the circle once it's drawn, you may be relegated to remain on the outside never able to get in. It was evident that if I was to obtain further information about the Reeser case, I needed to develop new sources. New sources presented themselves.

In 1975, Police Chief Handley had told me that "All of the personnel involved in the original investigation have retired from the department and are not available for comment." Five years later, I located one of those retirees, apparently the last St. Petersburg policeman alive who had direct involvement with his department's best known case. And he was willing to share his recollections with me.

I met retired officer Walter G. Tipton at his home in North Carolina. He vividly remembered his first rookie assignment on the SPPD. It was, he said, to stand sentry outside the Allamanda Apartments and "to make sure nothing happened. They didn't know *what* they had there!"

How had his more experienced colleagues on the SPPD responded that long-ago July week to the case which fate dealt them? "There was no speculation at all. *We* didn't know! That's what medical examiners are for!" he chuckled.

"Now one interesting thing about the whole thing was the *scorch line*. I suppose you heard about that?" continued Tipton, referring to the four-foot-elevation below which smoke failed to penetrate. "I know it was overwhelming! I probably had the closest contact of anyone left [alive]. Her body was almost completely disintegrated. The woman was *gone*! There must have been tremendous heat in there."

But that was not the case (about the heat, that is).

Tipton told me he had kept some artifacts from the fire—including her black satin slipper—in his desk drawer for years, but "I turned over all that stuff and my records over to the department when I retired."

Had he ever seen any photographs of the fire scene, photos the records division claimed no longer existed? "The photographs I had were the *initial* photographs taken at the crime scene," he answered. What did they show? "The *foot*! They showed the chair and where the wall switches were burned. *I just wish I had those pictures so I could show them to you!*" he said apologetically. "I still can't figure out how the heat spread *out* rather than up." And when had he last seen those remarkable photographs? "Last November."

So the nonexistent photos did exist, after all! At his request the SPPD had forwarded them, and he had sent them back *only nine months before*. What impressed him most was "that one foot left."

As I prepared to depart, Tipton recounted another burn case from his police career. Was it like Reeser? "Oh no!" he shot back without hesitation. "*Nothing* was like *that* one!"

(Twelve years later I telephoned Tipton again, to be told he not only had these elusive photographs back in his possession but was willing to show them to me because he respected the research I was doing. A few weeks later I was back on Tipton's front porch, studying intently sixteen different interior views of Mary Reeser's apartment. They documented the astonishing capriciousness of the fire's behavior: plastic switch plates melted above—but not below—the line of soot, linens and furniture so pristine one would

> The peculiar phenomenon of the light switches and receptacles being melted on the walls was uniform throughout the apartment....
> That struck me as having some information for us, if we knew how to analyze it. Was it lightning? Was it a short caused by her burning out, or something? We don't know.
> —Nelson Aters, fireman at the scene; personal interchange (Jan. 27, 1995)

never know an inferno had roared but inches away, a toothbrush bowed over by heat in the bathroom several yards away at the far end of her apartment...)

It was all so very, very peculiar. And about to become even more Alice-in-Wonderland-like.

PROFESSOR KROGMAN REVIEWS THE REESER CASE

When beginning our research of the Reeser case in the mid-1970s, the one person associated with this incident that I *most* wanted to talk to was Wilton Krogman. If he was still alive, had he come to a resolution in his own mind about his most notorious case?

I wrote to him at the University of Pennsylvania, where he had taught... and back came a reply! Excited anticipation of learning informative new insights changed quickly to consternation as I opened the envelope.

"I reject SHC/PC," Krogman's firm response began. He said *True* magazine, which published in its May 1964 issue an often-cited article by Allen Eckert about SHC, "attributed quotes to me that I never made. I wrote a letter of protest to the Editor which was never acknowledged. I do not wish, *in any way*, to be associated with this sort of sensationalism."

Stunned, I wrote back to reiterate my desire to compile a thorough and historically accurate record of this event. I requested examples of misquoting by Eckert. No response. Meanwhile, I discovered the article Krogman himself had written about Reeser—and it contained statements *identical* to those Eckert attributed to the forensic anthropologist.

The high strangeness of the Reeser case had not ended in 1951.

Then *The Philadelphia Inquirer* (June 15, 1975) ran a profile about Professor Krogman, and touched briefly on Mrs. Reeser. "At no time has he seen any sign that the body itself, which is basically a bag of salt, can actually catch fire," reporter David Bolt said of Krogman. "The case has troubled him through the years."

Without question, this expert on human incineration had significantly modified his views of the Cinder Woman. Why?

Bolt's article provided the impetus to reapproach Krogman, and eventually the eagerly sought and anxiously awaited interview was scheduled.

Krogman welcomed me graciously into his book-lined office located, conveniently, only fifty miles from my home. He delighted in discussing all aspects of his illustrious career. Save one. The Reeser case. But at last, he did speak about it. Reluctantly. Hesitantly.

"Alright, now there's another part of this story," he said of Mrs. Reeser. "These bones—these *remains*, sorry—allegedly had been sent to the FBI for further investigation. *They never got there!* So the next thing I know, the case was closed.... I think the case was closed with undue haste. That's all I can say."

The Reeser case evokes contradictions—and lingering enigmas.

The bone detective had another bombshell to drop. "My feeling about the Reeser case is, the set-up in her apartment was staged," he said forcefully. "And this is my professional opinion: in my opinion those who saw the bones—the remains—were not competent (a) to say they were human, and (b) to say they *were* those of Mrs. Reeser. It's all circumstantial. Now I'm not suggesting there's a plant. I'm just saying—"

Krogman paused, apparently surprised by the expression on my face. My reaction went beyond stunned. Flabbergasted would be more appropriate. Notwithstanding the international news coverage plus weeks of intensive investigation by St. Petersburg authorities, the Reeser case was a setup? A hoax?

Was he playing a joke on me? Being facetious? No, Krogman was quite serious.

This meant the good professor was admonishing everyone not to accept anything *he* had written twenty-four years earlier; not to believe *he* had inquired at all into the events at 1200 Cherry Street N.E.; not to recognize the training, competence, reputations, and just plain hard work of scores of on-the-scene investigators and technicians,[8] including himself. At no time had anyone involved in the investigation offered evidence for the scenario Krogman now proposed. Heretofore, that had included *himself*.

> The longer the time goes by, the less relevant the statement.
> —Patrolman Donald Smith,
> Susquehanna Township
> Police Department,
> on the reliability of recollection

In fact, he was saying that no one at the scene was sufficiently competent to identify as human a *foot* inside a satin slipper!

In fact, Krogman's own professional expertise had previously established the murder-elsewhere-and-salting-the-scene scenario he now advocated as irrelevant.

Even if criminals (there would have had to have been several) had accessed a crematorium next door to the apartment building (if there had been one next door, which there wasn't), did they have the time to abscond with the widow, cremate her, return her ashes to her room, and make it look like she burned *in situ*, then clean up all trace of ashes in the hallway and other evidence of their crime? And do all this without waking Mrs. Carpenter at the other end of the hallway?

This scenario is far more outlandish than admitting to SHC, a prospect that raises none of these new incomprehensibles.

Flabbergasted would be too mild to express my reaction, now. Nothing short of Krogman combusting in front of me could have astonished me more.

Krogman continued by recounting his numerous experiments on burning bodies—including crematoriums that utilized fans to create strong drafts to accelerate combustion—and of the one inviolate fact that, for him, had no exception: "There were *no* bone fragments I couldn't identify as human. Some were very small—they had shattered in the heat—but my God, I could tell them! They were calcined, but they weren't ashed—powdery ashed, as you say.... You end up with some bone fragments, not bone dust.... *You don't get powder! Period!* You don't touch it and have it go to powder."

Yet officials discovered very little bone matrix when examining Mrs. Reeser's death chamber. What they found was *exactly* what Krogman himself said was there, what he had never seen produced naturally before: *an adult female reduced almost entirely to powdered ash, including her skeleton!* Produced without the assistance of a crematorium.

The more he talked about his career, the stronger he made the case for Mrs. Reeser dying abnormally, by means unknown.

"But there's another thing, Arnold, I have to tell you," Krogman advised. "I have learned over the years to discredit so-called eyewitness accounts. 'Ya know, I seen that and, by God, it fell to pieces!' Sure, a body that's badly burned: the bones are shattered by the heat—*but not consumed!*"

Again, he now refuted his own published commentary. He now repudiated the SPPD's description of Mrs. Reeser's minimal remains. He now rejected his own comparison of Reeser to the collapsed-to-ash-when-touched bodies in earlier SHC cases, which, he noted, had "parallels so striking as to almost suggest intentional reduplication."

How were these earlier chroniclers—several of them physicians, you'll recall—able to describe so precisely an incident that would not occur until 1951? Was the medical coverage spanning three centuries of ashened bodies actually a conspiracy by clairvoyant medical illuminati for unknown purposes? Surely no one believes that. It is more rational to admit that history *was simply repeating itself* in St. Petersburg.

Krogman had said all he would about this, his most disturbing case. But how to explain his unexpected turn-of-mind?

Nostalgia is not always faithful. Sometimes, the haze of time draws a curtain of discreet forgetfulness. Sometimes, a penchant to deny the obvious and blithly repeat the big lie reconstructs a past misremembered. Consolation.

> Nothing is impossible; there are ways that lead to everything, *and if we had sufficient will* we should always have sufficient means. It is often merely for an excuse that we say things are impossible.
> —La Rochefoucauld

Pennsylvania State Fire Marshal Fred Klages taught me in his arson investigation course that "you have to eliminate a lot of possible causes to discover the *true* cause."

In Mary Reeser's death, a dropped cigarette and many other possibilities had been effectively dismissed years before on practical or evidentiary grounds. By none other than Krogman's own expertise.

Hence, precision dissection of the Reeser case points to spontaneous (perhaps preternatural) human combustion as *the* solution to the mystery story beneath Chestnut Hill Cemetery's tombstone. Indeed, Krogman should be prepared for this determination. He, himself, had affirmed that the improbable is not impossible; therefore, possible.

Nothing could be more true in the continuing saga of the mystery of Mary Hardy Reeser, alias the Cinder Woman.

ENTER THE NAY-SAYERS

This saga took another twist when Joe Nickell and John Fischer co-wrote for *The Fire and Arson Investigator* (June 1984) a debunking of everything mysterious about Mrs. Reeser's death.

They asked forensic anthropologist David Wolf for his opinion regarding statements about her shrunken skull, arguably the most controversial aspect of an already highly controversial case. Wolf told them the roundish object found in her apartment *might* have been "a globular lump that can result from the musculature of the neck where it attaches to the base of the skull," which *might appear* to a nonexpert as a "shrunken head."

Nickell and Fischer overlooked Wolf's qualifiers (appropriate, given that he was speculating about an artifact he hadn't seen) and unequivocally dismissed the shrinking of Mrs. Reeser's head as preposterous; then concluded that what "actually happened" was that "she became sedated from the sleeping pills" and dropped her cigarette upon "her own considerable body fat," resulting in "the rather complete destruction of her body which—while unusual—is nevertheless understandable."

Rather complete destruction? Understandable? As American actress Ruth Gordon once observed: "I think there is one smashing rule: 'Never face the facts.'"

If Nickell/Fischer are right, then why, pray tell, were St. Petersburg officials so baffled? The human-candle theory had been around long before 1951, and Mary Reeser wasn't the first fire death to occur in their jurisdiction. Hers was just the first to so thoroughly nonplus the professionals.

And why, forty years later, do crematoria operators spend $50,000-plus per unit and hire technicians to operate this expensive equipment, when for the cost of a five-cent cigarette and a few hours of patience they could accomplish the same result? The answer is, they *can't!*

Dale Auer, president of the Cremation Society of Pennsylvania, explained the operation of his crematorium. His brick-lined retort burns 1.5 million cubic feet of natural gas an hour, with an average cremation lasting two and a half hours (of which no less than ninety minutes is at 2600°F); large intake fans provide the "tremendous amount of fresh air" required; an afterburner, which must be at least 1800°F hot before the retort is even lit, further reduces the body; special filters then trap the noxious odor of burned flesh. Yet the fire-baked bones *still* need to be pulverized, first to "pebble-sized fragments" by a rolling process and finally in another machine "to the texture of flour."

Yet rotund Mrs. Reeser—in her apartment *devoid of all this modern, sophisticated hardware*—achieved the same end! Remarkable.

Dale Auer taught me one more lesson based on his twenty-five years of experience. "When a person has a lot of body oil, you have to be concerned about that," he explained. Why? Because, he said, overweight,

> Cremationists have much to worry about these days: shrapnel from exploding pace-makers, gooey silicone from melted breast implants, toxic mercury from dental fillings, bodies exuding debilitating odors that sicken Emergency Room doctors.
>
> But finding a cadaver fully ashed in a retort, within a few minutes, is not yet a worry to worry about.

fatty bodies "take *longer* to burn!" Just the opposite of what human-candle proponents argue on behalf of their theory.

And Mary Reeser *was* overweight. On this point, there is universal agreement. For just about every other aspect surrounding the last moments of her life, disagreement rages on.

Oh yes, about that telegram whose delivery led to discovering the macabre? It contained good news: everything had been arranged for Mary Reeser's return to Pennsylvania. It was a trip she never got to make, alive.

7

THERMOGENESIS: THE (QUANTUM) PHYSICS OF FIRE WITHIN

You've got to be prepared to step outside the currently accepted ideas, outside of the mainstream.

—Stephen W. Hawking,
"the most brilliant theoretical physicist since Einstein"

"I've never seen or heard of anything like it. There is no clue," arson investigator Edward Davies conceded during his examination of Mary Reeser's apartment as he sought answers to explain her death. In his befuddlement and frustration he off-handedly quipped, "We do not know that it *was* a fire..."

With that remark, I think Davies came closer than his contemporaries to grasping the *true* nature of the burning mystery in their midst. Because aspects of the widow's death truly spoke against the behavior of fire as it was understood in 1951. As it is understood by most people today.

What *is* fire anyway? Science defines fire as rapid self-sustaining oxidation accompanied by the evolution of heat and light. In most cases, that means a simple thing has to take place. Oxygen from the air must combine with some material that can burn. This reaction produces heat. Rusting of metal is a very slow combustion; more rapid reaction can yield intense glow and open flame combustion; very fast reaction generates explosive combustion.

People are most familiar with Class A fires, combustion that produces ashes and can be extinguished with water. Class A combustion byproducts are flame, heat, smoke, and fire gases, such as carbon monoxide (CO) and carbon dioxide (CO_2).

Fire has been known to man for about four hundred thousand years, according to science historian Jacob Bronowski, and is man's most valuable natural tool.

Yet it may surprise you to learn that, in this age of post-industrial hypertechnology, fire is not fully understood. Secrets still hide within the flickering flame of a candle; within the electric-blue jet of steel-cutting fire. The phenomena of ignition and propagation of flame, the complex interactions of the molecular events underlying the combustion process, and the evolution of postcombustion particles are aspects of every fire that elude full comprehension. Fire remains a mysterious and dangerous enigma.

Before criticizing anything, one should have a basic understanding of the subject. So when SHC debunkers categorically say people cannot spontaneously combust, it is incumbent on them to *first* prove that they know more about fire than do laboratory scientists who *specialize* in studying fire's behavior. Experts such as Professor Paul Garn, Ph.D. in chemistry, who said when appointed to direct the University of Akron's new Center for Fire and Hazardous Materials Research in 1984: "There is so much that we don't know about fire."

Especially about spontaneous ignition, that biological (metabolism) or chemical (oxidation) reaction that creates its own heat to produce combustion.

John E. Bowen wrote in the Australian Fire Protection Association's *Bulletin* (July 1984) that "The word spontaneous simply means that the process arises from internal forces or causes, that it is unplanned and that it is a natural process." That, alone, allows the scores of burnt humans already chronicled to be included as a subset of combustibles which are spontaneously ignitable.

Yet the *Fire and Arson Investigator* (December 1985) does not include humans in its listing of seventy-two items prone to spontaneous ignition, even though it mentions such rare exotics as sperm oil and powdered zirconium. Nor does Feller's *International Bibliography of Burns* (1969) include SHC in more than 150 citations dealing with the extent and depth of tissue burning.

The resistance, the vituperative opposition to spontaneous human combustion stems in large part to the word spontaneous. And to a lesser extent from a perception that SHC is an oxidizing combustion... what everyone knows as the common, ordinary, garden-variety, run-of-the-mill, what-happens-when-you-strike-a-match type of fire. It's the type of combustion that occurs in crematoria.

I could fill the remaining pages of this book with quotations by SHC commentators who deny, decry, and debunk SHC, such as Joe Nickell's admonition that anyone who defends SHC "will end up with egg on your face." However, this has no more relevancy to the facts of SHC than marshalling a thousand flat-earthers would have on changing the roundness of this planet.

Some people change reality to match perceptions; some people change perceptions to match reality. The perception that combustion is solely based on the classic Fire Triangle—which says burning happens only in the presence of adequate "heat" and "oxygen" and "fuel"—is, while accurate to a point, nonetheless far too restrictive for the study of SHC. Just as it has proven too limited in science's quest to understand fire itself. The Fire Triangle has recently been replaced by the Fire Rectangle. The new side—an additional avenue for sustainable combustion—is a "chemical or uninhibited molecular chain reaction."

> Well, it's kind of like how many fireplaces do you have to sit in front of to know that Santa Claus isn't coming down them?
> —Barry Karr, CSICOP executive director, explaining why thirty cases can disprove SHC; in *The Hartford Courant* (Oct. 31, 1994)

Besides broadening the popular conception about the forces underlying fire, this new element is crucial to investigating just what Davies's do-not-know-that-it-was-a-fire fire can be. To think of light puncturing metal would be crazy to the person holding a flashlight, but light waves made coherent and amplified in a laser can pierce plate steel in a second.

Combustion, as it relates to Mary Reeser and others, should no longer be thought of in terms of oxidizing fire per se but as release of potent energy. To broaden the range of consideration for and to defuse the prejudice against spontaneous combustion within people, I propose an alternative meaning for the SHC acronym: *SuperHyperthermic Carbonization.*[1]

COUNTING THE CALORIES OF SHC

While waiting for Professor Krogman to take a phone call during my interview with him, I noticed a placard prominently displayed on his office wall: *"The man who follows the crowd, will usually get no further than the crowd. The man who walks alone is likely to find himself in places where no one has ever been before."*

The "crowd" has walked away from SHC, saying it doesn't happen. So let this signpost be our philosophy, as we begin exploring in earnest some possible ways by which Mary Reeser and others managed to accomplish the seemingly impossible.

Up to this point, research has shown that the victims were almost completely dehydrated by a released energy that burned their tissues and bones to dust. The question this observation raises is: How much energy does this require?

The answer is relatively simple.

The human body is mostly water. Blood is 90 percent water, and even fibrous muscles are 80 to 90 percent water. Estimates for the total water content in a human range from 70 percent to 95 percent of body weight. Therefore, by taking the mean of this range, Mrs. Reeser, who weighed 170 to 175 pounds just prior to her death, contained about 140 pounds of water.

And her body was transformed *almost completely to ash*. That is, her body had *dehydrated*—meaning the approximately 16.5 gallons of water in her not only had to reach the boiling point, but had to turn to steam. Accomplishing this would produce 28,000 gallons of steam in her small apartment, and require 34,385,867 calories of energy.

This is no small feat. Roughly equivalent to eating 169,000 standard-size Hershey chocolate bars instantaneously!

The next question is: What can trigger an energy discharge *capable* of dehydrating a 170-pound Florida retiree in a short period of time, or for that matter, a slightly less heavy physician in Pennsylvania?

An answer to this second question is relatively *not* so simple. It involves the science of particle physics, and it goes to the very core of the makeup of our bodies; indeed, the makeup of the universe itself. Getting to the bottom of this question goes to the very foundation of the principles of physics as understood today... and a bit beyond.

> The true delight is in the finding out, rather than in the knowing.
> —Isaac Asimov

My investigation of the physics of SHC began in Spring 1975, when it was suggested to me that Mary Reeser was, in some way, a victim of a "freak accident" involving a high-speed, high-energy, and "infinitely small" particle that fissioned or fused with an atom in her body to unleash tremendous energy and a chain reaction which then dehydrated and disintegrated her body "in a split second *or less*."

What does modern physics say about such an esoteric concept regarding Mrs. Reeser's fate? Surprisingly quite a bit, though some background into the very strange world of modern theoretical physics is first necessary.

UNRAVELING THE NATURE OF MATTER

The more science unmasks the world, the stranger the world looks.

Once upon a time long ago, Aristotle taught that the universe was composed of four simple elements: air, earth, fire, and water. The Greek philosopher Democritus instructed his countrymen that Aristotle's elements were, in turn, formed by the smallest thing in the universe: the *atom*, that which is "not cuttable." The Roman poet Lucretius lauded in verse the atom for solving many perplexing mysteries. Upon this particle defined as indivisible, the world—as viewed by man—was constructed; the atom was fundamental, at the heart of soft clouds, hard stones, flowing water, and blazing fires.

Atoms are small: twenty-five million strung together would measure just one inch. For some twenty-four hundred years these tiny atoms ruled the world, and things ran pretty smoothly and understandably. Except for a few people who inconsiderately burned themselves to ash, life was familiar in the everyday world governed by atoms and Newtonian physics; so predictable that physicist Albert A. Michelson (1852–1931), who won the 1907 Nobel Prize in physics for measuring the speed of light, declared that since all the key physical laws had been discovered, there remained little more for his profession to do!

The twentieth century would soon prove the foolishness of another expert who put limits on knowledge. Ideas about the fundamental components of the universe began to change. And change *dramatically*.

Scientists realized the atom is not indivisible; the universe, not so apparently simple. Rutherford showed in 1911 that atoms have an internal structure of protons, neutrons, and electrons. French physicist Louis deBroglie declared in 1924 that electrons behave like particles and *at the same time* like waves. Matter kept getting still more complex.

Then physicists recognized that not only the atom but the world it structured was terribly complicated and, worse, unpredictable. In 1925, German physicist Werner Heisenberg established a scientific revolution with his *principle of indeterminacy* (which states the position and velocity of an electron cannot simultaneously be measured exactly), then he devised the mathematics of quantum mechanics to explain this brave new (and strange) world. (Lots of Nobel Prizes were garnered in the process, too.)

Heisenberg's theory proposed that every substance that radiates energy emits light as discreet units of energy. Einstein called these units "quanta." Quanta led to quantum theory, which now underpins not only physics but biology, chemistry, geology, and many other sciences. "It was one of the greatest intellectual revolutions of

> Anyone who is not shocked by the quantum theory does not understand it.
> —Niels Bohr, physicist

the twentieth century," wrote physicist Leon M. Lederman, director of the Fermi National Accelerator Laboratory (Fermilab). It is a revolt that few people understand, but it makes clear (to theoretical physicists anyway) that the "real" world is in truth elusive...a set of probabilities...a realm consisting of wave-particle duality, randomness, chance, uncertainty, entanglement, and potentiality, where (egad!) the mere act of one's observing influences what will be seen. It is a realm where, at the quantum level at least, anything can happen anywhere at any time.

It is perfectly appropriate—yea demanded by quantum physics—to rephrase

Hamlet's soliloquy as "To be, *and* not to be." Things have gotten complicated inside Democritus' uncuttable atom.

And a new kind of uncertainty has become a core principle of science.

CAPRICIOUSNESS WITHIN THE PERPLEXING ATOM

Several decades ago, one physicist offered a theory I think is applicable to SHC.

In 1944, Russian-born physicist George Gamov postulated a variation on the fundamental law of thermodynamics known as "the principle of increasing entropy"—that is, disorder. As a consequence of disorder, heat cannot of itself pass from a colder to a hotter body. Gamov proposed, in his *Mr. Tompkins Explores the Atom* (1944), that it is conceivable entropy *can* reverse itself. Instead of dispersing, molecules could accidentally swarm together and leave the rest of the room in a vacuum; heat would be concentrated amid surrounding coolness. Or a group of molecules might fall into an introverted swirl of collisions, concentrating energy at one point. These "statistical fluctuations of destiny"—inhomogeneities—would be extremely rare, he believed, yet molecules coalescing in a room might spontaneously boil a highball. If a highball, than a person? Mrs. Reeser?

Going beyond the laws of thermodynamics, which were quantified a century ago, the new realm of quantum theory (which describes the infinitesimally small to the incomprehensibly huge) has officially made everything random and malleable. This made Einstein himself uncomfortable. Of quantum theory he said, "God does not play dice." Yet he could not prove its equations wrong.

Intuition and fantasy now reigned alongside hard science.

Frederick Reines and Clyde Cowan Jr. combined some of each quality when co-confirming in 1956 the existence of a theorized particle even smaller than an atom's components. Postulated in the 1930s by Pauli, physicist Enrico Fermi termed it the "little neutral one"—a *neutrino*. Neutrinos have done a lot to upset the simplicity thought to be inherent in matter, and they are important to the concept we are working toward in explaining SHC.

A neutrino is an ancient, ethereal, elusive, and ghostly particle you can neither see nor feel. Scientists have called it "the nearest thing to nothing, being distinguished from the other elementary particles of matter mainly by what it does not have and by what it does not do." Birthed prolifically in the Big Bang when this universe began, neutrinos are so ephemeral they almost never leave any trace of their passage;[2] so small they can zip through three hundred light-years of solid lead before colliding with one atom.[3] They stream at nearly the speed of light through the cosmos—*hence through your body*—at the rate of sixty billion per square centimeter (the cross-section of a sugar cube) every second... which means your body is bombarded by neutrinos up to *10 trillion times a day.*

Neutrinos are, literally, a penetrating experience! And perplexing. Reines has said a neutrino might behave in a way comparable "to seeing a dog change into a cat and then back into a dog as it ambles down the street."[4] And to think that I, and probably you, had been taught the universe was orderly and made perfect sense!

The exotic, fasting-moving neutrino had its turn as the basic building block in the universe's playpen. Now this so-called "hot" subatom is, at least in the eyes of physicists, evolving into even *smaller* components that do more unimaginable things. The realm of subatomic photons born of quantum theory made things very, *very* strange. Physicists said so themselves. As playwright Tom Stoppard said in *Hapgood* about the dilemma facing physics: "When things get very small, they get truly crazy."

Physics became like a ruptured dam, overflowing with a cascade of confusion and elementary particles that were anything but elementary. Quantum mathematics and erudite conceptualizations by nuclear physicists spawned a blizzard of rogue particles assigned curious names and curiouser properties. A swarm of antimatter, ultra-tiny leptons, and pointlike quarks, which come in a variety of "flavors"—by 1981 six flavors had been discovered; two more ("vanilla" and "chocolate") were theorized and still missing ten years later!*—with anthropomorphic properties like "color," "strange," "gentleness," and "chimeron," now comprise a bizarre zoo of subatomic shrapnel whizzing through the universe. To keep everything caged in this menagerie, physicists devised the sticky gluon—itself held together by another unstable particle called (I kid you not!) a *glueball.*

Confused? So were many physicists, who by 1975 found themselves in a wonderland where anything that could be imagined seemed to be confirmed soon afterward in atom-smashing particle accelerators. Some physicists, not necessarily amused, quipped that "the theorists may have gone off the deep end."

> The full structure of the world is richer than our language can express and our brains comprehend.
> —Peter Coveney, on quantum physics and Chaos theory; in *The New Scientist Guide to Chaos* (1992)

---

* With the detection of the "top" quark by researchers at Fermilab in 1994, author Gordon Kane reports in *The Particle Garden: Our Universe as Understood by Particle Physicists* (1995) that the so-called Standard Model of particle physics seems complete. M'thinks the model has its shortcomings—

"Nothing stays simple for very long in particle physics," Lederman would later confess. The search for the simplest thing of all—one *fundamental* particle upon which everything is built—had almost become a joke. But few physicists were laughing. (Still, they collected Nobel Prizes.)

Particle physics indeed has room for speculation.

TAKING THE QUANTUM LEAP WITH MRS. REESER

"Will simpler structures from which quarks are made soon be proposed? Is it possible there are no elementary particles at all?" Lederman asked in 1978. It is this goal of finding the last box in an endless series of Chinese boxes that drives the minds of leading particle physicists today.

The key to a Grand Unification Theory—GUT—for energy appears to be a matter (so to speak) of smaller size and higher energy. How *small* and how *high*? And what does all—if any—of this have to do with Mary Reeser? Actually quite a lot, because the answer relates to SHC.

When I met with Walter Tipton to examine his photographs of Mrs. Reeser, I asked him what he saw when he looked at them. "A foot," he replied. "Anything beyond that is speculation."

Let's speculate.

William R. Shearer, Ph.D., when chairman of the Department of Chemistry at Dickinson College, worked with me in applying the rules of quantum mechanics to the baffling case of Mary Reeser. He was intrigued by the suggestion of a subatomic explanation for SHC, and applied quantum theory to calculate that a single photon able to trigger the burning of those 34,385,867 calories which dehydrated her body would have a wavelength of 1.20×10^{-31} *centimeters*.

How does that compare to particle wavelengths in the electromagnetic spectrum presently defined by physicists? It is "a value *way above* the energy of the highest energy gamma rays," remarked Professor Shearer. In fact, it's higher—which is to say smaller—than gamma rays by a factor of 20; that is, 100,000,000,000,000,000,000 (1

> There is not a reputable chemist in the world who believes in human spontaneous combustion...
> — science writer and CSICOP luminary Martin Gardner, rebuking *Science Digest* (January 1982) for publishing our article on "Human Fireballs" three months earlier.
>
> In *Nature* (Nov. 26, 1992), Gardner repeated the charge, broadening it to include all physicists. We are very impressed that Gardner can contact, and elicit an opinion from, *every* chemist and physicist in the world.... How'd he miss Professor Shearer?

followed by 20 zeros) times smaller than a gamma ray. Smaller than the wavelengths that carry your FM radio broadcasts by a factor of 37.

Mrs. Reeser's body certainly dehydrated; on this point there can be no argument. If quantum theory is valid for this application, then the culprit in her death was not a cigarette, not a match, not an electrical short-circuit in her apartment wiring, not a murderous cadre, not a lightning bolt, but possibly *a hyper-minuscule energy mass that exceeds today's physics by a factor of 20!*

It appears Mrs. Reeser's combustion was not only quirky, but possibly quarky.

In 1975, I coined a term for this energy unit derived from quantum theory: the *pyrotron*—an energy pattern that interacts with a vengeance to thoroughly "burn up" (that is, disrupt atomically) whatever it impacts.

Compared to the size of currently defined subatomic units, already inconceivably tiny, this new hypothetical particle is minute beyond comprehension. The pyrotron is 775,000,000,000,000,000 times smaller than a quark. Expressed in more visual terms, you could squeeze almost 800,000 trillion trillion pyrotrons inside this letter *o*. And if the pyrotron were enlarged to the size of a pinprick (about the thickness of a $1 bill), that number of dollar bills[5] would stack about 8.5×10^9 light-years high—twice the distance to the Large Magellanic Cloud, the nearest galaxy to the Milky Way! To call the pyrotron *exceedingly small* is an understatement.

Pyrotrons are as penetrating an experience as they are small. To a pyrotron, the quarks in your body look like galaxies. To a pyrotron, your body is so full of wide-open spaces that the chance of an atom inside you getting pricked by a pyrotron rocketing through you is, shall one say, astronomically small! But once in a rare while, the odds catch up.

And subatomic hell breaks forth, as a freak accident rapidly unfolds. The pyrotron rockets into a quark and, in smashing it, triggers a complex subnuclear chain reaction within the body.

The precise development of this personal Hiroshima will vary, depending on the momentum of the pyrotron; the point of impact on and size of the subatomic particle impacted; and the location of the collision within the body. Generally, the energy released in the initial collision would cause neighboring atoms to spontaneously speed up, enlarge their diameters, and thus increase the probability of more interactive collisions. As the

> If hypothesis does not anticipate truth, it sometimes leads the way to it, and, like the scaffolding to a building, it helps us in keeping pace with the more solid edifice we are erecting.
> —Dr. Thomas Newell, "Observations upon Spontaneous Combustion of the Human Body"; *Midland Medical and Surgical Reporter* (vol. 4, May 1829)

transfer of energy continues, some over-energized atoms would at this point react with the oxygen in the respiratory system of the body. (Or in the atmosphere.) So there could be complete combustion (oxidation) in the immediate vicinity. The likeliest point of origin would be the bulkiest area of the body: the torso. The farther removed from the center of the point of the first collision, the lower the thermal level would be, resulting in less complete combustion (that is, nuclear disintegration) and creating partial dehydration, distillation, or other limited decompositions in the body.

The rapid distillation of decomposed fats (mostly hydrocarbon) and glucose ($C_6H_{12}O_6$) with the liberated carbon coming in contact with steam vapor condensed on the walls, explains the oft-reported presence of the "oily soot" encountered in many SHC cases. It explains why victims experience absence of struggle or pain; why they so often exhibit perplexingly varied degrees of disintegration; why their extremities (feet, hands, legs, head) are often discovered intact; why adjacent objects are rarely damaged (the "fire" being an internal atomic disruption incommunicable to, or having very limited effect on, energy patterns external to the human biosystem).

And how much energy does this unbelievably small pyrotron pack as it jets neutrino-like through your body, the planet, indeed the galaxies themselves? Quantum theory again provides an answer, and what an answer it is! *The pyrotron's energy is 9.6×10^{26} electron-volts* or 10^{18} GeV (billion electron-volts).

"That's an *incredible* amount of energy!" said Priscilla Laws, an expert on ionizing radiation and Ph.D. professor in the Physics Department at Dickinson College. She double-checked the mathematics. No mistakes. "Such energies do not exist on Earth," she added emphatically. "At least not man-made, they don't!" Professor Laws paused, thought further, then commented about how neutrinos can zap in from outer space and pass through the Earth without the neutrino's trajectory being affected; except rarely, when a neutrino collides with a particle and sets off a nuclear chain reaction...

Professor Laws was describing exactly the concept behind my little pyrotron.

THE "HOT CHEMISTRY" OF COMBUSTION

The enormous energy released in a pyrotron collision would generate awesomely high temperatures, albeit briefly, just as do particle accelerators. We leave behind the common sense of ordinary chemical combustion reactions and enter a new realm with its own conundrums: the world of "hot chemistry."

According to researcher Richard Wolfgang in *Scientific American* (January 1966), the classical reactions between atoms and molecules at "ordinary temperatures" do not occur at extremely elevated energy levels. Hot chemistry allows more elementary particles (ranging from nucleons to molecules) to interact, and the type of reaction changes from the molecular collisions of classical chemistry to the processes defined by quantum physics. "Hot atoms," explains Wolfgang, "can react by all possible modes, regardless of their energy threshold. Hence totally new reactions paths are found."

In experiments confined to a range of 10^2 electron-volts. Wolfgang pointed out that the resulting "hot" atoms collide to yield unexpected chemical changes with hydrogen atom reactions. (Hydrogen is, of course, one of two elements comprising water.) Molecules heavier than hydrogen are "less probable" to rupture their bonds in a hot chemistry reaction than is hydrogen, Wolfgang discovered. But, he added, "On the other hand, if a reaction does take place, the immediate product is quite likely to be left with enough energy of excitation to break a second bond."

Hence the greater the excitation energy, the more molecular chemical bonds will break apart and the larger the number of different molecules that can be splintered, thereby increasing the chances for a self-sustaining chain reaction. The key, then, is energy threshold.

First-order calculations indicate the suspect pyrotron that could have killed Mrs. Reeser had an energy potential of almost 10^{27} electron-volts. This *more than fulfills* the energy needed to split apart all the water molecules in her body; more than fulfills physicist Stephen W. Hawking's estimate that the grand unification energy "would probably have to be at least a thousand million million GeV." That is, 10^{24} electron-volts. (By comparison, the limits of current particle accelerator energy technology is 2×10^{12} electron-volts at Fermilab.)

Therefore, the new "limit" on particle energy as defined by the pyrotron is more than the *square* of what a physicist today (1991) can hope to find experimentally—truly a quantum leap (pun intended) in physics.

Convergence: The Pyrotron Meets the X Particle

"We have all these unanticipated mysteries," remarked Michael Tannenbaum of Rockefeller University about the new discoveries in particle physics in the 1970s. That's true about most of modern science, too. But particle physics is heady research, designed and engineered by brilliant men and women toting Ph.Ds. who spend billions of dollars hunting down things billionths-of-an-inch small (but much *larger* than a pyrotron). Was I out of my league in finding, so to speak, the mathematical existence of these pyrotrons? And finding it, no less, through the study of an event said to be impossible?

Unbeknown to me then, the universe was harboring a surprise up its quantum sleeve. Confirmation of the pyrotron theory—and confirmation of the incredible breadth of application for SHC research—was found, several years later, in the work of Harvard University physicist Howard M. Georgi III, Ph.D.

"There can be nothing simpler than an elementary particle," Professor Georgi began his article titled "A Unified Theory of Elementary Particles and Forces" in *Scientific American* (April 1981). He then postulated the *final* box in the Chinese puzzle of ever-shrinking subatomic units: the existence of "just one kind of elementary particle and one important force."

It was the key to GUT. It was what Einstein had sought and failed to find. Georgi estimated the unification of particles and forces would happen at "about 10^{15} gigaelectron-volts"—a scale at which all matter is *impermanent*, he believed.

The unified theory predicts "new phenomena that cannot be deduced from earlier theories" said Georgi, but has one problem: "it is most unlikely such an energy will ever be attained in the laboratory." Probably not in a laboratory... but his figures are *right in the ballpark with the pyrotron*, the hypothetical cause for some (if not all) incidents of SHC. In fact, at 10^{27} electron-volts the pyrotron has room to spare by being a thousand-fold more energetic.

And SHC, so far as orthodox science is concerned, is *certainly* a new phenomenon.

Georgi postulated further that the ideal structural arrangement for his fundamental unit—he named it the X particle—would be two superimposed triangles forming a Star of David. (The significance of triangles in a very different yet related context will reappear when we return to the macroscopic world of strange combustion.) And when each particle is displaced vertically by a distance proportional to its electrical charge, the third-dimensional pattern forms a cube, which in turn "suggests some underlying connection between them," Georgi concluded. (Metaphysicians will recognize the beauty and intrinsic significance of a unified field theory that incorporates at its microcosmic heart triangles and cubes.)

Calculating the scale at which leptons and quarks could be unified by the X particle was next on Georgi's agenda.

Along with researchers Helen R. Quinn at the Stanford Linear Accelerator Center and Steven Weinberg of Harvard University, Georgi determined X to be about 10^{-29} cms. "The unification scale of 10^{-29} centimeters is an extraordinary small distance," Georgi wrote in understatement about his hypothetical particle. (Minuscule as it is, Georgi was pleased: had the scale been 10^{-33} centimeters, he said, the universe could cease to be.)

The pyrotron's size of 1.29×10^{-31} centimeters is a mere a hundred times smaller than Georgi's X particle, yet a hundred times larger than the limit at which the universe theoretically would never exist. In dealing with matter so small, the closeness of these values *arrived at from two avenues of research about as opposite as one can imagine* is nothing short of remarkable!

Georgi made another point, of interest here. "X particles can do something no other particles can do: they can transform a quark into a lepton or a quark into an antiquark," declared the Harvard physicist. "This process puts into question the very stability of matter." And does not the matter comprising a human body become unstable and breakdown in SHC?

He proceeded to speculate that if two quarks approached to within 10^{-29} centimeters of one another, an X particle could be exchanged to unleash a catastrophic chain reaction in which hydrogen atoms would be annihilated to give rise to gamma rays or high energy photons. Said the physicist: "The end result is that a hydrogen atom decays into a state of pure radiation. This process represents a conversion of matter into energy far more efficient than that of nuclear fission or thermonuclear fusion."

It seems the *perfect description* for what might be happening in cases of classic SHC.

Also noteworthy to this discussion is the physicist's supposition that it is the hydrogen atom which is the likely target for initial annihilation into pure radiation. Since the human body is mostly water and every molecule of water contains two hydrogen atoms, hydrogen atoms present the largest of targets. Annihilation of hydrogen atoms would split the water molecules apart in a lightning-quick subatomic exothermic chain reaction, instantly dehydrating the body.

Like with Mrs. Reeser, for instance?

(Consider also that many SHC victims imbibed heavily. Maybe alcohol somehow causes the hydrogen atoms in the body's water and glucose to energize slightly[6]—hence expand—thereby making these people a "bigger atomic target" and more susceptible than their nonimbibing friends to the flyby of a pyrotron. This could lead to a chain-branching of elementary reactions in the body and the creation of atomic hydrogen, molecular oxygen and hydroxyl radicals.[7] This chain branching will occur at 2600 to 3600°F; at only 1420 to 2150°F if the molecules are decomposed by electrical discharge. Both temperature ranges should be *easily* attained in a pyrotron impact, which instantaneously increases molecular motion—heat—and generates a bioelectrical aspect to the reaction within the body as well.)

"The energy needed to create real X particles may be forever beyond the capabilities of manmade machines," acknowledged Georgi, "and indeed there may be no process anywhere in the universe today that can generate such a high energy." Yet already we have scores of events represented by Mrs. Reeser which, if calculations are correct, do indeed generate evidence of such a high energy and give testimony to the *continuing existence* of Georgi's X particle zapping through the spaces of the universe and in our bodies.

I just happened in 1975 to give it a different name: the pyrotron.[8]

PUTTING THE WRAP ON THE PYROTRON

In particle physics research, it's been said "the *real* fun is discovering unexpected things."

The theorized pyrotron—an "unexpected thing" 10^{20} times smaller than the shortest gamma rays—may indeed inaugurate *some* cases of SHC as it zips neutrino-like through matter and occasionally triggers subatomic (and then atomic) disintegration.

Since neutrinos apparently do not unleash macroscopic disintegrations (such as burned-up bodies) when they collide with matter in people, why then should pyrotrons do so when they are infinitely smaller? The answer lies in *energy potential*. The pyrotron is trillions of times more energetic than a neutrino; hence, as demanded by physics' conservation of energy maxim, its impact would be trillions of times more powerful.

One argument against pyrotrons even existing, let alone triggering human fireballs, is the belief of astrophysicists that such a fundamental, high energy particle—*the* fundamental particle in all creation—could only have existed for the first nanosecond of the universe's Big Bang formation, after which it instantly decayed into lower energy leptons and quarks. I disagree. I feel that the ultimate particle *still* traverses the universe, leaving as a calling card of its passage the left leg of Dr. Bentley and the left foot of Mary Reeser after kindling a micro-Big Bang inside them.

The instrumentation does not yet exist (and may forever be beyond technology) to detect or replicate within accelerators such a minute, energetic particle. As Hawking wrote in his 1988 book, *A Brief History of Time*: "a machine that was powerful enough to accelerate particles to the grand unification energy would have to be as big as the Solar System—and would be unlikely to be funded in the present economic climate."

Should the pyrotron-X particle provide evidence of its existence through the disintegrated human bodies that occasionally confront science, then there is little point in spending multibillions of dollars to construct the Superconducting SuperCollider accelerator.[9] What is far beyond the SSC design-criteria to quantify has *already* revealed itself through the ashes of SHC. (Funding for the SSC *was* eventually cancelled, for reasons not related to pyrotron-X.)

In the future, when remains such as those of Mrs. Reeser are discovered, if nuclear fragmentation is involved I would anticipate the presence of remnant radioactivity. Though firemen are normally neither trained nor equipped to even consider radiation levels on a fire call, there are two relatively easy procedures to test this prediction.

First, ashes from the victims should be placed on photographic film and wrapped in opaque paper for about ten days; when developed, the film would show luminous areas if the ashes were radioactive. An electroscope properly calibrated to ionizing radiation levels could determine the amount of radioactive material in the human ash and provide data for further calculations on the energies involved (and indicate whether investigators need protection against possible post-SHC radiation decay).

Another approach would be to use a scintillation counter to measure any energy release as light flashes during the loss of atomic excitation energy. To be fruitful, this method would have to be employed *immediately* at the fire scene; any delay might exceed the brief time for de-excitation of the atoms and molecules. (Light-producing excitation in the human body is already documented by both spontaneously combusting—and noncombusting—people, as you will soon learn.)

Exploration of the quantum world continues. The unexpected is not only anticipated but eagerly sought. As Samuel Ting, the MIT scientist at the forefront of current particle physics research, told *Time* (April 16, 1990): "I will only consider our experiment a success if we discover something really surprising—new types of quarks, for example—that would explode the standard theory."

To scientists investigating the frontiers of the GUT standard model, I submit for your consideration the pyrotron and Mrs. Reeser. Time and technology may one day prove me right (or wrong). Some day, perhaps, the awesome energy that incinerates people can be harnessed nonfatally to provide society with *a nonpolluting energy source* that will never be depleted. Science may never know with certainty whether the pyrotron exists. But as Princeton University neutrino researcher John Bahcall told *Omni* (June 1979) about the quantum world: "To a physicist everything is possible until proved otherwise."

In that realm of the possible, people continue igniting as did Mary Hardy Reeser. And they do so for *some* reason.

8

THE PARADE OF PERSONAL FIRE MARCHES ON

We have no evidence of the human body suddenly bursting into flame. It has been built up by people with very rich imaginations... the whole thing is a fairy tale.

—Paul Kurtz, CSICOP Chairman,
advice to the author (1986)

Curiosity is a characteristic of a vigorous intellect. If things remain mysterious it should be not be for want of trying to fathom their secrets. Science fiction laureate Arthur C. Clarke has regularly turned his keen intellect to contemplate the marvels of forteana. Of those that withstood his rigorous scrutiny, he declared SHC to be the "quintessential" example of the rarest type of natural phenomena. In his forward to *Arthur C. Clarke's Mysterious World* (1980) he wrote this about SHC: "There seems no way in which this particular mystery can ever be solved, without a great deal more evidence—and who would wish for *that*?"

One needn't wish. History provides many examples if one seeks them out.

December 16, 1952; Muncie, Indiana. Mrs. Lola McLaughlin "apparently" (and officially) died of asphyxiation even though she, found lying on her living room sofa, was covered with third-degree burns. Her canary and dog asphyxiated too in the localized fire. She didn't smoke, said *The Muncie Star* (December 17, 18, and 19, 1952).

February 1954; Springfield, Massachusetts. Catherine Sutton dies at home on 2132 Main Street by burning suggestive of SHC, said *Fate* (November 1954).

Early 1956; Wichita, Kansas. Authorities find Mrs. Eva Ola Godfrey, seventy-six, burned from head to foot. Paradoxically, the flames that killed her neglected to singe her hair and clothing.

April 28, 1956; Benecia, California. Sam Massenzi stood outside 141 East F Street while his friend went in to change clothes. After waiting thirty minutes, Massenzi checked on his friend, Harold Hall, fifty-nine, who was now frightfully burnt but alive on the kitchen floor: arms, face, and chest virtually charred, the clothing on his back untouched. Hall, unable to explain why he burned, soon died. Fire Chief Thomas Geifels flatly ruled out gas leaks and ruled the fire "unexplained."

December 1956; Honolulu, Hawaii. In his home at 1130 Maunakea Street, crippled Young Sik Kim, seventy-eight, was discovered wrapped in flames; flames so hot his neighbor Virginia Cagat couldn't approach him. Others were also repulsed by the heat; by the sight. When firemen arrived ten minutes later, the localized blaze had died down and was easily put out. Now came the uneasiness: how to explain his half-consumed body burned along with the overstuffed chair he sat in. The fire, which also destroyed a nearby rack of clothes, a venetian blind and a curtain, only slightly scorched the wheelchair on which his feet rested.

> An objective reading of the 'evidence' leads one to the conclusion that either it [SHC] is a true anomaly which suffers from inadequate documentation or it is simply a manufactured mystery generated out of imaginative interpretations of fire deaths which may have been unusual but which had conventional causes.
> —Jerome Clark, *Unexplained!* (1993)

December 18, 1956; Roxbury, Massachusetts. In her kitchen, Mrs. Catherine Cahill, seventy-eight, sits in a rocking chair and quietly burns to death. According to next day's *Boston Sunday Globe*, she didn't smoke and firefighters found no source for the fatal fire.

March 1, 1957; South Bend, Indiana. The South Bend Tribune (March 2, 1957) announced that a "body fire" ignited around Mrs. Sarah Wilderman around noon yesterday. She screamed she was on fire. Moments later she had "100 per cent burns of her entire body." Only the soles of her feet escaped. She did not smoke. Her bathrobe is "believed" to have caught fire, but "baffled" fire investigators cannot find out how. Said the newspaper, she herself had "ignited mysteriously"... and she died eleven and a half hours later.

Around September 30, 1957; Kensington, London, England. While bathing in her Queen's Gate Terrace apartment, Countess Maria de Serrant screams. Those who responded found the sixty-three-year-old noblewoman dead in her half-filled bathtub, hair smoking and body badly burned. *Not* your typical scalding water mishap! Her Majesty's Coroner Dr. Gavin Thurston ruled "accidental death due to shock from extensive burns." In a bathtub? If the countess had dropped an electrical appliance into the tub, no one noticed (or reported) it.

January 10 through 17, 1958; Glendive, Montana. The Charles King household awoke to a strange day on January 10 when Mrs. King walked into her living room at 8:00 A.M., and saw window curtains spontaneously burst into flame. At 8:30 A.M., the dining room curtains blew up in fire. At 4:30 P.M., a waste paper basket self-incinerated. Just after six o'clock, wispy smoke was traced to bedroom curtains aflame. During the next week five more inexplicable fires erupted, focusing on articles of cloth. Glendive Fire Chief George Smith ordered the removal of "all materials in which spontaneous combustion might possibly occur." (He neglected to include the occupants.) The precaution came too late for Mrs. King's daughter, who burned her hands fighting one of the blazes. Not SHC in the classic sense, but certainly a human burned by spontaneous combustion of the anomalous kind.

Mrs. Middleton Meets the SHC/PC Controversy Midway

January 29, 1958; Pimlico, West London, England. A sixty-nine-year-old hypertensive widow, identified by Coroner Thurston only as Mrs. E. M.—her full name is Mrs. Edith Middleton—burns up upon the hearth in her Hammersmith apartment. She was slim, lived alone, and in poor health, suffering high blood pressure and mild Parkinson's disease. She did not drink alcohol. Neighbor Annie Law had visited her around three o'clock in the afternoon and lit a coal fire for her before leaving. Between five and six o'clock, Annie noticed undue smoke and sparks spewing from her friend's chimney. Next morning another neighbor, concerned, discovered the deceased amid considerable soot. Firemen quickly found Mrs. Middleton's upper body lying inside her fireplace, where she was severely burned above her knees.

Reported Thurston: "The trunk, arms, head, and upper thighs were charred and calcined. A good deal of bony tissue was missing." Apart from the body, damage was limited to a small 1'5" by 7" hole burned through the floorboards in front of the hearth. An autopsy revealed she had died before the fire touched her. Death was attributed to "natural causes."

At first glance it's easy to think she was about to clean out the chimney and inadvertently lay atop a hot coal, had a heart attack, and slowly roasted to very well done. Indeed, officially her death was ascribed to "coronary thrombosis due to atheroma."

While it's certainly possible—some will say it's absolutely essential—that she met her demise this way, Fire Marshal Robert C. Meslin (whom you'll meet later) examined a photograph of the scene and concluded from his decades of fire-fighting experience that the plume of soot, which extended dramatically out into the room, was inconsistent with a flue draft. Instead, the V-shaped plume indicated to him that the fire had originated *inside* the woman, from whom combustion residues were then ejected into the room rather than *solely* up the chimney as one would expect from natural drafting alone.

To invoke SHC would be risky; not so, PC. Heed what Her Majesty's Coroner said about this case: "the body has burned in its own substance, without external fuel, and in which there has been a remarkable absence of damage to surrounding inflammable objects. Anyone with experience can recognize a difference between such a case, and the familiar ones of extensive burning in a fire, where the heat, at least initially, comes from outside."

Dr. Thurston saw striking similarity to earlier cases in the medical literature, and boldly called this death "undisputed" preternatural combustibility.

In 1961, a distinguished physician *officially* designates PC as a natural cause of death.

However, he disputed any indication of the paranormal (that is, SHC). He fell back on the explanation of body fat burning in the presence of air drafts, that old idea which had been proposed as early as 1838 by Beck and Beck in their *Elements of Jurisprudence*. Thurston added a new criterion for these inflamings: "all the cases have happened indoors, usually near the hearth."

That is wrong, as you'll see later in an episode from 1890, which (obviously) was in literature available to Dr. Thurston. Clearly there is something about this phenomenon that inhibits doctors and scientists from familiarizing themselves with its history.

THE YOUNG AND THE OLD

January 31, 1959; Laguna Home for the Aged, San Francisco, California. Orderly Sylvester Ellis waited for resident Jack Larber to drink a glass of milk, then left. Five minutes later he finds Larber swathed in flames. Since the victim neither smoked nor carried matches, a fire department official ventured that someone had swiftly doused the old man in lighter fluid and torched him—a grisly theory that investigation found insupportable.

Spring 1959; Rockford, Illinois. Four-month-old Rickey Paul Pruitt mysteriously burns to death in his crib. His crib, crib bedding, and clothing failed to burn. Unless counterpoint information is missing, Baby Rickey *certainly* wasn't smoking or playing with matches or dealing with any fire that called for common sense.

July 4, 1959; New York City. Edward Mottern is awakened in his home at 320 East 57th Street by smoke. It came from his wife, Virginia Mottern, burned up in a smoldering chair. Said *The New York Times* the next day about the fire's origin in the fifty-two-year-old woman: "not immediately determined." The NYFD hasn't told me if a cause was *ever* determined.

Summer 1959; Grand Rapids, Michigan. Dr. George Ruggy put his infant child to bed, then retired at 10:30 P.M., leaving his wife downstairs reading a book. About 1:00 A.M., passersby saw flames in the living room. They rushed in but could not save Dr. Lemoye Unkefer, thirty-six, who lay charred on the floor next to her book. Husband and child had slept through the fire that silently took her life.

Prior to 1961, probably in the United States, a woman known to Rev. Winogene Savage is discovered ablaze on the floor "with a strange fireball hovering over her blackened form." The preacher told *Fate* (April 1961) that buckets of water successfully extinguished the fire, but the married victim died anyway. Her clothing had burnt off, yet the rug beneath her was not scorched; nor was anything else in the room. The fireball simply vanished.

January 1961; Spokane, Washington. Charles A. McCollough dies in a fashion suggestive of SHC, suggests *Fate* (June 1961).

August 2, 1962; Lockland, Ohio. In this Cincinnati suburb Mary Martin, seventy-four, burns to death in clothing aflame in the absence of a fire source and other damage. Next day, *The Cincinnati Enquirer* called it a "mystery."

DR. GEE AND THE HUMAN-CANDLE EXPERIMENT

A separate mystery, and accompanying dilemmas, confronted Professor David J. Gee, M.D., a lecturer in the Department of Forensic Medicine at the University of Leeds. His interest in SHC sprang from hearing about Mrs. E. H., eighty-five, an Englishwoman in good health who lived in Leeds until she died strangely by fire in November 1963.

Her calcined body rested upright on a hearth-step. Her right leg was severed at mid-calf; its upper bone exposed and devoid of all flesh; the foot and its shoe unburned. Her left leg was indiscernible. Her torso was a burned-out hulk upon which rested a carbon-blackened head—appearing to have shrunken. She had been visited only three hours earlier, and her neighbors found no fire in the fireplace when they returned to find the gruesome scene.

Dr. Gee wrote up the case for *Medicine, Science and the Law* (1965), saying it was "curious" and "of some interest" because "I have only been able to find one other instance reported in this country in the present century" that paralleled such fiercely localized combustion.

Only two cases in sixty-five years? Oh well.

Like Liebig a century earlier, Dr. Gee turned to the laboratory for answers to what nature had accomplished.

He experimented. He rolled eight-inch slivers of human body fat in cloth, then held his flesh cigarettes over a Bunsen burner in front of an extractor fan.

He learned the fat would ignite "after about a minute"... and then "combustion of the fat and cloth proceeded slowly along the length of the roll, with a smoky yellow flame and much production of soot, the entire roll being consumed after about one hour."

He experimented again, using skin instead of cloth; same result.

The professor of forensic medicine was satisfied his experiments proved "the most reasonable explanation for the occurrence of the curious phenomena." In correspondence to me in 1975, Dr. Gee maintained this premise, and addressed the singular escape of the woman's foot: "The reason the foot remained intact is in my opinion that it was beyond the covering provided by clothing to the rest of the body and it was spared for that reason."

Is this plausible? To leap from charring in one hour an eight-inch sliver of fat over a Bunsen burner to ashing a whole adult in no more than three hours before a fireplace devoid of fire is, I submit, quantum extrapolation. Note: Dr. Gee managed to burn up only a sliver of fat, *not* an arm bone or pelvis. Think: how long would it take to reduce to ash a 170-pound widow (*skeleton included*) over a Bunsen burner! Another problem for his extrapolation is that his experiment produced yellow flames, whereas witnesses of SHC usually see *blue fire*.

> Dr. Gee could have benefited from reading about the experiment done by J. de Fontanelle way back in 1828.
>
> Incredibly, the myth persists that burning a cigar-sized sliver of cloth-wrapped fat equates to incinerating a 170-pound human. The fifth show of Arthur C. Clarke's series "Mysterious Universe" discusses several classic cases of SHC, then concludes with a laboratory technician showing how (with great difficulty) he can coax fat to slowly drip—it never ignites—out of a piece of pork suspended over a Bunsen burner. Gee—
>
> —The Discovery Channel
> (Jan. 26, 1995)

"At a fire scene," arson instructor Klages told me, "there's a reason for everything—if you're lucky enough to find it." I don't think Dr. Gee got lucky this time. The Irish Sweepstakes gives better odds than does his "resolution" for the fiery death of Mrs. E. H.

As Liebig had disproved the alcohol-impregnated tissue theory for SHC but not SHC itself, it seemed Dr. Gee called into question his own theory of a human fat-candle to explain SHC.* Success of the human-candle theory to

---

* In fact, he later did. Commenting in *Doctor* (September 5, 1985), Professor Gee referred to a man found burned in a British public lavatory: "He may have set himself alight with a cigarette and there was a suggestion he could have been carrying an incendiary device. But I am not entirely satisfied with this and I must say this case does intrigue me."

explain wholly incinerated people is about as likely as finding Tonya Harding's picture on a box of Wheaties.

And nature would continue offering its own disproof of this dubious theory.

CHALLENGES MOUNT AGAINST HUMAN FAT-CANDLES

December 4, 1963; Long Island, New York. In the town of Glen Cove, clothing burns off the body of Thomas Sweizerski, age sixty-six. Next day's *Long Island Press* noted there was no other damage or sign of fire in his room.

February 22, 1964; Clapham, London, England. A few minutes before nine o'clock in the morning, the local Fire Brigade arrives at a ground-floor apartment and begins a search for the occupant, a seventy-two-year-old Clapham woman whom they cannot find. Small wonder. In the tiny back room measuring six feet by eight feet are a few heat-cooked springs from a chair's cushion sticking out of a pile of ashes; overhead, the plaster ceiling shows moderate damage indicative of a flash fire. Aside from smoke residue, damage elsewhere is moderate to none. But where is the woman? Firemen finally realize they've been looking *right at* her, once they see two unburnt feet jutting beyond the perimeter of the gray and white calcined ashes. Beyond her feet in shiny black pump shoes, her left leg's flesh suffered third-degree burns until it gradually disappeared below the kneecap into the rubble; her right leg, pristine, abruptly vanished at mid-calf into nothingness. On the fire report, the nature of injury is listed as "Severe burn to body." (Kind of like saying Dr. Bentley died by asphyxiation!) Supposed cause of fire is simply "Unknown."

May 1964; Naphill, High Wycombe, Buckinghamshire, England. A thirteen-year-old High Wycombe lad is found upon the kitchen floor, dead by serious burns, in his home thirty miles west-northwest of London. The carpet beneath him was damaged, but there was "no other fire in the locked room."

Around June 21, 1964; Launton, Oxfordshire, England. Mrs. Beatrice Massey is seen "blazing like a torch" in her garden westward from High Wycombe. A passerby threw her to the ground and attempted to smother her flaming body. Her two children were in the house and heard nothing unusual.

March 26, 1965; Foster Park, California. On Camp Chaffee Road in this small town sixty miles west of downtown Los Angeles, Herbert Shinn, seventy, develops severe burns on his chest, back, and spine. He died thirty minutes later in the hospital. No known source of fire, said *Fate* (September 1965).

December 1965; Oneida, New York. Mrs. Katherine Elizabeth Chaires becomes a torch in her living room. Ed Wilcoxon, a tenant in her home, burned his hand trying to rescue her, reported *Fate* (June 1967). In 1982, I spoke to the Oneida Police Department, where Rosie Myers recalled the case immediately even though seventeen years had elapsed; she said Mrs. Chaires "burned in a chair." Former Oneida fire chief John F. Myers told me the deceased was a heavy smoker. "However," he pointed out in rejecting a dropped cigarette scenario, "the fire is very unusual. In thirty years I've never seen anything like it." Police Detective James Eddy said he too, seventeen years after the fact, remained "mystified" by Mrs. Chaires's demise.

Late May 1966; St. Louis, Missouri. Upon her burned bed lies Mrs. Ethel Woodward, sixty-eight, dead. She didn't smoke. Said the *St. Louis Post-Dispatch* (June 1, 1966) about the rest of the bedroom: "Firemen were clearly mystified as to why the damage was so slight."

Around November 16, 1966; Palo Alto, California. Baffled authorities find a bed destroyed by fire in the home of Marlin Stevens Smith, fifty-four. He lay nearby, dead. Unfortunately the *Palo Alto Times* (November 16, 1966) doesn't provide adequate information to determine whether this was a smoking materials accident, spontaneous combustion of the bed, or SHC with secondary ignition of a bed.

November 6, 1967; Harrisburg, Pennsylvania. Harry Boyd Halfpenny, eighty-two, a retired widower and a heavy smoker who lived alone a few blocks from the state capitol. Early that afternoon, he "died suddenly" in a tiny blaze amid a "hell of a lot of smoke" while seated in an easychair in the middle of his small living room. Officially, "asphyxiation and third degree burns over 90 percent of his body was direct cause of death."

In 1979, Harrisburg Fire Bureau candidate Gene Hoy told me his fire instructor had mentioned the case as a possible SHC. (Wow!) The instructor mentioned photographs too. I promptly called the Harrisburg Fire Bureau. Though a dozen years had passed, a secretary immediately recalled the episode. "Strange!" she said. "Yes, come down. You can see the fire file and the photographs."

I arrived at HFB within the hour. Only to be told by a senior officer (who also immediately recalled this incident) that the file didn't exist. Furthermore, he stated, that particular roll of film had never been developed. Instead, he insisted, it had languished inside a fire chief's desk drawer. For twelve years? Hmm...

I walked a few blocks to the office of the Dauphin County Fire Marshal. Introducing myself to Marshal Elmer Shover, I explained my interest in Halfpenny's death. Shover spun around in his chair, tugged on a filing cabinet drawer, pulled out a manila folder, and extracted a big black-and-white glossy. Another nonexistent photograph of anomalous combustion spontaneously appears.

It showed a pair of legs propped against the front of an overstuffed chair; cotton batting sticking out, unburned. The apartment walls were dark-sooted about four feet above the floor. Otherwise, the photograph proved how isolated had been the fire damage. An isolation restricted to Mr. Halfpenny, who was pretty much burned away.

I located a relative who agreed to discuss her deceased kin's personal and psychological history. Therein, perhaps, lay clues to what happened in his last moments of life. Unfortunately, a knock on her door for the scheduled interview was answered by the relative's husband, who ordered me away at threat of arrest and lawsuit. No insights, this day.

Sometime in January 1968, an occupant of a house in Ballinger, Texas—a rail town near Abilene—became a habitat for spontaneous human combustion, according to Persinger and Lafrenière's *Space-Time Transients* (1977). To which Gerry Hunt added in *Beyond Belief: Bizarre America* (1988): "The home is reported to have a history of SHC." Substantiation of neither individual nor collective claims can I find.

Around April 28, 1968; Preston, Ontario. Canadian Hugh McIlvenny, seventy-six, is burned and semiconscious after he crawled away from what at first resembled a normal fire. Damage totaled $500 and was limited to a chair, a four-by-six-foot patch of carpet, two heat-cracked windows, and room scorching. McIlvenny's face, hands, and feet were burned. Curiously, his socks were not.

April 7, 1969; Long Beach, California. At 5334 Cedar Avenue Mrs. Grace Walker, sixty, was found alive but burned over 90 percent of her body. She died en route to the hospital. She didn't smoke or carry matches. By the time I contacted the Long Beach Fire Administration Office, records for this case had been purged.

Around December 16, 1969; Toronto, Ontario. Elderly John Komar is found unconscious with "extensive burns of arms and back." No sign of fire beyond himself. *The Toronto Star* (December 17, 1969) called it "a puzzling case."

January 4, 1970; Kansas City, Missouri. Just before 9:00 P.M., Ed James checks into the Hotel Phillips at 106 West 12th Street. Within minutes he checked out via ambulance, en route to General Hospital with burns of feet, knees, chest, arms, and middle of his back. Strangely the clothes on his back were unburned, reported researcher Brad Steiger.

Around September 21, 1970; Poole, England. Six-week-old Simon Simeon burns in his blazing crib. Police, says the *Daily Express* (September 21, 1970), find no cause or culprit.

March 17, 1971; Chicago, Illinois. In a fire of "undetermined causes" William Sessions, sixty-nine, and the chair he sat in burn up. *The Chicago Tribune* next day said damage was a scant $200—barely enough to buy a new chair.

Around August 15, 1971; Orange County, California. An Anaheim woman, fifty to sixty years old and of low intelligence, obese, sedentary, and "sloppy and careless," burns up in her apartment's kitchen. The apartment was "not burned *anywhere*, just all smoked up," neighbor Lorena Gardenshire wrote to me. She said a cigarette was blamed officially, but hinted that something more sinister had happened next door to her.

Mid-September 1971; Shaldon, England. Suddenly a violent flash of lightning(?) "seared the whole village" and adversely affected the nerves and muscles of many residents, reported *INFO Journal* (no. 14). Betty Connell was among two women who found themselves "trapped" in a brilliant "tube of light." No indication that either lady was burned, but it might have been a close call as they stood temporarily immobilized inside their fantastic luminous confines.

Prior to 1972, somewhere in the Sunshine State, an anonymous Florida woman sitting in a wicker chair burns up—more accurately disintegrates—on her porch. Pennsylvania State Police Fire Marshal Fred Klages had seen a photograph of her legacy when attending a fire-fighting convention that year. "I don't remember if her head shrank," he recalled for me, "but her glasses were lying on the chair. She wore those old-style high-topped black shoes that lace up. About three inches of her legs extended above the tops of her shoes, and there wasn't anything else left in between. The chair remained—and wicker is *easy* to burn! By all rights for her body to be cremated, it should have burned the wicker chair."

A tip I received in 1992 suggested this event occurred in the Pensacola area. I called Pensacola fire official Jim Dixon, who said he had served since 1977 and didn't recall such an episode. "But wait," he urged, "here comes the retired chief!" Former fire chief Raymond Church did remember the case. "It was not here in Pensacola, however. A lady in a chair, her legs in front of it. Brought in by a training officer, Tommy Tompkins." It looked like classic SHC, Church said.

Somewhere, someone knows more about this case. I hope to hear from you.

Evidence from Photographs

A photograph, it's said, is worth a thousand words. Revelation can leap from photographs. They can offer the most convincing evidence for paranormally thorough human incineration, short of standing at the fire scene and watching it unfold. They also assure almost certain assessment of whether the incident likely is SHC or is caused by more prosaic means, to be uncovered as authorities pursue investigation of a case.

Proof of this point is to pick up the 13th Edition of *Glaister's Medical Jurisprudence and Toxicology* (1973), an internationally recognized reference standard for the science and philosophy of medical law and the pathology of toxins in the human body. In one chapter are many photographs of burn injuries. Three stand out as distinctly atypical.

One shows a presumably British woman who collapsed in her chair. *Literally* collapsed into her own ashes! It appears her lower abdomen ignited and rapidly consumed its way through her torso and legs. Her upper body, no longer supported by a backbone reduced to ashes, toppled onto the floor. Consequently her head and shoulders and arms remained intact. In fact, her head fell onto an electric heater that apparently was *not* on—her head not being the least burned. At her other end, on the floor in front of the chair, lay her feet inside fuzzy slip-ons. The slippers were pristine.

A second photograph depicts another anonymous, presumably British, pyrophoric woman who blazed on the floor in front of her small fireplace. Her stockinged legs end abruptly at mid-calf, beyond which a mound of ashes lies upon a *partially* singed newspaper. Characteristic of the conundrums of SHC, white linens resembling tea towels draped from a cabinet drawer directly above the victim plus fringe on a tablecloth only twelve inches away reveal no trace of damage whatsoever. Not even sooting.

Both deaths occurred before 1973. Attempts to identify who these photographs depict and where they died have failed, so far. As it is the practice of law enforcement agencies to purge their records after a few years, these women's identities may forever be unknown. I sense, however, that someone remembers having seen these startling scenes firsthand; word from you is eagerly sought, too.

Regarding *Glaister's* third atypical case, I'll address it after more of this parade of intimately personal fires passes in review.

One more point of interest about SHC photographic evidence, first. Cpl. John Balshy, who is the Pennsylvania state police's photographic expert, told me that during his twenty-five years of crime scenes investigation he had heard about SHC-like episodes in which fire marshals had ordered pictures *not be taken*!

It's always easier to reject a problem if one doesn't have to look at it, isn't it? Not having any disconcerting photographs lying about can make denial so much easier.

THE BURNING GOES 'ROUND THE WORLD

Oxidation fire pays no heed to national boundaries. Nor does the strange fire that plagues, even persecutes people. Brazil...England...France... Germany...India...Italy...Malaysia...Scotland...the United States of America...Canada...

February 17, 1973; Burlington, Ontario. Canadian Charles Foster, eighty-two, screamed when he discovered he was engulfed in flames. Another smoking materials accident? "Perhaps," said *The Hamilton Spectator* (February 19, 1973). And perhaps not.

Circa October 1973; Sao Paulo, Brazil. A "person-centered poltergeist" had been tormenting an unidentified family for more than six years, said researcher Guy Playfair in *The Flying Cow* (1975). After fire breaks out inside a closed wardrobe, the father called in an exorcist. The phenomena ceased for three days. Then a nightgown spontaneously burned to ashes, and the underside of the mother's mattress caught fire three times—the third time while she slept upon it!

There are spontaneous combustion of clothing that burn their wearers... and of mattresses that almost, and perhaps *do* incinerate those who sleep blithely unaware of danger lurking so near.

January 5, 1974; Hagersville, Ontario. In this small town on the Penn Central rail line twenty-eight miles southwest of Burlington where Charles Foster died eleven months earlier, Cecil House, eighty-one, burns to death in a chair. Said *The Hamilton Spectator* (January 7, 1974), the "fire extinguished itself." There was, the newspaper added, "no indication he was smoking." Cigarettes, that is.

January 15, 1974; Warrington, England. In the jail of this Lancashire town sixteen miles east of Liverpool, William D. Pindard, twenty-six, burns up so badly he arrived DOA at the hospital. He may have willfully set fire to the bedding in his cell, remarked the *Manchester Evening News* (January 16, 1974). Or: he or his bedding may have set itself aflame and burned the other.

February 6, 1974; Oakland, California. Pervis Mayweather, an elderly and retired black man, dies in his home; the "upper part of his body had been severely burned." The *San Francisco Chronicle* (February 7, 1974) concluded its coverage by suggesting that his "stove flared up, igniting his clothes." The *Fortean Times* (no. 6) suggested SHC. The Oakland Police Department kindly furnished me a set of photographs of the scene: they depict superficial burns only, showing this to be truly an accidental domestic fire and *not* SHC.

March 5, 1974; in the vicinity of Toronto, Canada. Authorities discover that Harold Potts has "suffered 90 percent burns over his body." They theorize that a pipe's ashes found in a wastebasket are to blame, says the *Toronto Globe and Mail* (March 6, 1974). They fail to explain why those ashes failed to ignite the waste basket and other readily combustible items while searing exclusively the seventy-seven-year-old man.

Around March 28, 1974; Blacon, Cheshire, west-central England. Mrs. Elizabeth McPherson, thirty-five, dies in an armchair fire. Localized. An inferno, it is said of the chair. Maybe the *chair* was not the fire source. Authorities find no cause, reports *Fortean Times* (no. 6)

April 26, 1974; San Francisco, California. The wife of an eighty-year-old Frisco man went to the garage, leaving her husband sitting on a sofa. She smelled smoke and went back indoors to find herself becoming a widow. Her husband died, unable to explain the flames that engulfed him in the few minutes he was alone.

August 26, 1974; Corby, England. In this tiny Northamptonshire town fifty miles east of Birmingham, Roland Davies fell asleep on his bed, then awoke to find himself "lying on a fiercely burning mattress," which burned him badly enough to require hospitalization. It sounded like Davies had fallen asleep while smoking, except London's *Daily Express* (August 27, 1974) said the bed burned "from an unknown cause"—which would rule out the obvious. Michael Harrison decided this was a case of SHC survival but, without more information, a nearly fatal spontaneous mattress combustion merits consideration too.

January 1, 1975; Yuma, Arizona. A "flash fire" in a motel here sears the lungs of Gene J–, who was in emotional turmoil at the time he checked in. Now internal physical turmoil compounded his distress, until he died eight days later. Authorities alleged a natural gas leak triggered the blaze, but the victim's sister told me that "no gas was piped to that room." She suspected SHC, a not unwarranted suspicion considering her brother was emotionally upset.

January 12, 1975; Miami, Florida. In her home at 1801 NW 2nd Court, Mrs. Esther Cooks becomes burnt over 80 percent of her body. What didn't burn was her face, her hands, and her clothing, reported next day's *Miami Herald*. The forty-two-year-old victim never cried out for help, and investigators were further baffled by the "thermal or maybe chemical" appearance of her injuries. For details of Mrs. Cook's curious combustion, see *Pursuit* (January 1976).

July 1, 1975; Lucknow, India. Another rash of spook-fires is this time attacking the home of a CID inspector, whose wife Shanti was not as lucky as the wife of the São Paulo family in 1973. "Help, help, I am burning!" Shanti suddenly yelled out this day. Her son threw a quilt on her. She burned on. When doused with water, she screamed "The evil spirit is burning me!" That night she died, leaving behind for her widower a mystery to test forever his investigative skills.

August 9, 1976; the urban complex of Teesdale in northeast England. An unidentified Middlesborough baby dies "in fire of unknown origin" said *Fortean Times* (no. 18). Perhaps subsequent investigation uncovered a conventional cause, but a disturbing pattern of insidious infant incinerations begins to belie the nineteenth-century dictum that only drunken, corpulent, old people can self-burn.

August 17, 1976; Somerset, England. Found ablaze in an armchair, an old man of Bath expires shortly afterwards; "cause of fire unknown."

August 20, 1976; again in Bath. A "mystery fire" is fatal to a girl in a Bath basement, recounted *Fortean Times* (no. 18). Speculation centers on whether her nightgown somehow ignited.

Less than a month later, Englishman William Seale meets his "mysterious fire death" in Brighton during the night of September 15, 1976. A neighbor, Miss Christine Harmer, said the sixty-five-year-old man "had chronic bronchitis and bad legs. He could not get around by himself.... I think he got fed up not being able to go out." She called him "a lonely man."

Loneliness...legs crippled...lethargy: the three "Ls" that appear again and again where mysterious fire deaths happen.

September 25, 1976; Scarboro, Ontario. Horace Butler, seventy-two, is discovered engulfed in flames. Nearby they found his sister, unconscious from smoke. Officials pronounce cause as "careless smoking." Damage is a mere $500, says next day's *Toronto Sunday Sun*.

Circa November 26, 1976; Thanington, England. The widow Dorothy Sample, eighty-two, relied heavily on neighbors for help. They could not help her this time; maybe she subconsciously planned it that way. She burned up so completely in this tiny Kent village southeast of London that authorities had to rely on her wedding ring found in the ashes for identification. Being a light smoker, it was "possible" that she had fallen asleep while smoking in her armchair, *The Kentish Gazette* (November 26, 1976) quoted the coroner saying, and her clothes "could have been ignited by standing close to the fire." He didn't sound very confident about this, though, and pronounced "Death by misadventure."

December 26, 1976; Lagos, western Africa. Correspondent Kayode Awosanya bylines a story in the *Nigerian Herald* (December 27, 1976) about the "greatest mystery" surrounding the previous day's deadly fire visited upon six family members in Iponri. On-the-spot investigators expected to find the family's wooden shanty burned to the ground. Instead, "the room looked unaffected by the fire"—two cotton mattresses were untouched, even, by fire. A crime? Collective SHC? Carelessness? The mother of the family, its only survivor, told authorities that someone had sprinkled petrol through an open window while they slept but, in an unexpected! twist, authorities implied "that this might not be true at all..."

December 29, 1976; Islesworth, England. Strange fires continue plaguing the British Isles, and now Mrs. Mary Norris in particular. In the Middlesex district of West London along the Thames, a fire had damaged the thirty-two-year-old woman's bedroom six months earlier. This day, a second fire quietly broke out without warning—not even her dog barked, unlike before. Neighbors, relatives, and the fire brigade were mystified. No one could figure out why the fire began, nor why she made no effort to escape the localized flames, which were extinguished "in minutes." The Slough *Evening Mail* (December 30, 1976) said Mrs. Norris had separated from her husband. Did distraught emotions play a role in this fire free of accelerant?

January 6, 1977; Moxley, England. A blaze of qualifiers surrounded the aftermath of the fire that killed Mrs. Clara Thomas in this town near Birmingham in the British heartland. Authorities investigated for a fortnight, then told the *Walsall Observer* (January 21, 1977) that all her clothing had "completely burnt" because her clothing "*may* have been inflammable." She had fallen in her kitchen "*near* the open coal fire"—no indication it was the source of her burns, however. In overlooking her scorched flesh, the coroner simply pronounced a "death by misadventure" due to shock.

Perhaps a determination of SHC would have stripped away all the equivocating uncertainties—

March 6, 1977; Edinburgh, Scotland. David Yool Abbey, fifty, president of the VAT Tribunal, dies in a small fire in his Lynedock Place apartment. Communicates the *Daily Telegraph* (March 7, 1977): "The fire started in a chair but burned itself out and it is believed he died from fumes." The victim, British fortean scholar Robert Rickard pointed out, was "a hated tax collector." More than a few Scotsmen, Rickard mused, believed that Abbey earned his comeuppance *spontaneously* as the recipient of overwhelming, enraged burning collective thought aimed his way.

AN UNCANNY OCCURRENCE IN URUFFE

May 11 and 12, 1977; the town of Uruffe, near Nancy in eastern France. On Wednesday, May 11, Mrs. Ginette Margerard Kazmirczak spoke to a friend who visited at 7:30 P.M. "She looked tired. Her eyes were tired," the friend later recalled. Two acquaintances knocked on her door at 8:20 P.M., and heard Ginette say "I am in the bathroom, washing myself."

Sometime thereafter, and before 4:10 A.M. the following day, this fifty-year-old teacher "goes up in flames" in her apartment on the second floor of city hall, in circumstances that will try French intellect.

Just after four o'clock on Thursday morning, one of Mrs. Kazmirczak's neighbors across the hallway was up early brewing coffee. "Do you smell that?" the neighbor asked her husband. "A strong smell of burning. Like something is cooking—"

The husband peered into the hallway. Smoke! "Call the fire department!" he screamed.

He knocked on Mrs. Kazmirczak's door. The door was very hot. He feared the worst.

Outside, Marc Yung, was walking down Church Street. "There is a fire at the city hall! On the second floor!" people were yelling.

Yung ran to find a ladder. He then soaked a cloth in the river to wrap around his face, and set the ladder to the second floor window. Up he climbed, furiously. The window was locked. He broke it in and entered the Kazmirczak apartment. He expected to find its short, 165-pound occupant somewhere in the darkness.

"The temperature must have been between 250 and 300 degrees," Yung will later tell TF1's program "Counter Investigation." In the soot-darkened air, he tripped over something on the floor as he headed to the kitchen to open a window and vent the smoke. He noticed a plastic dish brush on the wall. It was all melted.

"When I came back," he recalled in retracing his steps, "I saw the corpse in the entrance hallway. It had shrunk like a doll thrown in a fire. Small green flames were still burning on top of it. Right next to it, on a stool, there was a cage containing canary birds. They were dead but their feathers were not even blackened."

It is the first-ever report of green fire seen flickering upon a corpse.

Jean-Jacques Yung, brother of Marc and mayor of Uruffe, was next to enter this twilight zone a few minutes later. He was deeply affected by what he saw: "The body of the victim was almost a pile of ashes except for the legs and the right arm which were still undamaged. The shoes were laying some centimeters from the feet. The torso, the abdomen, the head, and one arm had almost disappeared."

A greasy black film from the burning of the body covered the ceilings throughout the apartment, and the walls over an area between four and six feet. Tupperware in the cupboards and other plastic items, says Jean-Jacques, were "either melted or bent." He finds the teacher's glasses on the floor; the metal frame was intact, though the lenses themselves had melted away. There was no generalized fire.

The mayor was perplexed: "the strangest thing was that there was no other source of fire except for the body itself. Nothing else burned, except the area of the wooden floor where the body had been laying." The flooring—inch-thick oak—was "totally destroyed" under her.

Firefighters arrived. When they broke down the door to the apartment, they observed that the woman's skin, viscera, and bones were so consumed they totally fell apart as the firemen entered. Yet around her lay pieces of the cotton dress she had been wearing; on her legs was intact hosiery; on her feet, a pair of white shoes "that the fire did not blacken."

So complete was the incineration of Mrs. Kazmirczak, that these white shoes provided the sole means of identifying the victim.

Though it was clear the fire was not spreading, firemen sprayed the scene with water anyway.

It is not merely the oddity of fire starkly isolated around her that makes this case especially intriguing. Already evident is a curious (one might say outlandish) set of parallels to another, better known incident. The stature and weight of Ginette Kazmirczak are comparable to Mary Reeser, before *and* after their individual blazes. Both had been widowed for several years, and both had been depressed in the days before their deaths. Damage is singularly localized around their bodies, save for plastics melted throughout their apartments. Ginette's shoes and Mary's slipper escape the fire and provide identification for the stumps of leg that extend from their footwear. Even times in the women's last night coincide remarkably, eerily. Both were last seen alive between 8:20 and 9:00 P.M. Ginette was found dead in her apartment at 4:10 A.M.; the clock in Mary's apartment had stopped at 4:20 A.M.

> Curiouser and curiouser!
> —Lewis Carroll,
> *Alice's Adventures in Wonderland*

Was nature striving to mirror one mystery in another, almost twenty-five years and twenty-five hundred miles apart? Maybe the resemblance is all just coincidence.

Regrettably, police and fire officials in Nancy failed to respond to my requests for information. A death like Ginette's is upsetting to agencies accustomed to neat, easy taxonomy. Maybe officials there would have been less disinclined to discuss her fate had they known that scores of their country(wo)men had died similarly in decades previous.

French journalist Emmanuel Pezé had better luck at getting information I could not. Mme. Kazmirczak's death became the anchor case in his 1986 book, *Le Feu Qui Tue* (*The Fire That Kills*). Upon his research, and information from French correspondant Michel Granger, have I reconstructed this case for you.

Clearly, this was no ordinary fire call for the town of Uruffe. (Nor for any town, I might add.) How to explain the stupefying selectivity of this strange blaze? The Court of Justice in nearby Nancy called for an investigation.

The presiding judge said he had no sympathy for sensationalized explanations: tales of UFOs and psychic phenomena being parlayed by the media would not be welcomed in his courtroom. Only an accidental cause will be tolerated.

Calling Mrs. Kazmirczak an accident was no problem, really. There was no evidence suggesting criminal entry, or evidence of how such a crime could have been committed if foul play *had* been the issue. The singular problem lay in ascertaining *cause* for the accident.

Professor de Ren, member of the Faculté de Médecine de Nancy and a specialist in medical jurisprudence, was charged with performing an autopsy on Mrs. Kazmirczak. Not only had the victim's entire abdomen disappeared in the cremation, he established, but there was not even minimal blood to be retrieved to test for carbon monoxide inhalation. "Thus," he pronounced, "I concluded that the body had been destroyed by the inflammation of her blouse."

(Whoa. Wait a minute. Objection! Sidebar, please. Judge, I ask that the professor be disqualified as an expert witness. For him to assert that a mere blouse is to blame for this human holocaust—and to have inflamed by itself, no less?—and that said blouse alone can do what a crematorium cannot do except at 3000°F for twelve hours, is, to be kind, preposterous. You agree. Thank you, your honor.)

The court's attention turned to an aerosol can of Timor insecticide found on the floor beside the body and below the mantle. Alain Kazmirczak said his mother, a nonsmoker, had a "psychotic obsession" about bugs in her apartment and used insecticides excessively. The aerosol immediately became the next focus for the accidental cause being sought. The theory: either the victim had lit her gas-range, which, in turn, ignited the aerosol she began to spray... or she willfully held a flame in front of the can and ignited both its gas and her clothing (the police reasoned the victim had "soaked herself" with the insecticide, believe it or not).

Incredibly, Professor de Ren defended this theory too! What no one thought to ask him in court was why the aerosol can had failed to explode in the heat of the human inferno, inches away? (I have been within inches of an aerosol can when it exploded in a small paper fire. Shredded metal blasting past one's face—and what would have happened had those few inches not intervened—is an image one does not forget!) Whatever the temperature level generated inside Mrs. Kazmirczak, it failed to radiate laterally. Like so many similar cases, the heat here was directed downward.

Wisely, others quickly pointed out fallacies in this theory.

Both Jean-Jacques and Marc Yung testified that on the fireplace mantle was a white circle the exact diameter of the can. This meant the Timor had been on the mantle *during* the fire, rather than dropped by the victim's hand after allegedly igniting her. (Probably Marc, or maybe a fireman, had knocked it off the mantle when stumbling through the apartment.)

The judge appointed local fire technician Captain Laurent to file a report. At the age of thirty-one, Laurent had already seen innumerable fires of all kinds. He realized straightaway that none had been like this Kazmirczak fire.

Laurent turned his experienced reason to the aerosol can theory. He was, after all, a specialist in aerosol fires. He quickly determined that Mrs. Kazmirczak had *not* tried to light her range, because the valve on its gas bottle was closed when the fireman arrived. That ruled out the suspected external ignition source. Soon he had other doubts.

Meanwhile, the dousing of Mrs. Kazmirczak with acid was being considered as a tenable explanation for the mini-inferno. Dr. Michèle Rudler, a specialist in toxicology at l'Institut Médico-Légal of Paris, told the court this was "both dangerous and difficult to do; too much acid needs to be used."

She, too, dismisses the official aerosol can theory as "practically impossible." Especially after she and Capt. Laurent conducted laboratory experiments which incinerated several rabbits laced with Timor, even to blasting one Timor-soaked hare with a blow torch. Each experiment dramatically failed to mimic Mrs. Kazmirczak's total calcination.

Dr. Laborier, a dental surgeon and expert for the Dijon court, was called to testify. His career included numerous fiery highway and railroad accidents. He explains that the teeth are the most heat-resistant elements in the human body; that in car fires where temperatures may reach 1000°F, significant quantities of blood and flesh remain; that limbs are the first body parts to be destroyed in a fire, with the torso least apt to burn up. Dr. Laborier told the court he did not believe in "spontaneous" combustion of a human, yet he could not hide his shock when faced with a human burning that blatantly belies his experience.

"Actually," Dr. Rudler later told Pezé, "it looked like an insane individual used a blow torch to burn the body during more than an hour."

The court asked Dr. Rudler and Capt. Laurent for a report. They concluded that (a) no accelerant was present to fuel the blaze in Mrs. Kazmirczak; (b) a heat source "close to 2000 degrees was necessary" in herself; yet because wallpaper twenty inches away did not ignite; (c) the temperature around her body "did not go above 200 degrees." This closed the inquiry.

On January 18, 1978, the Court of Justice issued its eagerly awaited finding into the fate of Madam Kazmirczak: "Cause of death—Unknown."

THE BRITS, THEY ARE 'A BURNIN'

Prior to 1978, an old Scot in Glasgow burned up in an old people's home. Within ten minutes he was gone, leaving behind "a pile of ashes" on the floor; black and greasy smoke; and a "strong smell of roasting." Whether there was a noxious or a not-unpleasant odor, was not reported. It was explained away as a heart attack and dropped cigarette. Science writer Peter Laurie, in *New Scientist* (March 23, 1978), counter-argued that this make-sense scenario just wouldn't do in this instance.

On January 6, 1978: Mrs. Inder Kaur, fifty, was found dead on the floor of her home in Bedford, a town fifty miles north-northwest of London, England. Said London's *Evening Standard* (January 6, 1978): "She had burn marks on arms and legs, although there had not been a fire in the house." Well, it seems there *had* been a fire in the house, a fire isolated to the deceased tenant only.

January 16, 1978: east of London, England. Mrs. Mary Watts, eighty-eight, "was found dead in a burning armchair at her home in Stoke Albany, Northants," relates next day's London *Daily Telegraph*. "She is thought to have dropped a lighted cigarette." Maybe. Maybe not.

Often I am asked, "Why is it that only *humans* suffer spontaneous combustion? Don't you find that strange?"

Yes, I find many things strange.

Like what happened in Jarrow, England, on February 6, 1978. Four-month-old Hayes yelps. Understandable, because Hayes is a golden retriever. Understandable, because when its owner and passersby rush immediately to aid the pooch they see "flames leaping up" from its belly. They can find no matches, no lighter and—worse—no culprit near the candescent canine, reports the Newcastle-upon-Tyre *Journal* (February 7, 1978). That is, no explanation. What they apparently did find was evidence for SAC—spontaneous *animal* combustion.

Robert Rickard noted wryly that this SAC occurred just after SHC writer Harrison announced that animals don't inflame. As one might infer from this fortuitous synchronicity, it's not nice to limit Mother Nature.

Shortly before midnight on Valentine's Day, 1978; Hove, England. One hour earlier, Mrs. Françoise Price, thirty-five, had gone to her upstairs room at the Lawns Hotel in this coastal town just west of Brighton on the English Channel. Her husband waited in the hotel lobby. Patiently. When his wife didn't return after an hour, Mr. Price became concerned. With good reason. He and a porter, and soon the fire brigade, too, were repelled by smoke coming from the Prices' room...coming from Mrs. Price and a mattress she had inexplicably pulled over herself. Both had charred throughout.

"Wife Dies in Blaze at Hotel" headlined an article in the next day's *Brighton/Hove Gazette* about Mrs. Price. Police Superintendent Bob Allen initially considered foul play. By the afternoon, Home Office pathologist Dr. Hugh Johnson issued a report which led Sussex central division criminal investigation superintendent Ian Eadie to dismiss foul play. Harrison interviewed a hotel chambermaid who said the victim was attractive, French, and "sometimes sad."

How the fire started and why the victim was cowering beneath a mattress in a corner of her room remained unresolved. Harrison (1978) attributed her death to SHC and noted that eighteen months earlier lonely William Seale had inflamed barely a mile from where Mrs. Price burned.

February 21, 1978, 2:00 A.M.; Bognor Regis, Sussex, some twenty-five miles west of Brighton on the Channel coast. Discharged the previous evening after a three-day hospital stay, eighty-eight-year-old Herbert Mill fatally burned in his Burnham Avenue home. Police Constables David Mills and Bob Smith discovered the blaze, but were driven back by its smoke and heat even though a neighbor, Mrs. Arnold, had given them wet towels to cover their faces in the abortive rescue attempt. Bognor Regis police ruled out criminal intent but hadn't found any cause for the fire by the time Brighton's *Evening Argus* ran its coverage at day's end. "He lived alone," said Mrs. Arnold of her former neighbor. "We don't know anything about him. We hardly ever saw him." Another recluse ends life by curious combustion.

March 14, 1978; Roxby, Lincolnshire, England. Around 11:20 this Tuesday morning, Mrs. Ida Gladys Carter, eighty-one, apparently widowed, paralyzed on one side since 1972 and confined to a wheelchair, dies. Dense smoke filled the room where, says the *Scunthorpe Evening Telegraph* (March 15, 1978), she was found "badly burned." Conversely, it's also reported that "The only fire damage was to her clothes." So was her flesh badly burned or not? Coroner Allan Collins concluded that burning "took place after death," the victim having "died from heart failure." Mrs. Carter had been sitting four feet from the fire guard when she tumbled off her wheelchair onto it, explained the *Telegraph* (March 16 and 17)...leaving one to ponder why it was not her head but apparently her clothing which suffered severest burning.

Early October, 1978; a scant six miles northwest of Bognor Regis. A Chichester woman is said by the *Brighton/Hove Gazette* (October 6, 1978) to have awakened "with her shoulders and back ablaze." Bedsheets were not singed; there was no odor of fire. The cause? "Unknown."

Early Sunday morning, November 12, 1978; Reading, England. The young daughter of Mrs. Lucy Gmiterek knocks on her neighbor's door and sobs, "My mummy has turned to ashes." Mrs. Gmiterek, forty-nine, a widow for six years, lay dead on the basement floor of her Baker Street home forty miles west of London. Said Mrs. Randall, the neighbor: "It was terrible in that basement. I don't want to see anything like that again." An examination conducted at Royal Berkshire Hospital on November 13 concluded cause of death to be "shock due to extensive burning." Police remained puzzled. Authorities issued contradictory statements, dutifully recorded by the *Reading Chronicle* (November 14 and 17, 1978). No official has been willing to sort out the confusion for my benefit. The widow's doctor stated that Mrs. Gmiterek suffered acute fits of depression and hysteria.

January 25, 1979, 8:15 A.M.; Hutton-le-Hole, a village in northeast England. "The body of a seventy-six-year-old widow was so consumed that it was impossible to say how she died," announced the *Yorkshire Evening Press* (February 15, 1979) after speaking with Ryedale's coroner Henry T. Blakeston. He was speaking about Mrs. Lily Smith, another widow who lived alone. Three men broke into her locked Greenfield Cottage home after postman Rex Rudding raised the alarm. Said one of the would-be rescuers, Hubert Ford: "I didn't see any flames. I searched and shouted round but couldn't find anyone."

Police Constable David Binder *did* find the widow's body—well, sort of a body: "two human legs laid on the floor" under a dining chair near the fireplace. Where the body should have been, on the chair, were charred bones. The widow was known to have an "infirm" leg; now only it and its mate survived Mrs. Smith.

Another smoking materials accident? The reclusive, security-obsessed widow did not smoke, said her son Geoffrey. And although she regularly sat in front of her small fireplace with her feet on the hearth, he said the "fire guard was in front of the remains of the previous night's fire"—thus dismissing a Kirbymoorside fireman's suggestion that a hot ember had fallen out of the fireplace. Coroner Blakeston returned an open verdict—admitting it was "impossible to say how she died."

I can tell him that this case fits every condition of SHC... but SHC doesn't happen, right?

Epileptic Annie Webb

February 2, 1980; shortly after nine o'clock in the morning. In the Welsh county of Gwent, in the town of Newport, on Corporation Street, a neighbor of spinster Annie Gertrude Webb noticed Annie's living room windows were smoke-blackened. No one answered his raps on her door. He fetched another neighbor, who had a key to Annie's home. They entered her ground-floor living room. Darkness and a strange foulness in the air greeted them. Annie did not.

When they opened the kitchen door to let in some light, they discovered an amorphous mass of ashes on the floor, ashes which led to a pair of human lower legs. Also, a right forearm and a blackened skull, the latter reposing on the tiled hearth of a gas heater. On the forearm's wrist they found a plastic hospital tag, intact. It identified the victim: Annie Webb.

Scenes of Crime officer John Heymer, of the Gwent police, took a personal interest in this incident. (You will learn why, later.) Heymer's fascination with the demise of Miss Webb preserved her ashes, so to speak, for posterity; otherwise it's likely her fate and her name, along with her few remains, would have been buried quietly and forgotten. To Heymer, goes the credit for what you will now learn about this seventy-five-year-old spinster's last moments.

Her house was tightly sealed, effectively draft-free. Cluttered into her small living room were several chairs, a stuffed armchair, and a wooden table covered with plastic cloth. Annie's feet lay under the table. Her lower legs were burned off across mid-calf, at an angle that followed the edge of the tabletop directly above. The tablecloth's overhang had been singed black, and a corner of the plastic cover overlay was partly melted by a low heat. Two wood dining chairs were tipped over. One chair, minimally charred, rested against a wall and above her burned away head and shoulders; the other chair, actually in contact with her right leg, was unscathed.

Just days before, Annie had been hospitalized. Now her doctors had no cure, no clue, for whatever ailment befell her after being discharged for treatment of epilepsy. All the pathologist stated was that Annie Webb had a carbon monoxide content of 15.5 percent in her blood—an extraordinary high level for a nonsmoker like Annie, thus indicating she had inhaled considerable smoke before expiring—and that cause of death was (surprise!) "burning."

Commented Heymer in his article "Fire in their bellies" for *Fortean Times* (June–July 1994): "Again we are faced with the paradox of a grossly incombustible human body being reduced to ashes in conditions so devoid of oxygen that the fiercely burning chairs of wood and plastic foam ceased to burn.... Of course, Annie's epilepsy and her body's proximity to a lighted fire [small gas heater] gave rise to the obvious conclusion that she suffered a fit and fell head-first onto the gas fire and so burned to death."

Heymer found many reasons to dismiss the obvious. Evidence suggested the gas flame had extinguished in the blaze's early stages due to lack of oxygen; hence the buildup of excess gas in the room from a coin-slot meter. No trace of her skin or hair was found on the grill of the heater, where her head allegedly had contacted the flame that ultimately incinerated all but portions of three limbs. A tea towel one foot above the heater should have at least been scorched, if not completely burned away; it was untouched.

> The poor old lady, she was reduced to ashes except for two lower parts of two legs. Well, this was amazing to me.... There's no doubt that it was a case of a human body catching fire, and it was called spontaneous human combustion. I have no doubt at all.
> —Inspector Colin Durham, about Ms. Webb; Arthur C. Clarke's *Mysterious Universe*, The Discovery Channel (Jan. 26, 1995)

Perhaps the most telling argument against the heater-as-cause conclusion was not mentioned by Heymer: the worst burning of Miss Webb occurred not to her head lying atop the heater's ceramic burner (hence in contact with an external heat source), but to her torso, which was two feet *distant* from the burner. This paradox of thermal transmission, unaddressed, remains unexplained.

Miss Webb's weird immolation (no one mentioned any great heat in her home) puzzled Inspector Colin Durham for years after he left the scene of her death. He told Heymer that, whatever caused ignition, he "was convinced" it was not the gas heater. And that seems to leave but one remaining source: the spinster herself.

More Imponderable Human Incendiaries

One month later, somewhere in the British countryside yet another elderly British woman leaves a legacy hauntingly akin to Miss Webb's. On March 4, 1980, the deceased was found lying in front of her small fireplace, her head and shoulders (as once they had been) atop the raised hearth. Extending away from the grate lay the rest of her calcined form, only two stockinged legs identifying this as until very recently a human being. Furniture all around showed no damage; not even a trace of blistered paint above the hearth.

At 3:30 A.M. on May 13, 1980, in that English city particularly prone to strange fires—Birmingham University student John Roberts screamed and collapses in the hallway outside his twentieth-floor dorm room, # C2. Fire Service College instructor Anthony J. McMunn wrote to me that the undergraduate had suffered "severe burns to the neck, back, chest, both arms and legs. He had received the burns whilst in bed asleep."

This appeared to be no ordinary, careless smoking accident. Roberts was a nonsmoker. Ambulance Control was alerted to respond to "a man with electrical burns."

McMunn said of Roberts' room, "electrical wiring was found to be correct." What wasn't quite correct was this: "the white bedsheet he was lying on had a scorched outline of his body." Not burned through, just scorched.

Some officials blamed a dropped cigarette. McMunn said Roberts vigorously protested this, telling the ambulance driver "I was lying on the bed and I awoke in flames." The victim of somnambulistic searing over 70 percent of his body then went into shock and died two days later.

Robert Holland, West Midlands Fire Service division officer, told me the forensics people disagreed with SHC, mentioning something about "body fat" to explain Roberts' burns. McMunn indicated the coroner, although admitting it sounded like SHC, ended up labeling it (you guessed right!) simply "Accidental death." This conveys the impression that all is well.

British writers Jenny Randles and Peter Hough examined this case in detail, considered all the conventional scenarios, only to conclude in their book, *Spontaneous Human Combustion* (1992): "The truth is that *none* of this makes sense."

McMunn concurred: "It's a real mystery. There was no evidence of a break-in, no ignition source, nothing to explain how the young man came to wake up in the early hours of the morning to find himself ablaze."

Sounds like this student passed the entry exam into the twilight zone of SHC, doesn't it?

August 4, 1980; Singapore, Malaysia. Whatever burned up the Birmingham student in his bed may explain what became doubly mystifying half a world away in the same year. Malaysia's *Straits Times* (January 27, 1981) recounted that a delivery man noticed smoke coming from a seventh-floor apartment. Upon entering, he discovered Mrs. Saemah Sapari aflame. He beat out the fire, took her to safety, then returned to the apartment where he heard a child crying. This time it was young Mohammed Sapari ablaze. Mother and six-year-old son died soon afterward. Before she passed away she said she had been watching television when her son's bed ignited, and she rescued him and extinguished the bedding. Whether this contradicted the delivery man's story or represented an incident before his arrival isn't clear. State coroner Seng Kwang Boon returned an open verdict, admitting he could not discover any cause of the fire(s) and their injuries.

Third week of October, 1980; Cassville, Missouri. Meter reader Howard Fulton noticed smoke inside the home of semi-invalid Elijah William Currin. As meter reader Gosnell had done almost fourteen years before in northern Pennsylvania, Fulton raised the alarm. And as I did for Dr. Bentley, I again introduce to the world another new case.

Firemen and Deputy Sheriff Max Fields arrived quickly and crawled through the hot opaque smoke on their hands and knees in an attempt to find Currin. Smoke ejectors were set up. As the sooty air cleared from the rooms, an unsuspected puzzle emerged from the haze.

The smoke had been emanating from the old man himself who, lying on the floor, burned to death in a fashion the authorities found mysterious and memorable. The extensive heat and damage to Currin "was out of all proportion to the fuel loading," said Fields, a volunteer fireman also. Fire damage limited itself to Currin; a burn six inches in diameter in the flooring beneath him; a melted plastic planter inches from Currin; and a plastic waste basket nearby that also melted but did not burn. Nor did the paper in it burn. "We discussed this among ourselves," Fields told me; "I still say it's strange."

Coroner Darr Williamson told me that Currin's legs had burned to the bone; that his torso had burned out; that "percent-wise, it was reduced some 25 to 30 percent. The rest of his body was burned but not reduced." The coroner also noted that Currin had apparently been filling a lighter when the blaze erupted. "But I can't conceive how this little amount of lighter fluid could so burn the body and not burn the house down," Williamson said, based on his twenty-four years' experience as a coroner and thirty-one years handling dead bodies as a funeral director.

He said he had watched cremations and had burned dead farm animals. Did the death of Currin stand out? "Yeah it's different, it sure is!" Williamson responded. "The overalls' buttons lying there on the floor, and the hardwood floor not really burned—yeah, it stands out in my mind. It sure does!"

Cassville Fire Chief James E. Anderson confirmed that Currin's leg had burned "to the bone... as if the flesh was sliced off with a knife." Deputy Fire Chief Bob Norman concurred with the coroner's assessment: "In regards to the intense heat, yes it was unusual."

An odor of burned flesh, fabric for fuel, plus presence of an accelerant (*however* it may have ignited) does a weak case for SHC make. But preternatural combustibility? Yes, indeed.

December 4, 1980; 11 North Street, Lockwood, West Yorkshire, England. In this village near Huddersfield, some fifteen miles southwest of Leeds, seventy-eight-year-old Mrs. Frances Kenworthy—a widow of one month—had the audacity to burn to death on her kitchen floor; audacious because, said Her Majesty's Coroner's officer Sgt. H. Bert Booth, "the clothing she was wearing was not marked."

The victim's nephew, Jack Quarmby, found her in the "extremely warm" room where he had last been seen her eighteen hours earlier. She was still wearing a nightdress, vest, white wool socks, plus one shoe. However: under these unscorched garments, radiation-type burns extended midway up both thighs to just below her navel... and upward from just below the breastbone to just below her nose. Similar burns were found on her buttocks, back, shoulders, and back of the neck; her neck-length hair was singed at the neckline only.

It was conjectured that she had collapsed in front of the kitchen's small gas heater—which Gas Board engineers "found to be in good order"—then momentarily regained consciousness, rolled over, becoming burned by radiant heat on her other side as well. Subsequent tests showed that close exposure to radiant heat would have scorched and ignited her nightgown's fibers, thus disproving the speculation.

The *Huddersfield Examiner* (December 5, 1980) noted a not uncommon trait for uncommon combustion: a small hole had burned in the hearth rug "but nothing to suggest a serious outbreak." Photographs given to me by the investigating officers prove her burns were extensive but not deep... and offer no hint as to how she could sear herself inside garments unburnt. Coroner Dr. H. J. Black performed the postmortem examination and recorded her cause of death: "Shock due to burns."

As to cause of the burns themselves, fire officers remained unable to explain how Mrs. Kenworthy could have such severe burns when there was no damage to the clothing which covered them. As West Yorkshire Metropolitan Police Sgt. Booth wrote to me about his first-hand observations at the victim's home, "I have worked for the past twelve years in the Coroner's Office and the case of Mrs. Kenworthy is unique."

March 22, 1981; Kent Road, Halesowen, on the outskirts of Birmingham, England. Caretaker Victor Webber unlocks the Leasowes Sports Center for business and smells smoke. The fire brigade discovered a mop ablaze in a broom closet. "Heaven knows how it started," confessed a befuddled Mr. Webber; "It looks like one of those cases of spontaneous combustion." Said a brigade spokesperson to the Wolverhampton *Express & Star* (March 23, 1981): "It's a mystery."

I open a new file dedicated to SMC... spontaneous mop combustion.

(And immediately the file grows. At the Dominion Golf and Country Club at Windsor, Ontario, all hellfire breaks loose in December 1941, as flowers in floral arrangements on dining tables instantly ignite and—this time—a *broom* is discovered ablaze in blue fire inside a closed closet. For more information, consult Frank Edward's *Strange World*.)

June 22, 1984; Berthelming, northeastern France. In this village in the Moselle near Nancy, Madame Francis Fétique, twenty-seven, pauses from working in her garden to take a drink of water and light a cigarette. Instantly, said *Le Républicain Lorrain* (June 24, 1984), she "was transformed into a human torch." A torch of frightening ferocity: everything within forty-five feet of her was reduced to ashes, including trees. Astonishingly, she wasn't killed in this monstrous fireball! The Sarrebourg Hospital administered first aid, then she was admitted to the Burns Unit at Merlebach, where specialists from Metz "were trying to determine what had happened." I've been unable to learn if they ever succeeded.

Late Christmas Eve, 1984; Newton Abbot, Devon, England. The husband of Mrs. Christine Middlehurst, thirty-six, discovers his wife sitting in a chair, aflame. He tossed water onto her, burning his hand in the process. Startled, the woman screamed and ran outdoors. Whether the screams were due to pain or surprise was not made clear by the Exeter *Express and Star* (December 27, 1984). The house on Drake Street was "virtually untouched" although burns covered 50 percent of Mrs. Middlehurst herself. Detective Inspector D. J. Westlake told the newspaper: "What happened is a complete mystery."

In 1990 Mr. Westlake, now Superintendent of the Devon & Cornwall Constabulary, said in response to my inquiry that Mrs. Middlehurst had a habit of sitting before a kerosene heater, where she was "found alight by her husband on 24 December 1984." Westlake visited her in the hospital and told me she had recovered from her injuries. Although the records of the incident had been discarded, he remembered there was "no evidence of any criminal offences being incurred and the cause of the fire was accidental and connected with the paraffin [kerosene] heater."

If this connection is legitimate, then Mrs. Middlehurst missed out on the distinction of having survived SHC because her burn injuries could be plausibly linked to a fire source external to her. At least she survived her combustion.

The same cannot be said for Mrs. Mary Carter, eighty-six and widowed, found dead in the hallway of her home at 22 Ivor Road in the Birmingham suburb of Sparkhill. (Sparkhill!) Sometime mid-day prior to three o'clock on Sunday, February 3, 1985, she became "extensively burnt." Patricia Gilbert made the distasteful discovery of what she termed "the object"—that is, her mother—and later stated "All I could see was two legs." Thus, reported the Wolverhampton *Express & Star*, firemen labeled this case a "fire riddle."

Had SHC befelled yet another Birminghamian?

Thanks to kind cooperation from Dr. Richard Whittington, Her Majesty's Coroner for Birmingham and Solihull, I have the inquest transcript for Mary Carter. And the riddle seems to have a solution."I think you will find that it is not quite as interesting a case as might otherwise have appeared," Dr. Whittington cautioned.

At the inquest, West Midlands police constable Moore testified that the fire brigade had found a candle holder by her side, and matches in the living room. (For once, the debunkers' often alleged "nearby candle" *is* present.) He concluded: "It is possible that she may have fallen against a candle."

Now if Mrs. Carter had indeed left behind but two legs, P.C. Moore's suspicion would be highly suspect. But much more remained of the victim than her daughter first noticed, enabling forensics officer David Parry and pathologist Dr. I. McKin Thompson to perform a thorough examination of her internal organs (which wouldn't be possible in classic SHC. Their findings led them to reason that the aged widow had fainted during congestive cardiac failure and that, due to absence of smoke particles in her stomach, she had died before burning commenced. Dr. Thompson pronounced her death as "Left ventricular failure. Ischaemic heart disease. Natural causes."

It appears that neither spontaneous nor preternatural combustibility is involved here. Sometimes, good investigation will resolve a seeming riddle. Sometimes, the riddle deepens the closer one looks.

A Case of Human-Baking in the Bakery

Barry Soudain: late of Folkestone, a town on the English Channel just down the coast from Dover. Unlike Mary Carter, he was neither old—age between forty-five and forty-seven—nor (technically) widowed, but he had lived a series of tragedies, the most recent the death of his live-in girlfriend. He coped using the bottle, his landlord stating that Soudain "never bought any food and spent all his money on drink."

After almost two years of this regimen, he suffered his last tragedy. Midmorning of Sunday, December 27, 1987, his landlord opened the bakery above which Soudain rented an apartment. It was not the odor of baked goods that greeted baker Reg Gower, but the smell of burning. He investigated. On the floor of the bakery's kitchen, devoid of smoke and fire, his search ended. There lay Soudain, strangely scorched to death.

Kent police Scenes of Crime officer, Detective Nigel Cruttenden, answered the call. Answers at the scene, however, were not easily forthcoming. The kitchen was compact, only about sixty-five square feet. Soudain managed to burn in an even smaller space, in the three feet between a sink and a cooker. His burning managed to restrict itself to an even *smaller* space, basically Soudain himself. Cruttenden noted that a plastic dustpan and brush within six inches of Soudain had not melted; nor had the polystyrene ceiling tiles. A tea towel hanging over the sink was unsullied. The slight melting of a plastic air vent next to the body and the superficial charring of the sink were attributed to Soudain's burning arms detaching from the body and falling against each item.

At the inquest, detective Cruttenden testified that Soudain was known to be a heavy smoker, yet no cigarettes or matches could be found in the kitchen. Hence, he turned aside careless use of smoking materials. He admitted the gas fire on the kitchen stove could not be absolutely ruled out as cause, but he was reluctant to implicate this source because "there were no remnants of clothing or ash to suggest burning."

Dr. Michael John Heath, a Home Office pathologist, found the blood-alcohol content in Mr. Soudain's corpse to be three and a half times the legal driving level. (Clearly he drank right up to the fiery end!) He also found flecks of carbon in Soudain's larynx. Cause of death was listed as accidental.

Obviously, Barry Soudain's bodily destruction was not as complete as in cases like Hannah Bradshaw, Mary Clues, Mary Reeser, Annie Webb, and Dr.

Bentley. But it was sufficient, nonetheless, for Detective Sgt. Cruttenden to contend: "It is inexplicable that with a body as severely burnt as this there was very little damage to the surrounding area."

All this left John Heymer, upon whose research the above is gleaned, to conclude: "As Barry Soudain was the only source of burning, he must have been alive whilst breathing in the smoke of his own burning."

9

BURNING DOWN NOT UP: "I THOUGHT I'D JUST DROP IN..."

> I will give the facts, and then leave it to the judgment of those who may be seeking evidence *pro* and *con* on this subject, upon which we have all read "but knew no more."
>
> —Dr. Floyd Clendenen
> on the Rooney case and SHC,
> *The Therapeutic Gazette* (1889)

The English language is replete with colorful colloquialisms and idioms. "The floor is now open" invites an exchange of ideas or debate. "I thought I'd just drop in" announces the unscheduled visitor who (unless he's a wayward skydiver) walks through your door. To take these expressions literally would be silly, of course.

But in a topsy-turvy world of unlimited possibilities, might it *not* be so silly to take these idioms literally? Might "I'm burning up" as an expression of anger or profuse sweating in some bizarre way get turned upside down and literalized, so that the speaker combusts in heat not rising and real fire not flaming toward the ceiling?

In a world that includes SHC, beware: if you live in a multi-story building, someone upstairs just might some day drop in unexpectedly. Aflame.

A BRIGHTLY FLAMING VISITOR DROPS IN

Around two o'clock one morning in 1780, a distraught lodger at the Five Pounds Alms-House in Limerick, Ireland, complained vigorously to the innkeeper, Mr. O'Neill. Bleary-eyed from being awakened so early and certainly none too happy, O'Neill undoubtedly thought his tenant was overreacting. He offered assurance (as innkeepers constantly do) that any problem could easily be rectified.

The lodger would not be appeased; *please* look! he must have urged. As sleepy-eyed O'Neill stepped into the complainant's room, he found the complaint was indeed justified!

A dead body lay incinerated on the floor. Overhead, he saw a body-sized hole burned in the ceiling through which the body had plunged, said the lodger, with "red as copper" flames. O'Neill immediately ran upstairs and burst into the room of Mrs. Pococke.[1] He found her not. Just a hole through the floorboards.

History, unfortunately, does not record what action the inn's management took regarding the complaint of a flaming corpse burning through a boarder's ceiling and perhaps almost landing on top of him in bed.

O'Neill—now very much alert!—and other lodgers aroused by the commotion took quick action to quench the fire about the hole. Then O'Neill set out to determine the cause of this fatal, strange fire. He could not find it.

"There was no candle or candlestick near the place; nor fire in the grate, except what was raked together in the ashes," said the *London Mirror* (September 22, 1838). "The room was examined, and nothing had taken fire but that part of the floor through which the wretched woman had fallen. Even a small basket, made of twigs, and a small trunk of dry wood, which lay near the hole, escaped; not being so much as touched by the fire."

The mayor plus several clergymen and gentlemen of Limerick inquired into the woman's fantastic fall. Rev. Wood, a Wesleyan minister in the city, reported his own observations in the *Methodist Magazine* (1809) and reconfirmed the above details in his personal communications to Dr. James Apjohn, "the highly accomplished Professor of Chemistry, in the Royal College of Surgeons in Ireland" who took an interest in such flaming enigmas.

The incident was confirmed by reputable gentlemen, but a confirmed solution to Mrs. Pococke's outrageous drop-in eluded them all.

Lest the skeptic contend this lady who "indulged immoderately in intoxicating liquors" was a flight (downward) of someone's fancy, a remarkably similar event had occurred on the opposite shore of the Atlantic Ocean nearly a decade earlier.

Hannah Bradshaw's Horrible Holiday

Colonial America in 1770 prospered as part of the British Empire, and New York City became a rapidly developing metropolis for commerce and commu-

skull. Nearby a candlestick contained part of a candle, its tallow melted, its wick "untouched by the fire, as also the screen which almost touched the hole." Holt's *Journal* noted that the candle "did not appear to have touched the body, or to have set any thing on fire."

A chair had its leg and part of the underside of its woven-grass seat burned, but only "as they were within the compass of the hole in the floor and no further." The white-washed plaster ceiling was now "as black as if covered with lampblack, as also part of the walls and windows," said the *Journal*. Heat had been so great that turpentine was extracted from the oak boards of the wainscot in the room. Yet, said Holt, when the room was opened "not a spark remained."

(Recall: a normal fire produces three times the damage above its point of origin than on the floor, because heat and flame rise. Miss Bradshaw's ceiling, however, was *not* destroyed.)

The newspaper conveyed (in archaic prose) the amazement of those at the scene: "It is extremely surprising that with such an intense degree of heat as was necessary for entirely to consume the floor, etc., with the body and to calcine the bones, the Fire should have stopped of itself without burning the house, or even scorching the screen—which is hardly conceivable if the Fire had blazed. And if it did not, it is equally inconceivable that the heat should have been so intense."

> Facts are stubborn things.
> —Alain René Lesage

The *Journal* pointed out that when the floor burned through, it opened the room to a powerful draft so that the fire "could not have stopped for want of air."

Had she not been alone that night, perhaps Hannah would have been seen glowing "red as copper" as she burned her way into her basement and into the history of bizarre blazes. Just as Mrs. Pococke would do a decade later.

Reports of two events so unexpected and strange yet so alike, ten years and a thousand miles apart, must raise questions. Questions for which there are no obvious answers.

JOHANN CASPER CHALLENGES SPONTANEOUS COMBUSTION

Johann Ludwig Casper thought he did have the answer to these two cases, plus scores of other thoroughly combusted humans, when he wrote *A Handbook of the Practice of Forensic Medicine, based upon personal experience* (1861). His solution was simple: deny.

"It is sad to think," he bemoaned, "that in an earnest scientific work, in the year of grace 1861, we must still treat of the fable of 'spontaneous combustion,'

nication. Its people were eager to read about current news, formative talk of independence, general deeds and curiosities; so publishing flourished. Once-plentiful newspapers from this era are today rare and extremely fragile, hence seldom read. In their brittle pages, though, awaiting modern eyes, lurk some very interesting stories.

One story most curious (then *and* today) can be found in John Holt's short-lived *New York Journal; or General Advertiser* for January 3, 1771.[2] Since then, this story languished in dark archives, never recounted in detail until now. In an original issue of Holt's broadside, yellowed and brittle and ever so gently held in my hands in the rare documents room of the New York Public Library, here is what I learned.

"On Monday Evening, the last Day of the Year, the following dreadful and surprising Accident happened at a house in Division Street, in this City," Holt's *Journal* began. In an upper room of this house lived Hannah Bradshaw, about thirty years old, a large and robust woman who liked to drink. On this New Year's Eve, around seven o'clock, she bid holiday cheer to a visitor who noticed that Hannah already "seemed to have drank a little too freely." The friend wished Hannah well for the New Year, said the newspaper, "little thinking, as it was so near, that she would not live to see it, that she was then on the brink of Eternity, and that before the Morning Light *her soul would be required of her.*"

Hannah was now alone to celebrate the arrival of this special midnight. Special is too mild a word; *extraordinary* would be more appropriate. As New York's clocks struck in another annual cycle, something unanticipated *struck* one of the revelers.

New Year's Day, 1771. An employee arrived at Hannah's home early in the morning, as requested. The door was bolted. Hannah did not answer. The woman kept trying to rouse a response until, at eleven o'clock, she asked a man living downstairs to help her crawl through a window to see if Hannah was too drunk to get up or if...

There, behind a screen inside the room, she beheld the most shocking spectacle imaginable. A hole four feet in diameter was burnt into the floor. On the ground, about twelve inches below, lay what was left of Hannah: "The flesh was entirely burnt off the bones of the whole body and limbs, except a small part of the skull, a little on one of the shoulders, the lower part of the right leg and foot, which was cut off at the small, almost as even as if cut off, and left lying on the floor; the stocking burnt as far as the leg and no farther." Her bones were "like a pipe stem" and "crumbled to dust between the fingers" when touched.

The floor joist beneath her incinerated shoulder was itself "burnt almost through" and at the edge of the hole, on the floor planking, lay a portion of her

a thing that no one has ever seen or examined, the very proofs of whose existence rest upon the testimony of perfectly untrustworthy non-professionals, upon newspaper paragraphs, all of which in their statements laugh to scorn every known physical law."

This German forensicologist would be right at home with many of his professional colleagues a century later, who give SHC a passing thought only long enough to reject it. For four pages he lashed out in exhilarating diatribe at anyone involved in promoting this vile fraud.

Casper defended himself, pointing out how often people had been terribly burned in great conflagrations. "But never in such cases—never," his words veritably screamed, "even when the body has been exposed for days to fire or incandescent heat, has it been reduced to ashes!"

Oh?

At last he concluded with this gem of nationalistic rationalization: "Moreover, it must be remembered, as an incontrovertible fact, that the mass of the French people are more credulous than the Germans, and this helps to explain why France has been the actual birthplace of '*spontaneous combustion*,' of which it is to be hoped that we shall hear no more in relation to the science of Medical Jurisprudence."

Therefore anyone who believes in SHC—especially the French—must be stupid if not lying, because the truly intelligent see it "as one of the silliest fables."

While it's easy to dismiss two or three episodes that belie common sense, Casper had access to scores of historical cases, yet he did not even allow for *one* exception to his rule.

So what do we make of Mrs. Pococke and Ms. Bradshaw? Casper, in effect, called reports of these women's deaths lies. I wish Johann Ludwig Casper could have spent Christmas 1885 in America with the Rooneys.

CHRISTMAS AT THE ROONEYS

Fiery deaths such as Pococke and Bradshaw have been "discovered, investigated and filed away without acceptable explanation," wrote Frank Edwards in *Strangest of All* (1963). "Yet from time to time they recur—just often enough to plague the experts who must deal with them." If ever there was a Ghost of Christmas Past worthy of recurrence, it visited the Rooney household on December 24, 1885.

The account that follows is compiled directly from original reporting by two rival, small-town newspapers, *The Ottawa Daily Republican Times* and *The Ottawa Free Trader Journal*.

Patrick Rooney and his second wife, Matilda, both seventy-two, lived three miles north of Seneca, a small rural town sixty-five miles southwest of Chicago. A prosperous farmer, Mr. Rooney kept "a little brown jug," which he refilled weekly with whiskey at the saloon of his son-in-law, Michael Murphy. The week before Christmas 1885 would see his jug's last refill.

On Christmas Eve, the Rooneys sat down at the kitchen table with hired-hand John Larson to share the merriment of the holiday over the jug of spirits. Larson had two drinks, then retired to his room over the kitchen and was soon asleep. During the night, Larson awoke gasping for breath. He thought he had a violent cold and went back to sleep.

Next morning, Larson went downstairs to Mr. Rooney's room. There he discovering the old farmer prostrate on the floor. Larson, thinking the old man had collapsed in a drunken stupor, knelt down and vigorously shook his employer. Rooney was dead. Larson then rushed into Mrs. Rooney's bed chamber—she occupied a separate room—but found her not.

Larson decided to take care of some farming chores. Then he trekked off to the Murphy home, not to relate Christmas cheer, but his suspicion that Mrs. Rooney had killed her husband and fled.

If only it had been *that* simple!

With Murphy accompanying him now, Larson retraced his steps back to the Rooney homestead, where the two launched a thorough search for Mrs. Matilda Rooney. In the upstairs bedroom the men discovered Larson's pillows were blackened with soot, undoubtedly the cause of the hired hand's nocturnal suffocating coughs. In the kitchen below this room they made their momentous discovery.

To quote *The Ottawa Daily Republican Times* (January 9, 1886): "Finally they found a calcined skull, part of a vertebral column and a handful of white ashes beneath the kitchen floor. There was a hole burned through the floor four feet in length and three in breadth through which the incinerated remains had fallen. The walls were blackened as if with lampblack, as was the floor and the woodwork."

No wonder Larson had at first overlooked his employer's wife! For practical purposes, the scene was a duplicate of Hannah Bradshaw's room.

"The fire burned a hole just large enough to let Mrs. Rooney's remains fall through," added a correspondent for the rival *Ottawa Free Trader Journal* (January 9, 1886) in verifying this scene. The hole through the one-inch thick pine flooring measured two and a half by three feet.

"The flesh was entirely consumed," the *Free Trader* continued, "except that part which covered the bones of one foot. The [left] foot was found just at the edge of the aperture in the floor, through which the body had disappeared.

The limb had burned slowly to the ankle—flesh and bone—and when the body dropped the charred bone snapped, and the foot, with the shoe intact, righted itself and stood up as if its owner had been burned at the stake. Nothing in the room was burned except Mrs. Rooney's body, the hole in the floor and the table cloth fringe." A kerosene lamp on the table was out, giving no hint of having contributed to the blaze.

I question the assertion that the limb had burned slowly. Surely Mr. Rooney, who seemed to be preparing to retire for the night, would have smelled something and returned to the kitchen. Instead he fell, still clothed, in his bedroom next to the kitchen, indicating either that he succumbed to toxic fumes or suffocated in an oxygen-robbing odorless fire. This means that Mrs. Rooney, overcome with equal swiftness by her burning, was probably standing upright by the table when she quickly exited this world through the floor of her kitchen. The foot, found the next morning, stood aside the hole in mute embodiment of, truly, the woman's last stand.

That a fire erupted and somehow consumed itself was not to be denied. Its localized nature and fearsome destruction of Mrs. Rooney was not to be denied, either.

The newspaper raised the issue of burning at the stake, so let's digress a moment to revisit what the medical literature has to say about this once popular mode of body dispersal.

To quote Smith and Fiddes' *Forensic Medicine, A Textbook for Students and Practitioners* (1955) again: "The amount of external heat required to incinerate a body is enormous, and even when bodies are placed on piles of wood and burnt, there is usually only partial carbonisation." They cited an instance of two murdered young men placed on a stack of cotton stalks thirty feet square and six feet high, the pyre then ignited. After the huge fire's prolonged combustion and frightful heat ended, strangulation marks in the neck of one and fragments of carbon in the bronchi of the other were identified. "This case shows the great difficulty of burning the human body," the two authors justifiably concluded.

Imagine: 5,400 cubic feet (153 cubic meters) of readily combustible fuel could only partially burn two men, yet Mrs Rooney's body *with not so much as her clothing and one tablecloth* to support combustion was reduced to ashes in less than twelve hours!

What force had unmerrily been at work that Christmas Eve in Seneca, Illinois?

One force considered, according to the local newspapers, was murder. Larson—despite being termed "an old man and apparently honest"—was nonetheless an obvious suspect. *The Ottawa Daily Republican Times* reported his exoneration—along with his death—a fortnight later, death being caused by what the coroner called "the poison inhaled that night, as supposed." Likewise, *The Free Trader* said "no indications of foul play have been discovered."

The other force considered was an innocent mishap that turned fatal: that Mrs. Rooney, as was her custom, had stood on a chair with candle in hand to wind the clock, and, being under the influence of liquor, fell down and her clothing took fire from the candle. This explanation undoubtedly satisfied everyone seeking a superficial solution. But the problem of a candle and a housecoat unleashing more flaming destruction than fifty-four hundred cubic feet of flammable cottonstalks is one the modern thinker must ponder.

Whether Casper ever heard about the Rooneys, I don't know. If he had, he likely would have ignored them or dismissed their fate as he did other similar cases to which he remained blind. Fortunately, another man of science—*true* science—dealt with the Rooney case open-eyed. Hands on. Up close and personally.

Dr. Floyd Clendenen[3] thought about this "mishap" when the evidence was freshest. As coroner for La Salle County, Illinois, he was telephoned the morning of December 25 to investigate the two deaths. He immediately set out for the Rooney farm and a Christmas he'd never forget. He realized upon arriving that an autopsy was not possible to perform, so he catalogued the woman's remains as they were retrieved from the crawl space under the kitchen.

"Upon removing them," Dr. Clendenen wrote, "we found the skull, the cervical, and half the dorsal vertebrae reduced very nearly to a cinder, also about six inches of the right femur, together with part of ilium [the broad upper portion of the pelvis] in about the same state as the vertebrae." Her right foot and shoe "were reduced to a complete cinder. The other parts of the body and clothing were reduced to a *very light* cinder, leaving *no* shape of former body."

Destruction was thorough yet limited. Marvelled the doctor, "The skull and hip bone were really the only evidence by which it could be told that a human body had been cremated there."

Cremated! That's what the coroner said. Yet the crematorium had been invented only nine years earlier by Dr. Julius Lemoyne in Washington, Pennsylvania. (Ironically, its inventor was the first person his invention cremated.)

Cremated? *How?* The Rooneys *certainly* had not installed one of the world's first crematoriums in their Seneca farmhouse kitchen.

Regardless, Mrs. Rooney's reduction had been nearly complete: her 160-pound body could be (and was!) scooped into a bushel basket. "The entire remains weighed twelve pounds," Dr. Clendenen recorded. Mrs. Rooney had lost 92.5 percent of her weight overnight.

The physician-coroner risked his professional reputation by testifying to these facts clearly anathema to the reigning medical doctrine as expounded by Casper a few years earlier: that "never, even when the body has been exposed for days to fire or incandescent heat, has it been reduced to ashes!"

Maybe Dr. Clendenen wasn't easily intimidated. More likely, he was unaware of Casper's diatribes and therefore didn't realize that he *should* be intimidated by this affront to the medical dogma. Whatever, he reported his "peculiar case" to the medical profession as he witnessed it firsthand—for which medical history should be grateful. He could have found solace in Charles Fort's remark in *Wild Talents*: "There have been many ailments and deaths of human beings that have not been satisfactorily explained in the medical terms that are just now fashionable..."

In 1993 I located, and visited, the Rooney farmhouse, still standing within view of Interstate-80 as it plows westward across the Seneca plain. And in what now serves as a sitting room, its current resident pointed out the corner where a pillar of fire transformed one human into medical history more than a century before. An eerie feeling.

Some years earlier, in 1987, Don Hetzner, a direct descendant of Mr. and Mrs. Rooney, had contacted me in his research of family history and its most famous incident. Hetzner said that for years afterward, some of his relatives questioned quietly among themselves "whether divine retribution due to drinking on Christmas Eve" was to blame for the Patrick Rooneys' deaths. For Dr. Clendenen, pronouncement was more clinical. He called it, simply, "A Case of Spontaneous Combustion in Man."

THE MAN WHO COULD NOT BURN STRAW—ONLY HIMSELF

In his harangue against spontaneous combustion in man and woman, Casper resorted to sarcasm to defend his position. Every child and every physician, he wrote in *A Handbook*, immediately recognized that belief in SHC is tantamount to belief in fairy tales: "The old drunkard was burnt, but the stool on which her charred remains were found sitting was uninjured... of an individual who had been seen the previous evening alive and healthy, only a few burnt fragments or a handful of ashes were found in the morning..."

Poor Johann Casper. He just couldn't keep the vagaries of nature from raising disquieting challenges to his convictions. Not that it mattered to him, probably, but it did matter to Dr. J. Mackenzie Booth of Aberdeen, Scotland.

Dr. Booth (M.A., M.D., C.M.) took a deep breath, thought again about what he had seen a month earlier, then stepped to the lectern of the Aberdeen, Banff, and Kincardine Branch of the British Medical Association on March 21,

1888. I suspect not a dram of fine Scotch whiskey was poured as he began to speak.

"On the morning of Sunday, February 19, I was sent for to examine the remains of a man, A. M., aged 65, which were found in a hayloft off Constitution Street. This man, a pensioner, of notoriously intemperate habits, had been seen at nine o'clock the night before to enter the stable below in an intoxicated condition," Dr. Booth began. Two children watched the drunken old soldier climb a ladder to the loft, then noticed the skylight there glowing briefly and becoming dark again.

Between eight and nine o'clock the next morning smoke was seen issuing from a hole in the roof of the loft. The stable owner rushed in and was horrified to see through a hole in the loft floor the remains of the old soldier perched on the joists above, and leaning against the wall. As there was no fire to extinguish, the stable owner summoned the police, who summoned Dr. Booth, whose close examination revealed details quite strange.

"What struck me especially at first sight was the fact that, notwithstanding the presence of abundant combustible material around, such as hay and wood," remarked the doctor, "the main effects of combustion were limited to the corpse, and only a small piece of the adjacent flooring and the woodwork immediately above the man's head had suffered." Some slates on the roof, apparently dislodged by heat or flame, had fallen upon him, and some flooring beneath him had plunged into the stable below. Yet straw all around him was not touched by the limited holocaust.

Closer examination revealed details even stranger. Dr. Booth described the condition of the corpse to his medical colleagues:

> The body was itself almost a cinder, *yet retaining the form of the face and figure so well, that those who had known him in life could readily recognize him.* Both hands and the right foot had been burnt off and had fallen through the floor among the ashes into the stables below, and the charred and calcined ends of the right radius and ulna, the left humerus, and the right tibia and fibula were exposed to view. The hair and scalp were burnt off the forehead, exposing the bare and calcined skull. The tissues of the face were represented by a greasy cinder retaining the cast of the features, and the incinerated moustache of still gave the wonted military expression to the old soldier. The soft tissues were almost entirely consumed, more especially on the posterior surface of the body.

Place yourself in the audience of the British Medical Association, and imagine one of your esteemed peers telling you he saw a body that was thoroughly burned to cinders *yet retained the full display of its anatomical form—right down to its facial expression and trim moustache!* Only where some roof slates had fallen onto the body was the soldier's appearance destroyed, said Dr. Booth.

More consternation followed. "From the comfortably recumbent attitude of the body it was evident that there had been no death struggle," Booth confounded his audience further, "and that, obfuscated by the whiskey within and the smoke without, the man had expired without suffering, the body burning away quietly all the time."

Dr. Booth decided this episode of a burned body dropping in through the stable loft warranted recording for posterity. He left to find a photographer, and then would resume his examination of the dehydrated corpse.

Unfortunately bystanders attempted to pick up the body and remove it. "The bearers told me that the whole body had collapsed when they tried to remove it *en masse*," he frustratedly told his audience. "So much for the condition of the corpse."

But *what a condition!*... a condition calmly and rationally attested to by yet another physician with firsthand experience. No armchair fantasizing here. Dr. Booth had seen it himself and conveyed in understated amazement the impossible scene his eyes and hands beheld in this unfortunate incident. *The body of Mr. A. M. was 100 percent dehydrated by a fire that did not destroy the form and shape of what it burned, and insulated itself from surrounding combustibles such as straw on which the body lay.*

Fortunately, should anyone doubt Dr. Booth's testimony (and surely there will be some), a W. Reid of Aberdeen entered the loft with his camera before the powdered corpse's untimely removal and snapped the shutter—thereby earning the distinction of obtaining probably the first photographic evidence for a classic case of spontaneous human combustion.

MORE HUMAN FIRE AUGERS

A hole burned through a floor is "normally an indication of an electrical appliance fire or a piece of furniture with foam upholstery," arson investigator Tpr. Bill Sweet told me. Rarely will even a flammable liquid burn through a floor, he stressed, "unless a hole is already there!"

In May 1953, thirty-year-old Mrs. Esther Dulin defied scientific (as well as social) decorum by not only burning up but at the same time burning *down* through her floor. So intense was the fire in her Los Angeles home at 210 West 110th Street that she and the chair she sat in were "virtually consumed," as both bored burning into the room below. Nothing else in the house was damaged.

If the scene resembled that of Dr. Bentley, Ms. Bradshaw, Mrs. Pococke, and Mrs. Rooney—and the description in *Fate* (December 1953) suggests it would—then Mrs. Dulin became another member of that select fraternity-sorority who burn themselves through their floors like fiery augers. A call to the Los Angeles Fire Department records unit in 1987 failed to uncover more information, because their files no longer dated that far back. "I can't think of any way to help you," apologized LAFD officer William Reid, "it sounds very interesting..."

On November 14, 1964, a sixty-year-old Welsh woman in a bathtub was found severely burned, reported *Medicine, Science and the Law* (1966). Not only had she managed to super-scald herself, she managed somehow to burn a hole in the floor alongside the tub into which her head then fell.

On November 14, 1987, the search for bizarre blazes took me to Ontario, Canada, where I met retired forensic pathologist Dr. Douglass Mills at his home. By phone, he had told me his career as chief pathologist at St. Joseph's Hospital in London, Ontario, included not one, not two, but *three* cases he believed had the characteristics of SHC. "Just like in Dickens' *Bleak House*," he affirmed.

This would be unprecedented! No wonder that I was especially eager to meet him.

After carefully listening to his narrative and being escorted to the hospital where he had worked to see the relevant photographs, I concluded that two of these incidents did not meet the criteria for anomalous fatal fires.

But his third case does. A crippled Canadian woman had been cooking a meal years before when, ostensibly, her clothing ignited. She ran to the door but "dropped in her tracks," said Dr. Mills, after which her head and upper body reduced to a char and the rest of her dropped through a hole created by her incineration. "Burning was limited to a three-foot circle. Paint on nearby chairs wasn't involved. Her bones even reduced to char," he said.

"But it's not spontaneous, it's not preternatural combustibility. It's a very specialized type of burning, involving special conditions."

Surprised, but thinking this a wonderful opportunity to hear the long-sought solution to so-called spontaneous/preternatural combustibility, I asked "What *are* those special conditions, then? And *how* did they ashen her?"

These questions Dr. Mills could not answer. He could only say the fire scene resembled *Glaister's* "typical appearance."

This brings us back to the third photograph I mentioned earlier from *Glaister's Medical Jurisprudence and Toxicology* (13th Edition). The unidentified victim—I'll call her "Glaister's female fire auger"—was probably British, and her awesome legacy sometime before 1973 attests to SHC of the hole-in-the-floor kind:

(1) Combustion was so fierce in the woman's midsection it *bored a hole about two feet in diameter* through a Persian rug and the floor beneath;

(2) Combustion was so localized that—in classic SHC style—it left behind the woman's extremities: two legs below the knees and one-half of one hand with distended fingers and thumb; her skull literally burned in half;

(3) Heat was so minimal that, beyond the hole in the floor, no ignition temperatures were reached; cloth (or silk) fringe hanging from furniture just inches above the woman's halved head show not the slightest trace of fire; nylon stockings on the legs were *wholly* intact below the knees, where the fire had abruptly stopped.

Glaister's gave this woman a ten-word obituary: "Almost complete destruction of body with relatively slight surrounding damage." The epitome of terse, accurate copy writing. What is missing are clues to who she was... and, of course, just *how* she died.

The "Births and Deaths Registration Act of 1953" in Britain retains the command of the "Coroner's Act of 1887" that a coroner investigate any "violent or unnatural death" when the victim "died a sudden death of which the cause is unknown." This woman's death surely warranted a physiciancoroner's involvement; hopefully a file will surface so that the anonymity that cloaks yet another testament to medical passivity will someday be stripped away.

WHEN OLD MAN ASHTON BECAME... ASHES

Not cloaked in anonymity—though those who are offended by these incidents might wish otherwise—is the fate of eighty-six-year-old Alfred Ashton, late of Southampton, England.

At ten o'clock on the evening of January 8, 1988, neighbors noticed smoke issuing from the Ashton home, and soon the fire brigade arrived to discover a room filled with smoke and extreme heat. Plus the remains of its occupant, now best described by the first syllable of his surname.

A blackened skull and two lower legs lay on either side of a small hole in the wooden floor, through which the rest of Mr. Ashton's thoroughly incinerated body fell. Investigating fire officer Roger Penney noted that Ashton had burned facedown, atop the electrical wiring that fed a nearby outlet into which was plugged a small single-bar electric heater. The heater was intact, but cold, because the outlet into which it was plugged was "off." Whether a firefighter had switched it off moments before or it had been turned off before the fire began wasn't considered at the time by Penney—an oversight he later regretted—because the electric cable powering the outlet had itself shorted out, burned through by the *downward* burning of Ashton himself.

What Penney did notice was that Ashton had not fallen on the heater. He also noticed that highly combustible materials only inches from the hole were virtually untouched... except for the surreal heat deformation of a TV's plastic cabinet on a countertop some distance away. Otherwise, the consuming flames had been restricted to the corpse and whatever was directly in contact with it. Penney noticed, too, that the burn line on the pants legs *coincided precisely* with the point where the flesh of his legs escaped being singed.

> I've been to many fateful fires over the years, but I've never seen anything quite like this. The only way I can explain it is to consider SHC.
> —Roger Penney, Hampshire Fire Brigade; investigating officer on the Ashton case.

The pathologist concluded the cause of death was due to burning, and added: "Except for the lower legs, the bony skeleton and muscular tissues had been destroyed." So nonchalant, yes? Three extremities lay severed beyond the radius of a fireball that completely and quickly engulfed the body in between, and nary an eyebrow was raised. The inquest returned an open verdict: "cause unknown."

I wonder whether this old man was a widower.

IN MARKED CONTRAST

One irony is how the demise of these human fire augers stands in marked contrast to other examples of fire fatalities portrayed in *Glaister's* manual on jurisprudence.

Depicted is one man who perished in a conflagration. His corpse is burned black, ash-covered, and curled up in the pugilistic (fetal) position characteristic of exposure to extreme heat. Yet his entire body is easily recognizable as male, and his corpse remained intact during its transport to a morgue.

Another photo in *Glaister's* compendium shows "an unusual case where death occurred from a combination of slow excessive heat and anoxia." Anoxia results when oxygen in body tissues is abnormally low; in other words, a condition in combination with slow heating that is conducive to smoldering combustion of the body. It is the very process invoked by SHC debunkers to explain away Mrs. Rooney, Mrs. Reeser, Dr. Bentley, and scores of others. This "baked and dehydrated" body is instead intact, easily identifiable; a large percentage of the skin even retained its natural pigmentation.

How *unlike* Mrs. Rooney, whose coroner said that her "skull and hip bone were really the only evidence by which it could be told that a human body had been cremated there."

Advocates who would replace the SHC explanation with slow smoldering fires that pervade anoxic bodies from the outside, must satisfactorily explain how this process can produce two diametrically and dramatically opposite outcomes.

While we are on the topic of categorizing fire fatalities, the editors of *Glaister's* 13th Edition devoted a full chapter to medico-legal identification. In its fifty-three-page discussion of techniques used to classify cause in death by burning, *not one listed technique* would aid an examiner assigned to place cause for Mrs. Pococke, Ms. Bradshaw, Mrs. Rooney, Mr. Ashton, *et cetera*—except to rule out each technique as inappropriate. What is left?

Perhaps the next edition of *Glaister's* will add "SuperHyperthermic Carbonization" to its list of techniques for identifying causes of fire fatalities. Then, at last, a coroner would have a respectable—and wholly appropriate—choice for assigning death when death results from super-hot internal thermogenesis.

10

SuperHot Humans: Stoking Up the Body's Furnace

> Expert medical testimony has its chief value in showing the possibilities of the occurrence of alleged extreme cases, and extraordinary deviations from the natural.
>
> —Drs. George M. Gould and Walter L. Pyle,
> *Anomalies and Curiosities of Medicine* (1896)

Thermogenesis is the term scientists use to describe the organic generation of life-sustaining heat. We humans are homeothermic mammals, possessing a complicated internal mechanism for maintaining our core temperature at or near a preset, optimal temperature.

The primary controlling thermostat in your body is the hypothalamus, near the base of the skull in the primitive portion of the brain. It senses the blood's temperature, then fires off biochemical commands that, along with physical responses, maintain life-sustaining heat. It also exerts control over visceral activity, water balance in the body, and sleep.

Should our bodies become either too cold or warm internally by just a few degrees, we may die. But so long as thermogenesis operates within an extraordinarily narrow temperature range around the so-called normal* setpoint of 98.6°F, people give not the least thought to the wondrous process that

---

* New research by doctors at the University of Maryland School of Medicine may make some traditionalists hot under the collar, but they find 98.6 degrees isn't normal at all; rather, they told the Associated Press (Sept. 23, 1992), "98.9 degrees in the early morning and 99.9 degrees overall should be regarded as the upper limit of the normal oral temperature range in healthy young adults." Standards do change.

keeps them healthy and comfortable; neither too cold nor too warm, neither quaking with chills nor drenched in sweat.

Only when thermogenesis goes awry does a person become uncomfortable and concerned and probably ill. Fevers result when the body produces more heat than normal (even though fever itself indicates that the body is responding perfectly to an internal distress). Fevers themselves, even as they attempt to heal the body, must operate within a narrow temperature range or additional suffering is likely. High, prolonged fevers can irreparably damage the body; even cause death.

The question raised by the scores of people who have apparently self-combusted is this: How *much* internal heat can a human being generate?

A HOT TIME IN THE OL' BOD' TONIGHT

Physicians fear brain damage will occur when body temperature soars to 104°F, and sustained temperatures over 109°F for as little as several minutes is generally considered incompatible with life. That's what the medical community believes. And in most instances their belief is bolstered by millions of case histories.

Then there are the exceptions.

Temperatures in the 107° to 111.4°F range are not uncommon for sufferers of typhoid and rheumatic fevers, for example. The current *official* holder for record high internal temperature is Willie Jones, according to *The Guinness Book of World Records* (1990). On July 10, 1980, Willie's temperature soared to 115.7°F, attested doctors who monitored him at Atlanta's Grady Memorial Hospital. He survived... and was discharged the next day with a normal baseline temperature!

However, *Guinness* could reassign Jones's title to a nervously hysteric woman treated for acute rheumatism: her temperature rose to 115.8°F, declared *The Lancet* (1878), and she insisted on leaving the hospital when her continued fever dropped to a *mere* 104°F! In turn, she is outclassed by another English woman's fever described to the Clinical Society of London in 1875: for seven consecutive weeks her body temperature measured not less than 108°F, and on four occasions it *skyrocketed to an astonishing 122°F!*

Equally astonishing (in terms of hyperthermia) is the Italian mystic and Capuchin monk Padre Pio (1918–1968), famed for his display of stigmata that spanned more than half a century but less well known for his bouts with agonizing hyperthermia. More than once the young Capuchin shattered clinical thermometers used to measure his soaring inner heat, reported the papal *Acta Apostolicae Sedis* (May 31, 1923). Biographer Rev. Charles M. Carty wrote that Padre Pio's temperature sometimes reached 122°F.

In 1881, a twenty-two-year-old hysteric woman bedridden at Guy's Hospital, London, blew off hospital thermometers until her physicians procured a horse thermometer capable of reading 130°F. Not surprisingly, the patient objected to the insertion of such a large instrument, but her doctors prevailed. "128°F!" exclaimed the doctors. The woman lived another five months. In 1880 a similar fate warmed up another woman to a temperature of 130.8°F, the *Medical Times and Gazette* reported.

Actually none of these contenders for Hottest Human Being is in the running. Their thermogenic achievements up to 130.8°F earn no more than a Bronze Medal in the human furnace olympiad, and pall before evidence contained elsewhere in medical literature long forgotten and certainly overlooked by *Guinness*.

At a meeting of the Association of American Physicians (AAP) in 1895, Dr. Jacobi of New York stood to report a case of extraordinary hyperthermia in a man—a fireman, no less!—who incurred severe injuries after falling into machinery. For five days his temperature averaged 120° to 125°F, said Jacobi; "*on one occasion reaching the height of 148°F.*" A collective gasp of incredulity must have swept among the AAP as Jacobi presented that statistic.

Faced with this "excessively high temperature," as Dr. Jacobi modestly described it, he did what a good researcher would: he maintained skepticism but didn't reject the evidence outright. He told his colleagues that he repeated his measurements, using different thermometers to guard against defective equipment. The extraordinary levels persisted, impossible though they seemed.

Probably to Jacobi's relief, another member of the AAP arose to offer corroboration even more astounding. Welch of Baltimore spoke of another closely monitored human whose natural body temperature, Welch affirmed, "reached as high as 171°F."

171°F! Makes you sweat just thinking about it, doesn't it? Unquestionably, this is in the realm of human *super*hyperthermia.

Since at least one human's body temperature is said to have soared a phenomenal 73 percent, rising from 98.6°F to a sizzling 171°F, is another increase of only 24 percent more so improbable to consider? Such internally generated temperature would now be 212°F, the boiling point of water...the point where bodily fluids could boil, body moisture could begin to evaporate, and flesh could burn from within. The point where internal superhyperthermia leads to superhyperthermic carbonization.

The search for an answer requires a brief foray into what is presently known—and not known—about human superheating.

MALIGNANT HYPERTHERMIA

When I began research for *ABLAZE!* in 1973, medical science had yet to name a mystifying ailment that infrequently plagued humans undergoing surgery. Occasionally patients would die, not from botched operations but due to a dramatic spontaneous rise in body temperature. The instantaneous increase of several degrees Fahrenheit proved feverish, not to mention perilous, to both patient and personnel. Baffled doctors and anesthesiologists, unprepared for this surprising development, did not know how to safeguard the lives of these patients who began to fry internally on the operating table.

Today this condition has a name: malignant hyperthermia.

Malignant hyperthermia is an inherited condition that usually occurs after administration of a general anesthetic, wrote Dr. Peter M. Waterman in the *American Journal of Ophthalmology* (October 1981). "For the prone individual," said syndicated medical columnist Dr. Paul G. Donohue in 1986, those gases "spark a metabolic inferno in the body. Temperature shoots up to extremely high levels." Quick treatment with the drug Dantrolene, he said, "effectively puts out the body conflagration."

But, confessed Dr. Donohue, "I cannot answer the jackpot question on everyone's mind concerning malignant hyperthermia: What causes it? We don't know what sets off the great temperature rise."

Stress seems to be one factor, said Dr. Waterman, who noted in rare cases that *anxiety* rather than anesthesia was the prerequisite for unleashing malignant hyperthermia's hypermetabolic state (a state that, incidentally, and of particular interest in the context of SHC, produces mind-numbing and flammable carbon dioxide as a byproduct).

Malignant hyperthermia syndrome is, while not fully understood, a phenomenon acceptable to the medical community because it's happened in the operating room under the eyes of alarmed surgeons. Could it explain the preceding anomalous hyperheating in humans... perhaps even SHC?

I put that question to Dr. Henry Rosenberg, chairman of the Professional Advisory Council of the Malignant Hyperthermia Association of the United States. "I have never seen or heard of such a case," he replied. "I don't believe that body temperature can be raised to the extent described by the mechanism underlying Malignant Hyperthermia."

Perhaps. But might it be that malignant hyperthermia, an acknowledged medical condition, has an anomalous aspect *not yet identified* because of its rarity; a powerful aspect which can lead to excessive hyperheating and the torrid burning of a person from within?

Tumo: Fuel for Superhyperthermia?

One of mankind's more spectacular yet lesser-known powers is *tumo*. It is the voluntary production of prodigious heat in the body.

Western medicine remains oblivious to tumo, even though its existence and manipulation have been acknowledged for thousands of years by Tibetan lamas (holy men) and Indian yogis in the high Himalayas. Early in this century, Madame Alexandra David-Neel wrote about the marvels of tumo in *Magic and Mystery in Tibet*, a chronicle of her fourteen years spent in this isolated land. "To spend the winter in a cave amidst the snows, at an altitude that varies between 11,000 and 18,000 feet, clad in a thin garment or even naked, and escape freezing," she marveled, "is a somewhat difficult achievement. Yet numbers of Tibetan hermits go safely each year through this ordeal... by generating tumo, a subtle fire which... drives the energy, till it runs all over the body along the tiny channels of *tsas* [veins, arteries, and nerves]."

Thus in the most severe weather did Buddhist monks and hermits maintain comfort by a supernormal fire generated within themselves. Noted Mrs. David-Neel, "The word *tumo* signifies heat, warmth, but is not used in Tibetan language to express ordinary heat or warmth. It is a technical term of mystic terminology."

As an example of the *g Tum-mo* yoga technique and the person who could control tumo through meditation, Mrs. David-Neel described a test administered by Tibet's lamaseries to their students. "The neophytes sit on the ground, cross-legged and naked. Sheets are dipped in the icy water; each man wraps himself in one of them and must dry it on his body. As soon as the sheet has become dry," she wrote, "it is again dipped in the water and placed on the novice's body to be dried as before. The operation goes on in that way until daybreak. Then he who has dried the largest number of sheets is acknowledged the winner." Some students, she was told, dry as many as forty sheets in one frigid night.

And American students cringe over *their* final exams?

More recently, Dr. Erwin J. Dingle watched lamas sitting around a frozen lake at 16,000 feet, drying the wet woolen blankets wrapped around their bodies within minutes. Dingle reported in *Borderland of Eternity* that one lama's tumo proficiency had not only been ceaselessly thawing his icy blankets for two days but also had been melting the snow in a ten-foot circle around himself!

Impossible? Dr. Herbert Benson of Harvard Medical School proved otherwise in February 1985, when two filmmakers working with his scientific research group documented a similar thermal feat at a Kulu Valley monastery in Northern India. In the cold night air, the camera filmed twelve monks enacting the ritual David-Neel had described years earlier, using wet three-by-six-foot cotton sheets.

In his book *Your Maximum Mind*, Dr. Benson described what happened next as the monks began *g Tum-mo* meditation to increase their body heat: "Although most people would have begun to shiver violently when exposed to such cold wetness, these monks didn't react at all. Instead, they sat calmly, and within three to five minutes, the sheets wrapped around them began to *steam!* The room filled with water vapor so that the lenses of the cameras became fogged over and had to be wiped off constantly. Within thirty to forty minutes, the sheets draped around the monks were completely dry."

> The history of scientific advancement is full of scientists investigating phenomena that the establishment did not believe were there.
> —Margaret Mead

The dozen monks twice repeated this feat. I, too, have seen this remarkable (by Western standards) achievement. It is awesome to watch steam boiling off drenched sheets warmed only by the energy of the naked bodies beneath and know that human talent (aka tumo), and not trickery, is the force at work.

Mayne R. Coe, an American medical researcher with a Ph.D. in biochemistry, posited a way to generate tumo that he thought was consistent with present physiological ideas. Coe, whose research into the paranormal often consisted of experimentation on himself, wrapped himself in wet woolen blankets in freezing temperatures and learned to dry them through cold shocks to his naked skin. He wrote in "Discovering the Yogis' Secret" that this technique forced his liver to release extra glycogen into his bloodstream, where it oxidized and created additional heat.

Coe claimed a world record for drying a wet woolen blanket in 18°F weather *eight times* within two hours. While his technique worked for him, it is at variance with situations observed in Tibet. There tumo practitioners do not shiver to release glycogen, but sit in stoic meditation; they do not shroud themselves in thick insulating blankets (which Coe deemed important), but wrap themselves in frail cotton sheets or *nothing at all* and endure, unperturbed all night long, temperatures that drop to 0°F (-18°C).

So something other than Coe's mechanism can produce these feats of superthermogenesis.

Dr. Benson postulated that a relaxation response was involved in the monks' physiological display of nonshivering superheating. "This involves the ability of the body under some circumstances to burn or metabolize a type of fat called brown fat," he explained. Scientists once believed that only certain types of nonhuman mammals, especially those that hibernate, could burn brown fat.

But now it appears that human beings may also have the capacity to generate heat from this fat, said Dr. Benson. "We hypothesize that the monks may have learned to do so through the use of generally unknown powers of the mind."

Benson's conclusion based on his tumo study of Tibetan monks offers another possible mechanism for SHC: the rapid metabolizing (burning) of brown fat. Superfast oxidation would liberate a flash of heat coursing through the unsuspecting person carrying excess brown fat and bake the body from the inside out.

You will recall that many nineteenth-century physicians believed corpulence to be a requirement for SHC. Maybe the victims they catalogued, along with plump Mary Reeser in this century, had in their overweight bodies an elevated level of brown fat—a condition no autopsy thought to consider. Combining an overabundance of brown fat with a dramatic surge of tumo— perhaps the victims suffered high fevers that also escaped mention—and, instead of sheets dried by steaming shoulders, we find steam-pressed sheets beneath burned-out human beings. Like those of Mrs. A. B., the Irishwoman found in 1808 lying incinerated on her unscorched sheets.

One need not be a Tibetan ascetic to generate such abnormal warmth. A few years ago, a young Connecticut female suffering involuntary extreme heating contacted John W. White, a researcher of human consciousness phenomena. Her skin, White wrote in *Venture Inward* (1990), was "so hot that she took icy showers and drove around in a convertible with the top down in winter, while wearing only a halter and shorts. These measures barely gave her relief."

A Chinese boy living near Nanjing, Jiangsu, acquired the nickname "fire body" because beginning in 1971 when he was five and lasting for at least a decade, his body heat was so torrid he wouldn't wear clothing and played naked in the snow. Elsewhere in China around 1981, eighteen-month-old Wu Xiaoli also constantly refused clothing; townsfolk in Miangyang, called her "fire baby."

When I discussed SHC on KYW-TV's *People Are Talking* in 1984, a woman in the audience remarked that her husband sometimes became terribly hot in bed. Program host Dana Hilger quipped that the lady should consider herself lucky! True, so long as her mate's somnambulistic thermogenesis doesn't overheat. Her husband is not alone in this trait; in a personal note, my own sleeping body sometimes generates heat that's been likened "to a fiery furnace." I sleep blissfully unaware amid projected warmth said to be unbearable.

Tumo, perhaps? As you are learning, SHC often occurs when its victims are asleep in bed or after they've dozed off in a chair. Sleep—or at least a relaxed mental state—seems conducive to generating thermal anomalies...and, perhaps, super-sweltering nocturnal nightmares from which a few people never awake.

Not to be overlooked is *carbunculus*—the curse of menopausal women—which are those damnable hot flashes attributed to rapid hormone decline plus the erratic metabolic behavior of the hypothalamus. Sensations of dramatic heat have prompted women to say their upper bodies felt like a roaring inferno burning them from within.

INCENDIUM AMORIS: INTERNAL FIRE OF ADORATION OR HYSTERIA?

One thermal phenomenon more outrageous than these dramatic examples of tumo is even less acceptable to mainstream medicine, perhaps because it has occurred outside the observations of an operating room. Totally unacceptable to some, because it is a bit *too outrageous*.

An attribute common to many of the 108°F-plus cases cited above is heightened, even hysterical, emotion. Once again we identify a link between abnormal bodily thermogenesis and a person's mental state, first noticed with the number of inflamed alcoholics. This link can be extended to encompass the emotionalism of spiritual ecstasy.

Historically, a rise in body temperature often attends intensely emotional spiritual fervor. Great mystics and devout ascetics, embodied by a power they associated with divine presence, have described being imbued with a holy love that so energized their bodies that in the coldest winter weather they'd throw off their stiflingly warm clothing, fling open windows to gasp for frigid air, and pant as dogs do to cool their bodies.

Richard Rolle, the fourteenth-century English mystic who suffered this plight, called it "Fire of Love"—a blazing heat inside himself that he attributed to profound adoration of his God. Its Latin equivalent, *incendium amoris*, has since been adopted by the church for this peculiar effect of religious ecstasy first named by Rolle. By whatever name, the phenomenon is an enduring enigma; its cause as mysterious—and often uncomfortable—for those who experience it as for those who study (or debunk) it.

A centuries-old question that history has largely chosen to overlook is: How hot can *mystical* thermogenesis become? Let's look briefly at a few documented examples to see if an answer is lurking there.

Religious passion so affected St. Stanislaus Kostka (d. 1568), said his biographer Father Goldie, that "he was obliged to apply cloths dipped in cold water to his breast in order to temper the violence of the love he felt." One cold night in the Novitiate's garden the rector asked St. Stanislaus what he was doing strolling alone there. "I am burning, I am burning," Stanislaus replied straightforwardly.

Four hundred years later researchers would see this seemingly mythical tale replayed in Tibet's highlands. Whether the intense heat of St. Stanislaus's inner love could out-perform (out-dry) modern monks practicing tumo meditation, no one will ever know. Nor will anyone know if Stanislaus *literally* spoke of "burning" inside.

His contemporary, St. Philip Neri (1515–1595), shared this questionable blessing, but only after his thirty-ninth year. Thereafter, said his biographer Father Bacci, this priest walked about bare-chested during the coldest winter nights. So great was St. Philip's fire that when he touched Cardinal Crescenzi's hand, he said "it burned as if the saint was suffering from a raging fever." On at least one occasion his supernatural heat "scorched and blistered" his throat. St. Philip bore his thermal affliction with humor, laughing good-naturedly at his countrymen who shivered in the cold, which he, an open-shirted old man, literally could not do, due to his inner fire.

Coe's woolen-blanket-cold-shock theory doesn't make sense here.

During the life-long ecstasies of the Venerable Orsola Benincasa (1547–1618), this Theatine nun repeatedly endured abnormal thermogenesis. Her body became so hot that when she dipped her hands in water to cool them, the water boiled and steamed in a vessel which itself became so hot that it was hardly possible to hold it. Sometimes white smoke issued from her throat, a phenomenon witnessed by many onlookers and said to be beyond question.

(In 1822 an unnamed Italian male, about age twenty-six, suffered intermittent fever accompanied by burning heat in his throat, reported the *London Mirror* for October 6, 1838. Inconsequential news, it would seem; *except* that at the same time his "breath smoked.")

For water to boil, the temperature at sea level must be 212°F. Therefore, if the eyewitnesses have their testimony taken at face value, one deduction is inevitable: that thermogenesis inside this woman produced a skin temperature of *at least 212°F*—215 percent greater than normal. Was Orsola Benincasa, who many times proved immune to fire, at other times living on the verge of spontaneously combusting?

So bizarre does this sound that one is apt to dismiss both Benincasa and the possibility I suggest. Is this a singular example of *extreme* heat attributed to incendium amoris, and hence in its uniqueness deniable?

No.

The body of St. Catherine of Genoa (d. 1655) exhibited two physiological miracles: stigmata *and* incendium amoris. (In this sense her life was a precursor to Padre Pio's three hundred years later.) Outbreaks of "extraordinary heat phenomena" tormented St. Catherine, from whose throat often came white smoke "as if ascending from a brazier lighted within." Her thermogenesis thus appears to be much hotter than that of Padre Pio; furthermore, it could be transmitted externally. Once, she was given a bowl of ice water to cool her overly hot hands. Plunging in her hands, the water boiled and the shallow bowl heated to scalding. During her last two days of life this warm-hearted saint became hot-blooded, literally. As she bled, her red-hot blood both heated the vessels (including a silver cup) used to catch it and scalded her skin. When she died, this "great fire which consumed her" enabled her corpse to retain heat a full day as nuns warmed their hands over her stilled heart.

While St. Catherine's piety forgave many people, what cannot be forgiven is any dismissal out-of-hand for this evidence of superhyperthermia (itself a form of spontaneous human burning). Nor can its possible link to classic SHC be overlooked.

What amount of heat did the flesh of these saints generate? Common sense would demand that for ice-cold water to boil by immersing a hand in it, the hand's temperature would need to remain for a considerable time at no less than 212°F. This suggests St. Catherine and Orsola Benincasa were *veritable living furnaces* fueled by incendium amoris (or something else) within them!

Impossible?

The Dominican nun Maria Villani (1584–1670) also lived the impossible. Besides being credited with a profound knowledge of mystical theology, in the words of her biographer Fr. Francis Marchese, Maria had a reputation as "a furnace of love, a furnace thrice-heated." Literally. She had to quaff three and a half gallons of water daily to cool or to rehydrate her body "continually consumed by an almost insupportable flame of love." Amazed onlookers in Naples attested that as she drank water they would hear "a hissing sound like that of water falling on a sheet of red-hot iron."

Venerable Serafina di Deo, a seventeenth-century Carmelite nun of Capri, in life "was consumed with a living fire and that her blood was boiling." Physicians affirmed that Serafina's "blood-boilings" lasted up to three days at a time. After communion, her face would glow with "a red flame and her eyes sparkled with fire"—as did Philip Neri.

Franciscan nun Francesca dal Serrone of San Severino had, like Maria Villani, a wound in her side. From this wound issued, and from her mouth vomited, blood "so hot that it cracked an earthenware vessel used to receive it, and had to be caught in a metal bowl."

Extraordinary hyperthermia is not limited to Italian nuns and mystics.

Canon Martinon placed his cleric reputation on the line when, standing before French Dominican nun Agnes of Jesus, he declared himself a witness to water being "poured upon her breast to cool her in her transports of burning fervor, and that the water sizzled like water poured onto red-hot iron."

In 1798 Anna Catharine Emmerich, called the "Living Crucifix" in her hometown of Coesfeld, Germany, suddenly emitted from her twenty-four-year-old head "a strong but not unpleasant glow of warmth." For fourteen years her life apparently was unremarkable... until August 28, 1812, when, in ecstacy, her bosom spontaneously burned upon itself the image of a three-inch cross, after which her skin blistered and oozed a colorless fluid "of extraordinary heat."

B. Bhavan described in *Sufis, Mystics and Yogis of India* (1971) a Sufi holy man who displayed ecstatic superheat. Such was his ecstasy that prior to dining his hand-washing became a sterilizing phenomenon: as "the water would fall on his hands it would pass into vapor." As a consequence, "his hand is burnt and patches and swellings immediately appear on it."

Does the Venerable Mother Beatrice Mary of Jesus, abbess at the Spanish convent of Poor Clares in Granada, portend the ultimate outcome (or threat) of incendium amoris? Between November 3 and Christmas Day, 1664, she experienced "on four or five occasions... one of those interior conflagrations which were occasionally characteristic of her ecstatic state," testified reliable witnesses. Particularly noteworthy are their statements that during these ecstatic interior burnings *her body spewed ashes*!

QUANTIFYING SUPERHOT BODY HEAT

These examples of superhyperthermia are a problem of the deepest and most complex nature, whether we look to the supernatural for an explanation or seek to quantify some abnormal physical or psychic force of which the world has hitherto been largely ignorant.

Dr. Norman Shealy, director of the Pain Rehabilitation Center (LaCrosse, Wisconsin) and author of *Occult Medicine Can Save Your Life*, has demonstrated that through biofeedback training a person can consciously control body temperature up to 104°F.

Swami Rama of Rishikesh, India, participated in biofeedback research at the Menninger Foundation in 1970. He demonstrated precise control over his autonomic nervous system and brain, which included a 10°F temperature difference willfully induced between two points on his palm (above the radial and ulnar arteries). Dr. Elmer Green concluded that Swami Rama used dilation and constriction of the arteries to control his blood flow, hence the temperature of his palm. A dramatic achievement, but hardly capable of steaming soaking-wet cloth in a few minutes.

SuperHot Humans: Stoking Up the Body's Furnace 163

In 1981, Dr. Benson connected thermocouples to three Tibetan Buddhist monks, and showed that the most proficient tumo practitioner could increase by 15°F (8.3°C) the temperature in his toe.[1] Benson concluded the monks were able to increase the flow of blood through a particular part of the body. (If so, the feat is all the more amazing because the normal body response to cold is to constrict blood vessels and lower body temperature at the extremities.)

These findings are remarkable and barely within the understanding of Western medicine. Such voluntary elevations of body temperature, while physiologically significant, nevertheless fall *far short* of attaining the 130° to 171°F body temperatures that medicine documents, let alone producing internal boiling and burning of bodily fluids and flesh.

However, these subjects may have simply lacked the motivation (or need) to raise their body temperatures higher through the conscious application of will. Nor can it be ruled out that Himalayan yogis, acting out of self-preservation, *choose* not to demonstrate an ability to generate much hotter tumo. After all, why burn oneself up just to confound a scientist—especially since there is a high probability that science would disavow the spectacle of blood bubbling and a self-igniting human being anyway?

Alternatively, brain injuries or extreme stress could be the trigger that sets a human being ablaze. Anesthesia can launch a biochemical chain of events that causes malignant hyperthermia. The involuntary heating of incendium amoris is often associated with spasms and convulsions. Much as injuries to parts of the brain create amnesia or photographic memory or localized motor paralysis, I sense that extreme emotional or injurious stress to the hypothalamus system in the brain can produce violent and puzzling thermal aberrations.

To determine the temperature at which water sizzles like being "poured onto red-hot iron," I conducted a simple experiment in the kitchen. A stove burner cooler than 140°F failed to sizzle water droplets, indicating that the skin temperatures of Maria Villani, St. Catherine, and Orsola Benincasa (among others[2]) must have been at least 140 to 150°F.

Since skin is a poor conductor of heat—U.S. Air Force personnel survived exposure to 500°F temperatures during experiments conducted in 1960—the interior of Maria's body must have been even hotter. *How* hot is impossible to estimate, of course. We do have the statement from Cardinal Crescenzi that he burned his hand when touching the skin of St. Philip Neri; plus numerous statements that blood coursing through a living person has been seen to boil—not radiantly warm but bubbling hot, hot enough to make metal almost untouchable.

These testimonies, however uncomfortable to the percipients and disquieting to medicine, should be taken quite literally: on the inside a human body can heat itself to the boiling point, perhaps *beyond*. I will have more to say about this later.

Physical phenomena are not restricted to so-called holy men. Ordinary people at times spontaneously achieve the same wondrous abilities that ascetic yogis and enraptured saints spend decades seeking to perfect. Spiritual revelation or/and miraculous physical phenomena can descend on the unsuspecting everyday man and woman, as the lives of St. Paul and Philip Neri testify. Excessive thermogenesis is one of many "wild talents" that, arriving unannounced and probably unwelcomed, can lead to amazement, consternation, sometimes discomfort, even death. Death by SHC, even?

Consider William Eckle of Rochester, New York, who died on July 10, 1912, the victim of what *The Daily Record* two days later called "an unusual condition." Journalistic understatement. Surgeons said that Eckle had burned to death internally due to prolonged body temperature of at least 108°F, resulting in his internal organs being "all withered and dried."

The surgeons' conclusion: *Eckle's own body burned him up alive, from the inside out*.

And doesn't that fit within the definition of SHC?

Hyperthermia has plagued mankind for thousands of years, inexplicably heating the agonized bodies of countless people. People *do* cause themselves serious, even fatal harm when biological thermostats go haywire and fry their bodies... perhaps burning a few thousand to ashes. It and tumo, incendium amoris, and SHC are probably related in some way through interaction of the sufferer's consciousness and biological processes. The precise mechanism awaits discovery.

Meanwhile, a few years ago a production staffer for *The David Letterman Show* told me (in all seriousness) that the show would happily invite me to talk about SHC, but only if I brought along someone to demonstrate making an ash of him/herself on camera. As the appearance would naturally be brief (not to mention probably fatal) for the guest who must ignite on cue, I turned down the opportunity. True story.

11

Kundalini, Chi, and Chakras: The Serpent-Fire Within

> We feel that holding back data is evasive and, in the long run, destructive to the spirit of science. If today's models prove inadequate, it is because tomorrow's will be better. Not only the researcher and the writer, but also the reader must here make a special effort to be objective.
>
> —Lee S. Sannella, M.D.
> *Kundalini—Psychosis or Transcendence?* (1976)

The preceding chapter took us beyond standard medical knowledge for body-temperature generation, just as SHC takes us beyond the powers granted to the whole human body by current medical models. I see no point in stopping the study of SHC and anomalous human thermogenesis at tumo and incendium amoris, however. So *many* more mysteries await. I will not hold back data.

In 1989, Teresa Miskell placed her hands on a friend to channel energy intended to heal an ailment. The power of her touch amazed both persons when, after several minutes, she moved her hands and discovered two perfect outlines of her palms and fingers on her friend's dress—as though a hot iron in the shape of her hands had scorched the fabric. Both ladies insisted these imprints were not due to discoloration from perspiration, but resulted from actual singeing of the fabric.

In 1963, a napping middle-aged professor awoke to confront the mystery of a three-inch blister spontaneously appearing on his thigh where his hand rested. Seeking the reason behind this self-induced partial SHC that his rational mind must have told him shouldn't happen, he began to study metaphysics and Zen. During a meditation four years later his body again blazed, with a difference. Dr. Lee S. Sannella told in his book, *Kundalini—Psychosis or Transcendence?* (1976), how the professor "became engulfed by a bright golden light that lasted several minutes. A few weeks later this occurred again."

Body aglow, preparatory to body aflame? Ignis lambens, as prelude to ignis incendium?

One day in 1968, a forty-four-year-old librarian rested her hands on a wooden table and entered a hypnagogic state in the library's relaxing environment. What transpired next could not be referenced in any textbook in her library. Said Dr. Sannella: "She awoke to find charred marks that deeply marred the table and corresponded to her hand prints." Her hands had become like red-hot pokers searing themselves into the tabletop, yet her flesh revealed *not the least injury!* We'd be leery about shaking hands with this woman—

Inside these three professional people, a paradigm-shattering force was at work. Limited externalized spontaneous human combustion; a body-engulfing halo of golden luminescence, termed *ignis lambens* by the priesthood; and palm-projected blasts of fabric-scorching and table-charring energy. And each time, the fire is preceded by a state of altered consciousness.

Thus is the range that spontaneous combustion by humans can produce. Indeed, hot human hands have burned into hundreds of items, many on display at the Sacro Cuore del Suffragio museum next to the Vatican.

Questions are raised, each demanding a special effort to be objective.

The most obvious one is *How?* Normal human thermogenesis surely proves inadequate to answer this question. Psychological and psychosomatic venues to explain away perceived internal heating as illusory fail, too, what with fabrics being scorched and wood charred. A better model is needed. Dr. Sannella believes the "better model" for his examples of limited SHC is Kundalini. I agree.

THE SERPENT FIRE WITHIN

If you know about Kundalini, then you already probably agree with Dr. Sannella. If Kundalini is a term unfamiliar to you, a brief foray into esoteric traditions that predate Western medicine by thousands of years will help you understand what Kundalini is, why it has been linked to these three cases of SHC, and how it can provide a general solution to the continuing puzzles of SHC.

As I was taught in elementary school science, our bodies are made up of atoms; or, if you're in the youngest generation currently being schooled, of quarks and leptons. In the end we are, however minutely subdivided, a special pattern of dynamic energy. Esoteric Hindu, Vedic, and Tantric texts say the body's subtle energy system is composed of *prana*. Prana is the biological quasi-electromagnetic fuel that energizes every tissue and cell. This vital, fundamental energy that is the real "you" has many names: prana, chi, ki, qi, the Holy Spirit, bioplasma... and a score more, depending on the culture describing it.

By whatever name, prana surrounds and permeates the gross tissues of the body, says Gopi Krishna in *The Dawn of a New Science*. Prana is "a living electricity, acting intelligently and purposefully, controls the activity of every molecule of living matter. It carries the life principle from one place to the other," he explains, and "energizes, overhauls and purifies the neurons and maintains the life-giving subtle area [soul] of the body much in the same way as the blood plasma maintains the grosser part." This energy is crucial to the homeostasis of the body. Disease, even death, results if a grossly imbalanced interplay of mind, body, and environment blocks prana's natural, unrestricted flow.

Kundalini, a Sanskrit word meaning "curled up," is one of the mechanisms for concentrating the movement of ultrapotent prana. Tradition says that Kundalini resides coiled in an area called *kanda* at the base of the spine. It awakens in one's life only infrequently, if at all; uncoiling naturally, it usually stirs gently, but might upsurge toward the brain like positive charges in a lightning bolt's return stroke.

Kundalini is the most powerful mechanism in the body because it is *both* energy and consciousness, according to personal communication from Gene Kieffer, director of the Kundalini Research Foundation. It involves the entire nervous system and brain. Hence, Kundalini is in every person and can affect every aspect of a person's being.

Prana and Kundalini are distinct from the tangible anatomy and bioelectricity that practitioners of Western medicine are accustomed to treating. "The average neurophysiologist still thinks in terms of molecules," says Kieffer, "which is hardly sufficient to develop an understanding of Kundalini. That is why the theoretical physicist is so much better equipped to comprehend the phenomenon. Kundalini is the 'channel' that connects the unmanifested universe, i.e., the forces contained within the atom, and the solid world of atoms, molecules, etc.

"People say that we get our energy from the air we breathe and the food we eat, and this is all true. But when Kundalini is awakened, there is a release of energy far in excess of what can be extracted from air and food. From where does this energy come? It must come from the inexhaustible fount within the atom. Science tells us that there is more energy contained in one cubic centimeter than is contained in 3 million pounds of TNT (for example). That cubic centimeter of 'empty space' exists within the human frame, of course."

Kieffer's statement about Kundalini meshes nicely with the concept underlying the subatomic pyrotron theory for Mary Reeser, whose body transformed its "empty space" into a few pounds of ashes.

Once Kundalini uncoils from the kanda it surges up the spine through a central *nadi* (nerve channel) called the *sushumna*, one of 720,000,000 nadis in a human body, says Swami Muktananda in *Kundalini: The Secret of Life* (1979). Two subsidiary channels, the *ida* and *pingala*, criss-cross the sushumna and help guide Kundalini to the top of the head and into the brain. The cold channel of ida (yin energy), traveling along the left side of the spinal cord, is called the "lunar nerve" by yogis; the hot channel of pingala (yang energy) is the "solar nerve" that travels along the spinal cord's right side. The ida and pingala together appear as two serpents ascending entwined around the energized sushumna nerve channel, poised to strike at the brain.

This powerful "fiery serpent" life-current is a vital, energized *caduceus*—interestingly the symbol Western medicine has adopted to represent itself as healer! (The serpent-caduceus is also a very old symbol: depictions can be found as early as circa 2000 B.C. on the famous libation vase of King Gudea of Lagash.)

Along its journey up the spine, Kundalini flows along specific pathways called meridians to strike the body's seven major power centers called *chakras*.[1] Generally, the chakras are said to be situated at the base of the spine; the genital area; behind the navel at the base of the ribs; near the heart at the center of the chest; the throat; between and behind the eyebrows at the pineal gland; and near the pituitary in the head.

These chakras—nerve plexuses—radiate colorful energies called auras and have been described by Eastern spiritual leaders for thousands of years and currently by countless clairvoyants. Recently they have been confirmed by quantitative electronic experiments conducted by Valerie V. Hunt, Ed.D., professor emeritus at the University of California (Los Angeles). She told me her research, reported as "Energy Field Studies—Electronic Aura Study" in *Project Report: A Study of Structural Integration from Neuromuscular, Energy Field, and Emotional Approaches*, concluded that "the relationship between emotional states and auric color should be viewed as facts and not subjective judgments."

Ideally, Kundalini energizes each chakra into accelerated activity that benefits the person. This "opening" of the chakra allows the next higher chakra to be activated in turn, until the whole body is aglow—sometimes literally, it's said—with heightened energy.

Two auric chakras and their emotional attributes are of special interest here. Kundalini's initial rise first encounters the lowest (*mūlādhāra*) chakra, associated with self-preservation; the drive for personal survival. The third (*manipūra*) chakra—the solar plexus—is located at the base of the rib cage and is associated with vitality and the drive for power. The solar plexus chakra, also associated with feelings and emotions, is regarded as one of the body's most powerful energy centers.

Kundalini can also travel beyond the chakras through meridians that connect the body's organs to acupuncture points on the surface of the body (the skin itself being the body's largest organ). The chakras are analogous to the central nervous system of the spinal cord; the meridians to the peripheral nervous system which extends (along with the autonomous nerves) beyond the spinal cord. In other words the entire body is wired to channel the Kundalini-accelerated flow of prana, just as it is wired with nerves to transmit sensory information.

The successful completion of Kundalini's natural unimpeded cycle of movement through all the body's chakras is enlightenment—self-realization.

Kundalini's activation can also be encouraged by specific physical and mental techniques. The *Mundaka Upanishad* and *Mahakala Nidhi*, among other Eastern texts, explain how a person can raise and direct this subtle energy through specific channels that pervade one's body to achieve transcendent consciousness. This manipulation is not an undertaking for the faint-hearted dabbler, though. Kundalini has a downside, the ancient texts warn. A warning based on millennia of experience and wisdom that says "Danger!"

A body and psyche unprepared for the power of awakened Kundalini can exhibit various spontaneous *siddhi* (supernormal abilities), which might lead to physical or psychological damage, temporary or permanent; minimal or severe. Involuntarily contortions, hot flashes, and a variety of psychic phenomena (including psychokinesis) are among the side-effects that can plague the body and torment the mind during Kundalini's rush, says W. Thomas Wolfe, a Kundalini scholar who described his own dramatic experiences of the serpent-fire in his book, *And the Sun Is Up: Kundalini Rises in the West* (1978).

Kieffer warns that if this heightened flow of new bioenergy is impeded, madness and worse is probable. "In the case of insanity and psychosis," he says in *Kundalini for the New Age* (1988), "one almost invariably finds the people affected complaining of sensation, burning, terrible light, fire..."

Wolfe recounts one outcome: "In some very violent cases, men would be spirited completely out of this world with an accompanying clap of thunder and a wisp of quickly disappearing smoke."

Hmm. Could we be onto something here?

HOT TIMES, KUNDALINI, AND CHI

The power of Kundalini moving through the body can be likened to electrical current in a lightbulb and water in a hose. If the bulb's filament is thick and offers great resistance to a weak current, the movement of energy is impeded and the bulb glows dimly at best; if the tungsten is fine, and the current moderate, the lamp blazes with light; if frail, and the current's amperage too great, the filament melts and the bulb is dead. Water under pressure produces violent whipping of a narrow, frail garden hose yet barely flexes a larger, stronger firehose.

Power that is properly balanced accomplishes constructive results; too much power moving through a channel too constricted, creates havoc. As Dr. Sannella completes the analogy, "so also does the flow of Kundalini through obstructed channels within the body or mind cause motions of those areas until the obstructions have been washed out and the channels widened."

Until that happens, the power channeled by Kundalini's twin serpents can produce a variety of physical sensations and phenomena. Most of these are beyond the scope of *Ablaze!* to explore, fascinating though they are. One Kundalini-related *siddhi* that isn't, as you might guess, is the supernormal heating of the body by means seemingly quite independent of the hypothalamus.

"Body temperatures that soar high above normal are generally attributed to fever, but in those cases the temperature rarely exceeds 104°F," replied Gene Kieffer to my questions about Kundalini and extreme human hyperthermia. "This leaves—in my opinion—only one explanation for temperatures that are far in excess of 104°F, and that explanation is Kundalini. If there is some other explanation offered by science I am not aware of it." As the director of the Kundalini Research Foundation, Kieffer has knowledge that merits earnest consideration.

Evidence for heating by Kundalini "fire" is richly varied and wide-spread. The !Kung people of the Kalahari Desert strive to arouse in themselves *n/um*, an energy apparently comparable to Kundalini. R. Katz, who studied this African tribe, wrote in the *Journal of Transpersonal Psychology* (November 2, 1973) that "*N/um* makes you tremble; it's hot."

Ajit Mookerjee, in his *Kundalini—the Arousal of the Inner Energy* (1989), calls Kundalini the body's "most powerful thermal current." In India, Swami Nargayanananda writes in *The Primal Power of Man* (1960) of Kundalini's ascent: "There is a burning up the back and over the whole body." The Indian saint Ramakrishna had spoken a century earlier about the "Spiritual Current"—his term for Kundalini, as quoted in Campbell's *The Mythic Image* (1974)—and the sensation it produces: "The place where it rests feels like fire..."

Oriental philosophy and martial disciplines speak of *qi*, the primal matrix of life energy. Chinese medical tradition polarizes this vital energy into *yang* (positive/masculine) and *yin* (negative/feminine), stating emphatically that it and its eight hundred channels are quite distinct from the body's electrical nature as identified by Western science. Furthermore, the Chinese say one's vital energy is affected by weather, solar and lunar phases, and cosmic radiation; plus internal moods, thoughts, physical and mental traumas, and acupuncture needles. They wisely view man holistically; interconnected to *all* aspects of his inner and outer environment.

"Qi is stored right below the navel, in the solar plexus area," David Cannon, M.D., told me after his trip to China in 1988 to study the marvels of Oriental medicine. He learned that qi channels a "morphine-like substance" released by the brain in conjunction with acupuncture manipulation of meridians. He said its potency could vary two-hundred-fold depending on external stimuli, and he watched Caesarian deliveries performed on mothers "fully alert and aware of what's happening," anesthetized by nothing more than two pinlike needles at each end of the incision!

Qi can also be amplified by a biological process called *qi gong*. Qi gong mirrors the Hindu attributes of Kundalini, one learns in Paul Dong's fascinating book *The Four Major Mysteries of Mainland China* (1984). One of these mysteries is exceptional human function (EHF). Studies conducted at Beijing's Institute of High Energy Physics and other academies in China indicate that qi gong induces EHF, producing the same range of anomalous biological phenomena that is associated with Kundalini.

Evidence of *thermal EHF* includes thermographs taken of the hand of Beijing qi gong master Dr. Zhao Guang, which document his ability to produce "marked increase in heat" in his palm over an eighteen-minute interval using only the willed movement of qi in himself. Additional support for qi's biological EHF heating comes from the respected scientist Stanley Krippner, Ph.D. Following his own trip to China to study EHF, Krippner stated in an afterword to Dong's book that qi gong does indeed produce "differences when body heat is measured and when electrical qualities of acupuncture points are measured."

"The vital element is hot," writes C. Luk in his study of Chinese Taoist tradition, *The Secrets of Chinese Meditation* (1972). One's qi when accelerated by qi gong, he continued, "may even become bright and perceptible to the meditator. In exceptional meditators it causes illumination of a dark room perceptible to others."

One recalls the glowing *ignis lambens* that the meditating professor saw around his body during meditation.

THE FIERY SERPENT RISES

Kundalini (or qi gong) is not unknown in the West, though the few cultural references to it are often veiled. Here the preponderance of evidence—what little has been recorded—is not traditional but personal, anecdotal.

In America, for example, Dr. Sannella's medical practice dealt with several thermal anomaly cases that he ascribed to Kundalini and meditation. In one instance, a healthy forty-year-old writer registered a 101°F temperature, dropped in a few minutes to 99°F, though his hand then quickly became 104°F. And in the case of a forty-one-year-old psychologist who worried that her excessive body heat—which others felt when they touched her—could be injurious to herself, Sannella concluded that her hyperthermia was Kundalini-induced "due to residual blocks and unresolved conflicts which were locked into her body."

The most instructive description of Kundalini I have found happening to a Westerner is offered by W. Thomas Wolfe. In the context of SHC, it is particularly fascinating.

Imagine living day-to-day oblivious to a quietly ticking timebomb of potent energy lying dormant in your body. That's how Wolfe lived, until he got the (literal) shock of his life in early 1975. He described his introduction to the serpent-fire in *And the Sun Is Up:*

> And then one day it happens. It is as though the body has accumulated so much energy that it can hold no more without bursting at the seams. And now, while the subject is resting or meditating, the rising of the accumulated energy takes place. In most cases, the surge is so powerful that the subject loses consciousness of his body while maintaining some sort of internal consciousness. During this period he experiences himself as a disembodied being of pure energy. Or perhaps he experiences profound bliss, or sees a magnificent, overwhelming vision of God—it is different for each subject.

Or perhaps, for any number of different reasons, the subject embarks on a one-way siddhi of supernormal thermogenesis that begins with a slight fever, then escalates to transcendental heating as red blotches form on the skin, culminating in the SHC of third-degree burns (or worse).

That did not happen to Wolfe, obviously, even though he said "sensations of heat... had ignited into a full-blown conflagration that traveled slowly but surely over my entire head, producing many unusual effects." How *close* he came to having that figurative imagery become literal is not known.

The many "unusual effects" he attributed to Kundalini within himself are particularly noteworthy in connection with SHC. First, his stomach "got very hot" and then his lower spine "*got very hot,*" he noted in his diary. This suggests that his Kundalini was initially blocked, causing overheating as it moved upward into his solar plexus chakra.

(Noteworthy is that in many classic human fireball episodes, the energy seems to originate in the victim's lower abdomen—site of the solar plexus chakra and referred to esoterically as the body's pranic furnace—and then radiates outwardly from there.)

After his Kundalini broke through and moved into his upper torso, he wrote: "*I noticed a minor stiffening of all torso muscles into a locked position, as though I was being constrained in a force field.*" This stiffening of the torso and locking of muscles is said to be rather common in mystics who excite their Kundalini. Earlyne Chaney described in *Remembering: The Autobiography of a Mystic* how her entire body became immobile as her muscular system rigidly locked during her own Kundalini movement.

This phenomenon can, in context, explain why SHC victims often seem to make no effort to escape the fire enveloping them. If Kundalini rockets upward faster than 300 feet per second (205 MPH), it would exceed the nerves' ability to transmit pain signals to the brain. The victim would *literally never feel* the fiery serpent moving up and outward from his or her solar plexus or rushing up toward his or her heart and head. What if one's Kundalini moved less rapidly? A frequent side-effect of Kundalini is a desensitizing or blocking of the body's transmission of neurological impulses, thus again offering an explanation for why a victim of SHC doesn't react to the *expected* pain of burning. And if intense pain transmission *does* register as a consequence of intense hyperthermia, the immobilized neuromuscular system precludes any escape. The victim *simply can't move* even if he or she tried.

Wolfe's diary of his Kundalini experiences addresses another of the serpent fire's unusual side-effects: "*Soon I noticed a very strong fragrance within the room, like a flower or a sweet perfume. A brief period of headiness accompanied the aroma. Later, at bed time, I noticed Kundalini tightening...*"

A subjective perception? A deranged mind's schizophrenic illusion? Wolfe insisted that his personal, redolent fragrance "was emanating from within, was quite real."

Absence of the noxious odor characteristic of burned flesh, sometimes supplanted with a "sweet, perfume smell" as Don Gosnell found in Dr. Bentley's apartment, is a frequent hallmark of classic SHC...and one of its most puzzling. It has perplexed investigators. It has allowed critics of SHC to smirk at this imponderable though often reported phenomenon, thus enabling them to debunk it along with the burned-out bodies they would like never to consider. The critics will have to show why Wolfe's fragrance born of internal fire is not real, and why Hindu texts full of descriptions of aromatic smells emanating from Kundalini-empowered persons should be likewise discounted.

Another notation from Wolfe's diary about what he called his "baptism by fire" is doubly pertinent.

One evening about six months after he began to experience Kundalini, Wolfe sat down after dinner in front of his television to rest. His mind was weary, his body exhausted. Soon he sensed the Kundalini rising within him, which he found to be not unusual after dining. His diary entry follows: "*I was almost asleep when I was struck—and struck is the right word—by a $1/2$ second surge of extremely heavy current on the top of my head in the crown area. It was like someone had hit me with a hammer. The blow was extremely forceful, shocking me into wakefulness. I couldn't help but think that a fuse had been blown somewhere. I wondered what part of me had been burned away.*"

First, we find a link between Kundalini's thermal aspect and relaxation. Is it not significant that most of the SHC episodes so far examined occurred when the victim was relaxed, asleep, or in a drunken stupor? Dr. Sannella reinforces this connection when he concluded that Kundalini's symptoms (including heat) "usually occur during meditations or times of rest, and cease when the process is completed."

The author knows a few friends who experience, occasionally, waking up feeling quite comfortable—until they *move* to get off their beds or sofas. With their stirring comes a sudden, instantaneous heat; frightfully hot sometimes, they say, and usually short-lived. But very unnerving.

Does the production of tumo-like super-fevers, isolated blistering of skin and limbs or, rarely, total SHC of the resting individual sometimes attend Kundalini's completion? Wolfe understandably admitted that his feeling of being "burned away" worried him, though I doubt he realized that a potential for SHC could have left him unable to continue his diary entries.

Second, we face the issue of Kundalini's intensity of activation. Wolfe might have been far more worried had he known what the director of the Kundalini Research Foundation told me. "Almost always there is considerable heat associated with Kundalini phenomena," revealed Gene Kieffer, "especially if the arousal is spontaneous and forceful. This heat can become so intense as to cause death..."

Fortunately Wolfe's encounter with Kundalini, though spontaneous, was neither forced nor intense. It made him think and feel that a part of himself was being burned away, but beyond changing his belief system forever it left no lasting physiological injury. But can it be proven that others may *not* have been so fortunate when sitting or lying down, relaxing or dozing just prior to a forceful, spontaneously catastrophic baptism by inner fire?

GOPI KRISHNA AND HIS FIRE OF KUNDALINI

The presence of an invisible, psychic anatomy within the human body was long ago identified by the healing professions in Egypt, China, India, and elsewhere, but has been vigorously rejected by the current allopathic philosophy of Western medicine. Belatedly (but fortunately) some Occidental physicians have begun to heed the teachings of their ancient predecessors, and now give credence to the existence and function of a subtle energy-body within (and indeed larger than) the physical body they are trained to treat.[2] Drs. Deepak Chopra, Larry Dossey, Richard Moss, Bernie Siegal, and Carl Simonton, among others, have challenged the tenets of Western medicine with research that confirms the power of thought to restructure one's ailing, out-of-balance body back into healthy energy.

Pandit Gopi Krishna was not a physician, but he captivated the interest of many men of science when he spoke about healing the body, the mind, and the consciousness of humanity. Born in northern India to a poor family, Gopi Krishna spent most of his life in Srinagar, Kashmir. Despite his humble background, he earned the honorary title *pandit* (meaning "learned man") and traveled the world as a—some have said *the*—leading spokesperson for the evolutionary potential within mankind. His rise from obscurity began when he experienced at an early age, and for many years suffered through, a strange and powerful force rising inside his body. It was, he said, prana made autonomically ultrapotent by Kundalini.

In his book *Kundalini: The Evolutionary Energy in Man* (1967), Gopi Krishna described his battles with Kundalini that made him feel that he was on the brink of burning up within. It was as though "a tongue or flame...a jet of molten copper...a scorching blast had raced through every pore of my body," he wrote. He remembered a guru's admonition that if the awakening occurred on the right side of the spine, then "the unfortunate man is literally burned to death due to excessive internal heat, which cannot be controlled by any external means."

Gopi Krishna was horrified. "What had happened to me all of a sudden?" he asked himself. "What devilish power of the underworld held me in its relentless grasp? Was I doomed to die in this dreadful way, leaving a corpse with blackened face and limbs to make people wonder what unheard-of-horror had overtaken me as a punishment for crimes committed in a previous birth?"

His inner torment was noticed by others, too. "Gopi Krishna's children told me that when the heat was intense, their father's cheeks 'glowed like charcoal.' Now, they were not speaking metaphorically," Gene Kieffer, a confidant of the

Krishna family, wrote to me. "Also, they said that for a period of time, Gopi Krishna was consuming more than 200 lbs. of sugarcane a day; that he was eating as much food in one day as 30 working men would consume in one month. We know that assimilating such prodigious amounts of food requires tremendous heat."

These visible and metabolic displays of awesome internal heat occurred when the ida and pingala channels regulating the Kundalini up the sushumna of his spinal column became imbalanced, Gopi Krishna realized. Therein lay his salvation, too. He remembered that cooling ida energy could counterbalance the fire of right-sided pingala. Facing what seemed to be imminent fatal hyperthermia, he focused his consciousness on cooling the area of his neuropranic spinal system where imbalance in the twin serpents had dammed Kundalini's natural flow at a closed (imbalanced) chakra. "In that extraordinary extended, agonized, and exhausted state of consciousness," he wrote, "I distinctly felt the location of the [ida] nerve and strained hard mentally to divert its flow into the central channel. Then, as if waiting for the destined moment, a miracle happened."

Heat, inner fire, and the possibility of sudden death! Another hint at SHC. The miracle for Gopi Krishna was that, by an act of will, he triumphed over his three-hour-long tempestuous inner trial by fire. He had survived.

I wondered if Kundalini could have defeated him, culminating in classic SHC had he not been able to defuse the hotbox of blocked Kundalini. Gopi Krishna knows! I thought.

In 1979, I had the chance to ask him. He reiterated that the "living electricity" of Kundalini is the fundamental bioenergy of life and re-emphasized the extreme danger it poses—"equivalent to letting a child play with Three Mile Island's nuclear reactor control panel." (It was the perfect, poignant metaphor, since the near-meltdown of TMI's reactor had occurred less than two months earlier near my home.)

I then posed this question: "Could you have become a victim of SHC had you not consciously been able to remove the blockaged Kundalini within you?"

He looked into my eyes as if to probe why someone would ask a question so uncommon. Then he smiled knowingly. "Yes," he replied simply. "I escaped death perhaps dozens of times, only a hair's breath away. Yes, I would *very easily* have become an example in your study."

Kundalini and SHC became inexorably linked by the pandit who wrestled for years with the inner fire-heat that attended Kundalini's movement through his body. Gopi Krishna had triumphed. It would seem not everyone does.

WARNING: THE KINDLING POWER OF KUNDALINI

Testimony and evidence attest that both tumo and Kundalini can generate inordinate heat in a person. Are tumo and Kundalini synonymous, then? I think not.

I think tumo's heat is induced by consciously accelerating the flow of prana (chi) in parts of or throughout one's body at the cellular level, whereas Kundalini is the discharge along the cerebrospinal column of a bioenergy having much greater potentiality. Tumo and Kundalini can both be activated by the body spontaneously, through specific body postures and breathing techniques, or by the will of heightened consciousness.

Tumo can be a byproduct of Kundalini's potency. Kundalini, however, is almost certainly not a byproduct of tumo. Kundalini is more specific, more powerful, more visually and spiritually illuminating than tumo. Consequently, for all these reasons, Kundalini can be more hazardous than tumo.

All responsible techniques to induce Kundalini recognize this and carry a warning: to coerce Kundalini is *dangerous*! Its awakening should only be attempted under the guidance of a knowledgeable teacher. "Deliberate practice of...forcing the Kundalini," says Dr. Sannella, "may cause premature and imbalanced [sic] release of titanic inner forces."

Titanic heat can be one result, Dr. Sannella affirms. So much heat, states Gopi Krishna, that a *sadhaka* (a spiritual aspirant practicing yoga) must be immersed for hours in water to absorb the Kundalini-fueled heat. As you know, Gopi Krishna knows! He had years of experience dealing with the fire that he believed could have caused him to self-ignite.

Unsynchronized activation of Kundalini's ida and pingala duality is one risk best to avoid. Gopi Krishna spoke about this to the All India Institute of Medical Sciences: "In the case of a morbid awakening the two movements do not start together, or there is imperfect coordination between them.... In such an eventuality the consequences can be terrible.... An impure, toxic current now flows through it [the body] creating a chaotic condition in the whole province of thought. The irritation caused in the brain cells by the contaminated pranic radiation is soon translated in the anarchy in thought and behavior of the afflicted individual."

An anarchy that unleashes within oneself potential storms of fearsome energy, including aberrant thermogenesis.

Imagine prana as a fuse, sheathed in a casing of ida and pingala, connected through small incendiary charges arrayed up your body to the seventh chakra blockbuster in the brain. Kundalini is the spark that ignites the fuse that detonates a series of chakras en route to the blindingly bright fireworks of a final explosion of blissful consciousness in one's head—*if* all goes well. But if the fuse's sheathing is frayed, the secondary charges faulty or defective or improperly wired, the whole operation might fail in one catastrophic fireball. In SHC.

The forecast for Kundalini rising, then, could read like this: Alert! Human heat wave imminent, followed by possible firestorm.

PETER VESEY: A WARNING UNHEEDED?

This forecast for Kundalini may apply to Peter Vesey, a contributing editor for *American Astrology* in the 1930s.

Vesey's private life was secretive; his demeanor often strange; his interests as odd as his astrological fiction was popular with the magazine's readers. Rumor said he was obsessed with the occult,* the cabalistic rituals, and the lives of medieval magicians. Rumor also said he researched spontaneous human combustion.

When his manuscripts abruptly stopped coming to the magazine, no one apparently wondered why until a letter arrived from a reader who lived in Vesey's part of the country. Attached newspaper clippings announced that Peter Vesey was dead. Bizarrely dead.

The newsclips stated that, one morning, Vesey had called his wife and son into their living room. He wanted to be alone. He asked his family to go outdoors and not return until one hour elapsed. Mrs. Vesey and her son dutifully walked outside. After the allotted time, they returned indoors.

On the floor of the living room lay the remains of Peter Vesey. Charred remains. A body burned crisp, a blackened cinder amid unscorched furnishings. At the far end of the room a modest fire in the fireplace warmed the room, but nothing between it and Vesey was burned. Authorities could not determine what happened to the widow Vesey's husband, said the newsclips.

A fictional drama? Or a real-life disaster that reads like fiction? Vesey was not the kind of man who confided his aspirations and inner feelings to anyone. Perhaps a despondent Vesey intended to commit suicide that morning and succeeded. Or perhaps that morning Vesey wanted to practice an obscure rite he had discovered in his studies, one that forecast an expressway into the unknown. And he found it: a secret on-ramp that accidentally (or intentionally) activated Kundalini. Forcibly unleashed, the serpent-fire raced through

---

* Joseph Cater (1982), in his effort to synthesize all physical and occult phenomena under one unifying principle, writes of SHC victims: "In most casess, the person was undoubtedly the target of a practitioner of so-called, black magic." Whether this might apply to Vesey—and to Grace Pett who, researcher Peter Christie (1981) discovered, was said by contemporaries to be targeted by neighbors' witchcraft prior to her fatally odd burn-up—I see no support for so general an assertion.

Kundalini, Chi, and Chakras: The Serpent-Fire Within 179

his body like a gale-driven firestorm. Vesey was unprepared to channel its gigantean power. Its journey was swift; its destructiveness complete. Vesey never knew what hit him from within.

Details needed to confirm this speculation are unavailable. An account of Vesey's death appeared in the tenth issue of Tiffany Thayer's *Fortean Society Magazine* (later named *Doubt*), and Gaddis attempted to corroborate the strange story (which he got fourth-hand) for his book *Mysterious Fires and Lights* (1967). He failed. I sought confirmation of it through various American astrological groups. I failed too.

I can therefore but agree with Gaddis's conclusion about Peter Vesey: "What was it he wanted to do—alone—that added his name to the long list of victims of mysterious combustion? I would like to know more—much more—about Peter Vesey." I would too.

Kundalini is a topic that fires the imagination... and possibly the human body. It offers a key—perhaps *the* key—to the evolutionary progress (or lack thereof) of mankind's consciousness, a topic largely beyond the scope of this book[3] but interrelated with SHC nonetheless, as I will discuss later.

I have examined the thermal and combustible properties that esoteric tradition has long ascribed to Kundalini, and found substantiation by modern empirical and experiential evidence.

As misdirected prana courses down the meridian of a resting librarian's arm and exits through her palms... as overabundance of Holy Spirit envelops a meditating professor in visible light, where once it blistered an acupuncture point on his thigh... as dammed chi crashes through closed chakras to unleash a cascade of sizzling power that incinerates a dozen sleeping women... as an impaired pingala assaults a soldier in a hayloft and spontaneously burns away the dross of his grossly alcohol-saturated tissue but not the straw beneath him...

12

Driving the Fire: Human Auto-Cremation

> But the true mark of a humane society must be what it does about *prevention* of accident injuries, not the cleaning up of them afterwards.
>
> —Ralph Nader,
> *Unsafe at Any Speed* (1965)

The 1990 Academy Award winning film "Driving Miss Daisy" never addressed a hazard to the motoring public that, if known about, would send chills through insurance company actuaries, among others: SHC behind the wheel.

Auto = vernacular for automobile; *auto-* = prefix meaning 'self'; *cremation*—the act and process of burning a human being. *Auto-cremation* = the burning of one's self...in an automobile.

The Blistering of Carl Blocker

Sometime during his journey on State Road 15 between Wabash and his home in Syracuse, where the northcentral Indiana terrain is flat and conducive to roadway reverie, Carl C. Blocker ran into trouble as he drove alone during the evening of May 3, 1951. A passing motorist saw a car on the road's bern and a glow nearby. A flare, indicating trouble? Horrors, it was a body. Carl Blocker's *body*. Burning!

For the forty-four-year-old Indiana businessman who owned several successful high-fashion dress shops, there would be no more profits, no more earthly riches. That night by the roadside his plans, hopes, and ambitions ended, consumed along with his body by a fearsome fire. It seemed Blocker had been struck down by a bolt of fire hurled from night's blackness.

Sheriff Cecil Reynolds and coroner L. W. Yoder immediately launched an investigation to find answers. Blocker himself could provide none. Found unconscious, he died that way in Wabash County hospital.[1] It was thought cleaning fluid used in his business had ignited under the front seat. However, no container was found; besides, Mrs. Blocker insisted that her husband never transported such chemicals in his car.

Ten weeks later, the circumstances of Blocker's burning remained a mystery. The sheriff and coroner had not been able to determine what happened to cause Blocker's death. Insurance representatives also came up empty-handed. In an official statement that presaged what Capt. Jake Reichert would say in the Reeser case only *two months* away, coroner Yoder filed "an open verdict of death by third-degree burns" and remarked that "no additional information has ever turned up to explain the fire which caused death."

Carl C. Blocker: another mysteriously burned human.

Vincent H. Gaddis, respected writer about mysterious events, remembered the incident clearly because it happened when he was a newspaper reporter in Indiana. He told me he had checked the facts and found a larger mystery. There was negligible damage to the car's interior—*except* in front of the driver's seat where the heat had been so intense it *melted* part of the metal instrument panel! This reinforced the suspicion that Blocker himself, as the driver, was the point of combustion—though no one could figure out why he began to burn.

"I'm sorry I can't help you more on this story," Gaddis apologized. The highway division of the Wabash Police Department chose not to reply at all to my letters. Hence, Blocker and Reeser share another trait besides the timely proximity of their similarly odd burnings: investigation into their deaths, and attendant unanswered questions, are closed to the public.

What triggered this fire? Why was the car seat so little damaged where its burning driver sat? Why did the blaze's heat travel horizontally, melting the metal dashboard, rather than vertically as fires are supposed to do? Did Blocker, sitting in flames, nevertheless have time to pull his car off the road and park it before leaping out? Was he already parked when tragedy *from something unknown* struck him down in flames from which he tried to escape and could not?

Spontaneous human combustion was considered. But, Gaddis revealed, "SHC was not referred to out of consideration for his family."

With the records of this case closed as tightly as Blocker's coffin, you will have to provide your own answer. Blocker's death was quick, but the subsequent investigation suffered a slow death of gradual indifference and silence.

ABLAZE ON AMERICAN BYWAYS AND BYPASSES

In 1951, about thirty-seven thousand automobile drivers died in the United States. I assume Carl Blocker was counted among that number. The highway death toll climbed as the years passed, as did the number of drivers dying in automobile fires of undeterminable origin.

An Idaho motorist in Wallace "burned beyond recognition" in his car that was unharmed, reported *Fate* (December 1952). The auto-incineration happened on Monday, June 30, 1952.

Southeast of Indiana in western South Carolina, Waymon P. Wood tried to bypass Greenville's downtown traffic and encountered a hazard of a different kind on Bypass 291. It was Sunday afternoon, the first day of March 1953. A passing motorist noticed a parked and smoking 1951 Nash automobile. He pulled over. Other passers-by stopped too, out of curiosity or concern. Smoke curling through narrow openings at the top of the darkened windows prevented a view inside. One samaritan reached to open the door and quickly withdrew his hand in pain. The handle was searingly hot. Unexpectedly, the Nash began to move. Lurching forward several hundred yards, it stopped momentarily, then drifted over a steep embankment.

A few minutes later the wreckage stopped smoking as firemen doused the vehicle with foam. Inside sat Waymon P. Wood, a fire-gutted corpse.

"Fire of undetermined origin," writer Frank Edwards reported of the investigation that followed. The heat had been so intense, said Edwards, that "plastic fittings had melted and the windshield had bubbled and sagged into the car. Yet the fire itself had been confined to the front seat."

Again, the authorities' investigation was fruitful only in producing unanswered questions about the fifty-year-old Greenville man's fate. The victim had talked cheerfully to friends only one hour before the highway holocaust. There was no evidence of suicide, of incendiary materials, of mechanical malfunction, of a leaking gas tank. Concluded Edwards: "A fiery fate of inexplicable origin overtook Waymon Wood as he drove along the highway that afternoon. What it was and what caused it will probably never be explained."

Another abnormal fiery accident occurred about fifteen miles southeast of Baltimore, in Arundel County, Maryland, in early April 1953. It appeared even *more baffling* to authorities than Wood's cremation-in-a-car one month earlier.

Here, Maryland State Police found Bernard J. Hess in his overturned car at the bottom of a twenty-foot embankment. The Baltimore man had a fractured skull. Therefore, the cause of death appeared obvious, the case routine. Then the coroner investigated. Routine quickly ceased. Although he found no trace of fire damage to the wreckage, the coroner discovered first- and second-degree burns covered two-thirds of the dead man's fully clothed body. Police failed initially to notice Hess's searing because...well, *because his garments hadn't been burned!*

Authorities concluded that Hess's severely blistered skin would make it impossibly painfully for him to dress himself after being burned. Contemporary reports do not mention officials finding any electrical or fuel problems with the car that would have caused his injuries.

Did Hess fall victim to foul play at the scene, as unknown assailants stripped Hess naked, doused him with unidentified chemical accelerants and lit them, then re-dressed and drove their victim to a location remote from the crime to push him over the embankment in his car? No evidence supported this. Did Hess succumb to SHC?

As Frank Edwards remarked three years after this incident: "The burns which played a part in his death constitute another mystery which remains unsolved."

AUTO-SEARINGS ACROSS NORTH AMERICA'S HEARTLAND

On Sunday evening, December 13, 1959, Billy Peterson incinerated while sitting in his garaged car in Pontiac, Michigan. Said the doctors who failed to save his life: "It was the strangest thing we've ever seen!" The reason for their mystification will be explored later.

During the night of November 19, 1960, five Kentucky men—Harmon Robinson, 42; Denver Yates, 25; Buddy Hopkins and Harvey Ratliff, both 24; Louis Hopkins, 16—all died inside an automobile parked on Greasy Creek Road in Pike County. Damage in the unwrecked car was restricted to its interior, except that part of its metal roof had melted in a circular pattern. A fire therein had dehydrated and fire-fractured the men's bones, deduced the Dr. Frank Cleveland of Cincinnati's Kettering Laboratory. Investigators, including coroner Raymond Call, noted that the position of the men's bodies indicated the occupants made no attempt to escape the inferno. It appeared these five men sat placidly in a parked car and waited, immobilized, for death to strike them down in flame.

If these good ol' boys had been out drinking that Saturday, were they *all* so inebriated that escape from a frightfully accidental blaze was impossible or undesired?

One can theorize that they collectively decided to commit suicide. But no accelerant was ever found, said the *Pike County News* (November 23, 1960).

One can theorize a coincidence: that all five men were physiologically synchronized to die anomalously by spontaneous combustion, and they all happened to be together when each one's biological clock sparked their independent personal infernos. However, the odds against five men being together to die by synchronized SHC-time clocks must outnumber the stars in the Milky Way.

One can theorize a literal fire from heaven.

Just before the witching hour of May 10, 1961, a "ball of fog, about three feet in diameter, and slightly elongated" descended rapidly out of the sky at a forty-five-degree angle and smashed into the hood of Richard Vogt's car as he drove a familiar highway between Osakis and Eagle Bend in Minnesota. Now engulfed in the unfamiliar, he felt himself and his car's interior begin to cook instantly from heat. Not wanting to be fried, he jammed on the brakes and jumped out. The car didn't ignite, though its metal remained too hot to touch for some time. Scientists working for Prof. W. J. Layten at the University of Minnesota could not explain the incendiary fog that Vogt encountered.

Nor were scientists qualified to explain the "molten substance" which fell from the Saskatchewan sky outside Yorkton on June 7, 1974, to melt a hole into a Canadian car's hood and its windshield before setting fire to the grass— but *not* to the traumatized youth at the wheel.

One can posit still another theory, which will be introduced later.

On Friday, October 16, 1964, Mrs. Olga Worth Stephens[2] of Dallas, Texas, became the tenth statistic for cremation-in-a-car in the years following Blocker.

Mrs. Stephens, who gave Clark Gable his break into acting, had recently arrived in Dallas "to receive treatment for high blood pressure," stated the *Dallas Morning News* (October 24, 1964). Instead, said the newspaper in reporting her death, which occurred the afternoon before, she wound up being treated for "burns received in mysterious circumstances eight days ago."

Mrs. Stephens had been waiting in her car while a nephew went to get a cold drink. Why the seventy-five-year-old former actress suddenly turned into a torch in a parked car mystified homicide detectives. Firemen were baffled too, after they realized that the *automobile itself was not burned*.

Since Mrs. Stephens's nephew had left her alone for only a few minutes at most in his search for refreshment, the fatal fire could not have been smoldering. It was *quick!* With no indication of or reason to suspect suicide, willful immolation was ruled out. With no combustible materials in the car and the car itself unburned, there seemed to be but one fire source left: Mrs. Stephens's own body.

And that was *impossible!* Or was it? The answer lay in Mrs. Stephens, the human torch. Unfortunately, she was as baffled by her predicament as everyone else.

COMPARING CONVENTIONAL AUTO COMBUSTION

Might these cases depict abnormal but not anomalous fire deaths in automobiles? Not readily. Automobile fires resulting from ruptured fuel lines and overheated engines produce consequences that are very different, as demonstrated by these typical examples:

- A loose hydraulic line in a 1973 Bentley sedan leaked gasoline onto the engine's exhaust manifold during a roadtest, transforming the $30,000 (1976 value) luxury car into a true hot rod. Though the mechanic escaped unharmed, we suspect the car's owner was burned up too—emotionally.
- A cocaine user committed suicide in his car by dousing himself with gasoline from two gallon cans found nearby. "It was a raging inferno," said Pennsylvania State Fire Marshal Fred Klages, "and the body looked like a burnt marshmallow." But he had not burned to a white powder as Dr. Bentley and Grace Pett had.

These examples, plus the Schuylkill Expressway's nine-thousand-gallons-of-blazing-horror accident referred to earlier, should prove, clear as crystal no-lead, that gasoline-fueled automobile fires can cause tremendous vehicular destruction; yet entrapped occupants nevertheless remain *relatively intact and identifiable*.

These are normal cases of deadly fire in a vehicle. Blocker, Wood, Stephens, and the five Kentucky men appear to be *anomalous* cases of unwanted fires in humans when in a vehicle.

EARLY CASES OF AUTO-CREMATION

Not all cases of auto-cremations are recent. An early example occurred on November 4, 1928, in Allenwood, New Jersey. About 5:30 P.M. the wife of Albert Van Alstyne noticed their car in flames. Inside, sat the body of her thirty-five-year-old husband. "No evidence could be found to explain how the fire originated," said *The New York Times* (November 5, 1928), "or why Alstyne had not been able to save himself." That officials with two fire departments identified no characteristics expected of normal car fires, hints that spontaneous combustion of *some* form played a role.

Another problematic death by fire in an automobile occurred inside a Model T during the 1930s in Cleveland, Ohio. Dr. Krogman called it his Case of the One-Legged Victim. It will be explored later.

The United States is not the only country blessed by (or scourged with) an abundance of mechanized transportation; nor is it the only country whose motoring population has been plagued by curious combustion.

Near Caughall Bridge at Upton-by-Chester south of Liverpool, England, Edgar Beattie died on April 4, 1938, in a blaze that engulfed his body inside his truck, partly overturned in a ditch. Chester Royal Infirmary pathologist Dr. W. H. Grace performed a postmortem examination on the twenty-five-year-old man, and said "the body was badly burned and the skull had been fractured by the intense heat." He *assumed* heat had to be intense, though by now you know that need not be so.

An inquest convened and quickly uncovered facts that made the common misfortune now appear less common. Young bicyclist Joseph Reynolds testified that Beattie had asked him the time. When Reynolds rode by again just twenty minutes later, he discovered the motor-lorry on fire in the ditch.

Cyclist James Williams also saw the fire. "It was burning furiously," he stated. "The cabin was like a furnace inside." He heard moans, and saw Beattie in the truck with his head upon the steering wheel. The driver's door he found too hot to handle. "I went to the other door, which I managed to open," continued Williams, "but the flames made it impossible for me to get near the man." The inferno thwarted his good Samaritan efforts.

Constable Bartley testified he found pliers on the truck's hood, although, he said, "there did not appear to be anything wrong with the vehicle." Its brakes were good; the transmission was in neutral; the "filler cap of the petrol tank was detached."

Deputy coroner R. Moore Dutton suspected a simple "spilled gasoline" explanation would close the inquest, for it seemed something had gone horribly awry during a repair (although Beattie was reputed to be a safety-minded driver with eight years of experience as a professional). "Would the spilt petrol [gasoline] easily catch fire?" he asked the constable. Constable Bartley pointed out that the fire had started on the side of the truck *opposite* the petrol cap; furthermore, he said "the petrol tank had not burst."

Hence, evidence seemed *not* to support a gasoline-caused accident. Since combustion had occurred *away* from the petrol tank and *inside* the driver's compartment, what could ignite this fiery hell-in-a-truck?

Static electricity is one possible consideration. The *National Safety News* (January 1943) reported that for trucks "static is really dangerous. Heavy vehicles will generate 40,000 volts or more." Husky dock workers have been knocked flat when touching static-charged trucks; tire manufacturers have reformulated tire compounds to minimize road static buildup. Perhaps massive static discharge arced through Beattie when he stepped into his vehicle and turned the ignition key to test his repairs and unwittingly ignited himself and his clothing.

Or might the source have been *inside* Beattie himself, another victim of internal *human* ignition who happened to meet his blazing fate inside a truck rather than on a chair or a bed? As deputy coroner R. Moore Dutton told the inquest: "all sorts of suggestions could be made about how the fire started, but there was no definite evidence on the matter." With that said, the jury returned that infamous verdict that explains nothing: "Accidental death."

According to the *Liverpool Echo* (April 7, 1938), the jury complimented Williams the bicyclist for his brave action and recommended road improvement. The jury was then dismissed. Case closed. Everyone was satisfied... even though the *Echo* reported "there was no evidence to show how the accident occurred."

And that leaves room for SHC.

A TRIPLE AUTO-CREMATION?

Three cases said to resemble Beattie's supposedly occurred on the highways of Europe three days after his demise.

G. A. Shepherdson, a building contractor, waved to workmen as he drove past one of his construction sites at Hessle, Yorkshire. Moments later, his car slowed and came to a stop by the roadside. In those fleeting seconds Shepherdson had "charred" to death, revealed the *Liverpool Echo* (April 7, 1938). Nothing in the car, save the driver, was even singed, asserted Harrison in *Fire from Heaven* (1978).[3] The date is Thursday, April 7, 1938.

George Turner of Upton-by-Chester was discovered inside his wrecked truck, everything intact except his incinerated body. The incident's location, not given, must have been within a few miles of Beattie's burning. The date, according to MacDougall in *The Sunday Star-Ledger*, is Thursday, April 7, 1938.

Willem Ten Bruik drove through the countryside near Nijmegen, Holland. Suddenly the Dutchman "burned beyond recognition" in his car whose "upholstery was not even smudged." The date, according to Frank Russell in *Fate* (1955) and other subsequent writers, is Thursday, April 7, 1938.

A triplet of SHC? The Turner and Bruik incidents pose special problems, to be discussed later.

AUTO-CREMATIONS KEEP DRIVING ON

In Arcis-sur-Aube, France, Léon Eveillé, thirty-nine, died in his car of "burns by abnormal flames" on June 7, 1971. According to the French fortean journal *Lumière dans la Nuit* (August 1972), officials "could not explain how a fire of this violence could have been produced, nor the curious effects which were factually observed."

Back in the United States cars and human cremations continued to be linked. Around sunset, on April 9, 1974, California police and firemen in Redwood

City (on the Bay southeast of San Francisco) found a male occupant inside a 1968 Lincoln Continental about three hundred feet down a grassy slope at Canada College. The car they identified; the victim, "unidentified"—burned beyond recognition. "It's the worst I've ever seen anybody burned," Redwood City Fire Captain John Mace, a twelve-year veteran of fire-fighting, told the *San Francisco Chronicle* (April 10, 1974).

Unfortunately Capt. Mace didn't say if the *outside* of the Lincoln had also badly burned. A rationalist would quickly claim the fuel tank exploded. But by now you should know that in a case like this more information is needed before jumping to the "rational" conclusion. Witnesses of SHC often see a dazzling blaze; yet when the fire ends, damage is mostly confined to a body burned beyond recognition.

Not all cases of auto-combustion end in tragedy.

Consider this episode taken from the files of the Peerless Insurance Company: "Tim Keyes of Salt Lake City, Utah, was fighting rush hour traffic when a fire broke out in the *back* seat. Keyes pulled into an automatic car wash, rolled down the windows of his car, put money in the slot and watched the flood of water put out the fire." A resourceful solution to a startling situation!

Mark Benfield had a close call of a different kind in September 1989. Seconds after he parked under some 100,000-volt power lines and got out of his car, it began to spark and buzz like a door bell. When he and a coworker checked the vehicle with a meter, they discovered the car was giving off 2,000 volts! "That was the first reading we got. Then the meter went shot off the clock," he told the *Nottingham Evening Post* (September 2, 1989). The power company had to turn off the power lines before Benfield could drive his car away. Had he not left his vehicle when he did, perhaps he would have been electrocuted; maybe even fried execution-style inside his electrified auto.

Something perhaps similar, certainly fatal, befell a British motorist at High Ercall on Saturday, November 22, 1980. Here, on the north edge of Telford, Shropshire, the badly burned man sitting behind the wheel of his automobile perplexed investigators. If a suicide, police were at a loss to account for the trifling fire damage elsewhere in the car. So reported the *Birmingham Evening Mail* (November 22).

Thoughts on Auto-SHC: The Driver, the Car, and Nature

In the absence of detailed information that could shed insight on these auto-cremations, we can but speculate.

It may be that an occurrence of SHC in cars holds no special mystery, apart from the human body erupting in flames of its own volition. Since persons self-combust in bed, on chairs, while walking or drinking or eating, it is statistically reasonable that the phenomenon would happen in a truck or car. (Indeed, it would be a mystery if *no* SHC-type fatalities occurred in motor vehicles.)

On the premise that statistical support should not excuse one from seeking the cause itself, further study (as Ralph Nader advises) might prove useful to motorists wanting to evade auto-cremation as vigorously as an on-coming vehicle. Two areas to explore for insight: (1) factors outside the vehicle, and (2) factors inside the driver.

Factors outside one's car that could initiate a driver's internal inferno.

Ball lightning is one external source known to burn people, and it could have been involved in at least some of this chapter's cases.

The theoretical pyrotron is another plausible candidate for *any* spontaneous combustion in a car. It is simply a matter of the atoms comprising any automotive material—seats, gasoline, passengers, driver—being split in a chain reaction caused by a subatomic collision. It would be a collision for which no safety studies have been conducted; no manufacturers' designs could prevent; no amount of government or traffic regulations will avert. In short, a pyrotron could trigger the fire anywhere, anytime in any automobile, and in any of its occupants.

Another outside source, already recognized to exist but heretofore unassociated with motoring hazards, is the planet's magnetic field.

Of ten auto-cremation cases for which we know the terrestrial magnetic levels, six occurred on days when geomagnetic disturbance exceeded the mean level for the month. That's a 60 percent correlation. Half of these cases—all men—occurred during *especially severe* magnetic weather (the strongest magnetic storm being rated 2.0): the "burned beyond recognition" American male incinerated on June 6, 1952, a 1.7 magnetic storm day; Alstyne burned on a Sunday coming off another 1.7 storm; Wood on an even stormier 1.8 Friday.

Does this reveal a heretofore unsuspected connection between automobile operation and planetary magnetism that is hazardous to motorists' health? A subtle interplay between car and driver and an intensified magnetic field that can spark sudden, deadly combustion? I believe it does.

The late biomedical engineer Itzhak Bentov developed a model that allows this to happen. In his breakthrough book, *Stalking the Wild Pendulum* (1977), Bentov linked magnetism, automobiles, and occurrences of spontaneous phenomena (particularly involving Kundalini) within the human body. Spontaneous flare-ups of Kundalini may be attributed, he said, "to periodic exposure to certain mechanical or acoustical vibrations in our normal environment, which will cause the sequence of symptoms to arise." These symptoms can be induced by "a pulsating magnetic field around the head. When people are exposed to frequencies of about 4 or 7 Hz. for prolonged periods of time, as may happen

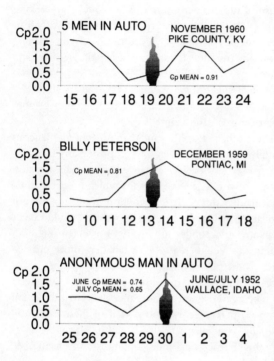

simply by repeatedly riding in a car whose suspension and seat combination produce that range of vibration," he pointed out, the "cumulative effect of these vibrations may be able to trigger a spontaneous physio-kundalini sequence in susceptible people who have a particularly sensitive nervous system."

Added Bentov: "The severity of symptoms is always proportional to the degree of stress encountered."

Driving is unquestionably a highly stressful task for some people. Naturally, physio-kundalini symptoms for them would already be pronounced. Expose them to additional physio-kundalini stresses from repetitive automobile vibrations and aggravated further by a disruptive magnetic storm, and their physio-kundalini symptoms rampage uncontrolled. One plausible outcome might be *nothing less than SuperHyperthermic Carbonization.*

In a personal letter to me shortly before his untimely death, Bentov agreed that all these factors together could culminate to produce an outcome of SHC.

Perhaps some of the seventeen congressional committees and five federal agencies that oversee American automobile safety will want to hold hearings about ways to minimize the build-up of magnetic fields in automobiles and safeguard against this sequence of cumulative stress upon unsuspecting motorists.

Another outside influence to consider, one beyond the Earth's magnetic field, is the Moon. The next time the Moon looms bright and crescentlike as you drive, it might be wiser to not take your eyes off the road to look at it but to not drive, period.

Assigning lunar influence to human behavior and physiology has always been (and will again be) controversial. Nevertheless, Florida psychiatrist Dr. Arnold L. Lieber and psychologist Carolyn R. Sherin, Ph.D., examined the relationship between human behavior and lunar phases. Studying 1,887 Dade County police cases spanning 1956–1970, they determined that murder was most likely to happen during the new moon and full moon phases. They found that chances for the same crime were again predictable in Ohio's Cuyahoga County, where murders increased twenty-four hours after new moon and forty-eight hours after full moon. Concluded Lieber and Sherin about these slight shifts in the time-crime-phase correlation: "This difference suggests that geographic location may be a significant variable, as it often is for geophysical and meteorological phenomena."

Recall the concentration of mysterious auto-cremations in western England around Birkenhead, Liverpool, and Upton-by-Chester: it is an area where *other* pyrophenomena have also blazed to confound science. (I'll have more to say about geography and combustion later.)

Lieber and Sherin also found a "definite relationship between phases of the moon and fatal automobile accidents, not to mention suicides and cases of aggravated assault." (As I'll document soon, incineration and persons who harbor suicidal tendencies have a definite relationship, too.) But is there evidence to link lunar phase, automobile accidents, and SHC?

Of the ten cases under consideration that are dated, four occurred in the new moon lunar phase (with the five Pike County men inflaming at new moon); two cases occurred in full moon phase; four occurred in the first quarter phase.

If cremations in automobiles occurred randomly, the probability of the fire in each seven-day lunar phase would be 25 percent. Instead, 40 percent of these cases coincide with new moon, 20 percent with full moon, and 40 percent with first quarter. *Four-fifths of the cases*—representing eleven of the fourteen fire deaths, a startling 79 percent!—coincide with either new or full moon. (Curiously the third quarter phase doesn't appear.)

What do we make of this nonrandomness? Science has nothing to offer. But then scientists have not quantified every influence upon the human body. Nor do they understand the myriad variations those influences can produce. (And, of course, some scientists steadfastly deny that the evidence you are reading in this book exists at all.)

"It might influence psychological factors," psychologist Sherin said about the influence of lunar phase upon human response factors. "Maybe certain people are more sensitive to atmospheric changes, moon changes, barometric pressure, weather conditions. Maybe some people are more delicately balanced."

Undeniably, certain people *are* more sensitive to all the conditions Ms. Sherin lists—including, I believe, moon phases. But how might moon changes relate to a condition so dramatic and powerful as SHC?

Symbolically, the new moon represents the initiation of a new cycle for human consciousness. Do some persons, when it is their time to leave their bodies and begin a new chapter in life's progression, decide subconsciously to depart the physical body by invoking or merging with powerful energies that would assure bodily death?

Or does the nonrandomness reveal an association that is more physical than psychospiritual: that astronomical conjunctions and oppositions and quincunxes of the Sun and Moon focus (or intensify) cosmic energies upon human energies, resulting in a spontaneously unleashed awesome power within a person who at that moment is hypersensitively vulnerable to minutely subtle disturbances in his psyche and unique body? Inside a person in peril at this moment to unseen (and unmeasured) external energies, the fire erupts.

Factors within the driver that could initiate an inferno.

Accident investigators, faced with otherwise inexplicable explosions and fires, have concluded that for reasons not yet understood, some people are *human* spark plugs. According to the *National Safety News* (January 1943), a Colorado oil company truck driver burned up three different trucks—not due to negligence, tire static, or employee sabotage but, rather, due to the driver's own "magnetic personality." These human capacitors, stated the National Safety Council, have been known to store up to thirty thousand volts in their bodies. Robin Beach, a professor of electrical engineering who collected examples of human capacitors, told of a motorist who unscrewed the cap of his car battery to check its water level. Checking himself first would have been wiser, because there was an immediate explosion as he touched off the hydrogen gas escaping from the battery of his recently parked car.

If the English lorryman Edgar Beattie was one such human dynamo, and if his truck's electrical devices were situated near the driver's seat, perhaps his fiery demise resulted from the sudden discharge of his excessive voltage and an ensuing explosion that started a fire inside the truck cab...inside himself.

Herbert Greenhouse catalogued many cases in which appliances malfunctioned when handled by particular individuals, including an automobile that would not run for its owner (no trouble for others) even though trained mechanics could find no defects in the car. Greenhouse called these people *PK-missers*. "A PK-misser always has car trouble," Greenhouse wrote in the magazine *Psychic* (May-June 1975). Then he described one notable exception: a woman who consistently started her car when it should have been impossible—her car had no battery! Electrical problems of a different sort, I suppose.

In addition to these human capacitors who walk around immune to the huge electrical charges accumulated in their bodies, others are demonstrably hypersensitive to electric currents. These people have a much lower tolerance of, and significantly higher sensitivity to, induced electrical fields. A friend of the author can sense when an EKG-current is administered, a degree of sensitivity her physician said was impossible until she identified each and every time the current was administered.

Could, then, a vehicle's 12-volt electrical system—shocking but not lethal to most people—be sufficient to trigger (or *contribute* to the triggering of) SuperHyperthermic Carbonization in such people? Perhaps Waymon Wood was among those who, if extremely sensitive to DC current, touched a faulty electrical connection and found his body reacting to sudden, shocking, fiery electrocution in his car.

It's been known for decades that an energy interaction can exist between man and the machines he operates; that certain equipment will malfunction in the proximity of certain humans. Scientists aren't comfortable acknowledging this effect producible by electric people. (Sound familiar?) Wolfgang Pauli, a Nobel Prize winner in theoretical physics, caused equipment to break down so many times when he entered a laboratory that the phenomenon acquired his name: *the Pauli effect*. Physicist George Gamov wrote in *Thirty Years That Shook Physics* that the Pauli effect is "a mysterious phenomenon which is not, and probably never will be understood on a purely materialistic basis." Which is to say there can be more than nuts and bolts when it comes to assessing the functioning of machinery and its interplay with humans

Might auto-cremation have something to do, therefore, with a motorist's emotional response to driving? Some normally unassuming people metamorphose into hot-tempered, mightily enraged drivers when they grab hold of a steering wheel. (I first drafted this paragraph in the late '70s, years before the use of gunfire to express dissatisfaction among motorists became a national pastime on American highways.) Might their psychologically overreactive mindset enhance the probability for a physio-kundalini blaze behind the wheel? The hyper-aggressive driver works himself into a cauldron of seething anger that builds until the emotions *literally backfire*—and one more motorist burns up in a crescendo of his own fiery rage.

Conversely, it may not be drivers' angry aggression that triggers this type of highway holocaust but the opposite emotional attitude: passivity. I point out again that the majority of SHC cases occur when a person is alone and in a situation conducive to melancholy, reflection, pensiveness. If you drive, you must know about "highway hypnosis"—that roadway reverie that lulls a driver into a tranquil, oftentimes *trancelike* mental state (a result of prolonged exposure to the mechanical/acoustical vibrations inside a car, according to Bentov). As I discussed previously, in that altered state of consciousness apparently lies the greater chance that SHC will blaze forth. Hence, the same brain-wave patterns that lead to human fire on a bed can also trigger SHC on a highway.

Certainly fatal fires inside an automobile do happen in ways far less esoteric than I am suggesting here. A cigarette dropped onto flammable seat covers; a ruptured fuel line; an electrical fire detonating gasoline in ways auto designers did not intend—and cars and humans will burn up.

Yet throughout the cases in this chapter there lingers a sense of factors present, not explainable by these acceptable scenarios. Officials repeatedly alluded to conditions that puzzled them but which might be readily explained *if* they were willing to open their case files to outsiders and *if* they had been prepared to consider that sometimes humans will spontaneously combust while sitting, while sleeping, while walking, while eating and... why not while driving, too?

SHC in automobiles may be just another form of accident injury occurring on the world's highways, awaiting recognition and public outcry to safeguard against it.

13

EYEWITNESSES: SEEING IS BELIEVING?

> No one can attest to what you saw or observed, and no one will know unless you tell. Thus, it is important to... accurately record the cause and circumstances of fire.
>
> —*Officer's Handbook for Determining Cause and Circumstances of Fire,* National Fire Protection Association (1974)

The strongest argument opponents of SHC think they have is that the phenomenon has never been observed.

"Why is it these things never happen when somebody's watching?" lamented Dr. Lester Adelson to CBC-TV's Hana Gartner in 1986. "Doesn't it bother you a little bit? It bothers me. We'd like to get some reliable witnesses..."

Debunkers assert that no one noticed the unavoided fireplace ember or the ever present smoking material that ignited the careless victim externally. Nor, they contend, was anyone present who could disavow that the victims' bodies had actually—allegedly—burned up in a smoldering combustion over many long hours.

You know already that history does not support their contention. A few cases *have* been observed. The fiery death of Ignatius Meyer in 1811, for example, and Abdallah Ben-Ali in 1839. Plus, there are those experiments by Dr. Rybalkin who, as detailed in *Le Revue de l'Hypnotisme* (June 1890), produced burns and blisters on patients purely by suggestion.

But, the debunkers argue, this is ancient history: these few accounts can be dismissed as misunderstood observations derived from unsophisticated thinking.

I hate to disappoint—no, actually I *delight* in disappointing—haughty intellectuals who think themselves above looking at and listening to evidence because they have a Ph.D. or... well, just *because*.

Eyewitnesses are about to speak. Listen to their words.

> What a mercy it would be if we were able to open and close our ears as easily as we open and close our eyes!
>
> —Lichtenberg

WITNESSES TAKE A STAND ON SHC

On June 10, 1828, Dr. Peter Schofield of Ontario delivered the first temperance address in Canada. He described an episode "under my own observation" that, he said, should give fair warning of the ultimate danger of immoderate drinking. It is important to hear what he spoke of that day.

One night at nine o'clock on a day not given, Dr. Schofield saw the town drunkard of Brockville who was, "as usual, not drunk, but full of liquor." Less than two hours later the town's blacksmith noticed a bright light in his smithy and ran to investigate. Inside, said Schofield, the smithy "discovered the man standing erect in the midst of a widely extended, silver colored blaze, bearing, as he described it, exactly the appearance of the wick of a burning candle in the midst of its own flame."

The smithy seized the drunkard, a man about twenty-five years old, and jerked him violently to the floor. The silver fire "was instantly extinguished." The blacksmith, acting instinctively and as a humanitarian, should perhaps instead have allowed the fire to take its natural course uninterrupted. The unidentified victim, his flesh sloughing off, shrieked and groaned and said he was "suffering the torments of hell"—although he complained of no body pain. On the thirteenth day he mercifully died.

Dr. Schofield had fortuitously been near the blacksmith's shop and, once called for, arrived at eleven o'clock. Of the victim, Schofield said "I found him literally roasted, from the crown of his head to the sole of his feet." Unaccustomed to hearing of a human burnt by argent flame, the physician examined the scene carefully. "There was no fire in the shop," he revealed, "neither was there any possibility of fire having been communicated to him from any external source. It was purely a case of spontaneous ignition."

Hear, again, the physician's own words about the fate of Brockville's youthful bacchanal: "It was purely a case of spontaneous ignition."

The words written by Dr. Etoc Demazy to *Archives Générales de Médecine: Journal Publié par une Société de Médecins* (1830) were in French, but the subject Dr. Schofield would have understood. For the first time in English, here is Demazy's forgotten story—titled "Combustion humaine spontanée"—about the last thirty minutes in the life of Madame Clotilde Herpain Guy.

During 6:30–7:00 o'clock Monday evening, February 2, 1829, this fifty-three-year old shoemaker was home alone in the Sarthe village of Ferté-Bernard, near Le Mans some 120 miles southwest of Paris. The weather was good; a northeast breeze cooled the evening. Mr. Guy returned home to find his wife—described as "hot-tempered, remarkably obese, ordinarily happy"—immobile in a chair. He thought she was asleep. Then he noticed a small fire...then "in an instant her clothes were consumed." Acting quickly amid dense smoke, he threw water onto her. This, said Demazy, surprisingly "increased the activity of the flames."

No less surprising was the ferocity of injury: "the head preserved but a part of her hair...neck, lips, eyes, ears were destroyed...torso and limbs were more or less entirely burned...a foot only and a breast had escaped burning." Monsieur Guy was now a widower.

Demazy declared that Mme. Guy had burned in her own body fat "as if impregnated with alcohol." Oh? He said so without benefit of medical evidence, because a local judge held an inquest without calling any of the village's three doctors, none of whom examined the victim. "It is grievous that no autopsy was conducted on the woman Guy!" decried Demazy.

It wouldn't be the last time such a scene would see proper investigation deficient.

On December 18, 1839, a pauper woman on Eldridge Street died about five hours after neighbors noticed smoke in her New York City apartment. Said the *New York Express* (December 18, 1839): "The woman was discovered lying in her bed—herself and the center of the bed in flames—but no traces of how the fire was communicated could be distinguished. It is, therefore, presumed that the miserable woman lost her life by that well attested, but very rare occurrence, spontaneous combustion."

October 17, 1844; Liverpool, England. As death neared for Mr. W. Mercer, forty-seven, he became insensitive; then suddenly a "red-hot coal-like streak" blazed from his mouth for twenty minutes until he took his last breath and died. No singeing of face or hair happened in this conundrum with multiple witnesses, among them medical personnel, said *The Lancet* (January 4, 1845).

Around ten o'clock on the morning of April 10, 1852, Englishman William Vodden stood on his farmland at Bedford, picking up stones. A few minutes after he had been seen looking for another stone to toss aside, companions heard a piercing scream. Turning in Vodden's direction, they saw their friend now standing wrapped in a pillar of fire. Twice they tried to rescue him, but the stubborn flames would not go out. After ten horrifying minutes, his wife succeeded in quelling her husband's blaze. Sexagenarian Vollen died within the hour.

(Four years and four months later, in August 1856, some forty unexplained fires flared forth in Bedford briefly. Was Bedford then free of odd fire until Mrs. Kaur inflamed in 1978?)

If the cases of the Canadian in Brockville and the Englishman in Bedford sound too absurd to believe, what should one conclude about the fate of Dr. Averet in Virginia on July 6, 1855? He was found "standing in the middle of the room, with his person enveloped in flames." He, too, burned with "an unusually brilliant light" whose color wasn't given, and died almost instantly by so-called "accidental burning."

For Christmas of 1869, or earlier, friends of a Louisiana man gathered around the fireplace at his home to celebrate the season. Their host, a drunkard, suddenly gave them a haunting, non-holiday image to behold. According to *Scientific American* (1870) and Walford's *The Insurance Cyclopaedia (1871–1880)*, blue flames shot from his mouth and nostrils... and soon he was a corpse. Merry Christmas, humbug!

During the afternoon of May 12, 1890, a hard-working "strictly temperate" forty-nine-year-old woman from Ayer died in this Massachusetts town, northwest of Boston, in a blaze that interested medical science. At least it interested Dr. B. H. Hartwell, who saw the five-foot-five-inch, 140-pound woman "in the actual state of conflagration." The *Boston Medical and Surgical Journal* published Hartwell's account:

> The body was face downward; the face, arms, upper part of the chest, and left knee only touching the ground; the rest of the body was raised and held from the ground by the rigidity of the muscles of the parts. It was burning at the shoulder, both sides of the abdomen, and both legs. The flames reached from twelve to fifteen inches above the level of the body. The clothing was nearly all consumed. As I reached the spot the bones of the right leg broke with an audible snap, allowing the foot to hang by the tendons and muscles of one side, those of the other side having burned completely off. Sending my driver for water and assistance, I could only watch the curious and abhorrent spectacle, till a common spading fork was found with which the fire was put out by throwing earth upon it.

Now Hartwell could examine the scene closer. "The flesh was burned from the right shoulder," he noted, "exposing the joint from the abdomen, allowing the intestines to protrude, and more or less from both legs. The leg bones were partially calcined. The clothing unburned consisted of parts of a calico dress, cotton vest, woollen skirt, and thick, red, woollen undergarment."

Another fire that horribly burned flesh but not garments.

Dr. Hartwell said this case was medically significant for three reasons:

> No one ever saw the incineration begin, proceed, or end.
> —W. Keysor, referring to SHC;
> *A Medico-legal Manual* (1901)

(1) "It is the first recorded case in which a human body has been found burning (that is, supporting combustion) by the medical attendant." That, as you know, is not true; medical personnel in Liverpool witnessed the fiery death W. Mercer in 1844.
(2) "It differs from nearly all of the recorded cases, in that it occurred in a person in middle life, not very fat, and not addicted to the use of alcohol." That was generally true.
(3) "It is interesting in a medico-legal sense. It proves that under certain conditions—conditions that exist in the body itself—the human body will burn, and this has given rise to the belief in the spontaneous origin of the fire."

Hartwell noted that the woman had set fire to a pile of stumps and roots some thirty feet away. As to injuring her, Hartwell concluded the blazing brush was blameless. This tragedy happened soon after a rainstorm, he knew; hence, the ground cover was wet and offered no means by which the brush fire was communicated to the woman. "As proof that the flesh burned of itself and nothing but the clothing set it afire," he wrote, "it may be stated that...there was nothing but charred leaves under the body; that her straw hat which lay several feet distant was simply scorched; that the wooden handle of the spade was only blackened."

Was it the *clothing* on her body or her *body* in her clothing that ignited this personalized conflagration? Dr. Hartwell had no answer.

In London's Bermondsey district, directly opposite the Tower of London, Mrs. Mary Brown, seventy, died on December 14, 1904. She was discovered standing in flames and "badly burned about the upper part of her body," reported *The Daily News* (December 17, 1904). Her death was blamed on the twentieth-century version of visitation by God: "Death by misadventure."

If people aren't safe from the fire when sitting or standing or sleeping in their own homes, where *could* one feel secure? Surely in a boat, surrounded by lots of water—

Yarmouth, England; the year 1938. On July 29 or a few days earlier,[1] Mrs. Maude Comissiong, fifty-four, was on a nautical outing on the Norfolk Broads with her husband and children. Unexpectedly before her family's eyes, she quickly burned to a charred corpse. The voyage back to shore must have been a hideous trip for those she suddenly left behind—because of what little she left behind in their midst. "I suppose her clothes caught fire," the investigating officer told the *Liverpool Echo* (July 30, 1930) once the boat docked, "but I can't understand how it happened." *If* foul play was involved, the family got away with murder. Otherwise, they got a nightmare. The case was adjudged accidental.

As the clock approached midnight on August 27, 1938, the fire again appeared in public. In a *very public* place—the Chelmsford Shire Hall in southeast England. Young and pretty Phyllis Newcombe, twenty-two, was whirling and twirling on the floor at the weekly Saturday dance with her fiancé, Henry McAusland, when she dance-stepped into combustion. Phyllis screamed. Blue flames were swirling around her body. In horror, everyone screamed. The unnerved McAusland tried to beat out the flames bare-handed. All attempts to rescue his fiancée failed, and "in minutes she was ashes, unrecognizable as a human being."

The *Essex Chronicle* (September 2, 1938) had this story first, one that didn't spread like wild fire through the British press. London's *Daily Telegraph* picked up the story on September 20, when it chronicled the coroner's court proceedings. Ironically, testimony began with a plea for an ambulance to serve the city of twenty-five thousand: it had taken the ambulance from a nearby brigade thirty-five minutes to arrive at the dance hall—as if a more timely response would have saved the poor girl. Not! McAusland testified Miss Newcombe's dress and body ignited "in seconds" and that no ambulance could have gotten there quickly enough.

What material was her dress? asked Coroner L. F. Beecles. "A frock modelled on an old-fashioned crinoline" came the answer. Why, crinoline will burn! He thus declared "the tragic results that we have heard" were caused by a dropped cigarette, which caused the dress (and not Miss Newcombe) to inflame.

Fortunately for justice, Miss Newcombe's father had a more inquiring mind than the coroner first demonstrated. Mr. Newcombe produced a swatch of fabric from which his daughter's dress had been cut. It *would* burn when lit by a lighter, he showed; but lighted cigarettes *failed* to produce the same, expected result. Coroner Beecles now had to admit, "From all my experience I have never come across a case so very mysterious as this."

A case that became even more mysterious, if possible. The *Essex Weekly News* (September 1938) also covered this fantastic story, though it appeared to be a wholly different story. The *News* had Miss Newcombe "descending the staircase and on discovering that her dress was alight rushed screaming up the stairs. Her dress immediately burst into flames and at the entrance to the ballroom she collapsed." On-lookers smothered the fire, limiting its damage "to her legs and body." Many days later, she died in the hospital of "hypostatic pneumonia and toxaemia secondary to burns." In the *News'* version, the coroner admits this was his oddest case, then appends this: "Obviously this is not spontaneous combustion."

Whether a fiery 'Jack the Ripper' was afoot in southeastern England this year, even surviving eyewitnesses to Miss Newcombe's fate might be at a loss to know. History, though, presents another case of very much the same fashion.

October 1938; a nightclub in Soho, the London district famed for its nightlife. Nightlife became nighthorror, then nightdeath for Maybelle Andrews, a nineteen-year-old out on the town with Billy Clifford. In midstep with Billy on the dance floor, she erupted in flames... flames shooting from her back, chest, shoulders, and hair. She died moments later. According to a personal communication from journalist Michael Harrison, the inquest was conducted by Coroner James F. Duncan, who concluded: "In all my experience I have never been confronted by a case as fantastic as this. I can find no logical explanation whatever as to why the deceased's clothing and hair caught fire."

Perhaps had His Majesty's Coroner known about the long history of SHC, he would have found the logical—though no less fantastic—explanation he sought.

On December 26, 1938; Ballina, Ireland. James Duncan, seventy-six, a non-smoker and not near any observed fire, became a pillar of flame in his bedroom.[2] "So fierce the fire," reported *The Western People* (December 31, 1938), that rescuers were unable to approach him until he burned himself up.

December 31, 1938; Wrexham, Wales. Mrs. Selina Broadhurst, fifty, inflamed in her room. Would-be rescuers again were repelled by the furnace blast of heat, said the *Liverpool Echo* (January 2, 1939). Said Eric Frank Russell who investigated: "Burnt to a crisp by her own clothes in circumstances where it was not possible..."

Around January 4, 1939; Peckham Rye (London). Put to bed, eleven-month-old Peter Seaton presumably lay fast asleep until a houseguest heard the toddler scream. Harold Huxstep ran upstairs to effect a rescue, and was stopped cold (or hot). "It seemed," he told *The Daily Telegraph* (January 4, 1939), "as if I had opened the door of a furnace. There was a mass of flames, which shot out, burning my face and flinging me back across the hall. It was humanely impossible to get Peter out."

How did a toddler manage this? London Fire Brigade Superintendent E. H. Davies conducted a painstaking search for the answer, any answer, and could give no answer but that there had been nothing in the nursery which could have initiated even a small blaze, let alone a holocaust of such horror.

Paradoxically, paint did not blister in this inferno whose heat halted Mr. Huxstep's heroics.

And circa 1961; in England's heartland. A man mowing his Midlands lawn "just burst into flames," Miss Charla Cohen recalled seeing on the evening local news. She told me that witnesses called it "classic SHC," but she could provide no more information. Can anyone?

THE FIRE SEEN IN OL' BAILEY'S BELLY

Witnesses to these unexpected, sudden, rapid, thorough and unresolved fires have seen intensely bright coppery red, silver, or white, but most often blue, flames. A normal oxidation-type fire characteristically burns with a yellow-orange-red flame. The blue hue suggests either a *very hot* oxidizing fire, or a gaseous or electrical or pranic-type discharge. As often seems the case, a fireball originates in the abdominal area where destruction is most fierce, then radiates outward until, at last, its energy abruptly spent, the extremities beyond the energy's abdominal-centered radius remain untouched as evidence that the ashes were once a human being.

Sometimes there are feeble fireballs; incomplete human cremations. Internal, nevertheless.

September 13, 1967; London, England. At 5:19 A.M. a group of early-risers walking to work noticed bright light emanating from an abandoned building at 49 Auckland Street in Lambeth. One of them telephoned in a fire alarm. The Lambeth Fire Brigade arrived at 5:24 A.M. In charge, Fire Brigade Commander John (Jack) Stacey. "We think there's a fire in there," one of the half-dozen onlookers said, pointing up to the first floor. In two minutes, Stacey's brigade had the situation under control, save comprehending the situation.

(After Ron Bentley of England brought this incident to my attention, I phoned John Stacey at his London home; he said he'd be delighted to discuss this case during his forthcoming visit to relatives in Daytona Beach, Florida. I immediately booked a flight to Florida. The resulting extensive interview is sharply condensed here.)

"When we entered the building," said Stacey, "he was lying on the bottom of the stairs half-turned onto his left side." He had died in great pain, reasoned Stacey, because his teeth were embedded into the solid mahogany newel post "and his knees were drawn up as though he was trying to bend the pain from his stomach."

The person about whom Stacey spoke was Robert Francis Bailey, age about sixty-three, a derelict and an alcoholic known to the local police. He had certainly chosen the abandoned building as shelter for the night. And it became his deathbed. Beyond that, certainty stopped. He either spent the night sleeping on the lower steps of a staircase or had begun to climb the steps when something impossible overtook his body.

The blue light that caught the eyes of the passersby, caught Stacey's own eyes next. "There was about a four-inch slit in his stomach, just about at this point"—Stacey placed his hand just above his navel—"and the flame was emanating from that four-inch slit *like a blowtorch!* It was a blue flame..."

A *blue* flame jetting from the abdomen of an aged drunkard? And *witnessed* by a fire brigade commander and his professional crew?

"The flame was actually coming from the body itself," asserted Stacey, "from *inside* the body. He was burning literally from the inside out! And it was definitely under pressure. And it was impinging onto the timber flooring below the body, so much so that the heat from the flame was charred into the woodwork."

Great balls of fire! the firefighters thought to themselves.

After the momentary shock of having to believe what they beheld, Stacey's men emptied a fire extinguisher on Bailey. The flame didn't go out. Another extinguisher was emptied. Still, the blue flame blazed! Several more extinguishers were needed to finally douse this stubborn internal fire. Now the firemen could pry open Bailey's jaws, and free him from his death bite into the stair post.

"When we put the fire out," Stacey continued, "we continued to investigate the cause of it." Stacey looked for the three elements that combustion classically requires: fuel, oxygen, and a heat source. "We had the fuel, which was obviously [alcoholic] spirit of some kind. But we didn't have oxygen"—from inside the body, he meant—"and we didn't have a heat source inside the body. *And that fire was coming from the body.* It didn't start outside. It started *inside!*"

> My personal opinion is that spontaneous human combustion... is based upon superstition exclusively and has no relation to reality at all.... Spontaneous human combustion... has to be condemned as totally impossible.... there is no fact in the world what so ever indicating that [the] human body would be able to burst into flames with or without other sources.
> —Dr. Mogens Thomsen, Ph.D. Chief Surgeon, Copenhagen Burn Center; president, Danish Society for Medical History; personal communication (Sept. 17, 1981)

> I am sorry to tell you that our different attitudes to the problems [of SHC] clearly demonstrate the distinct difference between superstition and natural science. So this will be my only comment.
> —Dr. Mogens Thomsen, Ph.D. personal communication (Nov. 8, 1981)

Likewise, Walter Kirk, a British crematorium superintendent, affirmed that "No doubt, the fire had come from within!"

Characteristic of SHC, the destruction was extraordinarily localized. And fierce. The photographs he handed me revealed the aftermath of the blue flame that burned outward from the vagrant's stomach. It consumed his abdomen plus the fabric of his light-weight suit-style coat—but *only* in front of the abdomen and on the back of the right sleeve where his elbow rested a few inches in front of his waistline. Because he was lying face down, the flame augured into the floor. A one-by-six-inch pine plank had charred, but just underneath his hip. His abdomen lay on the first step, its bare wooden corner now charred; burned moderately was the mahogany post. Other readily combustible debris, scattered only inches away over the floor, escaped without a hint of damage. In all, the fire confined itself to six square feet—the size of the hole Br. Bentley had burned himself through.

The color of the fire was just what history would favor: blue. Suggesting a gas torch, of a sort, and not the oft-invoked "candle effect."

If this case was one of classic SHC, history would also favor no noxious odor of burned flesh. What about odor? "No. No odor," Stacey replied to my question.

Evidence of accelerants was sought; none detected. Evidence for an ignition device on the premises was sought; none located. Evidence of foul play was sought; no evidence found.

Fortuitously, the inquest came before Dr. Gavin Thurston, whose career provided previous knowledge about paranormal human combustion. He found that the contents of Mr. Bailey's stomach had indeed ignited. No mention of how they took fire inside a human stomach.

Initially, cause of Bailey's death was listed as "asphyxia due to inhalation of fire fumes." In other words, Bailey suffocated on the fumes of his own combustion.

"I have my doubts about that," Stacey said softly, solemnly. "I have my doubts about that."

In the end, the inquest hearing ruled more honestly. The court ordered a new, official cause for Robert Bailey's death: "To remain unknown."

Stacey said this had been his first case of a body seen burning from the inside out; probably would be his last. He *hoped* so, anyway. It didn't make him comfortable, this relic of an unsophisticated past that wasn't supposed to happen then or now, in 1967. Yet he had seen it with his own eyes, and he would not deny it to himself or others. He said that as a professional, he would call the case as he saw it.

Everything—investigations and eyewitnesses—led Fire Brigade Commander John Stacey to the only conclusion supportable: *Robert Francis Bailey died by spontaneous human combustion.*

I differed with Stacey on one point only. He thought that Bailey must have been in great pain because his teeth had bitten deeply into the staircase post. I pointed out that in most SHC episodes there appears to be little if any pain felt by the victim. On the other hand, the victims often appeared to have been immobilized, as if an immense nervous or muscle spasm had locked their bodies into instantaneous pugilistic rigor mortis.

I thus suggested an alternative scenario: Bailey, sleeping off a hangover perhaps, was lying on the staircase; his head resting on the stairstep; his face against the stair post. When his abdomen initiated SHC, a sudden violent contraction in his neuromuscular system caused his neck to flex forward and his open mouth to snap shut around the wooden post with such force his teeth clamped into the mahogany. His jaws locked tight. Pain was not at issue. Though speculative, Stacey admitted this made sense. But then, everything in this case, beyond what he himself had witnessed, was speculation at this point.

Would Bailey have burned to ashes without the fire brigade's intervention? If Stacey's crew had been able to determine that Bailey was already dead, letting the fire proceed to its natural conclusion would have proven instructive. Whatever the end would then have been, the *outcome* of the Bailey case is known and cannot be changed. What *must* change is the belief firmly held that SHC is never witnessed by people who know what they are looking at.

Now, SHC stands as a phenomenon observed in progress by a reliable, competent, and no-nonsense fire professional and his brigade.

Weirdness at Woolworth

I have another incident to chronicle, one never reported in print until now. It comes from Cheryl Paserman, whom I met at a Jersey Society of Parapsychology symposium in 1976.

She said a friend's friend, named Irene (or Renie), had gone to a Woolworth store in New York City about three years before (circa 1973). Irene was "normally in a very distressed state," she said, "and felt her cohabitants were mean and cruel to her." This day, Irene felt particularly persecuted and paranoid; "definitely in a very extremely negative state of mind at the time." She went shopping for candles for possible use in a ritual because, Cheryl thought, Irene was "maybe into voodoo."

At the Woolworth candle counter, two sales clerks suddenly and anxiously pointed to Irene. "Your trouser leg is on fire!" one yelled out. The salespeople "brushed out" the *painless* flames—to use Miss Paserman's description as it had been told to her—and Irene fled the store.

Fabric aflame? Or *human flesh*?

Did anyone mention the flames' color? I asked Miss Paserman. "I think they were *blue*."

A Marvel at Mardi Gras

During my live appearance on Baltimore's *People Are Talking* (WJZ-TV) in 1985, a viewer called in to say that she had never felt so relieved in her life as she was at this moment hearing about people whose bodies self-ignite. That response caught everyone off-guard.

A crank call? I pondered during a commercial break. Or not? This viewer believed she had seen SHC. Was she making this up?

People unaware of the historical background of SHC would likely describe red flames, and that would tip-off her story as contrived. I asked her to describe the fire's color. "Blue," she said without hesitation; "bright, electric blue." It was the expected answer, if SHC was indeed at issue.

After the telecast, she agreed to speak once more, at length and on condition of anonymity, about the unsettling experience. It would be, said criminal justice major Miss D. W., an *exclusive* interview because she didn't want to have to think about the incident again.

She had been in New Orleans, in 1971, during Mardi Gras. One muggy afternoon, she casually noticed about 150 to 200 yards in front of her an "average size" man walking alone in the French Quarter. She glanced away for a moment. As she turned to face him again, a burst of light engulfed him. "He just disappeared in a ball of fire before my eyes!" she exclaimed. Horrified, she turned and ran, thinking herself crazy to see this. "Now I think I understand what I saw."

No one was closer to this man than she. No sound, no screams associated with the burst of blue light did she hear. No odor did she smell. No indication of writhing, panic, or pain in the man did she see. She told a few friends about her unforgettable Mardi Gras experience, only to be repeatedly laughed at. Consequently she began to doubt her sanity, until she saw me on *People Are Talking*.

Attempts to link her account with New Orleans Police Department "missing person" files have been unsuccessful. Due to the number of transients passing through the city during Mardi Gras, the police may never have known this pedestrian was in town; never knew the odd manner by which he left town, leaving behind a few ashes that scattered in the next breeze.

More "Misadventures" Seen

August 5, 1982; morning in Chicago's South Side. City News Bureau issues this bulletin: a Negro woman at 4052 S. Wells Street "burst into flames for no apparent reason, police said."

Winfield Gattlin claimed to have seen the woman walk past his parked car; then, now one block away, he watched her become an instant spire of fire. CNB soon issued an addendum which mentioned spontaneous human combustion. *The Paul Harvey Show* picked up the story for radio, and I picked up the telephone.

Brighton Park Police Violent Crimes Unit detective Daniel Fitzgerald said, "She was completely burned; I mean scorched. It was obviously not just a case of her clothing catching fire. We really don't know what happened here." Detective Nick Creszenzo and Sgt. Owen admitted in the early hours of their investigation to their puzzlement, too. Firemen thought at first they had responded to a rubbish fire, Fitzgerald said, so unrecognizable was the woman. Her purse lay untouched nearby.

Meanwhile, Cook County medical examiner Dr. Robert Stein announced his disbelief in SHC before beginning his investigation: "Such theories make nice fairy tales." Media telephoned me, eager for information about a hot story they knew little about.

The eyewitness could help sort out this confusion, I thought. Fitzgerald gave me Gattlin's phone number. It proved invalid. I was on the phone with Dr. Stein when Bob Boese of the Cook County Crime Lab informed him that gas chromatography had *just* identified traces of hydrocarbons on the woman's garments. To Dr. Stein, that meant accelerants and, therefore, either homicide or suicide. At this point, the media dropped the story like a too-hot potato. I pressed on.

Autopsy indicated the woman, in her mid-forties, had died twelve hours *before* the burning began—which is difficult to harmonize with the witness's statement. (So are a lot of other aspects of this incident.) I requested photographs to assess whether extent of injuries hinted at SHC. Marshall I. Considine, director of the Chicago police crime laboratory, explained that he could not release photos because of potential prosecution of a suspect. He added, "I will however state spontaneous human combustion was not involved." Plus, a physician involved in the investigation told my informant Mickey McWilliams that "there was decomposition in the organs" of the victim; that is, no deep body burning.

So: scratch one alleged SHC, pending revelations to the contrary.

A case whose outcome is less ambiguous transpired one month later in London, England. Hunker down, it's a doozy.

September 1982, daytime; Edmonton, north London. The weather is hot, humid. No breeze wafts through the open windows at the Saffin household, where Jean Lucille Saffin, sixty-one, and her eighty-two-year-old father swelter in the kitchen. She is behind and a bit to his left, sitting in a wooden Windsor chair.

> Nothing is too wonderful to be true.
> —Michael Faraday, physicist; discoverer of electromagnetic induction.

The old man is aware only of the heat until, out of the corner of his eye, he glimpses a momentary flash. He turns around to ask his daughter if she saw it. He sees his daughter engulfed in sinuous flame!

He sprang to rescue her. Jean Lucille remained seated, still sitting there with her hands in her lap as the flames swirled around her. She didn't make a move. Nor cry out.

Grabbing onto her, he yanked her from the Windsor—she didn't collapse but was able to stand—and dragged her across to the sink and started throwing cups of water over her. He shouted for his son-in-law, who was in the house coming down the stairs. He rushed in, got a saucepan of water and doused the hapless woman.

As the men battled to save Lucille, the son-in-law swore that "she had flames roaring from her mouth like a dragon."

Ambulance men took her to the hospital. In picking up their charge, they noticed there was *no* smoke damage in the kitchen; that her clothing was *not* burned. Only a portion of her red nylon cardigan only had melted.

I'll let John Heymer tell the rest of the story in his own words, shared graciously with me and published here for the first time.

"The surgeons obviously thought she had set fire to herself. And they wanted to know how. One of them stated that the burns to her face and her arms were full-depth burns—right down to the underlying fat. It takes some time to do that, you know.

"Now this woman sat next to her father whilst all of this was going on! And he also said, afterwards, that once he was aware of the flame, the flame made such a *roaring* noise he would never forget them. He was in the First World War. He'd seen flames enough, as he used to say to his other daughter, but he had never heard flames roaring like that!"

A doctor told the son-in-law he had to be mistaken, of course, about any *loud* fire issuing from Ms. Saffin. Apparently, he simply ignored the father's confirming testimony.

But the surgeons could not ignore the patient with face and hands burned away. At first when they spoke to her trying to figure it all out, her eyes would open and she was aware of them. But she never uttered a word, eventually lapsing into a pseudo-coma where she only responded to pinpricks. She died after eight days.

"The inquest was adjourned," Heymer continued, "to find out the cause of ignition. The police officer was sent down, and he went back and told them it was *spontaneous human combustion*. He could find no possible cause how she could have become a flame."

The police further stated that when she was admitted to the hospital, she was in no pain. "Now she had full-depth [third-degree] burns—her complete face was burnt off and her hands—and she showed no pain!" Heymer emphasized about a condition surely astonishing to anyone unfamiliar with the history of SHC.

He chuckled. After all, this kind of irony was familiar to him. "The only parts of her body that burned are the parts that weren't clothed. There were some minor burns to her body elsewhere, you know, one or two patches on her buttock and on her abdomen, I believe. But the burn to her abdomen could've been caused by her flaming hands which were in her lap."

Might Jean Lucille Saffin have burned to ashes were it not for the quick intervention of her relatives? No one will ever know.

Heymer, however, will contend that Ms. Saffin was "burning from the inside, below the surface.... 'Cause otherwise, how could she have burnt—her face and hands burn away—when her clothes didn't burn?"

Makes sense to me.

HUMAN COOKING AT THE COOKING SCHOOL

January 28, 1985, around 11:00 A.M.; Cheshire, England. At the Halton College of Further Education in Widnes, seventeen-year-old Jacqueline Fitzsimon had just finished a cooking class. With friends, she walked several hundred feet along the hallway and into a stairwell. A fellow student, John Foy, noticed smoke coming from beneath her clothing. He shouted the alarm: Miss Fitzsimon is afire! Nearby friends removed some of her clothing and beat out the flames. But she was already badly burned over eighty percent of her body and died fifteen days later from pneumonia while still in intensive care.

What happened to her, not surprisingly, remains controversial to this day. Had her clothing brushed an open flame in the gas cookers in her classroom and unobtrusively smoldered on her smock for several minutes? Had a cigarette thoughtlessly been dropped down the stairwell to fall upon her back or hair-sprayed curls? Had her back itself been the culprit?

What happened in the investigation of her injuries is just as controversial.

At the time the incident made headlines, I was the only nonpolice investigator who was speaking regularly to Cheshire Police Sgt. Martin Findlow, chief investigator in the case. In other words, getting information no one in the media was getting at the time, from the best source available.

In the space of "five-to-six yards and a few seconds," Sgt. Findlow told me, Miss Fitzsimon had transformed from a happy cooking student to what "looked like a stunt man." It took three staff members and two students to put out her flames, he explained. "We can't say how she caught fire," Findlow confided, saying he was definitely uneasy with the dropped cigarette theory being promulgated by other officials.

What about SHC? "It should be examined," Findlow believed. Cheshire fire prevention officer Bert Gilles agreed: "I have interviewed seven eyewitnesses. So far there is no clear explanation of the fire... We must look for other causes like spontaneous combustion."

Whoa! Officials, stumped, open-minded; willing to consider the bizarre because the bizarre might make the most sense.

Then I spoke to Police Constable Jenion, who had been assigned as acting coroner to the case. It was now April 19, almost three months after Miss Fitzsimon had died.

Are you conducting the inquest into her death? "Yep."

Would you give me information about her death? "Nope."

Have you completed your findings? "Nope!"

He would, he said, say nothing about the case. Then he reconsidered. "I can tell you this, however. You're all wasting your time!"

"You think there's a conventional—"

Jenion curtly interrupted my unfinished question. "You're all wasting your time. I'll tell you that now, for nothing!"

Acting Coroner Jenion was clearly irked by this death of Miss Fitzsimon, whether or not it held the specter of SHC. Fair enough. But when an official of the court categorically rules out an explanation that, despite its oddity, the police themselves are willing to consider, and does so before he hears all the evidence being assembled for him to rule on, I get suspicious.

That suspicion was confirmed. The investigation into her fate unfolded in convoluted fashion, masterfully reported with gusto by British journalists Peter Hough and Jenny Randles in *Fortean Times* (no. 47). Coroner Gordon Glasgow, now assigned the case, opened the proceedings by admonishing the jury (and the assembled press) to ignore any speculation or suggestion about SHC. Evidence would be withheld. Experts who had no firsthand connection with the case testified against SHC. Home Office pathologist Philip Jones flatly excluded SHC because, he said, "the necessary chemical reactions required were not evident"—a point on which I'd have wished for (and demanded, were I the coroner) *much more erudition!*

Mr. Jones, *what* are the necessary chemical reactions for SHC?!

Interestingly, Sgt. Findlow was apparently *not* called to testify. Nor, apparently, was Runcorn Station Officer Richard Pink, even though he too was involved in the original investigation and told me personally that "SHC is a likelihood."

A reconstruction was carried out using an acrylic jumper and cotton outer garments like Miss Fitzsimon had worn that fateful day. They did not ignite and burn, unlike the young lady.

The media then turned its attention back to SHC.

Predictably, experts turned against SHC with elevated vigor. Said one Cheshire fire brigade spokesman: "The suggestion in the newspapers that this could be a case of Spontaneous Human Combustion is entirely unfounded and merely a creation of sensationalism by a reporter to sell a story." To which the *Daily Telegraph* countered: "I can only say that our correspondent is quite certain that... the circumstances of this Cheshire incident were so mysterious that a full investigation was necessary, including the possibility of Spontaneous Human Combustion—"

"I have found no evidence of spontaneous human combustion," Jenion assured the inquest. I wasn't surprised. He advocated the make-sense latent-smoldering-of-the-smock theory, even as he admitted he had never seen a smoldering piece of cloth burst into flames. Seemingly unsure of his certainty about so quick a smoldering-to-flame, he asked: "it's common knowledge, isn't it?"

Student John Foy was one who vehemently disagreed. When Hough asked him his reaction to the smoldering smock explanation, Foy replied: "What a load of rubbish! When we walked past [Miss Fitzsimon] there was nothing, seconds later her back was a mass of flame."

Two other witnesses at the inquest supported Foy. Randles and Hough recount in *Spontaneous Human Combustion* (1992) that Karena Leazer and Rachael Heckle testified they had just passed their classmate in the stairwell when, said Ms. Leazer, she noticed a small ball of light moving downward toward Jacqueline's right shoulder and seemingly enter the girl's back, whereupon Jacqueline cried out: "It's gone down my back—get it out!"

Moments later, both girls affirmed, their friend erupted "like a stuntman on TV"—to quote another eyewitness's graphic depiction of the tragic occurrence.

Mr. Jenion can have his opinion against SHC, an opinion which dictated the inquest verdict: "Misadventure." Nevertheless, facts favor that Jacqueline

Fitzsimon is one more episode of SHC witnessed; witnessed, then denied; denied, then dismissed in a conspiracy of sanctioned confusion.

SEEN ONE MORE TIME?

On May 28, 1990, twenty-six-year-old Angela Hernandez was undergoing surgery at the UCLA Medical Center. Injured in an automobile accident, she was expected to recover without problem.

Until the sheets which covered her on her gurney on the operating table caught fire, that is. Or was it Ms. Hernandez herself who inflamed under the sheets?

Dense smoke forced the medical staff to flee the fire they could not extinguish. Neither could university police quell the flames when they arrived. By the time L.A. firemen rushed in—five minutes after the fire's start—the blaze had eased to a slow smolder, reported *The Los Angeles Times* (May 29, 1990). The patient was dead.

Curiously, the fatal fire produced too little heat to set off the overhead sprinkler system.

If a spark from a carelessly placed cautery had caused the tragedy, UCLA Medical Center officials won't say. None of the tantalizing rumors could be confirmed. The incident had its momentary flash of publicity, then just seemed to disperse like a puff of smoke wafting into nonexistence.

How many other people, in how many other jurisdictions, have left this world in a way similarly ablaze, bizarre...a way that is, sometimes, eyewitnessed but not reported for fear of ridicule? Fear of the unknown? People are not *supposed* to see their fellowmen leaving this world in fiery glory. One shall not see what cannot be.

CHAPTER 1: *Left*, more than winter's bone-chilling cold would send shivers through the small community of Coudersport—especially through Don Gosnell, who walked into Dr. Bentley's bathroom and saw Dr. John Bentley—the quintessential legacy of classic SHC. (Copyright © 1976/1993 by Larry E. Arnold.)

CHAPTER 2: *Below*, typical severe burning of a human being trapped in a conflagration; nevertheless, his corpse can be transported intact to the morgue. (*Glaister's Medical Jurisprudence and Toxicology*; copyright © 1973 by Churchill Livingston; reprinted with permission.)

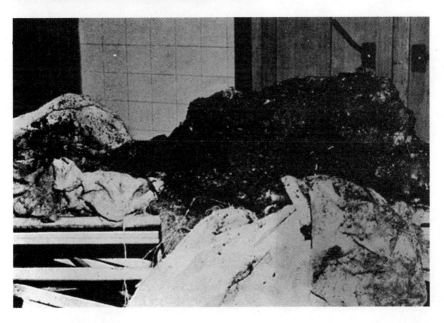

CHAPTER 6: *Above*, fireman Buddy Standish and a colleague had to *shovel up* Mrs. Reeser (*inset*) and her overstuffed chair, so thorough was their combustion in an apartment otherwise only minimally affected by a horrific inferno. *Below*, said St. Petersburg police officer Tipton: "I know it was overwhelming. Her body was almost completely disintegrated. The woman was *gone!*"—as her remains in the morgue attest. (Inset photo from author's private collection, originally published in the *Columbia* [PA] *News*; other photos used courtesy of Vincent Gaddis.)

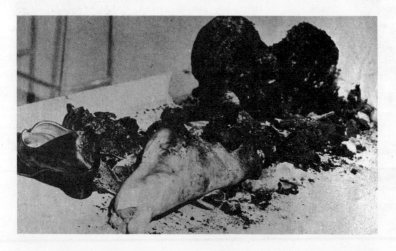

CHAPTER 8: One of three women burnt in fierce, localized cremations in homes where little else did—stockings on feet mark the amorphous mass to the left of the table as once human; a tea towel and combustible fringe inches away remain bright white. (*Glaister's Medical Jurisprudence and Toxicology*; copyright © 1973 by Churchill Livingston; reprinted with permission.)

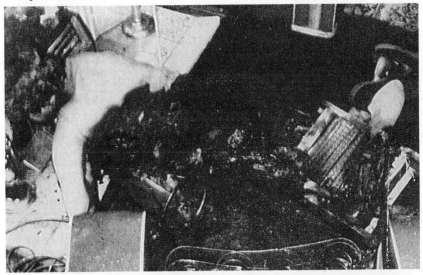

CHAPTER 8: Another of three pyrophoric persons fiercely burnt up—her head and shoulders collapsed into her burned away midsection as she tumbled to the floor. Her femur is exposed; feet nestled *unscathed* inside cloth slip-ons. (*Glaister's Medical Jurisprudence and Toxicology*; copyright © 1973 by Churchill Livingston; reprinted with permission.)

February 2, 1980; shortly after nine o'clock in the morning, spinster Annie Gertrude Webb burned up with gruesome wonder in Newport, Gwent, Wales. Her head is barely recognizable upon the hearth-step; her right hand, fist clenched, and right arm still wearing what appears to be a plastic hospital ID band, lies at an odd angle behind what should have been her back; lower legs in nylon stockings lay sprawled mid-calf under the plastic (barely singed) tablecloth. (Used with permission.)

March 4, 1980; another English lady bites the dust as she burns up, preternaturally or spontaneously. If fireplace embers simply are to blame, with hundreds of thousands of fireplaces worldwide why is it that firemen find this scene so uncommonly rare? Note the total absence of blistering to varnish and scorching of fabrics nearby. (*Glaister's Medical Jurisprudence and Toxicology*; copyright © 1973 byChurchill Livingston; reprinted with permission.)

CHAPTER 9: The world's *earliest* known photograph of SHC: Mr. A. M. burned to a dusty shell of his former self in a hayloft in 1888, retaining his features until collapsing into dust at the touch of his would-be rescuers from his unnatural crematorium. (*The British Medical Journal*, April 21, 1888.)

CHAPTER 13: A blue flame jetting from the abdomen of vagrant Robert Bailey was seen by London Fire Brigade officer Jack Stacey and his crew, proving that SHC not only happens but has been witnessed by knowledgeable professionals. (Courtesy of John Stacey.)

CHAPTER 15: Jack Angel lost his hand and forearm to an "electrical type" fire for which researchers could find no source. Perplexed doctors diagnosed the burns as "third-degree, internal in origin"—burns which also scarred his chest, groin, nape of the neck, and, incredibly!, fused vertebrae discs in his spinal column. In 1981 the author introduced this survivor of SHC—one of many—and his fascinating story to millions of viewers of ABC-TV's *That's Incredible!* (Copyright © 1982 by Larry E. Arnold.)

For several hours, Angel felt no pain in the hospital, despite the gaping burn wound in his right chest that physicians say resembles "an electrical burn ... *definitely* a high-energy source." (Courtesy of Jack Angel.)

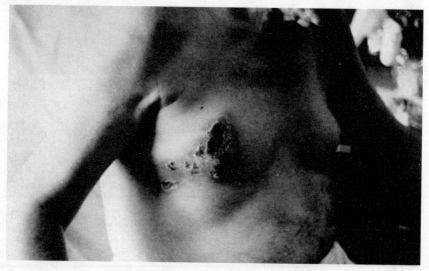

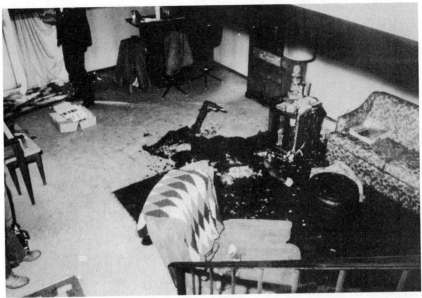

CHAPTER 20: Mrs. Oczki's former daughter-in-law first thought a minor chair fire had warmed the house and explained the small mound of debris she noticed in the lower room ... but on closer examination fire officials concluded the widow herself was both the victim of and the cause for this fiercely localized combustion. Like several victims of SHC, she was a widow; an invalid; and disinterested in living. (Above: courtesy of Frank Oczki, copyright © 1995 by Larry E. Arnold; below: courtesy Vincent Calcagno; used with permission of Frank Oczki.)

CHAPTER 24: Mrs. Helen Conway appears to have exploded in her apartment one Sunday morning in 1964. Most of the damage was confined to the widow herself. Firemen were dumbfounded, because a scene like this is not supposed to happen in a few seconds. But it did. (Both photos courtesy of Robert C. Meslin.)

14

SHC Fables and Select Myths

To treat your facts with imagination is one thing, but to imagine your facts is another.

—John Burroughs

"The laws of physics and chemistry have not been repealed. Spontaneous Human Combustion is a myth, a theory without facts. Its perpetuation is a stain on the science and professionalism of the fire analysis community." So concludes Patrick M. Kennedy, Ph.D., in a 1986 issue of *The National Fire Investigator*, the journal of the National Association of Fire Investigators.

As president of NAFI and a professional fire investigator who teaches fire education, Kennedy (who presumes expertise also in the highly complex and rapidly changing fields of physics and chemistry) desires to banish SHC to the world of myth. Dismayed that he and like-minded colleagues have been unable to bury the ghosts of spontaneous human and preternatural combustibility in the century since von Liebig, Kennedy laments: "It is the way of myths that they become self-sustaining. They develop lives of their own and once this happens, they are very difficult to kill."

According to Mircea Eliade, the University of Chicago historian of religion who is among the world's great experts on mythological theory, Western traditions have until this century treated myth only as "fable...fiction...invention." Kennedy is entitled to maintain his interpretation of myth, if this be his definition of myth, just as is the evidence entitled to speak its case.

Actually, Kennedy is partly correct. Many episodes alleged to be SHC *are* mythic, regularly foisted on the public by another explosive phenomenon known as the supermarket tabloid. Whenever myths of anomalous human burning are palpably false but presented as factual, serious research is threatened. Myth-slaying then becomes the worthy task that Kennedy advocates. So let's quickly slay some falsehoods that befoul SHC.

SOME TABLOID HOT TIMES

Father Domingos Fernandez spontaneously inflamed while exorcising Satan in Argentina in 1984... retired New York fireman George Mott "literally cremated in his bed" in 1986... French schoolgirl Marilee Mars "burned intensely for almost 10 minutes" while showering in 1987(?)... Herr Klaus Holtzberg blazed while doing battle with taloned, green demons in his children's bedroom in 1988(?)... gluttonous Maria Lissaro exploded with one big burp after devouring twenty-four pizzas at an unidentified bistro in Paris in 1988... two Belgians, Christian Ruys and his fiancée, Patrice Monami, inadvertently turned their public passion into a too-torrid embrace that shot flames fifteen feet into the air and "30 seconds later" left them forever co-mingled as ashes in 1989(?)... famed exorcist Father Luis Ponce "exploded in flames" spit by a "demon" inside a girl being exorcised in Barcelona in 1989(?)... hefty nine-year-old Alain Favier lost his French temper and blew up in a white-hot fireball "equivalent to three sticks of dynamite" that also killed five of his tormentors in 1989. Guess the tormenting isn't over until the fat boy explodes—

> You know what we're talking about. You've stood in the checkout lines. You've seen the tabloids. Heck, if there's even a spark of truth to those stories, people are bursting into flames as often as kitchen matches. And with the millennium closing in, this sort of thing can only be expected to get worse.
> —reporter Kenton Robinson,
> in *The Hartford Courant*
> (October 31, 1994)

There are more outrages.

In 1990(?), Stuttgart insurance executive Helmut Klommer is trysting with his secretary at the moment his wife directs hatred at him, so that his "head swelled up like a balloon" and exploded; an alleged police chief supposedly attributed this to "natural causes." Oh? A washed-up Budapest magician performs the trick of the century when he waved his wand and made his niece, Louise, "burst into flames and vanish." On the outskirts of Vienna Rev. Franz Lueger becomes his own fire-and-brimstone as he rails from his pulpit against evil, Bible in hand; he is instantly crisped, the book is not. Sunbather Sven Pehrson tans fatally on a Majorca beach—tans to ashes in a burst of heat rivaling the sun and injuring his new bride lying beside him. A New Jersey man, Herb Cook, already dead, ignites in "another case of human spontaneous combustion?" Deceased Carlos Aragon lies in his coffin "at a small church near Tarauac [sic]"—the town is Taraucá, in the rainforests of remote western Brazil—yet postmortemly becomes "an unusually well-documented case of spontaneous combustion of a human body" when in 1989 his corpse "vanishes into thin air!" before onlookers. In 1991(?), Swiss psychic Barbara Nicole engages in a "mind duel" with a rival and lost when the latter "blew Miss Nicole's head off with a burst of psychic energy from his mind to hers."

In 1994(?), a related fate befell Nikolai Titov at the Moscow Candidate Masters' Chess Championships, when "Hyper-Cerebral Electosis" overcomes his intense concentration and his brains blow apart "as if someone had put a bomb in his cranium," according to his blood-spattered opponent; a famed neurologist, Dr. Martinenko, asserted these "explosions happen during periods of intense mental activity" in people who are "literally too smart for their own good." The item made the Internet, along with a checklist of precursors for this rare malady; having any three of the seven conditions (practically ubiquitous to everyone) signals imminent HCE. The cure? Be dumb and don't think!

Finally, 1995 brings the advent of gross SHC contagion—"*First case ever of a human spontaneous combustion CHAIN REACTION!*" screamed the headline—as Marie de Challeurs began (in 1995?) to smoke, internally, at a lavish French dinner party; "she just sort of exploded," eyewitness Yves Gigot is quoted, but not before "a thin stream of fire" jumped from her to strike down Georges Casaubon, then the eyewitness's wife, Renee Gigot, plus four more guests at the ill-fated table.

These tabloid tales appear with unnerving frequency—the above are from the *Sun*, *National Examiner*, and *Weekly World News*—and invariably lack confirmatory data. With lurid headlines like "Woman wolfs down 24 pizzas—and blows to smithereens!" and "KA-BOOM! Famed psychic's head explodes" mixed among stories about a WW II bomber photographed on the Moon and the ongoing lives of John F. Kennedy and Elvis Presley, it is little wonder that a discerning public dismisses SHC, especially since mainstream journalism rarely gives any (let alone favorable) coverage to the fire within.

The threat of derisive ridicule for swallowing tabloid tripe as anything but fable has made life *very easy* for those who want others to disbelieve in SHC.

Most of these pseudo-cases originated in *Weekly World News*, which has in recent years published more of these stories than its competitors combined. *WWN* either has an exclusive hotline to SHC, or writers with hot imaginations. Either way, the result is another full-page story bannering a new twist on SHC. Correspondent David Welk announced in the *WWN*'s first issue for 1995, that of the 17,307 people who fell prey to the strange phenomenon worldwide in 1993, the last year for which figures are available, 12,073—that's 69 percent—were plumbers, explained one expert.

Lord help us all, especially plumbers, if there are 17,000-plus incidents of SHC annually! Not surprisingly, said "expert" is anonymous; as is the woman in an accompanying photograph who "exploded into flames in 1993"—a photograph electronically composed, clearly.

In every instance but two where local authorities responded to my requests for confirmation, *WWN* stories have proved (as you might suspect) to be outright fables.

One exception was Herb Cook, an eighty-one-year-old cemetery worker found dead in his smoke-filled home in Franklinville, New Jersey.

> TOP 10 SHC-Prone Occupations
> (with apologies to David Letterman)
>
> 1. Plumbers: 12,073
> 2. Elementary school teachers: 1,239
> 3. Cab drivers: 908\
> 4. Data processing clerks: 899
> 5. Dentists: 786
> 6. Clergymen: 473
> 7. Psychiatrists: 345
> 8. Police officers: 218
> 9. Actors, actresses: 189
> 10. All other occupations: 177
> —*Weekly World News*
> (January 5, 1995)

"Dead man bursts into flames! Is bizarre blaze another case of human spontaneous combustion?" headlined the *Weekly World News* for April 22, 1986. *WWN* reporter Dick Donovan called me prior to the story's publication, asking that I answer that question. I said I'd have to investigate, first.

Assistant Fire Chief Alf DeCesari confirmed the tabloid's account (amazing!) in that Cook had in fact died. He recommended I speak with Medical Examiner Claus Speth. Dr. Speth admitted the case was odd but saw no reason to invoke SHC. He reasoned that Cook suffered a fatal heart attack, fell onto a grate atop a gas heater located in a crawl space below, then for two days the heater cooked Cook as fat dripping from his roasting torso was consumed slowly—until police opened the door. Besides, Speth pointed out, Cook was not burned to ashes.

To the tabloid's supposition that SHC claimed this victim, the answer is no. Herb Cook roasted over a known external heat source and burst into flame due to a backdraft explosion in a carbon monoxide-rich environment. The circumstances were unusual but *not* attributable to SHC. Another case of tabloid SHC shot down in flames, though Cook did put the lie to the human-candle theory promulgated by SHC debunkers.

The other exception may be the best-documented SHC of the past decade, perhaps of all time. I'll tell you which one it is...later.

More Myths to Slay

"Tabloid SHC" does little to inspire belief in a phenomenon already too difficult for many people to believe. However, such myths are not relegated to the questionable journalism of tabloid editors. Reputable writers and scholars have erred too, as these examples show:

• On February 19, 1888, a "drunken old soldier" in Colchester died of SHC amid unscorched hay in a hayloft, according to Harrison in *Fire From Heaven* (1978), plus others who referenced him. I have found no corroboration in English newspapers for this event. Instead, this appears to be a case of relocating Mr. A. M. who burned 450 miles away in Aberdeen. Slay one myth.

• Strange flames engulfed Mary Hart "at Whitley Bay, Blythe, England" on March 31, 1908, according to writer Allen Eckert. Another SHC? Or problems? England has no shire named Blythe; Whitley Bay is near the *town* of Blyth, but the two communities are separate geographically (though linked by an unsuspected cartography to be examined later). Examination of *The Blyth News & Wansbeck Telegraph* (the local newspaper) at the Colindale Branch of the British Museum for the period in question uncovered no mention of a Mary Hart, dead or alive. After Eckert's article, Mary Hart gets entrenched in the SHC literature, as other writers cite it as evidence for mystery fire. A disturbing pattern. Until someone hands me the original news clipping, I'll say Mary Hart is a myth slain.

• Randles and Hough, in their book *Spontaneous Human Combustion* (1992), reassign Dr. Schofield's "silver colored blaze" incident of 1828 to the year 1847, and attribute it to the "first-hand experience" of a Dr. Nott, who wasn't there; Professor James Hamilton—his role in SHC will be introduced soon—of Nashville, Tennessee, is relocated to Nashville, Texas; three-day-burning John Hittchell becomes John Hitchen; young light-of-foot steeped-in-flame Maybelle Andrews inexplicably gets redated from 1938 to the "late 1950s" for her dance with fire. False, too, is their statement that *both* feet of Dr. Bentley "were untouched."

SHC-debunker Kennedy creates some SHC fables, himself.

• One involves a Mrs. Mary Carpenter who, he says, was allegedly consumed by SHC on a boat in front of her husband and children. A look through the *Liverpool Echo* (July 30, 1938) would have shown him the victim is Maude Comissiong and not a Mrs. Carpenter.★ Tsk tsk, Mr. Kennedy.

★ The burning lady-of-the-lake, Maude Comissiong, is also misidentified as Mary Carpenter by, among others, John Allen (1981); Gaddis (1967; 1968); Readers Digest's *Mysteries* (1982); Randles and Hough (1992).

- "Holes burned through the floor is a commonly mentioned situation" in the more than sixty-four cases of alleged SHC known to him, Kennedy contends. To the contrary. Although PARASCIENCE INTERNATIONAL's database of some three hundred SHC-style episodes is *much* larger than Kennedy's, as you have seen there are, at present, a mere ten examples of humans dropping in from above. *Not* a common trait—

If SHC debunkers expect to be credible, they must avoid their own mythmaking when rebuking SHC advocates for arguing upon what Kennedy himself calls "inadvertently misstated facts and outright prevarications..."

AN HONEST PROPOSAL PROVES WRONG

At 1:30 P.M. on December 5, 1987, firemen arrived at the scene of a lethal blaze in Greensboro, Caroline County, Maryland. Then they radioed for assistance, not to quell the fire—neighbors had already extinguished the small blaze—but to understand its larger question. How had this fire burned *only* Anthony F. Gianninoto, his clothes, and the flooring around him to the exclusion of everything else in the two-story, wood-frame house?

By evening, Maryland Deputy State Fire Marshal Bob Thomas thought that the fire could have been started by the gas stove, upon which soup was cooking, or by smoking materials. He also admitted to the media that the sixty-seven-year-old man might be a victim of a rare but documented phenomenon... "spontaneous human combustion."

Whoa!

For reasons I have never fathomed, this astonishing admission did not make the national news wire. Even the local newspaper devoted but eight column-inches of type to the story, initially on December 6 and in a follow-up on January 10, 1988, which said SHC had been ruled out.

When I received copies of those stories in April 1988, I grabbed the telephone and rang the Maryland State Fire Marshal's office. Bob Thomas returned my call, saying the state medical examiner had determined that Gianninoto died of smoke inhalation and burns. Perhaps so. The victim, wearing several layers of clothing, was not cooking food but was trying to light the stove when his clothes caught fire, officials concluded.

History gave me reason to be skeptical, however. Thomas helped put me in contact with the local fire officials in the case, Bel Air's fire marshal Ward and deputy fire marshal Dave Herring. They, in turn, were equally accommodating in inviting me to view the fire file and photographs.

At the Bel Air Multi-Service Center, I reached my own conclusion when seeing the photographs taken of Gianninoto and his kitchen. Yes, this fire *had* been localized. And no, damage to the body, by SHC standards, was *insignificant*. Only on the side of his chest had a small portion of his many layers of clothing burned. Inhalation of smoke and fumes from a carelessly handled match did indeed fit with the photographic evidence and other factors I learned about.

No new SHC to add to the database, this day, I thought. To raise the possibility of SHC was bravely commendable, given the circumstances encountered. But it was an honest mistake. On the other hand, given the excessive amount of fuel loading (clothing) wrapped around Gianninoto, the episode did reveal how *really difficult* it is to burn up a body.

THE PROFESSOR'S PYROPHORIC LEG

Early chroniclers of SHC contended that to experience SHC was to die. Then came apparent living proof to the contrary, in the person of a university academician. If SHC itself was not yet ready for slaying, at least its inevitable fatality now appeared to be a fable.

Professor James Hamilton, twice elected to the Chair of Mathematics and Natural Philosophy at the University of Nashville (now the George Peabody College for Teachers at Vanderbilt University), was one of its four outstanding scholars during 1826–1850. He was a man of high moral character and "quite observant of nature," his biographer Dudley Horton reported. As a member of the Temperance Society, the professor consumed no liquor "except 'gin and gypsum' for medicinal purposes." *The Peabody Reflector and Alumni News* (1937) said "it was his chief delight to acquire knowledge and to impart it to his pupils."

Ironically, the *one* incident that Professor Hamilton's life imparted to history and earned him fame beyond his reputation is overlooked by his biographer. As you can guess, it involved a freakish fire.

Walking home in the brisk winter air of January 1835, Professor Hamilton, probably thirty-nine years old, was suddenly afflicted by a stinging sensation in his left leg. "Directing his eyes at this moment to the suffering part," reported Dr. Overton in the *Transactions of the Medical Society of Tennessee* (1835), "he distinctly saw a light flame of the extent at its base of a ten cent piece or coin, with a surface approaching to convexity, somewhat flattened at the top, and having a complexion which nearest resembles that of pure quicksilver."

Dr. Overton told the medical society how his quick-witted patient, thinking to smother the flame by restricting its oxygen, "applied over it both his hands open, united at their edges, and closely impacted upon and around the burning surface." Whereupon, said Overton, by fortuitous chance or a triumph of rationalism "the flame immediately went out."

Professor Hamilton's silk and cotton drawers had a small hole burned through the fabric where the quicksilver flame had flared. But curiously, his outer woollen pantaloons escaped with merely a "slightly tinged... dark yellow hue" on the inner layer of broadcloth. It appeared the jet of flame originated underneath his pantaloons; beneath his drawers; inside his leg!

Consequently, modern writers[1] have stated that Professor Hamilton survived *very limited* spontaneous human combustion. He certainly belied the popular criteria for SHC: that victims were of the lower class; were dull-witted; were morally derelict due to alcoholic over-indulgence.

I must rain on their parade, however, because they all have overlooked an important additional source in this provocative incident: *The Transylvania Journal of Medicine and the Associate Sciences* (1836). After hearing of the professor's story, a *Transylvania* correspondent arranged to meet with him (good policy!) and, in so doing, learned the following particulars:

> The drawers worn by him, at the time of the accident, were newly made, and put on without being washed. The store at which he procured the materials for them had lately been painted, and the remnant of the goods from which they were made took fire, not many days afterwards, and was consumed, endangering the whole building. He [Hamilton] supposes, and there can be but little doubt of the truth of his opinion, that in painting the shelves some of the paint was dropped upon the cloth—a mixture of silk and cotton— and that from the action of the oil upon it, the combustion resulted, which has been repeatedly observed under such circumstances. A drop of oil adhering to his drawers, in due time, set them on fire; but the situation being one unfavorable to the process, the burning ceased when that part of the fabric imbued with oil was consumed.

Professor Hamilton did experience spontaneous combustion... of an external chemical contamination to his pants rather than internal human nature. He was never in imminent danger of becoming a human fireball (at least not at that moment). I'd wager, though, that he re-calculated the hazards of wearing virgin cloth, and until his death in 1849 never again donned pants without thoroughly washing them first!

Kennedy may be delighted to learn that another case of "half-truths" can be consigned to the slain myths that waylay serious investigation of SHC.

A Three-Pronged Trident of Fire?

Many centuries ago a popular belief was that what happened twice, happens thrice. The anonymous author of a 1570 Elizabethan manuscript put it to rhyme: "Multiplication is vexation, Division is as bad; The Rule of Three doth puzzle me, And practice drives me mad."

This belief has found its way into the lore about fire.

In his *The Golden Bough* (1922), Sir James George Frazer mentioned a myth of bonfires "lighted by threes, and the people spring thrice over them" on the island of Lesbos. Dr. G. Archie Stockwell wrote about SHC for *The Therapeutic Gazette* (1889), and remarked on the triplicate outbreaks of fire. "Strange to say," he noted before giving an example, "the vulgar [uneducated] hold that after a long period of immunity, one conflagration will be followed by a second and a third, ere another period of immunity occurs, and observation gives fair confirmation..."

Belief in fire's triplicate nature is established in modern fire-fighting, too, though without explanation. Upper Darby Township (Pennsylvania) Fire Marshal Robert C. Meslin told me that his three decades of investigating fires convinced him the Rule of Three has merit. "If you have two fires in a day," he said, "you might as well stick around—you'll have another. And *almost always* there'll be the third!" So well-known is this pattern, said Meslin, that one firefighter got himself in trouble when, after responding to two fire calls, he predicted there'd be a third fire soon. There was! His chief initially accused him of arson, said Meslin, even though belief in the Rule of Three deserved the blame.

It should not surprise, then, that The Rule of Three is found in the lore of SHC as well.

"Three Fingers of Fire and Three Odd Deaths" headlined columnist Michael MacDougall's essay for Newark, New Jersey's *The Sunday Star-Ledger* (March 13, 1966).

The first odd death, said MacDougall, occurred off the Irish coast aboard the tramp steamer S.S. Ulrich. At 1:14 P.M. local time on April 7, 1938, the vessel's helmsman, John Greeley, was found "a human cinder. There was no sign of fire. The deck, the wheel itself, even the shoes of the dead man, were unmarked. Whatever flame cremated John Greeley had come from nowhere, had vanished into nothingness." SHC at sea?

(Shoes anchored to the Ulrich's deck with no legs inside *is* reminiscent of the upright shoe with a foot but no leg found in Mrs. Rooney's kitchen.)

MacDougall compounded this mystery with another: "A few hundred miles eastward the police of Upton-by-Chester, England, were called to investigate a highway accident which caused George Turner to be terribly burned inside his truck." Was the fatality a motoring accident, or an accident inside Turner? MacDougall said it was no ordinary fire because the seat cushions hadn't ignited. "A broken clock on the dashboard showed that the accident had occurred at 2:14 P.M."—the "precise second" when Greeley incinerated one time zone to the west. "Coincidence?" he asked.

What happens twice happens thrice?

MacDougall had a third mystery up his literary sleeve: "at 3:14 P.M." at Nijmegen, Holland, a young man, Willem Ten Bruik, was burned to death in his little Volkswagen. The car itself was unharmed, but "the interior of the car must have been a touch of Hell since the boy's body was burned beyond recognition."

Three cases of SHC occurring hundreds of miles apart at precisely 2:14 P.M. Greenwich Mean Time on Thursday, April 7, 1938? A more *amazing story* Steven Spielberg could not have scripted!

"It was as though a galactic being of unimaginable size had probed Earth with a three-tine fork, three fingers of fire which burned only flesh. Fantastic? Yes," MacDougall answered himself.

Steiger and Whritenour embellished the story in *Flying Saucers Are Hostile* (1967), saying it "defies probability... without someone or something providing a catalytic agent that induced such ghastly self-immolation." Gaddis incorporated it in *Mysterious Fires and Lights* (1967; 1968). John Macklin repeated it in his own *Stranger Than Fiction* column (November 18, 1968), adding more details and affirming the precise moment of simultaneous incinerations was not 2:14 but "one-fifteen... on the afternoon of April 7, 1938."

Fantastic, unquestionably. But *believable*?

Despite each writer stressing the "precise" synchronism of the three-pronged outbreak, Macklin's clocks in the SHC triplet are all fifty-nine minutes ahead of MacDougall's timepieces. Hmm. Macklin headlined his column "Flames kill three continents apart"—despite the fact Holland, England, and the ocean off the Irish coast do not encompass three continents. Hmm. The automobile that evolved into the VW Beetle was called Kraft durch Freude (or KdF) in 1938, not Volkswagen; further, a German-born VW salesman told me the Volkswagen nameplate was not available to the public until 1947. How Dutchman Bruik could have been driving (let alone burn up in) a Volkswagen in 1938 is problematic. Hmm!

My examination of major 1938 British newspapers for April 7, 8, and 9 failed to locate a single mention of seaman Greeley or George Turner dying by any means. As for Bruik, the earliest reference to him I've found is a 1955 *Fate* article by Eric Frank Russell, citing "a translated report from an unnamed Dutch paper" and calling Bruik "a cinder" who passersby "lugged out of his car... little damaged; doors opened, gas tank intact." A correspondent in The Netherlands told me his search of Dutch newspapers disclosed no mention of Bruik's cremation.

It all was becoming *un*believable. Then it became downright impossible.

In 1976, British writer Michael Harrison announced two revelations about this trident of SHC in his first book on the subject. One revelation was startlingly provocative... at first. After placing the S.S. Ulrich at latitude 50° North and 120 miles from Cape Clear, Harrison turned to a map of Europe and discovered the three deaths "lay in relation to each other *topographically* [*sic*]." Precisely, said he, these men died "exactly" at the points "of an equilateral triangle, with sides some 340 miles long"—meaning Harrison had discovered *a huge equilateral triangle of SHC!*

He claimed 340 miles separated Greeley from Turner, Turner from Bruik, and Bruik from Greeley. A glance at any map shows that almost *twice* that distance had to separate Greeley from Bruik. Therefore, Harrison discovered (at best) an isosceles triangle. I pointed this out to him, and his geometry changed to "an isosceles triangle, with sides 340 miles long" in the 1978 edition of *Fire From Heaven*.

Though Harrison still provided no basis for pinpointing where in the ocean Greeley inflamed, let's assume (for argument) he is right on *this* point. A ruler and atlas in hand, I scaled the distance between Harrison's position for Greeley's ship and Turner's auto-SHC at Upton-by-Chester as 433 miles. If the isosceles triangle of SHC exists, 433 miles from Upton-by-Chester will bring one to the Dutch town of Nijmegen. Instead, The Netherlands is entirely overshot, and the Bruik-apex of Harrison's fire triangle falls well inside Germany. (The actual distance from Upton-by Chester to Nijmegen is 390 miles.) It's not even an isosceles triangle!

Harrison calls his discovery a "fact" based on cases for which I can find no confirmation in contemporary newspapers. Maybe I've overlooked something. Maybe Lloyds of London overlooked listing the S. S. Ulrich, too. But for now I'll err on the side of caution and consign this tale of three-fingers-of-fire-which-burned-only-flesh to the closet of SHC myths.

Have Governments Discovered SHC?

The odds which make the MacDougall-Harrison Triangle of SHC such a long shot are stacked even higher against two other episodes said to involve SHC.

Imagine that scientific research discovered a way to controllably kindle SHC. To the militarist it would be the perfect weapon: cleanup after engagement would be as simple and swift as sweeping up a few pounds of ashes per soldier. Now imagine that a research and development program actually perfected such a weapon during WW II and kept it a secret for over fifty years.

You have just imagined The Philadelphia Experiment, now firmly anchored in the mythology of secret advanced research projects and government coverups. Although it sounds like a story born in a tabloid, it has been given serious attention by serious writers and even Hollywood.

The Philadelphia Experiment is a tale about a destroyer-class ship, identified as the U.S.S. Eldridge, which allegedly, in October 1943, teleported itself from Philadelphia to the Virginia Beach coastline and back again in the most significant experiment of WW II. Imagine: the U.S. Navy with a dock-hopping ship!

The Experiment has been the subject of several books, two movies, and (not surprisingly) much debate. What interests us here is the least mentioned aspect of this wonder voyage: that two U.S.S. Eldridge sailors spontaneously combusted on deck after being exposed to the ship's space-and-time warping supertechnology. This dismayed everyone and frightened the crew, said the story's originator, Carlos Miguel Allende (alias Carl Meredith Allen). The duo burned for eighteen days in the resulting "Scorch Field," as Allende called it. At this point, according to Allende, the government halted the project.

No wonder. Watching men embraced by a flaming torch for two and a half weeks would terror-smite the most stalwart sailor!

Allende's last contention alone—of prolonged human combustion—would cast serious doubt on his entire story... unless perhaps there was a parallel elsewhere in history.

The *Annual Register for 1763* recorded the fate of one John Mitchell of Southampton, whose body "being fired by lightening [*sic*], continued burning for near three days, without any outward appearance of fire, except a kind of smoke from it." This is a retelling, under a different name, of John Hittchell's slow burn from natural forces for three days in seventeenth-century England.

Does this PHC of Hittchell (aka Mitchell) raise the possibility that a pair of sailors exposed to twentieth-century reality-bending technology could fry for six times longer?

Since the United States government has clandestinely exposed its civilian population to germ-warfare tests and innumerable toxins, this question is not far-fetched. So: was The Philadelphia Experiment a bizarre military exercise gone awry or a hoax?

I have endeavored to substantiate this story by several means. I searched Philadelphia newspapers for stories about a vanishing ship and crew. Nothing. Thrice since 1990 I met with Alfred D. Bielek, who claims that it was he and his brother who were operating the space-warping machinery aboard the U.S.S.

Eldridge when they and their ship teleported up and down (or into and out of) the East Coast on August 12, 1943 (instead of the oft-cited October date).

Al Bielek had no information to impart about the alleged SHC horror on deck. "Yes, that happened," he responded to my initial questions about the seared sailors, "but I didn't see it myself. One burned for eighteen days."

Bielek, who weaves a narrative of ever-increasing complexity that now includes ongoing manipulation by the United States government of the time-space continuum to thwart a takeover of Earth by extraterrestrials, is unable to answer what I view as a crucial question: Just exactly *where* in the Delaware River downstream from the sprawling Philadelphia Navy Yard was the U.S.S. Eldridge when it suddenly vanished... and reemerged? Even during WW II, this was a populated shoreline, and I find it incredible that in daytime in this busy waterway apparently no one noticed a destroyer's abrupt disappearance and reappearance.

Additionally, I have personal correspondence from one of Allende's neighbors, Robert A. Goerman, who states that Allende (aka Allen) had the reputation in his youth of being a notorious fabricator of fabulous tales and indeed proudly confessed to him that The Philadelphia Experiment was a grand hoax born of his own initiative.

If someday the Philadelphia Experiment should prove factual—and I am unconvinced it is anything other than modern myth—the United States Government has much more to answer for than the SHC of two of its sailors in 1943.

What I can say with certainty is that the Philadelphia area is a hot spot for anomalously burned people. Whether their number is increased by experimental research that smoked two sailors stationed at the Philadelphia Navy Yard in 1943, awaits better evidence.

The same need for better evidence can be said for the international conspiracy described in the book *Alternative 003* (1978), in which journalist Leslie Watkins purports that no less than three American and British scientists had been exterminated by a "hot job" to protect "the true story behind a shared secret that both America and Russia are ruthlessly obsessed with guarding"—a secret which includes a Mars landing by NASA astronauts seven years before Neil Armstrong set foot on the Moon!

Said to be executed by government-sponsored SHC technology were Mr. Peterson, a California aerospace engineer; Harry Carmell; and Sir William Ballentine, an eminent astrophysicist at Jodrell Bank, England. Sir Ballentine, according to a pathologist whom Watkins quotes, left behind in the wake of his murder in early February 1977 a shrunken skull. Shades of Mrs. Reeser, again? "The Americans and the Russians have certainly been experimenting along those lines, with a view to developing spontaneous combustion as a remote-controlled weapon," the pathologist explained, "but the results of those experiments have been kept secret."

Inquiries into the facts argue that *Alternative 003* is, despite Watkins's contention otherwise, an intricate myth spun around a fictional BBC-TV program of the same name, which was broadcast on April Fools day, 1977. One credible point Watkins does make is what he has the pathologist say about SHC: "There is still no known explanation for this phenomenon."

SHC-by-government-edict through sporadic testing of a doomsday weapon upon an innocent public is, one hopes, *not* the explanation.

Because some fanciful episodes invoke SHC and some deaths attributed to SHC shouldn't be, does that mean that SHC is always a half-truth; rumor; or outright prevarication, as fireman Kennedy avers when he speaks of the myth of SHC? Hardly.

As I view the evidence, the greater myth is that SHC *doesn't* happen. To conclude otherwise is irrationality on stilts.

That long-held myth is being demolished at last, I trust.

If this sounds like a very nonfortean statement to make of a world that forteans tend to view as universally eccentric, and especially if the assuredness of this statement makes you uncomfortable, I recommend that you skip over the next few chapters. More explosive evidence for the myth that SHC is only fable awaits, with new twists.

15

SURVIVING SELF-INCINERATION: THE ANGEL WHO BEAT HELL FIRE!

> It was driving us *crazy* trying to find out what happened. We exhausted all the evidence at our disposal. We just could not place approximate cause.
>
> —L. Paul Cobb Jr.
> attorney for Jack Angel

Do those individuals who experience the fire within always die? Dr. Joel Shew wrote under "Spontaneous Combustion—Catacausis Ebriosa" for *The Hypropathic Family Physician* (1854) that "in no instance has recovery been known to take place after the appearance of this most singular of all pathological states."

It would seem impossible to be otherwise. Classically, the person who becomes a human fireball leaves behind an extremity or two—feet, ankles, maybe lower legs, hands, perhaps a skull (sometimes said to be shrunken)—and a mound of sweet-smelling powdered bone and tissue burned more completely than a crematorium normally achieves.

But with SHC the impossible is commonplace.

In Atlanta, Georgia, in the fall of 1980, I gave a lecture about SHC. A woman in the audience told me afterward that she knew of a businessman whose roommate might have survived SHC. Pursuing this lead, I soon made contact with Jack Bundy Angel.* Six weeks later, I sat down with him in his living room in Georgia to learn about an incredible incendiary tale.

THE NIGHTMARE BEGINS

On Monday, November 11, 1974, Jack Angel set out to make sales calls in the motorhome bus that he'd converted into a garment showroom. He earned a lucrative $70,000 annually as a clothing salesman. He was successful, jovial, healthy, and happily married to a beautiful German woman many years his junior. Life was good, and the future looked heavenly for Mr. Angel.

The following morning, Tuesday, he had an appointment to meet with a buyer in Savannah, Georgia.

He arrived in Savannah around 9:30 Monday night, to discover his motel room had been given to someone else. Angel wasn't worried. "I called my wife and told her I'd be sleeping in the bus," he recollected, "and I gave a guy $5 just in case she'd call and need me for something."

Despite the cool evening, Angel felt hot. He dressed lightly in only a nightshirt and pajama bottoms and turned on the air-conditioner. It was an omen of tragedy to come.

Pushing aside racks of clothing samples, he sheeted the motorhome's sofa and lay down for the night... entering a nightmare he'd never escape.

Angel missed his Tuesday morning appointment because he overslept. *Overslept!*

"I don't remember when I woke up—whether it was Thursday or Friday—but when I came to, I finally woke up by myself in the bus." Subsequent research has shown he awoke around noon on Friday the 15th[1]—*four days later!*

Jack Angel arose from his Rip Van Winkle nap a different man. His right hand was burned black on both sides, from the wrist to his fingers. "It was just burnt; blistered!" he exclaimed. "Burned to a crisp—like I got hold of a hot wire, you know, and held onto it. And I had this *big hole in my chest*—exploded. It left a hell of a hole! I was burned down here on my legs and between my groin, down on my ankle, and up and down my back!"

Had the burning extended to the sofa he slept on? "No!" he exclaimed, mixing emphasis with disbelief that this could be so. Extended to his clothing? "I was wearing just a nightshirt and pajama bottoms. No, *nothing* was burned. There was no fire at'all! *Nothing!!*"

Feeling no pain, Angel got up, showered and dressed in the clothes he had lain beside his bed four days before. Unlike his body, Angel's pants and shirt were not touched by the mysterious fire; nor was the sheet on which he slept.

---

* I first introduced this case to the world on *That's Incredible!* (January 19, 1981).

Angel left the motorhome and entered the motel's cocktail lounge, staggering "like I was drunk," he recalled being told. Though he hadn't had an alcoholic drink in years, he felt parched and ordered a Scotch. "I thought maybe it would pick me up. But I never drank the Scotch—I'm not a drinker anyway," he stressed.

The waitress immediately remarked about her customer's right hand. "Yeah, looks like I got burned," Angel muttered, still only half-conscious of his injuries and insensitive to any pain. Minutes later, Angel felt nauseous and headed for the public rest rooms.

"Can I help you?" Bill Cribbs asked the man who stumbled past his office door at the Inn.

"I showed him how badly my hand was burnt," said Angel, "and about that time I fainted." He collapsed into Cribbs's arms. The next thing Angel remembered was lying on a hospital bed and an intern "with a pair of tweezers pulling skin off my arm."

Angel was in shock, unable to speak coherently. The physicians appeared shocked too, he noted. One doctor at Memorial Medical Center, Angel vividly remembered, "explained to me I wasn't burned externally, I was burned *internally*."

Around midnight, their bafflingly burned patient abruptly regained full consciousness. Then came what Angel called "excruciating pain." He began sweating profusely. He asked for, then demanded, answers to his predicament. None were forthcoming. "I couldn't get any answers from these people at the hospital," Angel's words still echoed with frustration.

He called his wife, telling her to examine the motorhome for signs of electrical faults. "There wasn't anywhere that you could find [burn] spots on the clothing," he said of his wife's search. "There was no evidence of any fire in that bus!" Once Angel had staggered away from it, that is.

Perplexed by his injuries and displeased with his treatment, Angel insisted he be transferred to a hospital renowned for burn treatment. Diagnosis remained unchanged, however.

"Mr. Angel presented himself to the hospital with a severe burn injury of the hand and minor injury of the chest wall," said Dr. David Richard Fern of the Veterans Administration Medical Center's burn unit in Decatur, Georgia. "This was a third-degree burn which damaged the skin severely and most of the underlying muscle of the hand, causing a total anaesthetic hand." Dr. Fern declared the ulnar nerve "completely destroyed," with the median nerve showing "questionable viability." The burning that caused necrosis in the victim's hand continued up the *inside* of his right forearm, as well.

Ironically, said Angel, *"There wasn't a hair burnt on me!* You understand what

I'm trying to say to you?!" (Remember Billy Peterson, whose body hair likewise was impervious to flesh-searing burns?)

Angel remembered with wry amusement the many doctors and interns who gravitated to his bedside "because this was such an unusual burn."

The hospital staff asked questions, among themselves and of their patient. Angel couldn't tell them how he had burned, because he had slept through whatever had happened to him.

His badly charred hand failed to respond to antibiotics and became septic. Faced with many months of painful skin grafting and reconstructive surgery the success of which could not be guaranteed, Angel elected to have his burned right forearm amputated. During the next fortnight his wounds healed with amazing swiftness, and on January 14, 1975, Angel was discharged from the hospital.

MR. ANGEL GOES TO COURT... ALMOST

Restoring normalcy to a life disrupted by what several physicians independently diagnosed as internal burning wasn't easy. Haunting questions remained. What caused the horrific burns that blackened his hand, ate into the inside of his right arm, and created explosion-like wounds in his chest without damaging any of the surrounding clothes? Why were his pajamas not charred? Why was there initially no pain?

Angel wanted answers.

Enter the partners of a Georgia law firm renowned for expertise and success in liability litigation: Ross Bell and L. Paul Cobb, Jr. They thought they had the answer: since Angel had been asleep and alone in his motorhome, certainly a design error or equipment malfunction caused the incendiary injuries, they told him. If Angel agreed to their representation, they would take the case on a contingency fee and file suit against the manufacturer. Angel would get his answers at the lawyers' expense, and get two-thirds of the settlement as a throw-in. He agreed. "We engaged in litigation right away," Bell said. They sought $3,000,000 in damages.

Now all the esteemed law firm had to do was pinpoint the cause.

Again, the motorhome was examined. "There was an Onan generator operating at the time he went to sleep," said attorney Bell, "and we addressed the theory that it was producing enough direct current that it might have shorted something out. But that proved not to be the case. That was our original theory. We went through many, many more."

The attorneys considered faulty wiring. "But that proved not to be the case," Bell admitted.

"We even checked into the possibility that lightning struck him," Bell continued, "because he had some burns around the lower part of his legs." Meteorological records mentioned no lightning in the vicinity during the fateful period.

They investigated whether overhead power lines had somehow conducted electricity to Angel through the vehicle's chassis. "We thought of direct current entering through his hands and discharging through some other joint," said Bell of their client's body. "But that turned out also to be wrong!"

The lawyers then hired Roy A. Martin Associates, a prestigious engineering and technology laboratory, to inspect the motorhome from top to bottom. They did, literally—eventually dismantling the vehicle right down to the wheelbase.

No evidence could be discovered to link the motorhome to its owner's baffling burns, and so the engineering lab *speculated* that Angel became scalded by a jet of pressurized hot water that gushed through a safety valve he had opened when allegedly adjusting the water heater.

"They told me that somehow I had been burned by steam!" exclaimed Angel, throwing up his shortened arm in exasperation. "Well, steam again is an outside burn. And I wasn't burned outside! I was burned internally. The thing didn't make sense!"

Nor did it make sense to Martin Associates, which admitted there was *no* evidence to support their scenario. "We had other engineers after that check on it," Cobb said in recounting his firm's efforts to make a case for their client. "And we came up empty."

The attorneys then tried hypnosis. Three times a noted Atlanta psychologist attempted to regress Angel, without success. The attorneys consulted with another psychologist. No success.

Confronted with a rapidly nearing day in court, Angel's attorneys became exasperated. "It was driving us *crazy* trying to find out what happened," Cobb declared. The fact that the engineers—on whose determination the case ultimately rested—admitted their "scalding water scenario" was nothing more than speculation left the lawyers empty-handed.

What began two and a half years earlier as a hot case for the legal team now fizzled before their eyes and their single-handed client.

Angel's attorneys withdrew the case from the court docket just one week before the scheduled trial, dashing all hope of a $3,000,000 reward for their heroic efforts. Why? To quote Cobb: "We just could not place the approximate cause... We never could come up with enough to keep us in court."

With legal counsel acknowledging defeat, the case of Jack Angel was closed. Officially.

COUNTERPOINT: A DISCREDITING NOT WORTH A NICKEL

The pursuit of spontaneous human combustion has more twists than a Brooklyn Bridge suspension cable, and, in this regard, the case of Jack Angel would not disappoint.

One of those twists came when Joe Nickell and John Fischer wrote an article for *Fate* (May 1989) entitled "Did Jack Angel Survive Spontaneous Combustion?" They answered a resounding no. Angel, they deduced, was "accidentally self-injured by a scalding, pressurized jet from his motorhome's hot water system—not by 'spontaneous human combustion,' a phenomenon whose existence remains as doubtful as ever."

Had Nickell the former private detective, and his associate, discovered facts that I and Martin Associates had overlooked; something obvious that should have led me, and Jack Angel's attorneys too, down the pathway of scalding water instead of internal burning?

That might be. I am not omniscient. But in repeating the scalding-water argument for their book *Mysterious Realms* (1992), Nickell and Fischer, who did not hesitate to quote generously my published research on this case as it stood in 1982,* impute that I am guilty of "superficial, even slipshod investigation" in the Angel case. Oh?

This damning allegation libels the quality of evidence I am presenting to you. Therefore, it is important to examine the investigation these two authors say they conducted into Jack Angel's burn injuries.

Earlier, you read about Nickell's techniques for a good investigation. Now you can judge for yourself if he followed his own guidelines in the case of Jack Angel's burns. I think the exercise will prove instructive.

Let the trial of the two investigations begin.

> INVESTIGATE: to examine the particulars of in an attempt to learn the facts about something hidden, unique, or complex, esp. in search of a motive, cause, or culprit.
> —*The Random House Dictionary of the English Language* (1966)

I reexamined my inch-thick file on Jack Angel, then telephoned Nickell to ask how he had reached such a different (and mundane) conclusion.

"When you start with the Jack Angel case, you know almost by definition the fact that the case is unusual," Nickell acknowledged. He then proceeded to demonstrate how he had applied his own techniques to the case at hand. I was unprepared for just how unusual this case would now become.

---

* *That's Incredible!*, ABC-TV, January 19 and April 27, 1981: "The Man Who Survived Spontaneous Combustion," *Fate* (September 1982), 60–65; "Jack Angel, SHC Survivor," *Fortean Times* (1983), 12–15.

How did Nickell explain Angel's burned-black hand, something not typical of a scalding injury? "No problem," answered Nickell; a "minor discrepancy" to be dismissed as inconsequential.

Had Dr. Fern told him he had changed his diagnosis that "an electrical burn... *definitely* a high-energy source" had caused Angel's injuries? No, Nickell retorted, he had not talked to Dr. Fern. Why not? Because Dr. Fern offered "non-expert opinion of what he alleged that he thought it looked like..."

A physician assigned to the burns unit of a large hospital is a nonexpert at diagnosing burn injuries?

Had Nickell analyzed the photographs of Angel's burns for evidence of scald injury? "I didn't know those existed," he confessed, quick to give himself an out by suggesting the photos "may be misleading because of contrast..."

Not true; contrast is razor sharp. Anticipating Nickell's allegation, I had taken these photographs a few days earlier to Dr. Donald R. Mackay, a surgeon who treats burn injuries at Pennsylvania State University's prestigious Hershey Medical Center. After carefully examining this evidence at my request, Dr. Mackay concluded: "This gentleman suffered these burns as a result of contact with an electrical source of some kind."

I pointed out that Dr. Mackay provided independent corroboration for Dr. Fern's firsthand determination of the type of burns Jack Angel suffered.

Nickell: "I'll find an expert to counter anyone you find..."

An interesting methodology on which to conduct an "unbiased" investigation.

> A skeptic is a person who, when he sees the handwriting on the wall, claims it is a forgery.
> —Morris Bender

Later, in *Fate* (December 1989), Nickell would demean this contradictory evidence to his scalding-water theory by referring to "the alleged opinions of two doctors"—that is, Drs. Fern and Mackay. In truth, the physicians' statements aren't "alleged opinion"; they *are* opinions, ones based on their medical expertise.

Further corroboration of the verbal testimony given by Fern and Mackay comes from the medical records dealing with Angel's hospitalization—evidence in my possession and unpublished until now:

"Third degree burn to the right hand" is what the Veteran Administration Hospital records say Angel was being treated for; a report filed and signed by none other than Dr. Fern. "Electrical burn, right hand" is what attending physician Dr. Paul Salter wrote on his report. "Patient suffered electrical burn of hand," wrote Dr. James Madden, the surgeon who conducted tissue examinations of Angel's necrotic right hand in early January 1975. Dr. William B.

Riley Jr. and Dr. M. Peters also assisted in Angel's treatment, and signed similar reports. Not one of these medical professionals found any reason to refer to their patient's injuries as scald-induced. One medical document clearly lists the nature of Angel's burns as "internal in origin."

Nickell confessed knowing nothing about this additional evidence.

Had he determined why the lawyers (with $3,000,000 at stake) found themselves unable to substantiate in court something so obvious as a faulty valve that allegedly led to the scalding of their client?

"Well, I don't know," Nickell responded hesitantly.

Had he even *spoken* with the attorneys? No, Nickell quietly replied. Hence he could not know that attorney Cobb had, in retrospect, accepted SHC as a reasonable solution to his client's injuries, based on best evidence.

Nickell and Fischer rely on the motorhome's water valve to spray Angel with scalding water, a theory I had considered and dismissed for lack of evidence. Had I missed something? I re-investigated this point, presenting it to five recreational vehicle technicians and/or dealers who together represented more than sixty-five years' experience servicing recreational vehicle water systems, including the exact model Angel owned. *Every one of them* rejected outright for technical reasons the Nickell/Fischer scenario. Two laughed out loud.

Even though Nickell instructs that investigation must be based on what he calls "best evidence," he conveniently ignored this important testimony when he later defended his theory for *Fate* (December 1989), even though he was made aware of it.

Where do these two armchair scholars lead the thoughtful? Consider the rationale behind the Nickell/Fischer alternative to SHC: that Angel allegedly opened his motorhome's hot water safety-relief valve and then *calmly stood there* while pressurized water seared his right hand and forearm from the inside-out, then raised his arm—*still holding the valve open*—so the alleged hot water jet could drill a hole into his chest, then pirouetted—*still damnably holding onto the valve*—so he could get "spattering" burns on the nape of his neck, his back, his groin, and legs before *finally*(!) letting go and returning to bed and forgetting that any of this happened.

Make sense? Only as the perfect *non sequitur*. To quote Nickell on the basics of investigation: "The burden of proof is on the advocate of the idea." So it is.

Had Nickell ever asked Angel if this was what happened? "No," replied Nickell, who then implied that Angel could be "lying" or promoting a "hoax."

Here's something else to test Nickell and his cohort. "It not only burned my hand and chest," Angel had told me matter-of-factly about whatever attacked his body, "but *also the cushions in the vertebrae of my back*!" The spinal column discs? "Yeah!"

As Angel had no spinal pain before this incident yet suffered considerable back pain thereafter, whatever it was that burned outward from his chest likely damaged his spinal column too. How Nickell and Fischer will explain spraying hot water as the means to fuse discs in one's backbone, one can only guess. Maybe as just another "minor" discrepancy?

ANGEL'S RETROSPECTIVE: REFLECTIONS ON A FIERY FATE

In *A Scandal in Bohemia* Sherlock Holmes, whom Nickell likes to quote, observes that "It is a capital mistake to theorize before one has data. Insensibly one begins to twist facts to suit theories, instead of theories to suit facts." If you believe that Nickell and Fischer conducted a solid investigation and are guiltless of theorizing against facts, there's a bridge in Brooklyn for sale.

In fact, Nickell and Fischer confessed in *Fate* (September 1982) that "We did not...actually investigate the Angel case...but we did privately discuss the case at some length." Imagine that! Had anything changed by the time I called Mr. Nickell seven years later, *after* he had published his scathing rebuke?

Your verdict, please.

Tested upon the fire of evidence, I say their theory evaporates like water spritzed onto a sizzling skillet. As attorney Cobb confessed to me in 1980

AVCO technicians and/or dealers, on the scalding water theory:

(a) Earl Witsil, service manager for Etnoyer's RV World with ten years in the business, on alleged pressurized jetting from the relief valve: "There's no reason for that."

(b) Dick Beaver, of Beaver's Trailer Sales, sold motorhomes since the late 1960s. Had he ever heard of a water valve pop off like Nickell described? "No.... The relief valve points 90 degrees down, *not out!* I haven't seen one that comes right out at your face."

(c) Dale Dentler, an RV service technician: "I don't see how that's possible. If there was that much pressure, the water input coupling pop-off would have popped out." Which it did not.

(d) Tommy Hippensteel, servicer of AVCO motorhomes in the early 1970s, about RV water relief valves: "All have basically the same design" and are angled to direct over-pressures down, *away* from anyone standing in front of the valve. I've *never* seen one that sprayed straight out." He added that the pop-off valve would have instead ruptured, something Martin Associates did not find. Or if the valve was truly defective, then the aluminum water tank would have exploded from a subsequent build-up of hot water pressure. "It would be very rare to be *scalding hot* water." Of Nickell's scenario, he remarked: "It seems strange!"

(e) AVCO dealer Wayne Schweitzer identified Angel's motorhome as a "Limited Edition" 1973–1974 model; he had sold one himself. "I don't know how he'd burn his hand that way," he said of Nickell's theory; "I find this hard to believe.... *Never!*"

—personal interviews (May 23, June 6, June 23, 1989)

and again in 1989: "We *never* could find out what caused it. *I would like someday to know.*"

And that brings us, and Jack Angel, back to the theory of SHC and survival.

What did Jack Angel *himself* think happened? "Hell, I haven't stopped thinking about it for the last five to six years!" he exclaimed. "I really don't know what the hell happened to me."

What did it feel like to be burned so inexplicably? "I don't remember any pain" for several hours after waking up, Jack told me. How long he had been painlessly burned before waking, no one could determine.

And *how* did it happen? "There are a *lot* of things that happen that you just don't know what," Dr. Fern told me as he reflected on Angel's fate. "Maybe it's a *miniature nuclear explosion.*" (The physician had unknowingly picked up on a promising explanation, explored in chapter 8.)

Jack told me he slept on his side with his arm usually tucked under him. Perhaps as he lay on his right hand, some force inside his right hand and chest "exploded"—to use Jack's own description—with secondary flaring up his spinal column and exiting at his neck, back and groin. Perhaps the searing force first exploded from within his hand, then burned up his arm and simultaneously into his torso. Perhaps his injuries will never be explained.

"I don't understand the whole damned thing," Jack Angel said as he waved the stump of his right arm in front of my face. "I don't know anything about human combustion theories. But something has to cause it, doesn't it?"

After years of confronting the mysterious incident that significantly altered his life (for the worse) yet stopped short of fatally transforming him into a human fireball, Angel has concluded that "The only explanation is that I was a victim of spontaneous human combustion."

It is an explanation that I, too, will continue to support—until a different scenario can be proposed that provides a less "lamentable" solution (to use Nickell's description) that encompasses, without discrimination, *all* evidence provided by Mr. Angel.

Is Jack Angel, then, the one person who literally escaped the popular contention that SHC is always fatal... *the* exception to the rule who lived to speak about his enigmatic self-inflaming? No.

16
"Light My Fire"... Beating Back These Innermost Flames

> As soon as questions of will or decision or reason or choice of action arise, human science is at a loss.
>
> —Professor Noam Chomsky,
> *The Listener* (April 6, 1978)

Joe Nickell, implacable, reaffirmed his disbelief in SHC and presented a challenge at the close of our conversation. "If you find you have some *real* evidence to the contrary," he said agreeably, "I'd sure like to see it. If you can prove me wrong, I'd like to know about it."

Challenge accepted.

Prior to Jack Angel, the most widely known case thought to be SHC survival happened to Professor James Hamilton. Remember, a con-

> Accept the challenges so that you may feel the exhilaration of victory.
> —Gen. George S. Patton

temporary document refutes this interpretation. Professor Hamilton never had to do battle with, nor triumph over, SHC.

But I'm not ready to throw in the towel yet.

There is Irene (Renie), who likely did *not* die as the result of her flame-gushing leg upon which she fled the New York City Woolworth store. This is, admittedly, a weak case to satisfy the standards of evidence espoused by Nickell.

Not to worry. In fact, I'll make *his* challenge more interesting. Not only do strange fires happen, as firemen and medical professionals admit, strange fires sometimes *persist* in haunting certain people.

Here, then, are more instances of truly remaining calm under fire. Persons who have survived a personal plague of pyrophenomena. Repeatedly.

• One day in 1943, Paul V. Weekly rented a hotel room in Sioux City, Iowa. During the night his foot began to itch, and he threw back the covers to see what was the matter. A lurid blue glare from flames leaping around his toes illumined the room, according to *Doubt* (June 1943). Weekly "smothered" the fire (if fire it was) with his bedsheets, and went back to sleep. Later that night his itchy foot reawakened him. Once more, Weekly had to contend with a hot foot... in the same successful fashion as before. After which, he slept without further incident.

Some people will sleep through anything! Nothing definitive can be said about this astonishing tale, except that Paul Weekly must surely be in the running for the *Guinness* title of Remaining Calm Under Fire.

• Surely another contender is Susanna Sewell of Maryland. In 1683, she entertained—perhaps horrified—her houseguests with her (almost) unique ability to project blue flames from her body, mostly from her fingers. Her wild talent projected from beneath her garments too, it was reported in the *British Medical Journal* (August 12, 1905), and her fingertip sparks said "to be of fire" could be transferred to others.

• And what does one make of Marguerite-Frédérique-Catherine Heins, a seventeen-year-old fraulein in Hamburg whose hand suddenly ignited on January 21, 1825? Her fingertips spewed blue flame, attested *Heckler's Annalen* (August 1825) about this incredible case. Hospitalized for three and a half months(!) until the fire went away, she then went back to work.

Clearly the fire with its blue flames, which had proven to be fatal in cases heretofore witnessed, could be survived as well. Survived repeatedly.

• Sometime prior to 1839, a unidentified friend of W. B. Carpenter experienced a fate similar to Fraulein Heins. Professor Carpenter, instructor of medical jurisprudence at London's University College, a Fellow of the Royal Society, and an eminent physiologist, believed in SHC with good reason. Not only had his Society chronicled several cases, but as he wrote in his *Principles of General and Comparative Physiology* (1839), he had a friend whose hand "twice at distinct intervals emitted a flame which burned the surrounding parts." Carpenter speculated that combustion of phosphorus explained the predicament.

• In the 1960s and early 1970s, according to Montague Ullman in *Research in Parapsychology* (1973), observers of the Soviet psychokinetic agent Alla Vinogradova attested that "sparks have been seen at her fingertips while she is working..."

• In the 1970s, another Soviet telekinetic medium, Nina Kulagina, convinced many who observed her that she channeled an energy that could burn her from within. The distinguished parapsychologist Professor Genady Sergeyev told London's *Sunday People* (March 14, 1976) about his experience with Ms. Kulagina: "On several occasions the force rushing into her body left burn-marks up to 4 inches long on her arms and hands... I was with her once when her clothing caught fire from this energy flow—it literally flamed up. I helped put out the flames, and save some of the burning clothing as an exhibit."

An exhibit that reinforced the ever-growing evidence that humans, when properly energized, can repetitively become an ignitable hazard to themselves.

REMAINING REALLY CALM UNDER FIRE

• October 1980: Peter Lynam Jones of central California was sitting on the side of his bed. His wife, Barbara, stood beside him. "I looked down and smoke was billowing from his arms as though something was on fire," Mrs. Jones told me. "We both started frantically trying to put it out. Suddenly it was gone."

There was no odor. No heat. No tissue damage. Just "gushing smoke."

"*What was that?!*" Mrs. Jones implored, after gaining a modicum of composure. "Beats the hell out of me!" her husband retorted.

Nonplussed, Peter went out that afternoon for a drive. With both hands firmly on the steering wheel, shirt sleeves rolled up, he was stopped at a railroad crossing mulling in his mind the bizarre way his day had begun, when... to his utter astonishment *it happened again!* Pale blue-gray smoke spewed from both his arms, quickly filling up the car's interior with a pallid haze. This time, the smoke had a distinctive "metallic taste."

> SHC "as a scientific fact, is not entertained.... The burning of the body cannot be accomplished without contact with fire."
> —*Burns. A Reference Handbook of the Medical Sciences* (1901)

Then as abruptly as it began, his smoking ceased again. *"Weird"* thought Peter Jones to himself. (Perhaps Orsola Benincasa and St. Catherine of Genoa had the same thought when, some four hundred years earlier, whitish smoke issued from their throats.) He did not tell his wife about this second smoking outbreak for many months.

If anyone is in the running against Paul Weekly for Remaining Calm Under Fire, it is Peter Jones.

Mrs. Jones said her forty-six-year-old husband drank heavily at the time. He was *not* drinking prior to either of these two events, however, though one could forgive him if he did so afterward! Her husband was also "violently outspoken," she said, and harbored "total hatred" toward her teenage daughter. An incendiary combination, perhaps?

I have since spoken several times in the 1990s to the Joneses, preparatory to producing segments about SHC for American television. Their congeniality indicates life has become much less stressful, and they now look forward to traveling (hopefully *sans* smoking arms) in their retirement years. That one day in 1980 *remains* vividly remembered, however.

"It was just the weirdest thing! Just bizarre!" reflected Mrs. Jones. "It's never happened again." I remarked that she must be relieved about this. "I sure am. And so is my husband!"

TWICE MORE THEY REMAIN CALM IN THE FACE OF FIRE FROM WITHIN

Again it happened—*again* twice!—in 1980. Again in a woman named Norris. This time, in the heartland of Illinois.

• "I am not sure that my experience will be of interest to you, but it was so puzzling to me that I'd like to share it with you," wrote Mrs. Elizabeth K. Norris after seeing me discuss SHC on *That's Incredible!*

Age sixty-five at the time, her story began on Mother's Day, May 11, 1980, around nine o'clock in the evening. She and her husband sat at the kitchen table working the Sunday crossword puzzle as they unwound after a day spent with grandchildren, when she faced a very different type of puzzle:

"One of the puzzle clues made reference to a type of gin. I went to check the ingredients on a bottle of gin we had in the bar. Just as I returned to the kitchen I was astonished to see a thin stream of smoke rushing down my left arm. I screamed, 'I'm on fire!' and rushed to the kitchen sink to douse the fire, although I saw no flames, just smoke. At that moment I had the taste of ashes in my mouth. I spit what appeared to be ashes into the sink. After a quick drink I used a Kleenex to wipe out my mouth. That was the end of the experience."

(Remember the Venerable Beatrice Mary of Jesus, whose "interior conflagrations" ejected *ashes* from her body?)

Mrs. Norris's husband immediately checked for his pipe, thinking that burning tobacco had fallen from its bowl onto her arm.

"He found it cold in the ash tray across the table from where I was sitting," she said. "He did, however, see the gray ashes on the Kleenex. I must admit that having seen a segment on *That's Incredible!* which told of cases of SHC, that my first reaction was that I was about to become a victim. However, nothing developed beyond that at the time."

Until twelve days later. "I did have a second experience with the stream of smoke down my arm and the taste of ashes on Friday, May 23," she wrote.

"This second experience was only for a split second and was not observed by anyone but me."

In a two-week period, odorless smoke had twice gushed from beneath her blouse sleeve. "It was just like a white plume, the width of a pencil or finger," she said of the smoke in a follow-up interview. "It came down the length of my arm, down along the side of my body. It was *definitely* coming from underneath the sleeve of my blouse."

Her garments were not singed, though they did get soaked as she frantically doused her arm each time with water from the kitchen sink. She suffered no physical injury nor a third recurrence of the smoking arm.

"My mother was not given to fanciful imaginings by any means," her daughter Marybeth Beechen informed me. Mrs. Norris, she added, had a "general take-charge attitude toward life" and "sought harmony in all relationships" before—and after—"her experience with what we believe was the start of human spontaneous combustion."

Perhaps her predisposition to happiness and harmony extended to the fire she found within herself, and twice warded off any serious consequences.

• The dilemma of double SHC occurred next to a Virginia woman who had an interest in spiritual studies and in raising her Kundalini. Had a bad hair day, too. Twice. In March 1986, the hair on the head of Susan C– took fire, each time preceded by light smoke. In neither instance was an external source of flame anywhere seen. All that Susan would tell me about her hair-raising/singeing experience was that she felt Kundalini was involved. She refused to say anything more.

IF IT PLEASES YOU, TWICE IS NOT ENOUGH (PART I): TONG TANGIANG

What happens twice, happens thrice. Firemen believe this about normal outbreaks of fire. The maxim is no less applicable to non-normal fires...and survival of them.

• Young Tong Tangiang was four years old when, in April 1990, a one-inch square hole burned through two layers of his clothing. Rushed to a Hunan Province hospital, the lad started smoking again, *spontaneously igniting himself four times during the next 120 minutes!* His right hand, armpits, and genitals burned, reported the *China Youth* News (April 30, 1990).

Another of those Chinese children only completely comfortable when not clothed, Tong was placed under the care of Dr. Hsing Peng who believed a strong bioelectric current intensified by stress or nervousness explained Tong's torture. The boy's mother added to the mystery by stating her son had set fire—not by playing with matches, but by his mere proximity—to a mattress, and narrowly missed igniting his grandmother's hair when she got too close. Fire-starting Tong suffered social stigma too, as friends avoided him. Said Dr. Peng, "No one wants to get burned and you can't blame them."

Tong Tangjiang: a victim of multiple spontaneous combustion *of* and *by* a human.

Young master Tangjiang is not unique, however. Now the evidence becomes really spectacular, and it comes from right here in America.

TWICE IS NOT ENOUGH (PART II): JOE NUZUM

British writers Janet and Colin Bord end the chapter about mystery fires in their superb book, *Modern Mysteries of the World* (1989), by pondering a possible synchronicity: "People whose presence starts fires, fires that plague homes, people who burst into flames and are totally consumed... Is there any link between them? That is the burning question to which we do not have the answer."

I believe I do.

Days before this book was set to go to press, I had the good fortune to speak with two gentlemen who spoke of unforgettable—and pertinent—experiences unknown to the public-at-large. Until now.

• Joseph A. Nuzum, thirty-seven, says he has had an unnerving relationship with fire since five years old. "It began with a charring effect when I was small," he revealed. "I got into trouble playing with matches—because I was fascinated with fire. Then papers ignited spontaneously by themselves, and I'd just sit there mesmerized. I had a fascination with fire. Still do."

That fascination with fire, including its spontaneity, led to life-long physical expressions of psi phenomena. Bending/breaking metal... handling hot coals, molten metal and glowing casings in his hands with impunity at the foundry where he once worked... moving objects remotely by psychokinesis (PK)... causing ice cubes to inflame... and telepathically extinguishing lit cigarettes are among this unassuming, sensitive Pennsylvanian's supernormal accomplishments.

Since the mid-1980s his feats have been studied and documented by Dr. Berthold E. Schwarz, one of the most brilliant minds engaged in researching fortean phenomena today. He told me that Nuzum indeed "has the ability to experimentally induce fire from a distance."

To learn more about this, I flew to the University of Nebraska in May, 1995, to spend two days with Nuzum and his wife, Sandy, plus Dr. Schwarz. There I witnessed firsthand some of these wild talents—including Joe's successful attempt to affect telekinetically a lit lightbulb some twenty feet distant.

Within a minute after he focused his intent upon it, the bulb loudly *exploded* in a flash of intense light.

To chronicle everything I learned that weekend is beyond the scope of this book. Even a partial listing of psi-induced spontaneous combustion (SC) claimed by Joe is daunting. But so provocative are they, that a representative sampling is worth the effort.* Here goes:

• January 5, 1990: Joe focuses on a cardboard safety-match, which suddenly flies off a countertop; minutes later, he smells smoke and discovers nearby a paper bag ablaze.

• March 14, 1990: Schwarz discovers that a new 9-volt radio, intended as a PK target for SC by Joe, is melted internally; wiring insulation melted; paper schematic diagram charred. Physically, Joe was nowhere near it.

• April, 1990: Joe aims his incendiary gaze at a tissue in his home; meanwhile, at a friend's house approximately eleven miles away, a hand towel instantly burns "to a crisp."

• September 30, 1990: Schwarz, in Florida, mails keys wrapped in a plastic bag inside a cardboard box to Joe in Pennsylvania. When the package arrives, some keys are bent; one is charred. Three of Schwarz's business cards in the box are charred, too; the bag is melted.

• December 29, 1990: Joe is psi-experimenting on a pack of cigarettes. The pack moves; inside the pack, one cigarette chars.

• 1991: at the home of Ray, a friend of Joe's, a towel draped over a refrigerator door suddenly ignites, burns a hole through itself, and chars the door. Ray telephones Joe, and learns that at that exact moment Joe was experimenting with spontaneous combustion.

• June 15, 1991: Joe debriefs Schwarz on how he achieves pyrogenesis. "I transform all the energy in the body into the hands," he explains; it is then explosively directed, like a martial arts practitioner (which Joe is) does when breaking boards. "Spontaneous combustion is exciting, a powerful agression, fire within. It is immense."

• 1992: Joe attempts another SC experiment... simultaneously, miles away, a mop begins to smolder in the kitchen at Ray's house. (Remember the 1981 case of the self-combusting mop at Leasowes Sports Center?) Wet or dry, this mop's burning was witnessed in progress.

• February 17, 1992, and beyond: the repertoire of this pyromeister expands to include psychokinetically lighting candles, plus a Butane lighter without flicking its flint; psi-dousing candles' flames from across a room; melting

---

* This list is compiled from personal interviews plus Dr. Schwarz's historic paper, "Observations on Presumed Psi-Induced 'Spontaneous Combustion'," in *Journal of the Fortean Research Center* (May 1995); available through FRC, PO Box 94627, Lincoln, NE 68509.

filaments in lightbulbs cradled in his hands and in flashbulbs at a distance; combusting sails on model boats in the middle of a pond; autonomically unscrewing a lid off a jar of apple juice, followed by the juice smoking until it ignites in twelve-inch-high blue and orange flames; forming sparks for several seconds on the blade of his Cris (ceremonial dagger); charring Play Dough; igniting packs of matches between his fingertips; sometimes charring paper inside glass tumblers, other times burning it with open flame; catching on fire the walls of mobile homes. Wood blocks char to smoldering charcoal; granulated sugar burns bluish-orange until it is a black gook. Dollar bills—another target of his piercing pyrokinesis—char inside overturned aquariums and in the clenched hand of Dr. Schwarz. Gasoline-soaked paper bursts into flame at Joe's intent, to the consternation of his wife, who yelled that she feared "trouble with the Fire Marshal for burning stuff at the wrong time."

That would not remain the only concern about the pyrophenomena surrounding her, however. Her husband's pyrokinetic propensities were about to be experienced up close and *very* personal.

• January 2, 1994: the Nuzums had been arguing this Sunday, prior to Joe attempting a new PK experiment which involved Sandy manually lighting a candle. Suddenly, the unexpected. "The flame *jumped* off the candle, onto her hair, which caught on fire. Her woolen sweater too!" confessed her distraught husband. Sandy Nuzum, facing a calamitous variation of SHC, slapped herself free of fire. Then erupted a firestorm of another kind, as Joe faced Sandy's understandable anger. One can but imagine a judge's reaction if charges of spousal abuse been filed based on this incident—

Three months later, fate turned the tables.

• April 5, 1994: shortly after nine o'clock this Tuesday evening, Joe felt the top of his head warming. Then hot. Then afire. Looking in a mirror, he watched as "bluish flames moved around the top of my head." No odor. No pain. Just the horror of seeing his hair singeing before his eyes. It was a short-lived subset of SHC: Spontaneous *Hair* Combustion.

• Just before midnight, April 6, 1994: Sandy noticed the singed dark hair on Joe's head. Before he can explain, his right index finger begins to tingle, then ignites! "It was blazing, bluish flames, about two and a half inches high," Joe remembered vividly. After a few seconds it was over, to the couple's mutual astonishment and relief.

I recall what history says happened to Susanna Sewell more than three hundred years before; more recently, to the toes of Paul Weekly. Outrageous hearsay; hoax; not to be believed? In 1994, another human finger spews blue fire; spontaneously; painlessly; without injury. It tingles. It self-extinguishes. It is witnessed. It will happen again, likely. History repeats.

Remember Hayes, the golden fire-from-its-belly retriever in England? Well, Nuzum's pyrokinetic talent may have some relevence, indirectly.

• Sandy Nuzum has seen her husband zap crickets "which go up in a puff of smoke." She described an incident prior to their marriage, when both noticed a big spider in her bathtub: Joe had the compulsion to stare at the arachnid which, they admitted, started to smoke and "just burned up." A mosquito became "a whisper of smoke" after Joe, some eight feet away, said he focused on it. Joe recounted walking in the woods in late Spring 1994, and noticing a crow some thirty-five feet overhead: abruptly the bird burst into flame and plunged to earth trailing smoke. "It was real," he said. "You could smell the burned feathers." Four additions to the catalog of spontaneous animal combustion, if you accept the Nuzums' testimony.

(I point out that Joe exhibits deep concern and reponsibility about what his psychic nexus might induce, these animal ignitions aside. He is careful, not cavalier; aware of the ramifications to damage he could cause.)

Debunkers masquerading under the guise of scientific skepticism will say all this is circumstantial or delusional, and proves nothing.

Really? Much in science is accepted on far less evidence than Joe Nuzum provides for psi-induced spontaneous combustion: eyewitnesses, photographs, videotape documentation—and the darling of Establishment science, repeatability.

"At the least," Dr. Schwarz concluded in his monograph to the Fortean Research Center about his decade spent with Nuzum, "it can be said that it is highly likely that the reported incidents of SC happened exactly as described and many or most or all should be accorded the status of facts no matter how difficult for the reader to accept or to integrate these with his/her usual orthodox understanding of science. These facts beg for attention and explanation."

I asked Joe Nuzum if he himself could offer insight into human spontaneous combustion, his own or others.

How many items total have you spontaneously ignited? "I couldn't begin to tell you," he confessed. (Dr. Schwarz has catalogued no less than ninety-nine incidents.)

How would you describe the fire that blazed around the hair of your wife and yourself? "Glowing blue color. Cool hot." And painless, naturally.

Any forewarning that this hair-raising SHC loomed imminent? "No. It was surprising to me."

Why did it stop? Did you somehow mentally douse the flames? "Like sometimes something will happen afterwards, like a distraction, to cancel it out." He said that's what happened in both of these hairy outbreaks. Many times, though, momentary amnesia blocks his memory of what transpires and how it comes to a halt. (Have you, too, noticed how often victims of SHC seem nonplussed, even oblivious to their predicament?)

How do you think you trigger your pyrokinetic talent? "I feel like I'm a vessel, pulling energy in from outside and projecting that energy, directing it," he answered. "I focus on a paper clip, to cause it to become cherry red. I cause wires to heat up inside a light bulb. Some of these things are difficult," he added modestly. "I'm in constant learning myself. Every combination is different. Every action is different. It may appear to be the same, but it isn't."

He believes innumerable factors influence pyrokinesis, from what one eats to the magnetic fields inside and around oneself. Nonetheless, he believes "The intent of the heart is the most important."

Finally, what's a day in the life of Joe Nuzum like? "Just getting through the day!" he grinned. A day that at any moment could prove wondrously challenging (disquieting) to those who would limit human capabilities—

TWICE IS NOT ENOUGH (PART III): PETER CALHOUN

For some, exploring the range of human potential is a conscious career choice. Peter Calhoun has made that choice. As a college-level teacher and spiritual counselor, musician, former Episcopal priest, co-founder of Atlanta Institute of Metaphysics, and shaman, he has sometimes found himself as "a mediator between the human world and the natural forces and at times a conduit through which the natural forces can flow."

Occasionally, those forces will manifest conundrums of combustion.

Like Nuzum, Calhoun tells me that "All my life I've had a strange experiences with fire. As a child, fires sometimes broke out around me for no apparent reason. For example, several times, upon starting an automobile, the carburetor would burst into flame." Since then, he has explored consciousness, meditated with Eastern sages, and studied shamanic knowledge.

However, as he began the sixth decade of life, his childhood connection with primal fire took on new dimensions in the broadest sense.

• In 1986 and again in 1989, Calhoun was among some two dozen people engaged in spiritual work in the Canyonlands of the American West. Both years, firewood gathered earlier in the day for use in a sweat lodge instantaneously ignited only minutes after it had been laid in a ceremonial manner. "Suddenly the flames were just there!" Calhoun said, recalling the first time it happened. "No smoldering. Just burst into flames. The reaction was almost a nonreaction," he remembered about the group. "Like, why *shouldn't* this happen?" The second time, not only did the campfire pyramid arrangement of twigs burst into flames, so did two-inch-thick branches ten feet away. "We were truly awed by this theophany!" Calhoun confessed, "because the second time was no random event. This time, we consciously invoked the fire."

The same elemental force has been used by Calhoun as a powerful tool for healing others in certain extreme or unusual circumstances.

• In one incident during Fall of 1987, he was counseling a teenage Colorado student hooked on cocaine. After the session seemed unfruitful, Calhoun told his associate that the young man needed to have his own inner fire awakened. Said Calhoun, "I'm going to call the fire spirits to light a fire beneath him." He did a brief ceremony.

The maxim Be Prepared For What You Wish For, would never be more true.

The outcome was instantaneous. Completely unexpected. When the Boulder High School student descended the staircase to leave, "the crotch of his pants burst into flames! He yelled, beating at his pants until the flame was extinguished. At no time would the fire have touched his skin," Calhoun stressed. "But *he* didn't know that, and the whole thing had a powerful effect on him—a little like shock therapy.

"It *worked!* By this time, I had learned that a literal manifestation, not just a metaphoric expression, is always a possibility, though remote."

Call it the Law of Unintended Consequences. Drug abusers might find the prospect of such dramatic intervention treatment an effective deterrent to continuing their addiction or, even, to begin use of controlled substances—

"As you can see, my experiences of spontaneous human combustion are not the negative kind." Calhoun prefaced the incident he was about to tell by saying, "One of the most recent was quite dramatic!"

• Early Spring 1989; in the Appalachian highlands of western North Carolina. Peter W. Calhoun was preparing to journey north to conduct educational workshops with a group living in an area where many people were hostile to alternative spiritual ideas. One evening, he felt their distant negativity focused on him. "A Native-American friend and neighbor was visiting me, and I asked him to work with me on a method I had used other times for protection."

The two men began a fire ceremony in Peter's living room, using a Four Directions spiritual ritual and four candles. In the room, devoid of drafts, three flames remained still; only the flame of the north candle flickered wildly.

Calhoun explains what happened next: "I invoked the help of the fire elementals. And *I* burst into flames!" Initially, the fire centered around his gut (solar plexus). "My natural reflex was to slap it out. Instead it ran down my arm, then my other arm. I felt no fear. I knew without knowing that I had total protection. It was not ordinary fire, but the *elemental force behind fire* that enveloped me."

Spellbound, Calhoun sat rigid and watched the fantastic golden-red fire swirl around himself, "$1/8$ to $1/4$ inch above my wool sweater.... At times it would shoot out from my body feeling like liquid fire—yes, that's what I would call it—*liquid* fire going up my spine."

At this point, Calhoun continued, "my greatly alarmed friend began frantically slapping me on the back—but the fire continued to spread, going up my neck and head. I attempted to calm him, saying it was okay, not very reassuring advice, considering the situation. A few seconds later, it simply burned out, all negativity having vanished."

Hallucination? Ponder, then, this fact. Globs of flames that dropped off the dazzling inferno engulfing his fire-immune body, burned holes in the carpet around him.

Dramatic, is understatement. It was SHC experienced; viewed from the inside out; witnessed by another. And survived. The only thing missing was videotape footage... and an explanation.

Was there any physiological damage? "No," he answered.

Heat? "At no time was there heat!"

"To me," explained Calhoun, "there's not any great mystery here, in that I have a strong belief in the elemental forces behind the flames. When one's inner fire has been activated," he elaborated, "it can actually raise the ambient temperature in a small room and cause persons to feel they are literally burning up." As one Colorado high school student could profoundly attest to—

"Every shaman knows that fire is by far the most unstable of the elements," Calhoun stresses, "hence the most difficult to control." Dimensions of human fire-starting are not understood, obviously, and he strongly recommends against experimenting with spontaneous combustion, even for good ends.

ONCE *Is* QUITE ENOUGH, THANK YOU

As incredible as these episodes are, no less amazing is that Weekly, Smith, Jones, Norris and, repeatedly, Nuzum and Calhoun reacted so calmly. That, understandably, has not always been the case.

• On July 15, 1681, a Friesland man fared not so fortunately as did Paul Weekly. According to Dr. Bäumler in his nineteenth-century treatise on spontaneous human combustion, *Studiorum Suorum Primitias*,[*] this resident of northwestern Holland cried out "I burn! I burn! I burn! Oh my captain?, come help me!" As he pleaded for help from Captain Douve, his foot *burned to powder*.

---

[*] The original source is probably Stephen Blankaard's even rarer *Collectanea medico-physica oft Hollands Jaar-Register* (1683), Century VI, 212–217.

Deep burns lacerated his body, his head subsequently swelled, his ears and nose shrank, and metal buttons melted on his clothing. He died five days later.

There is a rare malady called erythermaglia, in which a person's legs and feet become blotchy, purple blistered, and swollen—and they are fiery to the touch. Are there even rarer cases of erythermaglia *in the extreme*?

• September 7, 1822; Laégnon, France. "The fingers of Pierre Reyneteau caught fire with a blue flame that he could transfer to his pants," declared Grabner-Maraschin in *Gazetta di Milano* (April 7, 1823). The fire was not extinguished by water but by prayer, a novel safeguard that could not assure instantaneous healing. That took two months to accomplish.

• April 19, 1827; Puy, southcentral France. Very sober eighty-year-old Monsieur D– attended to his chores in the cathedral and then retired for the night. At 9:30 P.M. he became disturbed to see his fingertips burning with "a fire that progressed rapidly" and "burned like candles, and the flames were blue." He ran to a neighbor's house, where he plunged his flaming hands into cold water. Witnesses there remarked on a "vapor of smoke" emitted in the aftermath of this phenomenon of hands spontaneously burned red. Twenty-eight days later his hands had healed, according to Dr. de Brus in *Archives Generales* (March 1829), but Monsieur D– kept a pan of water by his bedside...just in case.

• August 7, 1844; Trinidad. A West Indies Negro didn't self-combust but he did exhale a "flame of bright colour," which was witnessed by a Quaker and a servant, reported Dr. William Huggins to *The Lancet* (September 6, 1845).

• Within the two years prior to 1890, a twenty-four-year-old Welshman in Newchurch attracted the interest of physicians when blue flames erupted from his body; flames that then burned him. In the event others might be interested in what transpired in this tiny town in east-central Wales, McNaught reported the incident to the *British Medical Journal* (1890).

• Prior to 1895, a Mr. Holcombe was traveling through China, where, according to DeGroot's *The Religious System of China*, he witnessed a miracle of the unexpected sort. Holcombe told of noticing a child staring at an elder Chinaman's queue (the braided pigtail fashionably worn by men). He turned his gaze to see what was of interest. Said he: "the man's queue faded from sight, leaving only a [*sic*] odour of burnt hair." Another example of SHC— spontaneous *hair* combustion—of the sort to be experienced by Susan C– a century later in America.

• July 14, 1894; Sea Cliff, New York. Sitting in her Long Island home, Miss Goldie Newlin suddenly felt "a sharp blow on her left foot." The foot turned black-and-blue, the result of ball lightning that struck her and produced "a stinging sensation at the same moment" in a man in an adjoining room, said the *New York Herald* (July 16, 1894). An SHC near-miss?

- In 1934, a different part of the human anatomy came under fire for forty-two-year-old Anna Monaro. Her affliction was flaming breasts! The hospital staff at Pirano, Italy, led by Dr. Protti and five other men of medicine, marveled at the "violet-blue light" which crackled from her bosom. As she lay on her back, the heatless flames sometimes reached eighteen inches high. Fifty-four frames of movie film captured this puzzling phenomenon for 3.6 seconds; otherwise, it was well-documented in *Sul Fenemeno di Pirano* (1934).

> Miracles happen, not in opposition to Nature, but in opposition to what we know of Nature.
> —St. Augustine

Dr. Protti postulated a "radiant power of her blood" as the mechanism behind her luminous mammaries, whose glow became enhanced when heart rate and respiration doubled. That meant metabolism—that is, combustion—was increased, as well. Other observers, in a diagnosis no less bizarre than the event itself, proposed to explain it all as psychosomatic! Propriety prevents me from saying more—

- In late December, 1938, fire struck down Mrs. Harriet Garner in her home at Walton, Liverpool, England. Her son testified in court that he found his mother aflame; that no other fire was in the room; that nothing he knew of could account for the blaze. Mrs. Garner, who survived, was equally mystified; she told her son, "I am totally unable to explain what had happened."

- March 28, 1953; Silver Spring, Maryland. Veronica Rae Klenke, eleven, was in her third year of playing the accordion when screams interrupted her practice. Her father rushed upstairs to find *flames* playing around Veronica. Before he could tear the instrument away and toss it out the window, his daughter suffered second- and third-degree burns over 35 percent of her body and he had second-degree burns over 25 percent of himself. Veronica said she had just strapped on her accordion when "jets of fire began squirting from it"—though Deputy Montgomery County Fire Chief Roy Warfield could find no reason for it to do so.

"Self-igniting instruments are rare," Frank Edwards deduced in *Strangest of Ali* (1962).★ I wonder if it was the young musician *herself* who ignited and shot flames from her abdomen, and then through her accordion.

- A fate that reads like science fiction struck British actor Derek Boote on October 13, 1974. The next day's *London Daily Express* said he had donned a monster costume for a children's "space comedy" thriller at BBC-TV's Cardiff

---

★ Alvin Moore, in his *Mystery of the Skymen*, places this event in Kensington, Maryland, on February 28.

studios in Wales and awaited his cue. His screams weren't part of the script. Nor was what followed. Fellow actors discovered Boote engulfed in flames in his dressing room, quite *unable to explain* his frightening predicament. Doctors pronounced him "critically ill," though he eventually recovered from his unscripted nonfiction bout with the unknown.

• Hot pants were hot fashion in 1977. Around August 2, eighteen-year-old Miss Sally Flack was wearing "a real hot number" when, said the *Daily Mirror* (August 3, 1977), she jumped off the Brighton bus moments after she boarded it because her fashionably hot pants had suddenly flamed. Or had *she*? Unaware of her plight until another passenger pointed it out, she managed to put herself out (so to speak) by vigorously slapping her thighs and legs. She was not hurt. The Brighton Council said "cause of the blaze is a mystery." In an attempt to defuse a lawsuit, the Council suggested "a cigarette end may have got caught in Sally's shoe." Then again...

• In 1978, the nonsmoking, well-adjusted, teetotaler seventeen-year-old daughter of W. Geerlings developed afflictions in her right hand. These included, her father wrote to me, "a severe electrical discharge" that enabled her to throw sparks when facing metal objects. Circumstantial evidence indicated the sparks were two- to three-thousand volts, he explained. For two years the phenomenon recurred. Then it waned. Following several traumatic episodes in the Australian girl's life, her electric fire reappeared in December 1982. During the next five months, wrote Mr. Geerlings, "she had several pin-point fires in the right hand, until May the 8th, 1983, at 4:00 A.M. when she came to show me how she had...a severe burn across her hand of approximately 2″ wide." Skin grafting surgery was performed. Still, her right hand continued to produce "spot burns" having one-eighth-inch to one quarter-inch diameters. The medical fraternity's answer? "Psychiatric Treatment," said Geerlings. He and his daughter refused. The medical fraternity should have its *own* head examined for making such a diagnosis.

• An unnamed woman aged twenty-nine awoke one morning in 1979 to see "flames spurting from her right arm, her neck, and her stomach," claimed writer William Michelfelder in *True Outer Space & Paranormal World* (Fall 1981). "No source of flame" existed in her small cottage near Dearborn, Michigan. According to Michelfelder, the victim subsequently became "an anonymous patient in a mental hospital, right arm amputated, lost in a schizophrenic fog..."

• Disco was hot in 1980. Around November 15, Vicky Gilmour, nineteen, was enjoying herself at a Darlington dancehall in Durham, England. A hot dancer, Vicky danced too hot this night—in the powder room of the disco "she burst into flames." Friends extinguished the inflamed dancer, reported

News of the World (November 16, 1980), but she would "need extensive skin grafts to her face and body."

A dropped cigarette was blamed. So, cigarettes were *intentionally* dropped on her dress and—in a repeat of the two 1938 dancehall episodes—they could only cause the garment to smolder. Said a spokesperson for the Department of Trade: "We've known for a long time that cotton can catch fire. But for a dress to go up in flames like this is a mystery."

And isn't that what *Ablaze!* is about?

Noted, in passing, is a statement by author Ajit Mockerjee about the activation of Kundalini: that music and dancing can arouse the thermal potential of this potentially dangerous bioenergy.

• A thirty-one-year-old Canadian woman, Miss Marilyn O–, lived alone in an apartment on the twentieth-floor of a high-rise in southern Ontario. An intelligent, thoughtful woman who worked as a medical secretary, she was a nonsmoker and a social drinker only. She had been depressed by the death of her mother six months earlier and now questioned her religious faith, but otherwise life continued uneventfully. On June 19, 1980, she went to bed at 2:30 A.M. And awakened at her normal time of 6:45 A.M.

There normalcy ended.

During those 255 minutes of sleep, her upper left thigh developed "a two-inch pencil-like red streak." She felt faint and dehydrated, rather than rested. The redness intensified.* She phoned the company doctor, made an appointment to see him, then left for work. (One wonders what her employer would think of the excuse "I won't be coming in today because my body's burning"?)

To her consternation then horror, the reddening dramatically worsened. Her left eye swelled shut. A circular burn developed, then blistered, on her left calf just below the knee. By the time she saw the doctor, she was vomiting, her left leg was numb, and the red streaks had blistered. The doctor sent her to Humber Memorial Hospital, where for the *first time* that day she felt pain. Red marks now appeared on her right leg. "People said it was like looking at raw meat," she remarked.

Physicians at the hospital told her that she had second- and third-degree burns on her legs. They could offer no explanation.

---

* Randles and Hough, citing me as their source, state in *Spontaneous Human Combustion* (1992), that Miss O– "awoke at 2.30 A.M. with major burns on her thigh and abdomen." Likewise, the Reader's Digest *Facts and Fallacies: Stories of the Strange and Unusual* (1988) says she "woke to find severe second- and third-degree burns had developed on her thighs and abdomen." Not so.

"I'd have gone straight to the hospital if I knew I'd been burned!" Miss O– exclaimed. She insisted she was not the type of person to inflict self-injury and knew no reason why her body would be nightmarishly burning itself.

Her brother, a policeman, checked out his sister's apartment for evidence of fire or an intruder who could have burned her. He found nothing amiss. Burning was limited solely to Miss O– herself.

Five months of agonizing skin grafting was needed to repair the scarring caused by an unidentified blaze inside her body, to which bed linen, night clothes, and work attire were immune. Marilyn was terrified her body would reignite on the first anniversary of her self-searing. It did not.

"It's still a mystery," she admitted when I called back years later to see how she was doing. "We really don't know what happened." She scoffed at the Humber doctors' suggestion that she burned herself chemically. "What I'm telling you, which is what I told them, is *exactly* what happened! It's *strange*!" Indeed.

• Once was also enough for U.S. Navy airman Jeanna Winchester. On October 9, 1980, she suddenly burst into yellow flames while riding in a friend's car in Jacksonville, Florida. The car crashed as the driver tried to beat out the flames. The fire went out, probably of its own accord. Miss Winchester suffered severe burns over 20 percent of her body; specifically her *right* shoulder and arm, neck, side, back, and across her stomach and breast.

Commented patrolman T. G. Hendrix, whose investigation found no spilled gasoline or other accelerant in the car: "The white leather seat she was sitting on was a little browned and the door panel had a little black on it. Otherwise there was no fire damage. I've never seen anything like it in twelve years in the force."

Airman Winchester told San Antonio's *The Light* (November 16, 1980) that she couldn't remember anything between riding uneventfully in the car and waking up in the hospital. "At first I thought there had to be a logical explanation," said Miss Winchester, "but I couldn't find any. I wasn't smoking anything. The window was up, so somebody couldn't have thrown anything in. The car didn't burn. I finally thought about spontaneous human combustion when I couldn't find anything else."

• Sometime in 1980 or the late 1970s, according to personal correspondence in hand, a young wife named Mary was sitting in the back seat of the family car, enjoying the scenery of central New Jersey. She lovingly cradled her infant child in her arms, while her husband drove. Suddenly, tranquility was shattered by an explosion "like a boom!"

Mary's husband frantically turned his head to see what had happened behind him. There, in astonishment, sat his wife *with part of her neck burnt*. The burn extended from the back of her neck's left side to her Adam's apple. Whatever its cause, the explosive combustion left her shoulder-length blonde hair unsinged and the baby she held untouched. Her physician, after hearing her bizarre story and examining her injury, blamed the seatbelt.

That no one in the car mentioned burnt seatbelt webbing was irrelevant to rationalizing away a situation otherwise too frightful to contemplate.

• On May 3, 1981, a thirty-seven-year-old mother in Wath upon Dearne, Rotherham, ten miles north-northeast of Sheffield in Yorkshire, is said to have burst into fire when striking a match. Damage was restricted to herself and her clothing. And she lived, say Randles and Hough (1992). Though sounding unsure about the episode, they do assert that "theories about hairspray causing the flames to rapidly spread have been mooted."

• Sometime prior to October 1981, an anonymous male foreman for an electrical firm was at a job site in Waterbury, Connecticut, when he suddenly keeled over unconscious. Said Sal Trapani, who lived in nearby Seymour and kindly brought this episode to my attention: "As his men reached to aid him, they were repelled by intense heat, as if approaching an open oven." Trapani said police and the responding ambulance crew packed the superheated foreman in ice "within minutes." His terrific thermogenesis caused blindness in one eye and restricted vision in the other, said Trapani, but spared his life.

• On October 10, 1981, forty-three-year-old Mrs. Barbara Greene complained to her daughter that she "felt sizzling all day." Apparently she had been, literally. On this Sabbath evening, as she undressed in her Brooklyn apartment, she discovered her brassiere and bodysuit had fresh burn holes in them. "No harm to body or over-blouse," her daughter told me. One just never knows *what* kind of surprises a day can hold, does one?

• Something like the day in 1982, when a comic book caught fire in the hands of Benedetto Supino as he sat in a dentist's office in Formia, near Rome? Life soon got more unnerving for the ten-year-old Italian and his family. One morning he awoke to find his bedclothes on fire—and himself painfully burned. He stared at a plastic object held by his uncle; the object inflamed. "I don't want things to catch fire. But what can I do," he lamented to the *Sunday Mirror* (August 21, 1983). What he did was produce electromagnetic effects that caused $6,000 damage to electrical machinery in his father's workshop.

His uncanny abilities came to the attention of top Italian scientists. Dr. Giovanni Ballesio, Dean of Physical Medicine at Rome University, commented: "It is wrong to call him an 'Electric Boy' because he really doesn't possess any more electricity than anybody else." Well, he must have possessed *something* different because even the worst mischievous adolescent couldn't do what this lad was unwittingly accomplishing! Professor Mario Scuncio of the Tivoli Social

Medicine Center declared young Benedetto to be "perfectly normal." Come again? Archbishop Fagiolo skirted the issue of defining normal versus paranormal, in warning that "Neither must his extraordinary powers be considered miracles."

Learned men of science and theology had spoken. And solved nothing. Young Benedetto continued, for some time, to cause furniture to smolder in his presence; to scorch the pages of books when he touched them; to knock out electrical circuits with his proximity alone.

• In early February 1985, a sixty-one-year-old pensioner at the Innsbruck University Hospital "suddenly spat fire" and burned his face. Nearby patients intervened quickly to what the *Soester Anzeiger* (February 15, 1985) understatedly called an "enigmatic explosion" from his mouth, and prevented further injury to his body. The pensioner remained personally responsible for lingering damage done to his psyche—

A HUMAN TORCH WALKS THE STREET

In most of these instances the individuals survived their bizarre internal burning with minimal discomfort, a characteristic of SHC. Indeed, the absence of pain in these inflamings—so atypical of what one experiences when burned by a hot skillet or blazing log—is one more mystery to solve.

A less fortunate survivor of inner fire is British teenager Paul Hayes—surnamed Castile by the *National Enquirer* (July 23, 1985) and Castle by other sources—who got mugged, by the unknown, as he walked a quiet road in Stepney Green on Saturday night, May 25, 1985.

"My God!" the nineteen-year-old lad suddenly screamed inside his head. Flames from the waist up licked at him as though his next step had plunged him into a steel mill's cauldron. He was ablaze!

"It was indescribable...like being plunged into the heat of a furnace," he recounted of his ordeal. "My arms felt as though they were being prodded by red-hot pokers, from my shoulders to my wrists. My cheeks were red-hot, my ears were numb. My chest felt like boiling water had been poured over it. I thought I could hear my brains bubbling." The lad seemed to be undergoing spontaneous ignis amoris, unleashing *raging flame!*

"I tried to run, stupidly thinking I could race ahead of the flames." (It is the worst possible strategy in a *normal* fire of external origin.) He thought he had been doused with gasoline and set afire. But there was no one near him, nor evidence that someone dumped napalm onto the hapless pedestrian from overhead. "I drew my hands to my face, pressing my palms tightly against my eyelids—desperately hoping they would be a barrier of defense against those awful flames."

In despair he dropped to the pavement and curled up in a fetal position, a heap of searing flesh. His life flashed before him. He prepared to die.

It was all over in about half a minute. Hayes found himself *alive*, not dead! As mysteriously as it began, the flames and smoke vanished.

He lay still, terrified he would reignite. Then he hesitantly dragged himself upright and assessed his condition. "I was numb in some spots," he noted, "white-hot in others." He stumbled around the corner and into London Hospital, where the medical staff stared wide-eyed at his lower face grotesquely burned and his shirt scorched all over. They must have become *more* wide-eyed when their patient recounted his ordeal.

"I'm still recovering from burns on my hands, forearms, face, neck and ears," he said about a month later. "But miraculously I wasn't seriously injured. Maybe I'm the luckiest man in the world to be alive. I'm also the most puzzled. What happened—and why did it happen to me?"

Regrettably, I have been unable to contact this lad firsthand.

DR. SULLIVAN SURVIVES A STRANGE SEARING

The same question also confronted Commander Edward Joseph Sullivan, M.D.C.M. in 1970, though at the time he didn't care if he ever got answers.

Contrary to the assertion by early writers that SHC befell only those of society's uneducated lower class, Ed Sullivan is a thoughtful, endearing, highly-educated professional; and yet...

"In my field, occupational medicine," Sullivan wrote to me, "there are many things hard to believe, but which turn out to be true." During the night of April 24, 1970, the hard-to-believe-but-true happened to him.

Some biography, first.

In a sense, Edward Sullivan's observation about occupational medicine epitomized his entire life. Born in 1930, he nearly died in his first week due to medical malpractice. Despite a difficult childhood, he sped through school and obtained his undergraduate degree in pre-medicine at nineteen. After earning his Doctor of Medicine and Master of Surgery degree from McGill University, he enlisted in the U.S. Navy to serve tours in the jungles of Eastern Luzon and as a Senior Medical Watch Officer.

Private practice followed. It was a time, he said, when "I was lied to, cheated and robbed by most of those I dealt with." Always a loner, he contemplated suicide.

In spite of these crises he married his nurse, and life became brighter though professionally unrewarding. He did pathology residency for a year, returned to active duty in the Navy, and then the couple moved to Cincinnati where he began his occupational medicine residency. There his wife became ill.

He loved her deeply and harbored considerable rage because of the many times he felt she had been mistreated by others. "As I gradually learned the facts," he revealed, "I became madder and madder at those who had made my wife a target. As one who had chosen the way of the warrior, I could expect war. But she didn't deserve it."

On his way to residency class on April 24, he visited her in the hospital and deduced she was again not being treated properly. Later that morning, a call from the hospital informed him that he was a widower.

"My wife had become my whole life, and with her gone, life was something I no longer wanted," he admitted to himself as he returned to his suite in the Vernon Manor Hotel in Cincinnati. He felt cheated again. He was angry, despondent, and suicidal.

In his hotel room, he decided he would persevere through a few more days of a life that now thoroughly disgusted him, and then end it all. He made the arrangements for his wife's funeral. Exhausted and numb, Dr. Sullivan went to sleep around 6:00 P.M.

"My next memory is becoming aware of a phone ringing, and trying to fight my way up out of whatever I was in to answer it. I gradually came to realize something was wrong. I was very weak, and could see but dimly. There was still some light coming through the window, so I figured I couldn't have been there long."

In fact, it was 6:00 P.M. on Saturday. He had been deeply unconscious for about twenty-four hours! (During this time he had a classic near-death experience, suggesting that he may have momentarily died.[1])

"I tried to get up," he recalled, "but could not. I was lying face down, with my right forearm under my chest. I gradually became aware of increasing pain in the right forearm. It dawned on me that something was wrong besides fatigue. I started to yell, and eventually someone in an adjoining room called the desk, and someone was sent up to investigate. I had trouble speaking. My lips were crusted and my tongue had a thick coating. My arm was gotten from underneath me, and I looked at my hands. Both were frozen in tight flexion. Sensation was gone beyond the left wrist and from shortly beyond the right elbow. Then I noticed the blisters. They were filled with dark fluid, and involved both hands and forearms. My left cheek was stiff and numb."

(Sixty-eight months later, Jack Angel would be able to empathize with Dr. Sullivan's unnerving state of body and mind: it would be Angel's *right* forearm that burned black also.)

The hotel employee left, and Sullivan was alone again with his thoughts.

"I tried to open my hands. They would not open. My right forearm was beginning to hurt—it felt as though it were in a vice. I tried to move it, but it was under my back, and I couldn't get it free. I brought my left hand up to where I could see it. It must have weighed a hundred pounds.... It was all blotchy. My vision cleared a little more. The blotches were blisters. I focused on my tightly clenched, locked left hand. It too, was blistered. One great amber fluid-filled blister ran along the side of my frozen thumb.

"Oh my God! Had I been in a fire?" he questioned of himself. "Burns were supposed to hurt, but my left forearm didn't. The hand itself was numb, as free of feeling as though it belonged to someone else, and was mine no more. My right forearm was hurting more, but I still couldn't free it from behind my back. I stopped trying and lay there, trying to figure out what had happened."

Hotel employees had called their tenant's physician. Dr. George M. Lawton now arrived, examined Sullivan, pronounced "electrical burns" as the injury, and called an ambulance to take his patient to the nearby civilian hospital.

"Oh my God!" gasped one of the doctors as he looked at their new inpatient. If his injury distressed a surgeon that much, Sullivan thought, he didn't want to see himself. More doctors surrounded his bed, looked over his scorched limbs, and talked among themselves. Sullivan tried to convince them he was painfully thirsty. They said he could have nothing to drink, because he had extensive electrical burns.

"I tried to explain that I hadn't been in contact with electricity," he said, "but was assured that I had had a head injury and didn't remember." He insisted that was false, also.

"It made no sense, but nothing made sense," he realized.

After a month under civilian care, Sullivan was transferred to the 4th District U.S. Naval Hospital in Philadelphia, Pennsylvania. The staff there viewed their new patient as a medical mystery, as well.

"As little as I knew about what had happened to me that evening when the compelling fatigue had claimed me," he said, "those on my 'case' obviously knew still less. Assorted consultants came to see me and formulate theories, or shrug their shoulders and pass on. Eventually, the original diagnosis of electrical burns and head injury was discarded, but alternative theories bloomed briefly only to be discarded in their turn."

By July 1970, sensation had returned to his left hand, though motion of his thumb and index finger was limited. Not so in his other arm. "Sensory loss and pain in the right forearm remained essentially unchanged," he noted, "and muscles had shrunken. The hand remained locked in a fist." The median nerve was caught up in scarring, and hand muscles were diagnosed as "dead" by a surgical specialist. (This is *not* the kind of damage that results from lying on one's forearm for a day, as some aides suggested had been the cause of his burns!) Not until fourteen months after his injurious twenty-four-hour nap did flickers of motion return to his crippled limb.

Being accustomed to adversity, Sullivan persevered again, this time living through his injuries. He successfully applied his professional expertise to a regimen of personal reconstructive therapy and regained most of the use of his hand.

He remained haunted, however, by that one episode that lay beyond his medical knowledge. *Why* had this happened in those twenty-four hours asleep? He remembered the advice one of his better professors at McGill University gave prior to an exam: "Start at the beginning, continue until you reach the end, and then stop."

If ever there was a victim qualified to analyze a personal burning predicament, it was Dr. Sullivan. He replayed the episode over and over in his mind, starting each time at the beginning to figure out the reason for his knurled fingers, crippled hands, sunken forearm muscles, and scarring on the knees.

Had he drunk himself into a stupor? No. Had he stumbled, fallen upon an electrical appliance or outlet and knocked (or shocked) himself into temporary amnesia? No. Had he been the victim of foul play; of hyperthermia; of leaking gases? Again, no supporting evidence. Had he been drugged? Absolutely not! Was it Volkmann ischemia, as some colleagues had diagnosed? No! vowed the patient. Best evidence said the injuries were electrical in nature, yet there was no satisfactory external source to produce them.

As much as he tried to follow his professor's advice, Sullivan said "I find it difficult to determine beginning or end." Until he realized *the end could include spontaneous human combustion.* Then, only then, did he feel he had his answer.

In June 1991, he called to say that "Something curious is going on. My hands and forearms will get red and quite warm. Skin around my eyes is itchy and black—I look like the Lone Ranger! Redness, heat, shedding of outer layers of skin—on the face particularly. Am I trying to make an ash of myself in *another* way?" he joked.

Sadly, Dr. Sullivan died by massive coronary two months later. "He didn't feel any pain," his daughter-in-law Bobbi Curry told me; "his face looked very peaceful." Edward Sullivan was cremated, in the conventional way.

For this gentle man, and my friend, the search for an explanation to life—and its one very personal burning mystery—had at last come to a peaceful resolution.

INFLAMING ON THE WAY TO THE WEDDING

I met Jonathan Karrer at the 1985 Fortfest, a conference sponsored by the International Fortean Organization. He had heard about my research and sought me out for answers to a personal mystery that was baffling to him.

In June 1981, while driving to his brother's wedding in metropolitan Washington, D.C., he felt his chest suddenly "pricked like by a venomous snake." The sharp pain gradually intensified. Upon arriving at the church he removed his blue T-shirt so his cousin, a medical student, could examine his chest. "It looks like you've been burned severely!" his cousin remarked.

"But I don't know *how*," Karrer countered. "I've been with you nearly all day!" The only sensation he could connect to the burn was the painful 'prick' where, now, he noticed a slight dark spot on his T-shirt. His white tuxedo shirt was not discolored.

Karrer continued his recollection: "The burn got worse and, needless to say, I was *very* uncomfortable.... I was in agony.... Halfway through the ceremony, about three to four hours later, I thought I was catching fire—that's how intense and painful this was!" He *literally* sweated through the wedding. At the reception, said Karrer, "I was writhing, and drank vodka straight" to dull the pain coming from the crimson area on his chest, now the size of his hand. It worsened by the minute.

Next morning he awoke from a deep sleep to a *new* pain two to three inches below the original one. His chest haad developed a *second* burn, though, said Karrer, "it turned out to be smaller, about the size of a silver dollar."

One week later, his skin blistered in these two areas. One month later, he suffered an anxiety attack: "I was vibrating at such a rapid speed I thought I'd explode!" An EEG revealed a brain-wave pattern apparently so abnormal, said Karrer, that "doctors didn't want to discuss it."

I asked him about his emotional state prior to the June nuptials. He acknowledged being under tremendous stress; his emotions pent-up to the point where Karrer felt "something *has* to explode." Karrer half-jokingly called himself a "self-arsonist."

Not jokingly, it seems he *was*. His heightened stress and bottled-up emotions perhaps could no longer be constrained...and burst forth through his chest as a physical burn-out that blistered flesh and discolored his T-shirt but not the white shirt he wore en route to the wedding.

Assessing a Psycho-Emotional State as Precursor to SHC

It's easy to see why Commander Sullivan and Jack Angel and Jonathan Karrer were obsessed with figuring out what happened to their bodies. It's even easier to understand why other survivors of partial SHC, like Marilyn O– in the Ontario high-rise, would be psychologically and emotionally traumatized after their searing ordeal with the Unknown.

On the other hand, Peter Calhoun saw his encounters with eerie flames as intellectually and metaphysically engaging. He chatted about instability in a person's aura or electromagnetic field, or of concurrent irregularities in the electromagnetic field in the proximity of a pending human fireball. "I think we have that combustible aspect in all of us," he philosophied. "And if that's unstable enough, spontaneous human combustion could be one of the things that could happen." He recalled instruction given him by a yoga teacher in India: "If you're going to bring up all that power, you're going to have to have plenty of insulation! And that insulation is your consciousness."

A compelling metaphor for generating, and shielding, spontaneous combustion.

There is another consideration though, no less easily grasped. Might a person's psycho-emotional state of mind *prior* to burning have any bearing on whether SHC afflicts one person and not another?

Can Dr. Sullivan, as both a physician and one of the few survivors of SHC, offer insight to a possible causal relationship? As already explored elsewhere, one's thoughts and state of consciousness do affect the physical condition of one's body.

I asked Dr. Sullivan to recall how he felt just prior to that fateful night in a Cincinnati hotel.

"Indeed," he replied, "it *was* an interesting situation for students of psychology and emotions. I had lost everyone and everything that meant anything to me at all, and I was a *container of rage*."

If ever there was an example of the vernacular "Burning up with rage," Dr. Sullivan said he personified it that day. It was as though, at some level of his consciousness, he had willed himself to die—his body choosing execution by bioelectricity—and then his consciousness decided to rescind its self-electrocution order once it had already been implemented.

Certainly a connection between emotional state and human spontaneous fire can be made in this instance.[2] If you are one who has a fiery temper—who is prone to spontaneous and uncontrolled expressions of anger—it would behoove you to temper your emotional stance, unless you don't mind the risk of becoming a pile of ash or surviving partial immolation by psycho-emotionally induced internal thermogenic immolation.

The SHC that happened to Dr. Sullivan affected him in more than the obvious, physical way. He noted a significant change in his persona, as well. He was now less stressed; *more* reassured about the struggle of living. Triumph over SHC had truly enhanced his life.

"My 'startle reaction' seems diminished to absent. While I can recognize risks and identify things I'd prefer not to do," he chuckled, "I tend not to notice *fear* as I did. As I've said, half-jokingly, 'How can you kill a ghost?'"

Dr. Sullivan is one of those very rare human beings who triumphed over the usually fatal consequences of spontaneous human combustion. More rare, he could make light of it.

17

BEAMED UP AND BURNED OUT: CLOSE ENCOUNTERS OF THE COMBUSTIBLE KIND

> It may be that if beings from somewhere else would seize inhabitants of this earth, wantonly, or out of curiosity, or as a matter of scientific research...
>
> —Charles Fort, *Lo!* (1931)

December 4, 1988; 5:20 A.M. Southbound on a side street, police officer A. P. Jones prepared to turn right onto U.S. Route 22 six miles east of Pennsylvania's capitol. Suddenly a burst of blue-white light, "so bright I couldn't see the stop sign," enveloped his car. Unable to proceed, he instinctively tugged his service cap snugly over his forehead and stepped from his car into the blazing light. And into the unknown! Officer Jones had just entered the Combustible Zone.

Shielding his eyes with a mock salute that extended the protection offered by his cap's visor, Jones looked up at the painfully bright light. He was not prepared for what his squinting eyes beheld: a UFO hovering motionlessly—despite 40 MPH wind gusts—some 150 to 200 feet above the intersection.

"It was circular, like a frisbee," Jones observed, "pretty big, 75 feet in diameter. And it made a weird humming noise, like from a science-fiction movie." The air was charged with electricity.

For ten minutes he watched it rock back and forth like a falling leaf not falling, then it sped eastward and disappeared.

Jones drove to headquarters and meekly hinted to his fellow officers about the disconcerting incident-on-the-way-to-the-office. To his amazement, one of them offered independent corroboration for his encounter with the bizarre. Said Jones, who had been an Army aviation mechanic, *"That craft was not from this planet!"*

During the next twenty-four hours Jones developed nausea, severely irritated eyes and an inability to focus his vision, and *burns*. His hands, ears, and face (where not shielded by his cap's visor), plus a V-shaped pattern on his neck and chest (where exposed by his unbuttoned shirt) looked "like going to the Bahamas," quipped Jones, noting that this was early winter in Pennsylvania. "My face felt real hot, like a 104-degree temperature."

The police nurse said it looked like Jones was under a sun lamp too long. When he told her about his UFO encounter, she said she knew of a similar incident in the neighborhood that had produced the same phenomena.

Jones's colleagues teased him about how he could afford an overnight Bahamas vacation. The joking ended a fortnight later when the lobster-red color left his face, though his chest and ear burns remained ten more weeks.

A Chronology: Possible Close Encounters of the Combustible Kind

Burns resulting during close encounters with UFOs is not a laughable subject, however. Neither is SHC. Officer Jones was lucky; others have fared worse. Encounters can be, have been, more serious; even deadly. And stranger than many ufologists have dared to imagine.

What follows is a chronology for what I call *Close Encounters of the Combustible Kind* (CE-CK).

• In April 1993, a Marrianne Leith of Hyde Park, New York, wrote to describe a series of "strange encounters" in her life: traumatic sightings of UFOs, electrical surges through her entire body, spontaneous scars over half her body, an ability to "make electricity go wacky as I go near it." Scary and confusing, she said. "I know it sounds crazy, but it's *true!*" There was one more oddity. "I get burns, all over my body. What's wrong?"

I'd like to know too. The confluence of several potentially interconnected phenomena is intriguing. Was surviving partial human spontaneous combustion among her plights? I can't say. Ms. Leith deactivated her phone and moved away before I could contact her.

• During April 1989, a mother and her son were driving near Langenburg, Saskatchewan, Canada, when a bright object descended from the sky, paced their car, and baked it. And the lad. The doctor who treated him called the injury to his singed arm a radiation burn.

• One episode with a variety of unusual aspects occurred in October 1985 to Penny Petty, a Southern girl twenty-five years old. At the time Penny was staying with Anthony and Lynn Volpe, UFO researchers in Ivyland, Pennsylvania. One morning at 4:00 A.M. she burst into the Volpe's bedroom, exclaiming "There's something outside my bedroom window with blinking lights making a loud humming sound!" With wondering eyes, all three went to the bedroom window, but saw nothing.

The next night, again at 4:00 A.M., Penny awakened the household with a terrified shout: "That thing is outside my window again!" The Volpes investigated and found only blackness outside the window. It happened three more nights in a row. Lynn Volpe tried to understand Penny's recurring dilemma, while wondering if their house guest was hallucinating.

Penny's skin was burned, though she had not been outdoors much, and besides, this was Pennsylvania, where weather in late October is not conducive to tanning. "Yes, I was sunburned," she told me, "but that was the least of my worries." In addition to burns on her chest, neck, and face, she found a needle-type puncture in the center of her chest. Then came recall of, as she said, "being taken aboard an alien space craft" and "a bright light shining down" upon her. Interestingly, Penny's young daughter had been dreaming of an "operation" being performed on her mother inside a UFO.

Skeptics can argue these UFOs and dreams were subjective, but her burns were objective; as were severe bruises "like from fingerprints" on her body. Within a few days her sunburned skin began to peel as if she had been to the beach in July, although it was now early November.

• At Weston Hill, Plymouth, a thin beam of light from a UFO overhead struck young Denise Bishop as she stood outside her south England home on September 19, 1981. It immobilized her for 30 seconds, she testified, and produced a scar on the outside of her left hand. A doctor who examined her injury gave his opinion, cited by Timothy Good in *Above Top Secret* (1988), "that the scar was similar to a burn from a laser burn."

• In Australia's New South Wales on April 17, 1981, around 10:30 P.M., an Aussie driving a Morris noticed a bright light overtake him from behind. It burned his leg and melted his car's tape deck. He and his car, he told friends, seemed to float upward amid "the intense heat." Afterward, he suffered recurrent nose bleeds and his urine changed color. As UFO researcher and NASA aerospace engineer John F. Schuessler said sarcastically: "Seldom does someone make up these symptoms!"

• One well-documented, ominous encounter occurred just north of Lake Houston, Texas, on the night of December 29, 1980. Vickie Landrum, fifty-seven, and her seven-year-old grandson, Colby, were being driven home after a Monday night bingo game by Betty Cash, fifty-one, when their number came up. It didn't win them bingo. A bright flame-belching object that Vickie described as "like a diamond of fire" and the size of a Texas water tower blocked their travel on the otherwise deserted Cleveland–Huffman road (Route 1485) around 9:00 P.M. Vickie thought the world was ending and in a sense, for them, it was.

The cool night (it was about forty degrees) became scorchingly hot, as heat flooded the car's passenger compartment and the dashboard warmed. Betty and Vickie got out of the car and watched the UFO glowing 150 feet overhead. Colby screamed for his grandmother to get back in. She did. Betty stayed outside, looking up. "The heat was tremendous," she remembered. "It felt like I was burning up from the inside out." When she tried to reenter the car, its door handle was so hot she had to wrap her hand with her leather coat to prevent burning herself.

For fifteen minutes the unknown device hovered above them, spewing flames from its underside. Then it left. Mrs. Cash floored the accelerator, only to round a bend in the highway and reencounter the UFO—and a sky full of two-rotor helicopters, either CH-46 Sea Knights or CH-47 Chinooks, plus a large flying crane and smaller, faster gunships. She and Colby counted no less than twenty-three of them.

By the time the trio arrived home, their enjoyable, *normal* world had already been shattered. Vickie complained of an aching head and feeling sick. Worse was but hours away.

The next day, Vickie's face erupted with big, red, swollen bumps the size of fifty-cent coins; "like a severe case of sunburn," she said. Colby complained of burning up and begged for water. She and Colby suffered nausea, diarrhea, and stomach pains, plus permanent damage to their eyes. Betty, who had a longer direct exposure to the UFO, saw her eyes swell shut and spent the better part of January 1981 in Houston's Parkway Hospital being treated for nausea, diarrhea, hair loss, red blotches that became huge water blisters, and acute radiation sickness.

"The doctors and nurses kept asking me if I was a burn victim," Betty said. In one sense she was.

"There are two possible explanations for this case," John Schuessler told me, and NBC-TV's *Unsolved Mysteries* (February 6, 1991), after he investigated this case. "It was an experimental craft of some kind by probably our government. The other, it was an unidentified flying object, possibly extraterrestrial." By 1992, Schuessler told me at a U.N. conference on UFOs that "I've found nothing to support the claim that it [the diamond UFO] was one of ours run amuck. Not a shred of evidence." Which leaves the ET option—

Mrs. Cash's doctor attributed her burns to radiation but had no idea how she could have encountered the harmful dose. Officials at nearby Army and Air Force bases first denied the presence of helicopters at their facilities, claims later proven false. Betty Cash sued the federal government in 1984 for $20,000,000, believing, she said, "that the government had something that got out of control." A federal judge dismissed the lawsuit.

> I wonder which government agency has files covering up SHC.
> —Lynn Volpe

Schuessler writes in the UFO newsletter, *Houston Sky* (October/November 1994), that a "*smoking gun has been identified that will soon be exposed.*" I eagerly await its revelation in his forthcoming book, *Night Terror in Texas: The Cash–Landrum Story.*

Meantime, I can say that Betty Cash and Vickie and Colby Landrum all suffered unearthly burns that night on a remote stretch of Texas highway. Either they produced inside their bodies spontaneous combustion plus hazardous dosages of radiation... or inadvertently they shared a rendezvous with a fire-spewing UFO whose presence was known to (and denied by) the United States government.

Could such burn incidents, in which the UFO had *not* been observed, be characterized as SHC? I asked Schuessler. "I wouldn't argue against that, no."

• Five months earlier—Thursday, July 24, 1980, 2:20 A.M.—and a continent away, Francisco Calle, fifty, was descending the stairs of the Roma Hotel in Rio Ceballos, Argentina. He heard a hum, then saw a ten-foot ball of red light with girdling yellow bands shooting out sparks a few feet away. Whether dubbed an Unidentified Flying Object or Unidentified *Flaming* Object, its heat burned his face and arms and clothing "as if scorched by an iron."

• For three nights in January 1979, Texan Greg Gregson dreamed about a UFO. On Wednesday night, January 24, the dream became a nightmare of the tangible, scorching kind. En route home near Tyler, Texas, a bright green light enveloped his car as what he called two "ships" hovered in front of his windshield. Five hours elapsed he could not account for. Soon a diamond-shaped burn five inches across appeared over his sternum and lingered for months. Dr. Richard Niemtzow discovered what he called puncture wounds "about the size of those caused by a hypodermic needle" on Greg's shins; these too lasted for months.

How a reddish diamond would get branded into a Texas lad's shirt-covered chest was never explained conventionally.

- In May 1978, a fiery oval disk hovered above Kerman, California. Patrolman Manuel Amparano stopped to gaze at it, an instinctive response he soon regretted. The UFO shot out a "bright bluish beam of light" and then shot upward. Standing in that beam—the target of it?—was the hapless policeman. Hospitalization followed for treatment of "third degree burns and facial blisters," according to *Specula* (April-June 1979). Observers remarked that the officer's injuries "resembled microwave radiation burns."

- In 1976, Cecillio Ferrera died in central Brazil after his incinerating encounter with an unknown aerial object. According to journalist Robert Pratt, his death certificate stated simply that "burns to his body were the cause of death."

- South American Sebastian Acevedo was a bit more fortunate in his close encounter with the unknown. Acevedo was himself the target of a searing ray from a UFO shortly before sunrise on April 14, 1974. Acevedo described the ray as "very hot. It was terrible, the heat that that thing was giving off. I don't know how to describe it," *Official UFO* (July 1976) quoted him, "but the heat was like an oven in steel mills." Apparently he escaped with nothing worse than a frightful memory burned into his brain.

- On January 8, 1974, Andrew Jackson Huckabee, seventy-six, walked along a road outside Ramer, Tennessee. He thought he was alone, but something stalked his footsteps. When it struck, Huckabee screamed. *His clothing had burst into flame!* Engulfed in fire, he collapsed on the road. After being found, he was rushed to Jackson-Madison County General Hospital where baffled medical personnel asked questions about his predicament. Huckabee was no help. To his dying breath he could not explain how he suddenly became a human torch.

What fearful force encountered this elderly man as he walked a Tennessee highway that night? The police were as baffled as the hospital staff: none could place probable cause for the weird combustion. It seemed as though an invisible *finger of fire* had sinisterly shot out of the sky...

- On the night of October 3, 1973, Eddie Webb, forty-five, was driving his truck near Greenville, Missouri, when a luminous object approached from behind and announced its intent to pass with a blinding white flash. "I'm burned! I can't see!" he screamed to himself. Webb had just survived a Close Encounter of the Combustible Kind. What made the episode more than a passing encounter with an uncourteous motorist—unless you factor in a UFO hot rodder—was that Webb's eyeglasses melted on his face.

- Fortean scholar John Keel tells in *The Eighth Tower* (1975) about a young male in Ulster County, New York, who sometime before October 1973 also had a CE-CK. "Man, it was throwing off some heat," Keel quoted the youngster about the low-flying UFO. "It felt like my sex organs were burning up." Keel noted that microwaves affect the eyes and the testes in particular. Whether this turned out to be a close encounter that destroyed (future) life, it inevitably left an indelible life-long impression on the unlucky boy.

- In 1971 in southern California, Trevor James Constable and a friend were operating weather-engineering equipment designed to attract a specific type of atmospheric radiation. It attracted a disc-shaped UFO. Constable's friend saw it first. Lucky for him, it wasn't the *last* thing he saw.

"*Crack!* A bluish bolt of energy, lasting only an instant, hit the tube he was touching and traveled up his arm to his skull. He was partially paralyzed for a few minutes," Constable described in *The Cosmic Pulse of Life* (1976), "and for that terrible instant he felt every nerve in the right side of his body irradiated by the bolt." Another friend had been permanently crippled by a similar incident in 1954, said Constable.

Blue fire from the sky; irradiation; paralysis of the right side. Had the UFO not been noticed, who would have known the reason why?

- February 7, 1969; the Pirassununga area north of São Paulo, Brazil. Scores of witnesses watched Tiago Machado walk to within thirty feet of a metallic-appearing UFO on the ground and watched it shoot forth a beam of light at his legs. The nineteen-year-old teenager was paralyzed, and his legs were burned "bright red and became painfully swollen."

- Less than a year earlier, at dusk on March 14, 1968,[1] Gregory L. Wells was walking the short distance between his grandmother's house and his parents' trailer in Beallsville, Ohio. It was a stroll the nine-year-old would never forget. Piercing screams brought his grandmother and mother running into the yard, where they found Gregory rolling on the ground—his jacket ablaze. They beat out the flames and rushed him to the hospital. Greg had become a *ufo*: an unfathomably flamed object.

Recovering from second-degree burns, he explained what happened as best he could to the curious authorities. Halfway home, he said he had noticed a large, oval, stationary, luminous red object just above the trees. "I wanted to run or scream," he said, "but suddenly a big tube came out of the bottom, which moved from side to side until it came to me, and a beam of light shot out." He turned to flee but could not outrun the bright light/flame that struck his upper arm, spun him to the ground and set his jacket aflame.

That was Gregory Wells's story, impossible and imaginative though it sounded. One supported by reports coming into the police station that a cylindrical unidentified flying object at low altitude had been slowly moving over the Wells's neighborhood, producing electromagnetic interference and black-ing-out a street light. As Greg carried no matches or lighters, officials, including the county sheriff, could find no earthly explanation for the boy's seared flesh, which remained scarred for three months. The Ohio Bureau of Criminal Investigation analyzed the jacket and could only report there was no radiation.

"No ordinary explanation for the sudden fire could be found," reported Major Donald E. Keyhole, director of the National Investigations Committee on Aerial Phenomena. Since no charges of false pretense were filed against the lad or the other witnesses, we are left with one conclusion: that Greg's story was believed and that he survived a CE-Combustible Kind.

• August 13, 1967; Pilar de Goias, Brazil. During the afternoon, Inacio de Souza, forty-one, stumbled onto a landed UFO surrounded by three aliens (*not* the kind who walk across borders without visas). De Souza could have stumbled away but made a fatal decision. He fired his gun at the figures. In return, according to Brazilian investigator Nigel Rimes's account for *Flying Saucer Review*, de Souza was hit by a "beam of green light." It burned "a perfect circle 15 centimeters in diameter" onto his chest. After suffering eight weeks of excruciating pain attributed to acute leukemia, the Brazilian died.

• May 20, 1967; at Falcon Lake, on the Manitoba–Ontario border. For thirty minutes, Steve Michalak watched two circular thirty-five-foot metal saucers maneuver near the lake. One landed. When he saw a portal open, he moved closer. Too close. The UFO shot out "gas or flame" that ignited his shirt and burned a *checkerboard pattern* onto his chest. Treated at Misericordia Hospital, Michalak remained ill for weeks. He lost twenty-two pounds, his blood lymphocyte count dropped briefly, and what appeared to be hives appeared on his body.

Five months later, a burning sensation returned along with large red spots on his chest; then every 109 to 124 days the bizarre malady reappeared. The Mayo Clinic isolated a "foreign substance" in the man's blood, John Keel stated in *Mothman Prophecies* (1975) without being more explicit. Dr. Horance Dudley, former chief of the Radioisotope Laboratory at the U.S. Naval Hospital in New York, speculated that Michalak received a short, intense blast of X-ray or gamma ray radiation. Whatever it was, it spontaneously burned him. By 1992, he had apparently recovered fully.

• Twenty-four weeks earlier another mystery combustion occurred amid a classic UFO flap. "Potter County teemed with UFOs of various sizes, forms and colors" during 1966,[2] reported local historian Robert S. Lyman Sr., in his *Amazing Indeed* (1973). One woman "nearly crashed into a saucer in the center of the road," said Lyman; she jumped from her car and began running away, only to find herself sitting back in her car miles away from where her terror began.

A victim of missing time/space—of whom there are hundreds, according to researcher Budd Hopkins in his book *Missing Time* (1981)—the woman at least lived through 1966 in Potter County. One person there who didn't was Dr. John Irving Bentley... but you already know about him.

• On July 1, 1964, the burning-from-above aimed itself at an anonymous female motorist in northeastern Georgia. Here, some unspecified aerial device swooped down to deliver third-degree burns upon her body. Furthermore, bags of groceries in her car caught fire. Sometimes, the fire of spontaneous combustion leaves toast along with toasted people to mark its presence.

• Two days earlier, on June 29, 1964, Beauford E. Parham left Atlanta to return home to Wellford, South Carolina. Somewhere in northeastern Georgia, probably on U.S. routes 23 or 123 (perhaps at the spot where two days later another human *plus* groceries would strangely burn), he and his car collided with the unknown.

"I saw the object so clearly I believe I could build one of the things," he said later of the "giant top" that swooped thrice over his car. After its third pass the UFO disappeared, leaving behind an odor akin to embalming fluid; a car roof covered with oily spots and blistered paint; a car radiator and hoses now riddled with holes; and nasty burns on Parham's arm.

He filed a report at nearby Warner Robins Air Force Base. The USAF told Parham he had been frightened by ball lightning. Explained an indignant Parham: "Lightning wouldn't rotate at the top and the bottom, as this object did." What he also saw—and felt—was the relationship between a UFO (whatever its nature) and spontaneous burning of human flesh.

• June 2, 1964; Hobbs, New Mexico. Eight-year-old Charles Keith Davis waited for his grandmother outside the DeLuxe Laundry. The three-foot distance between them was sufficient to safeguard Mrs. Frank Smith from the fate that befell her grandson at four o'clock that afternoon.

Suddenly a whooshing sound and a blackish ball of fire covered the youngster. She reacted immediately. "I grabbed him and tried to smother out his hair, which was on fire," she said. Laundry employees supplemented her efforts, and Charlie Davis eventually recovered after burns treatment at Lea General Hospital, where he told perplexed doctors and police that his injuries resulted from "a fire that came out of the sky."

• One week earlier, an egg-shaped UFO over Albuquerque painfully burned ten-year-old Sharon Stull on April 28, 1964. Her doctor diagnosed first-degree burns on her nose and under her eyes, plus membrane inflammation of both eyes. The incidents provoked Albuquerque Police Chief A. B. Martinez to issue a warning about UFOs: "they should be treated with respect and caution."

Charlie Davis and Sharon Stull already knew that. What no one knew was *exactly what* launched these unprovoked incendiary assaults on two New Mexico children, or why in the month following young Miss Stull's harrowing encounter she gained thirty-five pounds (a 41 percent increase!) and spurted from four-feet-eleven-inches to five-feet-two-inches.

• On Saturday, October 12, 1963, Eugenio Douglas was burned when a red beam of light from a UFO struck him near Valdes, Argentina. Douglas's burns resembled those caused by ultraviolet-type radiation.

• Shortly before the witching hour on May 10, 1961, motorist Richard Vogt had his unnerving thermal encounter with a "ball of fog" described earlier. Was his encounter another example of a potentially injurious link between UFOs—at the least an Unidentified *Foggy* Object—and aberrant combustion phenomena? Scientists working under Prof. W. J. Layten at the University of Minnesota couldn't explain Vogt's eerie encounter with "incendiary fog."

• Remember the five Kentucky men outside Pikeville whose fire-fractured bodies never made it home on Saturday night, November 19, 1960? Their car's nearly melted roof but uncharacteristically unburned lower panels suggest one more explanation not considered earlier. Investigators should have looked heavenward for the cause of this strange collective cremation, because it appeared the Kentuckians had been *struck from overhead by searing heat*.

C. W. Fitch had the same idea when he wrote to the APRO *Bulletin* (July 1963): "If we consider the evidence in this case along with repeated reports of glowing objects following cars, causing intense and sudden rises in temperature in the interior of vehicles being pursued, we begin to see the correlation of data which has led researchers to conclude that at least some UFOs are dangerous, and that the occupants are, at best, not concerned with the welfare of humans."

• The suspicion that Fitch and I share regarding Pikeville *can be confirmed* in an incident that mimicked it five years before. December 5, 1955; near Williston, Florida. Deputy Sheriff A. H. Perkins and Patrolman C. F. Bell began to bake in their rapidly overheating patrol car; their clothes became "intolerably hot and their limbs were virtually paralyzed" as six UFOs passed low over their location. Had the UFOs not moved on, perhaps Williston would have lost two officers to another burning mystery.

• In 1952 scoutmaster D. S. Desvergers warned three scouts to watch from a safe distance while he cautiously approached a glowing circular object hovering about thirty feet above the Florida palmetto undergrowth. His curiosity should have earned him a purple heart merit badge. From a "turret" atop the object a ball of red fire drifted toward him, changed to mist, and knocked him unconscious. When his boy scouts returned with a Florida highway patrolman, they found Desvergers' face burned, forearm hairs singed, and holes burned in his cap. USAF Captain Edward Ruppelt, as director of Project Blue Book (a government-funded catalog of UFO incidents), personally discovered that grass roots at the site beneath the UFO had been scorched while the grass above ground was undamaged; this was duplicated in a laboratory at 300°F over a gas burner, but no one could guess how that had been accomplished outdoors.

What happened to Desvergers? One moment he's in command of his scouts; the next, he's burned and unable to command anything. What happened to the motoring quintet on a Kentucky back road? One moment they are enjoying a scenic drive; the next, their bodies are candidates for SHC. And what of Huckabee? One moment he's enjoying a walk; the next, his body is fatally ablaze with flames for which a conventional source is absent.

But what about *unconventional* sources for combustion? Say, unidentified flying objects?

THE UFO/ufo DUALITY: MAKING SENSE OF THE MYSTERY

The body of evidence for UFOs and human combustion phenomena is too substantial to scorn and too rich in potential to disregard.

Too substantial, even, for some professional firefighters to disregard.

The late Charles W. Bahme, former Los Angeles Fire Department deputy chief and nationally regarded expert on disaster control, collaborated with William M. Kramer, a district chief with the Cincinnati Fire Department, to write in 1992 the second edition of the *Fire Officer's Guide to Disaster Control* (Fire Engineering Books and Videos). This comprehensive tome is used by the Federal Emergency Management Agency (FEMA) in its National Fire Training Academy Open Learning Program. It outlines traditional firefighting fare for rescue workers, plus one very untraditional primer: the ABCs of a first-ever disaster plan for public emergency personnel who find themselves facing a threat from UFOs.

Bahme's interest in UFOs ignited during the "L.A. Air Raid" of August 26, 1942, when, as a young Navy fireman, he was among those on the Western front who had the personal experience (the best kind!) of watching some twenty unknown skycraft maneuver "at incredible speeds while gun crews along the coastline pumped more than 1,400 rounds at them." Added Bahme: "The official explanation—weather balloons—was never taken seriously."

For four decades thereafter, Bahme studied the subject of UFOs as he advanced through the U.S. Naval Reserve (retiring with the rank of captain) and the fire services. He reached a conclusion startling for someone of his professional background, one he wasn't afraid to state:

Thus, UFOs may not only have the power to control some of our military and industrial establishment's highly technical scientific hardware, they may also possess the ability to impose pain and control over people who attack them, even to the extent of "liquidating" them in one way or another.

So what is a fireman to do? "In the absence of overt acts indicating hostility, there may be no danger in approaching a UFO with a positive, solicitous attitude of wanting to be of service," Bahme and Kramer advise—notwithstanding the fact they had just cited several episodes of people being burned by proximity to UFOs. Of more concern to Bahme and Kramer is their own government: it would be "inadvisable to make personal contact" with UFOs, they warn, because federal law (14 CFR, Ch. V, Part 1211) authorizes NASA "to quarantine under armed guard any object, person, or other form of life extraterrestrially exposed."

In one sense, these cases are not SHC *per se* because an external source (whatever its ultimate origin) for burn injuries is identified. On the other hand, that source is itself unexplained. Replacing one unexplained with another doesn't resolve the quandary, but does get us one step closer to understanding what might be happening.

UFOs may have been involved in other SHCs but—like Iraqi ground troops in the 1991 Gulf War who found themselves getting blown up without warning by bombs from unseen stealth F-117s—no one saw them overhead. In that sense, no less than two dozen persons have been subjected to and suffered curious, abnormal, indeed inexplicable thermal irradiation or burns in conjunction with UFOs of *some* kind.

Unfathomable flaming objects (human) offer significant insight into the seven basic theories proposed to explain UFOs, based on the connection between them (whatever it may ultimately prove to be):

(1) *Misidentification and hoax*. Many UFO sightings are misidentified known objects; *some* are hoaxes. However, it is implausible that people burn themselves in geometric patterns or, as in the Cash–Landrum case, expose themselves to cancerous levels of burning radiation just to validate a hoax. Those who would sweep UFOs under the carpet by debunking UFO sightings, must *first* defend their stance against a serious challenge to their smugness: the seared human flesh that coincides with some UFO close encounters.

(2) *Psyche/visionary encounters*. UFOs are events of the psyche projected from one's individual unconscious to form a visionary hallucination that is shaped by the reality-construct defined by the collective unconscious that is prevalent at the time, a view advocated by Jung in his seminal work, *Flying Saucers:*

A Modern Myth of Things Seen in the Sky. The problem here is the same one confronting the misidentification/hoax advocates: how to explain the physical evidence associated with some UFO sightings, which can include *burning* of soil and witnesses' bodies?

The only way out of this predicament that I see is to invoke a psycho-(pyro)kinetic aspect to the visionary hallucination: that is, the witness's psyche projects a force that registers on radar scopes, upon automobile wiring, in depressions at so-called landing sites, and on skin as blisters, burns, and worse. This theory, if one accepts it, *must* entwine unidentified flying objects and unfathomable flaming objects—and consequently confirms the reality of SHC because it demands that burns originate within and manifest from the psyche of the individual without contacting any known external fire.

(3) *Extraterrestrials.* Humans are coming in contact with real, physical spacemen arriving here from elsewhere in the astronomers' visible universe. Exobiologists like Carl Sagan have a problem with this theory, because they think interstellar distances preclude inhabitable worlds from venturing to Earth. This reasoning, of course, mandates that Alien technology and physics are restricted to the same advanced—or primitive—level as twentieth-century human engineering. To be kind, a myopic assumption.

(4) *Ultraterrestrials.* Aliens who contact humans are *truly* alien, arriving from other dimensions and other times to interfere momentarily with mankind's collectively agreed-upon reality. Charles Fort referred to this charming concept as "super-geography." If an item—Alien entity, candle flame, exploding fireworks—from another dimension or parallel universe suddenly "beamed up" (or through) and materialized inside an Earthling, might not the Earthling's human body combust, even explode from this ill-placed ultraterrestrial interchange?

The beauty and charm of super-geography is that it is speculation; hence impossible to deny without appearing closed-minded. The horror is that events occur in nature that seem best understood as super-geographical—

(5) *Natural phenomena.* Fireballs, ball lightning, will-o'-the-wisps and ignis fatuus can undoubtedly explain some UFO sightings, even those that leave behind the physical trace evidence of burned skin. This, then, could place UFOs into the misidentification category.

But what of energy patterns indigenous to the Earth but *unidentified*? There is photographic evidence for UFO-style *organisms* existing in the atmosphere that theoretically could trigger SHC. (For the adventuresome, you'll get the idea by reading Trevor James Constable's *Sky Creatures* and *The Cosmic Pulse of Life*.)

(6) *Earth-based technology.* UFOs are developed by government agencies acting behind the tightest security imaginable; technology that makes the

F-117A Stealth fighter look like the Wright Flyer at Kitty Hawk. There is support for this theory. (On October 2, 1990, Cable News Network revealed a new generation

> Truth is so precious that she must have a bodyguard of lies.
> —Sir Winston Churchill

of USAF "pulsar" triangular aircraft thought to be capable of hypersonic flight; after a few months of government denial, U.S. Secretary of Defense Chaney not only acknowledged this aircraft but canceled its funding.)

How better to explain the twenty-three helicopters that swarmed around the UFO that burned Betty Cash and her companions? Accepting the Cash-Landrum account means a U.S. government-of-the-people agency is injuring and burning its own citizens for reasons only a CIA or Dr. Strangelove mind-set could embrace.[3] One problem for this theory is that it can't (easily) be applied to early (pre-1950?) UFO sightings.

(7) *Other.* That the UFO phenomenon is a defensive artifact of the collective human consciousness, created to protect the psyche from something even more disturbing than the prospect of abductions and burnings by extraterrestrials.

BLAZING A NEW TRAIL: MORE UNFATHOMABLE FLAMING OBJECTS

Evidence for UFOs is too plentiful to be dismissed, no matter how seriously it undermines the status quo about mankind's delusion of omniscience and being alone to dominate the universe. SHC, certainly one of nature's great unsolved mysteries, is the ultimate burn phenomenon.

What might a cause-and-effect be that links these two profound enigmas, the UFO to the mysteriously burned Earthling?

Science measures things by frequency; by intensity. Frequency is what separates ice, water, and steam; intensity determines the speed of transformation from one phase state to another. A radar set safely pinpoints distant airplanes, but can burn its operator's chest if the power of its frequency is increased too much. Light, if scattered, harmlessly illumines book pages at night; when coherent, it can laser holes through plate steel. UFO energy sunburns people, sometimes in checkerboard patterns; when increased in frequency or intensity, sometimes they get worse than sunburn....

Were Mr. Huckabee and Dr. Bentley singled out by (unseen) UFOs, which, as with Greg Wells, fired at their unsuspecting targets a beam whose energy was so intense its impact left behind fatally seared flesh, smoldering ashes, and mystery?

Let's look at additional evidence.

- Around 2:00 A.M. on November 4, 1957, two sentries at Fort Itaipu in Brazil observed a brilliant light rushing downward at tremendous speed upon their military installation. It was a Monday morning the sentries wished they had stayed in bed. The object decelerated until it enveloped them in its orange glow. A painful humming noise soon broke the silence of their fear, then blistering heat engulfed them.

"To the terrified sentries it seemed they were actually on fire. One, overcome by the intense heat, fell to his knees and collapsed," recorded Maj. Donald Keyhole in his book, *Aliens from Space* (1973). "The other soldier, screaming from pain and fear, threw himself under a cannon for shelter." Soldiers ran to their battle stations, but three minutes later panic abated and conditions returned to normal after the glowing UFO streaked skyward. According to Dr. Olavo T. Fontes's original account published by APRO,[4] not surprisingly "The Fort Itaipu affair was kept secret."

This unprovoked "attack" by a UFO on two Brazilian soldiers—victims of SHC resulting from a CE-CK with a UFO—has an obscure and fascinating equivalent half the world away and thirty-five years earlier.

- The *Watford & West-Herts Post* (April 30, 1964) newspaper published a letter from Lawrence Bradley, who recounted an astonishing episode from the 1922 Irish Civil War. Local militia, defending a County Donegal cave, opened fire on what they first thought was an approaching enemy vehicle and provoked a kind of fire-fight not even practiced during the Vietnam war fifty years later.

Wrote Bradley: "the object retaliated by firing jets of flame at the cave. The defenders had to withdraw in face of the fierce heat... so that it was a case of facing the flame throwers or suffocating to death." The defenders decided facing the enemy was a more valiant death than ignoble suffocation, so they rushed toward the cave's opening. "When they ran out," Bradley continued, "they saw the flame-throwing object ascending into the sky... circular in shape and bright in appearance, as if made of aluminum. I daresay some of the men who saw this strange phenomenon are alive today and can vouch for this story."

In the era of Sopworth Camel biplanes, it's safe to say none of the factions in the Irish conflict possessed weaponry like this! Though it would be another twenty-five years before Kenneth Arnold sighted his historic UFOs over the American Northwest, by the early twentieth century their predecessors had already become a potentially combustible combination.

- One year before Arnold's sighting introduced the phrase "flying saucers" to the world mindset, Joao Prestes Filho, forty, returned home around 7:00 P.M. on March 5, 1946, after a day spent fishing near his home in Araçariguama, about thirty miles west of São Paulo, Brazil. As he opened a window in his house a brilliant UFO flashed before his terrorized eyes. Knocked to the floor by its force, Prestes collapsed into incoherence...and much worse. One can only imagine how the local medical staff responded to the gruesomeness that followed, as described by investigator Professor Felipe Machado Carrion for *Flying Saucer Review* (March-April 1973):

> Prestes' insides began to show, and the flesh started to look as though it had been cooked for many hours in boiling water... The flesh began to come away from the bones, falling in lumps from his jaws, his chest, his arms, his hands, his fingers, from the lower parts of his legs, and from his feet and toes. Some scraps of flesh remained hanging to the tendons, and none of those present dared to touch them. Soon every part of Prestes had reached a state of deterioration beyond imagination. His teeth and his bones now stood revealed, utterly bare of flesh.

Equally astonishingly, Carrion wrote that "at no time did he appear to be feeling any pain." Prestes died six hours later—his flesh falling off as though his body was decomposing alive.

Prestes' fate—whatever exactly caused it—remarkably echoed the SHC-style death of Father Bertholi in Filetto in 1776. And almost two centuries after Bertholi's bizarre fate, an eerie *double SHC* episode—an extraordinarily rare event—occurred with a UFO connection.

• December 11, 1979; on the outskirts of Bodfish, a small town some thirty miles northeast of Bakersfield, California. Melvin and Naomi Anderson enjoyed living in the remote southern foothills of the Sierra Nevada range. That remoteness revealed a sinister side the couple had not anticipated.

They would later recall that around two o'clock that afternoon, both of them lost consciousness in their trailer—at least that's what they initially *remembered* remembering. When they regained awareness of their surroundings, the couple found themselves on the floor of their trailer, unable to move; found themselves missing two days of time; found themselves haunted by unexplained physiological symptoms; found themselves burnt.

As Melvin lay immobilized on the floor, unable to move for another twenty-four hours, he told his wife how he felt he was "burning up inside." Already he had burns on the outside of his body: a bad burn *to the bone* on his left hip, a one-and-a-half-inch burn on top of his head, and more raw swollen burns on his heels. The most severe burns were below the waist and across his back. Naomi, lying helplessly nearby, had burns on her thigh and left hip.

Neighbors discovered the Andersons on day three of their ordeal, and soon the couple was hospitalized for treatment of "third-degree high radiation burns."

How did this unassuming couple in the rural Sierra Nevada foothills find themselves so oddly and painfully burned? No cause in their trailer could be identified to explain their injuries. Were they, then, the victims of an amazing *double case* of spontaneous human combustion that, even more amazingly, they survived? It appeared so.

Then came a twist that shed insight into their unearthly inflaming. As they recuperated, the Andersons both began experiencing an unnerving recall: of a UFO hovering around their trailer, bathing it in light, and suddenly finding themselves inside the craft. Then they were back on the floor of their trailer, burned.

It made no sense to them, yet with trepidation they shared their recollections with a trusted friend who realized their tale was not unknown to the literature of abduction ufology. However, thought the friend, the burn aspect was abnormal.

But *was* it? Abnormal, yes. Unprecedented, certainly not.

The Andersons's plight, like that of Gregory Wells, Francisco Calle, Steve Michalak, Inacio de Souza, Manuel Amparano, and Joao Prestes before them, and officer A. P. Jones since, raised again the disquieting horror that some humans in the presence of UFOs find their flesh suddenly burned or bursting into flames. And if a UFO had not been involved in each of these episodes, we would be facing overwhelming data on behalf of SHC. In a sense it is SHC because, as I said, the source of combustion is certainly unquantified and unrecognized by science today.

Another intriguing recent development that reinforces the UFO and ufo (SHC) relationship.

- As chronicled by Raymond Fowler in *The Andreasson Affair* (1979), Betty Andreasson Luca recounted under hypnosis ten years of encounters with extraterrestrial beings. In one session, she exclaimed over and over that she had become unbearably hot in the presence of these beings. In front of startled investigators, she began to pant and writhe in agony as she relieved hypnotically that moment, screaming that she felt as though she was burning up in a barrage of light rays and burning ashes which emitted a sweet smell of incense.[5]

These images may have had symbolic or mystical meaning for her, yet it is *noteworthy* that classic SHC does leave behind a sweet, perfumelike smell amid burned ashes. I asked her to elaborate about this Phoenix experience:

> It was more alive than, like, you are standing here. Somehow, it was real hot. And the rays of light kept getting bigger and bigger, shooting out in all directions. And then, the heat was extreme! And then I felt something penetrating every cell of my body, like tiny

stingings. At a point it got *so* hot, I think I must have passed out. The next thing I remember, is seeing a pile of ashes. And there was still some burning coals with it. And there was a cold, cold wind. I felt very chilled. That's what happened: the heat, then the cold.

Was it a physical heat, not a metaphor, I asked? "Yes. I physically felt it."

I wonder if Mrs. Luca's words might be describing the early stages in the evolution of internal spontaneous combustion, an experience that for her fortunately never fully manifested in this reality.

• In recent years, dozens of villages in Brazil have been plagued by what Robert Pratt, a respected journalist for forty years, calls "a terror which comes from someplace other than the Earth which produces injuries and death. I call it *a* dark side of the UFO phenomenon."

In one of the most provocative (and disquieting) lectures I've attended in years, Pratt detailed for 1994's "The UFO Experience" conference★ why he has concluded that "strange things are happening down there that I don't hear about elsewhere." Strange things, like people being engulfed by winds "like a hurricane going in circles" then lifted off the ground in the presence of what the locals describe as a "ball of fire." Strange things, like human combustion.

The owner of a large ranch in northeast Brazil described (in terms reminiscent of Betty Luca's phoenix experience) his encounter as "Hot...It burned my arms, and hurt very much." A woman in Colares was hit by a red beam from one ball of fire; consequently, said Pratt, she "has three scars on her chest to this day." Señor Louis was paralyzed for three days after his too-close encounter with a domed object, and was hospitalized in a catatonic state; doctors discovered "his hair had been burned off without singeing the scalp." Pratt interviewed four of eight doctors who treated Louis: "None could explain Louis' condition....As far as I know, Louis completely recovered from his experience. But not everyone does."

The mouth of the Amazon River has been a real hot spot, so to speak, for UFO activity (Pratt says thirty villages there have had recent CEs, one for an entire year) and for close encounters of the combustible kind. There, in 1978, a Colares woman was burned fatally (though she didn't die until 1981). In 1993, again in Colares, a thirty-two-year-old missionary and a forty-year-old domestic worker died, a month apart, as a result of their CE-CK episodes; according to Pratt, these women were "both burned on the throat and the chest, as were most of the other people the doctor treated."

★ Sponsored by Omega Communications (PO Box 2051, Cheshire, CT 06410-5051).

In 1986, not far from Colares, two persons on Crab Island were discovered "badly decomposed by heat" during a time of numerous sightings of balls of fire in the sky; cause of death "not known" in either case. Same area, different time: a "ball of fire" seriously burned three men out cutting wood on Crab Island; one of the woodchoppers died. One of the two survivors, Edmundo, had a gaping electrical-type burn injury on the side of his chest—a spitting image of Jack Angel's chest burn suffered in Georgia in 1974!

In all, said Pratt, "I know of about ten deaths that have some connection with UFO close encounters."

I asked Pratt if any of the Brazilians mentioned a metallic taste or ionization in the air around their close encounters and if burn tissue pathology had been done on the victims. "I don't know of anything like that," he replied. "A lot of the burns, by the way, look like sunburn—where the person is running away from the UFO and is exposed to light from the UFO." Sunburn that can prove fatal quickly and leave gaping chest wall injuries is not a normal sunburn, to be sure.

Of the many episodes involving UFOs and combustion of humans in modern times, one stands out—if true—as particularly disastrous.

• For several nights in June 1954, eleven-year-old Laili Thindu and his shepherd companions listened to the pounding of their neighbors' drums in the African village of Kirimukuyu on Mt. Kenya. The drums' rhythms announced a wedding about to take place in the mountainside hamlet. One night they watched something besides their flock: strange lights soaring above this sacred peak in central Kenya. When bright beams flashed from these unidentified lights, the shepherds were startled; when the drums became silent, they grew concerned.

The next morning Laili learned that "all the dancers, all the children, all the livestock—the entire population of the village—had been seared to death by terrible streams of light from glowing objects," report authors Brad Steiger and Joan Whritenour in *Flying Saucers Are Hostile* (1967). "It was not until Laili Thindu ventured to Nairobi that he was able to tell his story to someone who recognized the tale for what it really was: the annihilation of an African village by a UFO."

BEAMS OF COMBUSTION

UFOs...beams of energy, flashes of light...intense heat, open flames... human targets. SHC.

Short of displaying a retrieved UFO in Macy's Thanksgiving Day parade, there is no stronger evidence on behalf of super-advanced aerial technology than the burns upon people who have suffered encounters of the too-close kind. Disbelievers in intelligently controlled UFOs must come up with an alternative reason why scores of people worldwide, for generations, would—*could*—subject themselves to painful, sometimes fatal burning if the (usually) saucer-shaped sources they say are overhead truly do not exist.

These burn phenomena cannot be created by an energy that is merely ultraviolet. Although some symptoms do resemble sunburns, UV rays will be stopped by clothing—and Michalak, for instance, was burned *through* his garments. It must be more.

One suggestion is *microwave* energy. Based on incidents like Cash-Landrum, where lesions and symptoms of radiation sickness develop soon after exposure to the beam, a radioactive component is possible.

Witnesses to these beams sometimes report electromagnetic aberrations in nearby electrical equipment. So who can predict what effects might be transmitted into the human bioelectrical system?

Ultrasonics must also be considered a component, since ultrasound can vibrate water molecules until they boil and become steam.

Whatever frequencies UFOs and their beams project, the scorch, burn, and radiation injuries associated with their close encounters are

> I have my own theory. I think it's attributable to walk-in aliens who have taken over human bodies, then have been spotted by their own, other-dimensional police, and Zap! Blue smoke and ashes.
> —reporter Deloris Ament, about spontaneous human combustion; in *The Seattle Times* (April 1, 1990)

real. Many of these innocent victims lived to tell of their horror, though not to understand it. Others died, surely never knowing what hit (or lit) them.

A close encounter with a UFO can ignite more than wonder about man's place in the cosmos... it might, sometimes, kindle a very hot time indeed, one not wholly of this world that humans like to think of as their own. Close encounters of the combustible kind offer some of the strongest evidence that we humans are not alone in a universe we as yet do not comprehend.

18

Death-Fire: Searing Suicides and Combusting Corpses

> The act of dying seems to be associated with some other event, perhaps pharmacologic, that transforms it into something quite different from what we are brought up to anticipate.... Something is probably going on that we don't yet know about.
>
> —Lewis Thomas, M.D.,
> essayist and medical philosopher

What is the destiny of the animating life force after its release from the body at death? Does it go to the grave along with the corporeal form? Does it reside for eternity in radiant Heaven or fiery Hell as the result of a one-time incarnation on Earth? Or does it survive as a soul, having consciousness and maybe an option to reincarnate into another physical body?

Regardless of the answer you favor, there is one point on which nearly everyone agrees: bodily functions cease after death, rigor mortis ensues, and decay transforms the body to biblical "ashes to ashes and dust to dust."

This popular assertion is wrong. Among a variety of postmortem phenomena is the fact that corpses can *self-heat*, sometimes becoming hotter than when the body was alive:

- Venerable Antonio Margil, a Franciscan missionary to Mexico, died there in 1726 at age sixty-nine. From the moment of death his pale face became rosy and his flesh continued warm until his corpse was consigned to the tomb.

- Dead for three days when laid in his coffin, sixty-year-old Franciscan monk Andrew Ibernon's flesh was "still warm and soft, and all the sinews and muscles flexible, just as if he had only expired the moment before."

- Warm-hearted, water-sizzling Dominicaness Maria Villani continued to astound, even in death. Two surgeons cut into her corpse nine hours dead and found new mystery when it and its unbeating heart discharged smoke and heat—heat so intense it forced one surgeon to retreat. Upon attempting to extract her heart, it proved so hot it burned him and "he was compelled to take his hand out again several times before he succeeded in effecting his purpose."
- Not to be outdone in the realm of postmortem thermanomalies was Serafina di Dio, whose corpse in 1699 remained—despite March weather's chilliness—perceptibly warm for thirty-three hours after death as nuns warmed their hands over its hot, unbeating heart.

Now add to these mysteries of postmortem hyperthermia yet one more: that bodies dying—or dead—can self-immolate.

A TRIAD OF POSTMORTEM COMBUSTION

Dr. René Moreau described for the *Ancien Journal de Médecine* (1786) an incident from 1644 when mourners were astonished, perhaps appalled, to see "flames from stomach" of a dead Frenchwoman in Lyon who was being buried.

The *St. Louis Daily Globe-Democrat* (December 4, 1875) ran a filler from the *Dover Herald* about a "supernatural" event at the wake, a week earlier, for marble-cutter Patrick Savage, who died of consumption:

> About a dozen friends and neighbors were engaged in conversation in the room where the corpse was, when, as the story runs, and as each one of the party is ready to solemnly affirm, there suddenly emerged from the side of the wall of the room a number of small jets of what appeared like gas flame of a golden color. These jets moved along and arranged themselves in a circle or wreath about the head of the corpse, and continued to burn for half a minute, changing from their golden color to a dark blue hue. The party was much too terrified to make any investigation of the remarkable phenomenon, though it was generally accepted as the work of some supernatural agency, and as portentous of good to the spirit of the deceased.

This might be easily dismissed as too unbelievable, were it not for the similar outburst of blue sparks that would terrorize Mrs. Ona Smith in 1922.

Hereward Carrington recounted in *Death: Its Causes and Phenomena* (1977) a recent necropsy he knew about with an intriguing phenomenon. A girl who gorged herself on chestnuts died of acute indigestion because she had not chewed them thoroughly, wrote Carrington. In preparing the child for burial, neighbors discovered the body suddenly appeared to be aflame.

"While the heat from the bluish flame which enveloped the body was quite perceptible," Carrington narrated, "it was not sufficient to burn the body or even set the bed on fire; and yet, when the corpse was removed from the sheet on which it had been placed, it was found that the latter was scorched in such a manner that the outlines of the human fire could be plainly distinguished."

Blue flames; scorching material; soaring hyperthermia in a *corpse!*

Ironically, Carrington had disavowed SHC in the preceding paragraph of his book. What to do? He conceded that "in certain cases the body may acquire preternatural combustibility." Where was the external ignition source required for PC of this young lady? Carrington found none. This scholar on death had seen, it seems, even if he wouldn't admit it, none other than spontaneous self-heating inside a person deceased.

BETTY SATLOW: "GHOULISH FIRE IN A CLOSED COFFIN"

"It was like something out of nightmare theater: A fire inside a casket bearing the body of a woman awaiting burial."

So said the *San Francisco Chronicle* (December 26, 1973) of what could rank among the ten most bizarre events of 1973. For it wasn't so much the interior of the coffin that burned, as the *corpse* it contained!

Betty Satlow helped her husband Sam operate a tavern in Hoquiam, a small town on the Pacific coast of Washington state. On Friday, December 7, 1973, Mr. Satlow walked into his garage and found his fifty-year-old wife dead on the seat of her car. Hoquiam Police Chief Richard Barnes could find no evidence of foul play, only intoxication. Coroner Harold Schmid declared that carbon monoxide poisoning contributed to this fatality, but listed cause of death as simply "undetermined." Her body was taken to a mortuary, prepared for burial, and given a rosary service on Sunday.

But Mrs. Satlow's body, now readied for final rest on her third day dead, was not about to lie still.

Smoke issued from the mortuary. Soon, firemen discovered the blaze was not just inside the funeral parlor... it was inside the Satlow coffin... inside the late Mrs. Satlow herself!

The lower portion of the casket was closed, but the upper lid was open. Here the firefighters found the lady's body "completely consumed to the hips," said Chief Barnes. The police chief was baffled, reported *The Oregonian* (December 20, 1973). He had the coffin sent to the Treasury Department's laboratories in

Washington, D.C. There the case dead-ends. No agency involved wants to be troubled with replying to inquiries into this troubling case.

Again it is more convenient to forget than to confront, even though Chief Barnes had admitted in late 1973 that "We really need a logical explanation to put an end to so many wild, baseless rumors that are going around the community." He didn't elaborate on what these speculations were, merely saying of any theory about Satlow's immolation: "It's all conjecture."

Was spontaneous combustion *of a corpse* among those "wild, baseless rumors?"

BILLY PETERSON BAFFLES THE PONTIAC POLICE

"Impossible!" screams the skeptic. "Corpses *can't* burn! There must be some conventional explanation!"

Police Chief Barnes couldn't find one in Hoquiam, Washington, not even one so conventional (and obvious) as misapplication of embalming fluids. The experts in Pontiac, Michigan, couldn't either when they were confronted by—well, by the facts known about Billy Peterson:

- He was a thirty-year-old male.
- He lived in Pontiac, Michigan.
- He had been a welder for General Motors.
- He was alive at 7:00 P.M. on Sunday, December 13, 1959.
- He was dead less than one hour later.
- He was burned—somehow.

A hospital pathologist said Peterson died of carbon monoxide (CO) poisoning. The deputy coroner declared that Peterson died accidentally. A police detective suspected murder, then changed his mind to "suicide." The fire chief concluded the dead man was cooked—*after* dying—by extreme heat. Pontiac officials closed the case by pronouncing "Death by Suicide." However, said County Prosecutor George F. Taylor: "We haven't closed the case yet."

Why all this confusion among trained and competent professionals over a few facts? To have empathy for the consternation in which these officials found themselves, return now to that Sunday evening....

Billy dropped his mother off at 7:00 P.M. and drove the mile to his garage. He had been despondent over ill health and missed work, but was about to return to the factory in two days and, his family noted, Billy was "as jolly as could be" this day.

Forty-five minutes later, the Pontiac Fire Department was notified of a smoking car in the Peterson garage. Fire Lieutenant Richard Luxon and crew arrived to find a shocking scene. A flexible tube led from the shortened exhaust pipe into the car's interior. On the driver's seat sat Billy, dead. But dead with a difference: though his face and arms were vivid with burns, *nothing* of the car itself was ablaze.

Billy was rushed to Pontiac General Hospital anyway, where Dr. McCandless said the violent red color of the victim's blood indicated CO poisoning. Police, now alerted to the case, suspected foul play in the garage. Finding the car's missing tail pipe plus some flexible tubing in the garage, the police changed their speculation to suicide.

Meanwhile back at the hospital, doctors were exclaiming "It's the strangest thing we've ever seen!"

Why was the medical staff astonished? Though the United States has one of the world's lowest per capita rates for suicide (in contrast to one of the highest homicide rates), about twenty thousand Americans kill themselves annually. Peterson surely wasn't the first suicide ever examined by the hospital staff.

What mystified the health-care professionals was the nature and extent of Peterson's burns: his chest, back, and legs were *covered* with second- and third-degree burns; the left arm so badly seared that the skin rolled off; ears and nose were scorched. Yet his eyebrows and hair were untouched. And, his clothing was undamaged—not even his underwear was singed!

So the police returned to their murder theory. "Possible Torture Killing" headlined *The Detroit Free Press* on Monday the 14th. Physicians promptly said the victim *couldn't* have been undressed, burned, and then reclothed as the police now contended; besides, any fire *external* to his body would have burned the hair off his chest and arms instead of leaving it intact. No, that theory would not do.

Dr. Marra surmised that the exhaust pipe's heat ignited the upholstery, which in turn caused Billy's blue jeans to "become so heated that superficial burns of the skin resulted." Superficial burns would be comparable to first-degree burns; that is, reddening of the skin. Yet his colleagues spoke of charring.

Besides, how could hot blue jeans severely burn the wearer's chest and back? Furthermore, Fire Chief White said the hot exhaust had caused about $75 damage to the *right* front seat and that the blaze *had not touched Peterson*— or his jeans. No, the rationale of superheated jeans also fails to remove the perplexity surrounding Peterson's fiery fate.

"Can a man take fire in his bosom, and his clothes not be burned?" asked Proverbs 6:27. It seems the answer is yes.

I wrote Pontiac General Hospital, hoping to find *their* answer in the records on Billy Peterson. I received a form letter requiring "an authorization signed by the patient"—oh well, no one said this would be easy.

In a letter to *True* (August 1964), Dr. Tad Lonergan detailed his medical staff's reaction to what he called "inexplicable internal and external third degree burns" on Peterson's body. Contrasted to his staff's previous burns injury experience, he forthrightly said "this case was different. One could not account for the burns on his skin when the clothes were not even singed. Hence, a thorough investigation was launched. No explanation was available then, and as far as I know, none is now. I haven't seen a case like it since, and it is still baffling to me." Years later, Dr. Lonergan confirmed by personal letter that his puzzlement lingers on.

A propitious point at which to pause—

The experts in Pontiac failed to notice that Peterson showed symptoms of nuclear radiation burns, a fact that caught Paul Fought's attention. "This fact, of course," he wrote in *Fate* (March 1961), "brings us to an even greater mystery. Where was Billy exposed to radiation?" Fought had no answer for his question.

Evidence in Peterson's garage suggestive of suicide raises another question. Might Billy Peterson have been the victim of *belated* spontaneous human combustion?

Because there is no doubt Billy inhaled CO gas, a proposal made way back in the 1800s by Dr. Adrian Hava regarding humans is of interest: "that the accumulation of carbonic oxide [CO] gas was the prime factor in spontaneous combustion." Spontaneous combustion of a human, did the doctor say?

This physician performed experiments on animals to see what effect the gas had on tissue inflammability. Rabbits and roosters had a propensity to ignite in *bluish flame* after prolonged exposure, he found. But it required 169 days for the rabbits' hemoglobin to store enough gas for the tissue to ignite; eight months were necessary for the ill-fated roosters. Hava's experiments would *seem* to rule out bodily accumulation of carbon monoxide in Peterson's odd demise, since he was exposed to his car's exhaust for probably no more than thirty minutes. One could argue that Peterson ignited the accumulating gas with a pack of matches or a lighter, except that investigators found evidence of neither in or around his car.

The editors of Aerial Phenomena Research Organization's *Bulletin* offered another scenario, stating their "strong suspicion that Billy Peterson may have been burned by an ultrasonic scanner beamed at his car. But why? Therein lies the mystery."

To link Aliens with SHC, as APRO did, echoes the many cases in which humans suffered varyingly burned flesh in close contact with UFOs. Asserting that extraterrestrial voyagers would randomly assault this human with their "ultra-sonic scanner" may be an unjust accusation against unearthly visitors. (This assumes, of course, the Aliens exhibit a higher rationality and respect for life than many humans do.) But what if they did so with a purpose? That could resolve the problem—if one could just think of a reason.

Which brings us back to the theory of SHC of a dying body. Fire Chief White, when asked by Fought if Peterson might have succumbed to this extraordinary demise, replied: "I would not quarrel with the theory concerning Spontaneous Human Combustion.... I have never had any knowledge of this, but certainly would not care to say it was impossible."

"THE CASE OF THE ONE-LEGGED VICTIM"

More than two decades before Billy Peterson attempted suicide only to burn strangely, another case occurred similarly in Cleveland, Ohio. Wilton Krogman, who gained fame or (depending on one's view) infamy by reporting on the famous Mary Hardy Reeser death, found this one to be among his more intriguing experiences in forensic anthropology.

To summarize, police found a one-legged man burned to death in a Model T automobile. They attempted identification from a list of the area's amputees, unsuccessfully. Krogman's expertise reached a different conclusion: "I found *no evidence whatsoever* in the right pelvis of this man that said he had had an amputation. And I concluded that it [leg] had been *burned away!*"

I asked him if he had resolved how this man's leg could completely disintegrate.

"Yes," Krogman replied, "because it was nearest the gas tank." When this "*badly* burned" corpse was examined carefully, he gestured dramatically, "we found in his viscera—the belly wall had burned away—a gun. And when we put the head together, we found this—the gunshot wound. See? So evidently at the moment he fired a shot, he dropped a match in the gas tank. So that was the side that was completely consumed."

His narrative had taken on new significance. Not only was the man badly burned and his leg totally reduced to ashes, but now it involved apparent suicide as well.

There is a problem with Krogman's reconstruction of the event. An antique automobile dealer confirmed my belief that, while the gas tank in the Ford Model T is beside the driver's leg underneath the dashboard, the filler tube is *on the hood in front of the windshield*. Now consider the amazing dexterity of this soon-to-be-a-suicide man, once he has removed the gas cap. He must reach over (or under) the windshield's glass with a lit match, making sure the match is held far enough above the gasoline filler tube so that fumes won't prematurely ignite before he fires his gun. Then, as his body recoils sharply from the bullet's shattering impact through his head, the man nevertheless keeps a steady hand and dead-aim drops the lit match into the narrow filler tube where it finds its way into the fuel tank, whereupon flaming gasoline somehow *leaks* out below the dash—there is no mention of an explosion—and onto the man's leg, which then burns to ash right through the bone. See?

True, the auto was said to be ablaze. But as discovered with Billy Peterson—and as noted in other cremation-in-a-car cases—one should no longer assume this means the vehicle itself (and its fuel) is the origin of the blaze. Knowing what you know already about the normal resistance of flesh to fire, might not the ravaging flame have been birthed within the man's *own body* first?

This curious case could be another example of SHC accompanying suicide.

More Victims of SHC Suicide?

The unidentified Model T motorist in Cleveland, perhaps spurned in romance or dejected over an inability to recover from the Great Depression, may well have chosen to walk away from his personal depression by administering a bullet to the brain. Thereupon his body immediately inflamed—not due to a lucky drop-of-a-match-in-the-gasoline-filler-tube but because his physiology, shocked by this sudden trauma, responded by marshaling a *firestorm of energy* just as the body musters white corpuscles when invaded by germs.

A more thorough example of a firestorm of energy occurred more than a century earlier, just after its human host told acquaintances that they would see him no more.

Corpulent and intemperate sexagenarian Monsieur Vatim was "extinguished with difficulty by water" after friends found him aflame in Beauvais, France, on January 22, 1822.[1] With the exception of his head and one leg, and some organs that shrank, two physicians testified Vatim's body was totally consumed in less than eight hours. The scene was sufficiently noteworthy to be remarked on by *The Edinburgh Medical & Surgical Journal* (October 1823).

Vatim drank heavily during the week and, on the evening before the body was found consumed, bade farewell to a fellow-lodger and assured him they should not meet again. A Dr. Klaatsch of Berlin took a personal interest in these events, and procured a report from police investigators. He learned that a brass chaffing dish containing embers was found near Vatim. Therefore, he insisted to *The Edinburgh Medical & Surgical Journal* (1824) that "this case cannot be considered one of indisputable 'self-burning.' At the same time, we presume, there can be no doubt of its being one of preternatural combustibility; which, we suggested, was the limit of our belief with regard to all such stories."

History will never know whether Vatim planned to drink himself to death that night, setting himself alight before preternaturally burning up; or met a fate beyond belief.

Perhaps Vatim yearned to "end it all" in a flash? In the way that Alhaken ibn Itta is said by *Ripley's Believe It Or Not!* (23rd Series) to have done? This famed alchemist from Merou, in Khorasan, Iran, "was so skilled in his profession that when he committed suicide by immolation *not a trace of ash remained.*" Centuries later, one can but guess, and wonder, about his method.

What can be said, reasonably, is that both Vatim and Itta accomplished what no ordinary fire could: they reduced themselves nearly or wholly to ashes in (at least in Vatim's case) a room otherwise undamaged. Plus, they did it with the intent of never being seen again. Vatim succeeded dramatically, and ibn Itta perhaps even more so, and in their decidedly different deaths achieved a degree of historic immortality.

SUICIDE AND INCINERATION: ESCAPE OF THE LIFE FORCE

Biological processes that occur within a body after death differ markedly from an animated state. But exactly what happens to the animating life force at the catastrophic moment of life-to-death has been debated by theologians, atheists, scientists, philosophers, and mystics since communication began. What happens to the *physical* has been less prone to discussion, however, for here the evidence is usually perceived with man's five senses.

But the subjective and the objective sometimes intermingle. Duncan MacDougall claimed in the *Journal of the American Society for Psychical Research* (May 1907) that his monitoring of many dying patients gave him sufficient time to make careful measurements of a recurrent weight loss—about twenty-one grams—at the moment of death. MacDougall thought he had quantitative evidence for the existence of a soul, whose absence lightened the body.

Though his work has been harshly ridiculed, he may have succeeded in measuring what operating room physicians have only recently begun to acknowledge among themselves: light-forms rising from patients at the moment of their death on the operating table. In January 1990, surgeons from several nations convened in Amsterdam, Holland, to

> I do believe a soul goes on. I do believe there's more out there. And I sincerely *hope* a soul goes on.
> —Robert J. Wagner,
> widower of Natalie Wood;
> on *The Extraordinary* (1994)

share experiences their medical training had not taught them to expect: bright, glittering clouds of light, fluttering sounds, or rushes of frigid air which envelop the bodies of their patients at the moment of death. Whatever these phenomena represent, there surely will be more such observations reported and continuing controversy over just what is going on at this ultimate human phase-transition called death.

Another observation that MacDougall made is less controversial and readily confirmed by care-givers of patients who realize health is deteriorating fatally. For these persons who have prepared themselves emotionally and psychologically to die, separation from the body is expected; nontraumatic. But the decision to commit suicide can be a spontaneous action, permitting no time for the necessary bioenergy adjustments that are a prelude to death. Making the body suddenly inhospitable may force the animating energy to leave abruptly, resulting perhaps in another catastrophic event *within* the corpse: the arcing of bioelectricity which rages throughout the body. The rapid exit of the life force, fully vitalized a moment before, literally burns out the body from within.

The *abruptness* of the decision to self-destruct could be a factor in whether the suicide wholly inflames, or manifests a somewhat less dramatic fiery departure. That is, it may be the *brevity* of a suicide's premeditation that influences the body to fire up after the emotions sink into the darkness of gloom.

It would seem Billy Peterson acted impetuously that evening (though there is no way to document this), because he was said to be "jolly as could be" just before he made the tailpipe alterations that changed his car into a pseudo-crematorium. Was it his abrupt change in disposition, his own emotional outrage, that led his body to create its own thermal outrage?

Tantalizing support for this supposition comes from a case—lost to history until I rediscovered it—in an obscure Michigan medical journal. *The Therapeutic Gazette* (1889) tells about an eighteen-year-old Chenango County lad who shot himself in the late 1800s in central New York State. Dr. George O. Williams discovered the boy's corpse fearfully charred, the flesh split asunder by heat, the face "cooked." Yet planking upon which the charred remains lay was only "trivially damaged."

This youth had burned after a self-inflicted gunshot wound, Dr. Williams reported. But he could find no accelerant—only four pounds of clothing, a gun, and portions of the corpse. Dr. Williams was frustrated. The clothing was the only fuel source present to sustain a fire. The limited amount of combustibles, inadequate to so consume the victim, *plus* absence of adjacent destruction, posed an unsolvable mystery for Dr. Williams.

Resolution required a concept alien to nineteenth-century medicine in New York State: that the human organism consists of much more than tissue, bones, and circulating blood; that some unseen, undetected force inside the body may release a vengeful fury upon the person who attempts to prematurely kill himself. The concept is still alien to much of twentieth-century medical knowledge. Concepts, however, are changing—

"Into the Valley of Death" and Back Again—Sometimes

"Granting then, that such changes may take place in the human body, which permit it to be more easily burned," remarked Dr. W. H. Watkins in his 1870 article on preternatural inflammability, the occurrence of SHC "must be entertained, although science cannot account for the changes."

Since then, has science progressed toward explaining these changes, *especially* when corpses seem to self-combust?

Pioneering thanatologists in the last decade have begun to shed insight into what can occur to the physical body's life force at that mysterious moment termed death. Dr. Raymond A. Moody Jr., Elisabeth Kübler-Ross, Ph.D., and Professor G. Kenneth Ring, Ph.D., founder of The Center for Near-Death Studies, are notable among many researchers who have concluded that death is not a final, terminal experience but merely a passage into another realm of existence. (Eastern philosophy, mystics, and clairsentients have professed this truth for aeons, but Western medicine only now finds support for the ageless pancultural traditions concerning postmortem existence.)

Evidence for seeing beyond the veil of death dates back thousands of years. Today, it is called the Near-Death Experience (NDE). This ever more documented experience generally finds the individual hovering above his body and "watching" attempts to resuscitate it, hearing someone pronounce him "Dead!," then realizing that he still possesses a "body"—albeit quite different from the flesh-and-blood one just vacated. This ethereal body travels through an unfamiliar realm, sometimes alone, sometimes met by the images of relatives and friends who previously died; occasionally by a light-being who asks that the just-departed life be evaluated. A barrier representing the irreversible separation between physical life and the next life is not allowed to be breached, however, and reunification with the corporeal body occurs. Medicine has a miracle; the once-dead person, now resurrected, has a revelation.

Analyzing quantitatively the prospects for the continuation of consciousness after death and what awaits in this post-physical existence, sometimes yields insights that surprise the researchers as much as the NDE transforms the person who experiences it.

Dr. George Ritchie has spoken to me about his memorable experience of December 19 through 24, 1943, which mirrored the classic NDE pattern. His temporary sojourn into the afterlife was marked with vivid recollections of the events surrounding his body in the hospital, then of meeting a luminous being who showed him *everything* he had done in his life (not one secret was withheld, Ritchie says). Though surprised by this separation from his physical body, Ritchie learned it was not to be permanent. He was told, to his chagrin, that there was still work for him to do on Earth and he must return. Thus *four days* after he entered a coma, during which his doctors *twice* pronounced him dead, Dr. Ritchie arose Lazarus-like and sat up on the bed. The life-support equipment was no longer needed. Dr. Ritchie's soul had chosen to extend its life on Earth.

Many people who were interviewed by Dr. Moody and Kübler-Ross told of similar reluctance to return to the three-dimensional world. The reentry process occurred with no more than an emotional emptiness, these two pioneering researchers concluded. But is this readjustment to the physical always so serene? Could there be difficulties in certain circumstances? For example, might the act of *suicide*—a strong taboo in most human societies—create a unique barrier that prevents rehabitation of the body?

Ring surveyed thirty-six suicide attempts and learned that not one survivor recounted an unpleasant, "hellish" or "demonic" NDE.[2] However, Dr. Moody collected reports of NDEs associated with attempted suicides, and his findings contrasted with those of Ring.

For example, Harry Woods. Despondent over his wife's death, Woods shot himself. Dead for nineteen minutes, he returned to describe his resurrection experience in *Life After Life* (1975): "I went to an awful place.... I immediately saw what a mistake I had done.... I thought, 'I wish I hadn't done it'."

Robert Smith of Chickamagua, Georgia, was also not part of Ring's survey. This sixty-eight-year-old retired car salesman was pronounced dead during triple bypass surgery following a massive heart attack. "I saw a lake and fire and people doomed to burn in it for eternity," Smith vividly remembered. "They were fighting frantically to keep from being shoved into the fires. It was horrible. I was scared to death." But not scared enough to die, yet. He returned from the fiery portals of hell and redirected his life.

Dr. Rawlings, a clinical associate professor of medicine at the University of Tennessee (Chattanooga) and fellow in the American College of Cardiology, has resuscitated dozens of patients who experienced clinical death. In *Beyond Death's Door*, he wrote that about 50 percent of the persons revived from death told of going "to a place of great darkness, filled with grotesque moaning and writhing bodies crying out to be rescued" from a fundamental Christian's worst nightmare. His patients frantically sought escape from their terrifying plunge through a flame-swept tunnel and Dantean confrontation with tormented, inflamed souls.

Since the NDE itself varies so among its experiencers, one wonders if a special combination of forces might be playing out inside the mind and soul of the suicide?

Leading clinical researchers of NDE concur that people moving into (and out of) death achieve an altered state of consciousness in their momentary netherworld. In the trauma that is suicide, confusion and fear can reign; the normal course of events at death gets circumvented dramatically. Engrossed in despondency and repentance, the disembodied personality is likely to cling to the vacated body, even attempting to reincarnate the body just destroyed.

In the interim, however, decay has already begun a series of changes in the body's constitution. The two once-merged energy patterns—soul and body—are no longer compatible. As the soul tries harder and more frantically to force itself back into the devitalized body, a point is reached where—like a match drawn repeatedly across its striking surface—the energy created becomes sufficient to *kindle* the corpse.

And on the third day did Mrs. Satlow attempt to raise herself from the dead, and fail... fail except to kindle her corpse?

Another possibility depends on the emotions and belief construct of the individual now physically transitioned, plus the quickness with which the suicide seeks to reanimate his body. Was the person taught to expect harp music and tranquility in a heavenly afterlife, for instance; or the iniquity of eternal fire and damnation?

The suicide, in creating his own postmortem expectations of the latter afterlife, finds himself embroiled (and boiling) in the devil's flaming ovens, with all the fire and brimstone of Dante's *Inferno* to ensure just punishment. This belief, so vitalized by the intense sense of repentance and justice due, travels the weakened but still present energy link back to the corpse on Earth. The discarded body, now useless, is swept up in the energy field reality of its former occupant: it *burns* in a very real fire but with flames kindled from another dimension.

An "inflamed soul" leads to a flaming corpse. Is this how one can account for the strange scene discovered in Mr. Denny's bedroom four decades ago?

GLEN DENNY'S GHASTLY DEMISE

Across the Mississippi River from New Orleans, Mrs. Stalios Cousins sat watching the rain through the window of her apartment in Algiers, Louisiana. September 18, 1952, was one of those gloomy, depressing, and cheerless days when anything unpleasant would be expected. At one o'clock in the afternoon, the expected happened—unexpectedly.

Mrs. Cousins smelled smoke, saw that it came from the window of the apartment above hers, and called the Fourth District police headquarters just around the corner. They notified the fire department, and both agencies were on the scene in minutes. Firemen broke down the door to the apartment of Glenn Burk Denny, then rushed into smoke-filled rooms and stumbled over a body.

Lieutenant Louis Wattigny, one of the first members of Algiers' Fire Engine Company 20 to enter Denny's bedroom, saw a man lying in a mass of flames on the floor behind the door. The fire was extraordinarily hot! The burning body was odor-free!

Nothing else in the room was ablaze.

Police and fire officials studied the possibility of foul play in the death, because there was no evidence to suggest cigarette smoking had started the fire. Bloodstains were found on the kitchen and living room floors. But with testimony from neighbors that the forty-six-year-old Denny was a quiet man who never bothered anyone, murder was dismissed for lack of motive and no evidence of a struggle.

Three days later the police issued their report. It said Denny's death was "due to burns," and noted that several of his arteries had been *severed*. One arm, both wrists, and both ankles had been slashed. Carbon was found in his lungs, indicating he had been alive while ablaze.

Otto Burma, an investigator of fortean events, asked the coroner what caused the fire. The coroner's response, published in Burma's article for *Fate* (May 1953), was "that Denny had, after severing his arteries in five places, poured *kerosene* all over himself and ignited it with a match."

Hmmm.

The firemen detected no fumes, neither of kerosene nor even burning *flesh*. This wasn't the only oddity Burma found.

The victim, now classified a suicide, had been losing 1 percent of his blood every second from each wrist artery alone. Burma calculated that 30 percent of Denny's blood would have drained from his body by the time all lacerations were made. Death could be expected in seconds. Yet the victim is supposed to have struck a match—despite blood gushing from his wrists onto his hands—and ignited his clothing just to make sure he died?

Compounding the incredible is that Denny was *intoxicated* at the time, according to a friend who saw him alive "and shaking like a leaf" only a few hours earlier. Yet the coroner's office asked everyone to believe that Denny, drunk and *now losing 4 percent of his blood every second* from several severed arteries, managed to douse himself with an accelerant, hide the container where police and firemen would never find it, and finally strike a match to ignite his body!

Involved in the Denny investigation were a police captain, two homicide detectives, an assistant district attorney, two deputy state fire marshals, an assistant Orleans parish coroner, and a division director of the New Orleans Fire Department. This assembled firepower worked three weeks to prepare the official verdict on Denny's demise: "declared suicide, and closed."

Despite (or because of) all the problems created by the police report, the source of fire and the unburned condition of the apartment remained disturbingly unexplained—including the whereabouts of the container of kerosene that Denny allegedly poured over himself.

Conceded Wattigny about Denny's burning: "In all my experience I never saw anything to beat this."

Is it possible that Denny died from *self-induced* SHC created when his disembodied consciousness, filled with thoughts of Hell's fire awaiting suicides, tried fervently to reunite with the blood-gushing corpse lying on the bedroom floor? Or conversely, that his life force was so suddenly snapped free of the dying body it released a massive energy jolt, which burst forth as destructive flame?

Does either proposal strain credulity? No more than the official determination does, I submit. To consider the abnormal is not unreasonable, because these cases do not follow the *normal* modes of physical transition.

This brief foray into what happens to the bodies of suicides at and after death suggests the need for a new scientific discipline: *postmortem biology*. Of necessity it would surpass current forensic medicine, which only examines by comparative analysis the condition of organs that contributed toward the life force's exit from the physical body.

History offers many examples that show—sometimes via combustion—that a dead body does not always separate neatly, gently from the forces that made it a functioning, animated organism. Unquestionable evidence there is of life after death, at least for a *corpse*.

Unbelievable? "It goes without saying, however," Burma concluded about Denny's death, "that 'unbelievable' events occur as readily without our belief as with it."

19

Art Imitating Life: The Literary Side of SHC

And now, if you will excuse me, Connie, I propose to take a short stroll on the terrace in the faint hope of cooling off. I feel so incandescent that I'm apt to burst into spontaneous flame at any moment, like dry tinder.

—Lady Julia Fish, in
P. G. Wodehouse's *Heavy Weather*

"There is no science without fancy, and no art without facts," observed American novelist Vladimir Nabokov.

The study of SHC would be incomplete if it overlooked the historical richness the subject has contributed to the art of literature. Dickens, as you have seen, designed Krook's demise around this phenomenon. You may be surprised to learn that Dickens is not alone among the world's premier writers who have applied SHC to literature, where its legacy is even less well-known than it is in medical science.

Charles Brockden Brown (1771–1810)—arguably America's first novelist and the "Father of American literature"—incorporated SHC in his first novel and strongest romance, the classic *Wieland; or, the Transformation* (1798). This biography of the Wieland family debated moral responsibility and showed Brown to be a serious moralist who explored doubts about human reason through his own personal fascination for the darker side of the psyche. Brown also defended feminism in the late 1700s, proving him to be a man of radical and unconventional thought in his day (unlike Lair who chauvinistically believed women were the weaker sex and more prone to SHC).

His unconventionalism is evident in *Wieland*.

Brown injected into the Wieland family three "terror-compelling devices" guaranteed to horrify his readers in early America: ventriloquism, a father who murders his family, and spontaneous human combustion. My, how times have changed! Brown introduced his third terror first.

On the Wieland farm at Mettingen a few miles from Philadelphia, Pennsylvania, a hot sultry August evening portents a hotter event about to befall elderly Herr Wieland, the family patriarch who predicts that evil is near but cannot say why. He leaves his apprehensive family and walks to a small chapel where he had practiced unspecified mysterious rites, and apparently will again. Suddenly, a blast of brilliant "light proceeding from the edifice" is followed by an explosion.

Young daughter Clara rushes to the chapel, and recounts the gruesome scene:

> My father, when he left the house, besides a loose upper vest and slippers, wore a shirt and drawers. Now he was naked, his skin throughout the greater part of his body was scorched and bruised.... His slippers and his hair were untouched.... The preclusive gleam, the blow upon his arm, the fatal spark, the explosion heard so far, the fiery cloud that environed him, without detriment to the structure, though composed of combustible materials, the sudden vanishing of this cloud at my uncle's approach—what is the inference to be drawn from these facts?

Two hours later, Herr Wieland is dead.

Brown's graphic imagery must have truly terrorized his readers, in part because the colonists thought whatever appeared in print was likely to be true. In part, because they would have seen this event as apologue—a moral fable. Since the elder Wieland had come to America to be a minister and instead established a farm to make money, his terrible demise was just punishment for rejecting both God and religious duty. And should God so strike down—or burn up—one man for his abrogation of a holy trust, might not the same divine retribution befall *any* sinner?

Brown evidently realized his readers would be terrorized by this horror, yet skeptical of it. He has Clara, the personification of innocence, examine the scene and vouch that Herr Wieland died of his own fire due to natural causes: "Such was the end of my father. None surely was ever more mysterious."

Thus was America, itself young and innocent, introduced to SHC through the writing of its first novelist.

Brown said he had researched contemporary medical texts before dispatching his character by this strange and terrible fate. After stating four conditions popularly thought necessary for a human being to meet such a fiery end, he mentioned two more not identified by eighteenth-century medicine: a potentially *explosive* nature, and a *trance-like consciousness* that Brown assigned to Wieland just prior to his fatal blazing—a factor not emphasized again until the twentieth century when Charles Fort remarked on this feature of SHC.

How did Brown know to include these traits unknown to his contemporary medical scholars? Writer's luck? Or did he personally know of an SHC episode in colonial Philadelphia which shaped his story-crafting, a case now lost to history?

THE SATIRICAL SHC OF GOVERNOR KIEFT

Young Washington Irving (1783–1859) garnered international acclaim when he published at age twenty-six, under the pseudonym Diedrick Knickerbocker, *A History of New York from the Beginning of the World to the End of the Dutch Dynasty* (1809). Called "America's first internationally popular classic" and praised for its satire of politics and science, the book was so highly regarded that Charles Dickens reportedly wore out his copy from repeated readings.

Irving, like Dickens, detested moral deficiencies and officious bureaucrats, and also like Dickens, dispatched a despised character by means of the fire within.

Wilhelmus Kieft, the waspish governor of New York whose temperament led to his nickname William the Testy, exuded "fearful wrath" and moral dereliction that set the stage for this consummate politician's inevitable fall, which Knickerbocker described in the seventh chapter of Book IV:

> the incessant schemings, and projects going on in his own pericranium... did eternally operate to keep his mind in a kind of furnace heat, until he at length became as completely burnt out as a Dutch family pipe which has passed through three generations of hard smokers. In this manner did the choleric but magnanimous William the Testy undergo a kind of animal combustion, consuming away like a farthing rush light—so that when grim death finally snuffed him out, there was scarce left enough of him to bury!

While doing research for *A History* at the New York Historical Society, perhaps the young satirist uncovered the 1770 cremation of Hannah Bradshaw and thought it symbolized a just end for the contemptible self-aggrandizing politician he envisioned for his book. (Would that nature apply herself similarly in the twentieth century!)

Irving, like Brown before him, identified emotional disturbance—in this case Testy's inflammatory passion and fiery soul—as a prerequisite for SHC. He also used SHC as a device to warn about the danger of moral injustice.

JACOB FAITHFUL'S DOUBLE SHC TRAGEDY

English naval officer and novelist Captain Frederick Marryat (1792–1848) was a contemporary of Washington Irving, and he too took up the crusade of SHC as avenger of amorality.

In Marryat's novel *Jacob Faithful* (1834), the namesake character reminisced about his childhood spent on a Thames riverboat with his alcoholic mother—"a most unwieldy, bloated mountain of flesh," he called her—and a father who "passed most of the time in the cabin, assisting my mother in emptying the great stone bottle." One evening his besotted father suddenly sprang from the bowels of the boat and jumped overboard to disappear forever in the black sea. From below deck a "strong empyreumatic, thick smoke" issued forth, and Jacob waited for it to dissipate. Then he entered the hatchway:

> I descended the little ladder of three steps, and called "Mother!" but there was no answer.... Nothing was burning—not even the curtains to my mother's bed appeared to be singed.... I remained for more than a minute panting for breath, and then ventured to draw back the curtains of the bed—my mother was not there! but there appeared to be a black mass in the centre of the bed. I put my hand fearfully upon it—it was a sort of unctuous, pitchy cinder. I screamed with horror—

Thus did young Jacob Faithful become an orphan.

Again an unnatural fire utterly destroys mortal flesh but not the fragile veil ever so tenuously preventing immorality from tainting the world of the just. Again the innocence of youth is cursed forever by the ignoble lives of doomed elders. Forsaken by adults satirically surnamed Faithful, who personified abrogation of parenthood by their intemperately drowning themselves in alcohol, Jacob's pain is compounded by the stigma that he has been orphaned by a fate both incomprehensible and unacceptable.

Only later in life does Jacob, in maturity, grasp the fate of his mother: "Cases of this kind do indeed present themselves but one in a century, but the occurrence of them is too well authenticated. She perished from what is termed *spontaneous combustion*, an inflammation of the gases generated from the spirits absorbed in the system."

Years later, a pair of fire fatalities at the Rooney farm in 1885 paralleled this fictional incident on a Thames riverboat. One *could* say that Marryat's art precognitively imitated life, so closely did it resemble a peculiar death five decades in the future. Rather, it appears his childhood interest in the macabre, combined with his reading about SHC in *The Literary Gazette* (June 28, 1828) and a 1832 issue of *The [London] Times*, enabled him to write fiction that accurately anticipated the nonfiction horror destined for an Illinois farmhouse.

Fiction, when patterned on fact, *can* echo the world of real phenomena.

More years later, Marryat's graphic depiction of SHC caught the attention of Her Majesty's Coroner Dr. Gavin Thurston. He pointed out in a letter to the *British Medical Journal* (1938) that the novelist incorporated five of the six points deemed prerequisite for SHC: chronic alcoholism, an elderly female, little damage to combustibles in contact with the body, a residue of greasy ashes, and a lamp that "might have occasioned the fire." The sixth prerequisite, said Dr. Thurston, is that hands and feet often remain intact.

Marryat miscalculated the frequency of SHC by stating through Jacob Faithful that SHC is a "once in a century" phenomenon. Also, SHC is not universally attributable to alcohol-fueled gases in obese people. But Dr. Thurston himself erred in claiming that SHC requires an external source of ignition (which by definition wouldn't be spontaneous human combustibility anyway). Since Marryat didn't name an outside ignition source as the cause of Mrs. Faithful's self-conflagration, his fiction withstands Thurston's academic challenge to SHC.

FRENCH AND RUSSIAN LITERARY TRADITIONS

Considering the large number of SHC cases that have appeared in the French *journals de médicine*, it would be surprising if none of France's great writers utilized these baffling burnings in their writing. Two did.

Honoré de Balzac (1799–1850), a master of realistic writing who took great pains to craft precise details of character and setting, selected SHC to be the worst punishment that an intemperate husband can inflict on a suffering wife.

Combining realism and morality in his novel *Le Cousin Pons* (1847), Balzac has the shrewish Mme. Cibot hastily judge the punishment bestowed on an acquaintance. "That woman, you know," Cibot expounded in her broken dialect, "'as 'ad no luck because of her man, who drank everything in sight and who died of a spontaneous *imbustion*." To which Fraisier the ambulance-chaser, who rushed to the scene to cart off her ashes, retorted: "Providence moves in its own ways. It is not for us to question them."

Émile Zola (1840–1902), fascinated by science and today still considered one of the greatest European fiction writers, portrayed in his novels all that is low in society and brutal in man. So it is ironic (or is it?) that in Zola's *most* scientific work, *Le Docteur Pascal* (1893), SHC is allowed not only to be the fate of the drunken peasant-farmer Uncle Macquart, it is allowed to be witnessed in progress:

> At first Félicité thought it was cloth, the underpants or the shirt, that was burning. But doubt was no longer permitted; she was indeed looking upon the bare flesh; and the little blue flame was escaping from this flesh, light and dancing like a flame flickering across the surface of a bowl of blazing spirits. It was scarcely any higher than the flame of a night lamp, was quiet and gentle and so unstable that the slightest breath of air caused it to move about.... Félicité understood that her uncle was catching fire there like a sponge soaked with brandy.

When Doctor Pascal arrived at the home of the brandy-drinking Macquart, he found that "Nothing remained of him, not a bone, not a tooth, not a nail, nothing but this pile of grey dust that the draft of air from the doorway threatened to sweep away at any moment." Offering the perfect description of classic SHC—right down to its "little blue flame"—Zola has Macquart die "royally like the prince of drunkards, flaming up spontaneously and being consumed in the burning pyre of his own body...just imagine setting fire to oneself like a Saint John's fire!"

The Russian novelist and playwright Nikolai Vasilievich Gogol (1809–1852) embedded in the plot of his *Dead Souls* a passage that will sound familiar:

> "Last week my blacksmith was burnt..."
> "Did you have a fire, ma'am?"
> "The good Lord preserve us from such a calamity, sir! A fire would have been worse. No, no, sir, he caught fire himself. Something inside him caught fire. Must have had too much to drink. Only a blue flame came out of him and he smoldered, smoldered, and turned as black as coal. And he was such a clever blacksmith, too..."

This translation by Constance Garnett (for Penguin Classics) reads quite like the fate that awaited the man inside the obscure Brookville smithy in 1828—another happenstance of art imitating life and vice versa.

MELVILLE HARPOONS A CASE OF SHC

Herman Melville (1819–1891) is instantly recognized for authorship of *Moby Dick, or The Whale* (1851). He is less known for his use of SHC in a novel written two years earlier. In fact, so obscure is *Redburn: His First Voyage* (1849), based on his first round-trip voyage between Liverpool and America in 1839, that some lists of Melville's writings do not mention it.

In chapter 48, "A Living Corpse," the shanghaied sailor Miguel Saveda is so inebriated he is confined to his bunk deep in the ship's dark hold. At midnight of the first day at sea, an intolerable smell from the hold awakens every deckhand. Some blame it on a dead rat near Saveda's bunk, but no! it comes from Saveda himself. A sailor looks into Saveda's face:

> to the silent horror of all, two threads of greenish fire, like a forked tongue, darted out between the lips; and in a moment, the cadaverous face was crawled over by a swarm of worm-like flames... that faintly crackled in the silence, the uncovered parts of the body burned before us, precisely like a phosphorescent shark in a midnight sea.... the whole face, now wound in curls of soft blue flame, wore an aspect of grim defiance, and eternal death. Prometheus, blasted by fire on the rock.

Saveda is dead as a drowned bilge rat, and his body is ordered to be smothered in blankets and thrown overboard. "A few minutes more," said Redburn, "and it fell with a bubble among the phosphorescent sparkles of the damp night sea, leaving a coruscating wake as it sank. This event thrilled me through and through with unspeakable horror..."

Redburn is mystified, but his crew mates are not: "the sailors seemed familiar with such things; or at least with the stories of such things having happened to others. For me, who at that age had never so much as happened to hear of a case like this, of animal combustion, I almost thought the burning body was a premonition of the hell of the Calvinists, and that Miguel's earthly end was a foretaste of his eternal condemnation."

Saveda's bunk is nailed shut, and no one speaks of this again.

The flames' coruscating colors, the absence of pain in the victim's face, the emotions of witnessing and extinguishing SHC-in-progress all comprise a graphic potency unique to the treatment of SHC in classical literature.

What was Melville's inspiration?

Literary analyst William Gilman examined the novel and declared it a work of fiction with aspects of autobiography. Melville scholar Hershel Parker checked the log of Melville's own voyage to Liverpool but found "no spontaneous combustion of a dead body" and concluded that Melville wove the book's latter chapters largely from material lifted extant from other sources—yet "very few of the sources" have been identified. Actually, scores of pre-1849 SHC cases could have crossed the writing desk of Melville's mind, including Mr. W. Mercer from Liverpool(!) whose numb lips spit fire like a *"red-hot coal-like streak"*—a phenomenon reported by *The Lancet* (1845) and the *Transactions of the Liverpool Pathological Society* (1845).

Melville, widely read, might have come across some or all of these SHC episodes and retrieved them from the hold of his retentive mind as he sat down to masterfully craft the horror to be found in the ship's depths by Redburn (whose "autobiography" he composed in less than ten weeks). Maybe he tapped the collective consciousness from which gifted writers extract factual images and weave their fiction. Or maybe he patterned Saveda upon an actual event experienced firsthand; perhaps even witnessed on his own voyage though never entered into the log. Seamen of the nineteenth-century were a superstitious lot and would hardly have publicized an event of such ill-omen. As Redburn himself remarks of Miguel's fate: "strict orders were given to the crew not to divulge what had taken place to the emigrants; but to this, they needed no commands."

In this regard, fiction again accurately imitates life: SHC has always been a topic that provokes cover-up.

FICTION IN THE FICTION OF SHC

Dickens learned with *Bleak House* that controversy can be good for book sales, as most authors and publishers can attest. It was a lesson that Thomas de Quincey couldn't overlook when, in the 1856 edition of *The Confessions of an English Opium-eater* (1822), he listed SHC as one of the "Pains of Opium" (although the pain of SHC more often afflicts those who face its ashed remains rather than the victims themselves).

As the Dickens–Lewes debate over Krook's demise raged on, de Quincey wrote that a drugged stupor enabled him to recall a hidden memory that opium can cause an addict to cremate spontaneously—a terrible punishment he, curiously, had forgotten when writing the first edition of his memoirs.

True, selective memory is an opium user's stratagem. So is opportunism. When the anti-SHC argument of Lewes appeared to triumph against Dickens, de Quincey quickly disavowed SHC.

Thus, de Quincey gave himself away as the perpetrator of a double fiction: his own attempt to hitch a ride on Dickens' rising career, and his calling SHC a "popular fantasy" just shortly after having "remembered" it.

Across the Atlantic, an American began to churn out his own fiction like the stern-wheelers churning up the muddy Mississippi on which his fame would be anchored. Samuel Clemens's venerable *Life on the Mississippi* (1883) unfolds one of the "classic *literary* allusions to Spontaneous Combustion," says Michael Harrison.

The episode appears in Chapter LVI, when a citizen asks if anyone remembered when the harmless whiskey-sodden tramp Jimmy Flynn "burned to death in the calaboose?" Mark Twain replied that he had looked upon the pleading face of Jimmy and his death-grip on the iron bars, and for "every night for a long time afterward" that horror haunted him. Flynn had been jailed due to delirium. There, said Twain, the tramp "had used his matches disastrously: he had set his straw bed on fire, and the oaken sheathing of the room had caught."

The fiction here is that *Life on the Mississippi* describes SHC. It does not. It isn't even PC. Its only parallel to SHC is a haunting memory for those who saw Jimmy's alleged SHC. In real-life cases, that memory is commonplace. As it was for Don Gosnell, who, nine years after discovering Dr. Bentley's remains, told me: "I can *still* remember what I saw as plain as can be..."

Jimmy Flynn has none of the oddity of the real-life incineration of twenty-year-old Patricia Cummins, jailed in London's Halloway Prison on March 1, 1974. She would never make parole. Guards heard her scream, and "thick black smoke" gushed from beneath her cell's door. The fire, if that's what it was, was confined to one corner of what had now become a crematorium. The corner where Miss Cummins burned up. A cigarette was said to be involved but incapable of starting the blaze that left officials baffled.

Neither can Twain's fiction rival the comparable whole-body ashing of an Englishman in the Aldermaston jail. Years earlier, his crimes had led to his temporary incarceration in this Berkshire community's tiny outhouse-sized brick prison, some forty-seven miles west-southwest of London... a prison from which there would be no earthly escape.

SHC IN SCIENCE FICTION AND FANTASY

"Science fiction very often leads to science fact," Leonard Nimoy observed in PBS-TV's *The New Explorers* (January 18, 1995). Regarding SHC, it's the other way around.

Spontaneous human combustion has been adapted by twentieth-century authors, scriptwriters and filmmakers in a variety of science fiction and science fantasy plots.

SHC is *the* focus of Maby Ted White and Marion Zimmer Bradley's 1962 short story, "Phoenix." Max, possessed of vast paranormal powers over things that become meaningless in a world devoid of meaning for him when he loses his lover, wills his wild talent to immolate himself in a flash of flame... only to be immediately—ecstatically—regenerated as "a gestalt that was more than himself."

> Anyone who believes it's a quantum leap from science fact to science fiction is right—and wrong.
> —Jean Prescott, *Biloxi Sun Herald* (January, 17, 1995)

SHC factors prominently in Paddy Chayefsky's sci-fi novel *Altered States* (1980); in Pat Groverson's novel *Invisible Fire* (1981); and in Jeff Fain's occult novel *The Burning* (1981), based partly on those real-life outbreaks of spontaneous poltergeist-type fire that vexed the Charles Williamson household in Bladenboro, North Carolina, in early 1932.

SHC is lucidly depicted in Bob Shaw's novel, *Fire Pattern* (1984), in which a young girl seeks out her father, to convey a medical diagnosis that his recent abdominal pains were nothing more than some vague colic. In his room things were frightfully amiss:

> It was difficult to make out anything through the billowings of the curious light-blue smoke, but it seemed to Maeve that there was a blackened area of flooring near the television set. She went towards it, gagging on the sickening sweet stench in the air, and her hands fluttered nervously to her mouth as she saw that what she had taken to be a black patch was actually a large hole burned clear through the vinyl and underlying boards. Several floor joints were exposed, their upper surfaces charred into curvatures, but—strangely—there was no active flame. In the floor cavity, supported by the ceiling of the utilities room below, was a mound of fine grey ash.
> "Dad?" Maeve looked about her uncertainly, fearfully...

In terror she realized this was father. And thus, writes Shaw, did little Maeve lose her childhood innocence.

Shaw could have been describing with journalistic precision the story of Dr. Bentley or the female fire-auger in *Glaister's Medical Jurisprudence and Toxicology*.

If visualizing SHC through the written word has less appeal than watching it played out, fear not: scriptwriters for the small screen can accommodate you.

"Spell of Evil" (ITC Entertainment) was a 1974 made-for-TV movie that incorporated spontaneous human combustion.

Also in 1974, the mystery "Crackle of Death" featured Darren McGavin as a crime reporter hot on the trail of SHC.

Kolchak: the Night Stalker (1974 to 1975) devoted one episode to a frazzled newsman-with-a-nose-for-news-too-weird-to-print, wonderfully played by Darren McGavin (again) in the lead role, and his quest for an unnatural prowler who left behind ash mounds of cremated humans in unscorched homes in haunted Chicago.

The hit TV comedy *Barney Miller* (Four-D Productions) played SHC for laughs in one 1979 episode: a derelict is jailed on charges of arson because fires have spontaneously ignited five times in two years where he lived, and now a wastebasket in the police precinct office ignites spontaneously in his presence. "Sorry," Mr. Yago apologizes to the astounded Captain Miller, "I didn't mean to flare up. I am a combustible!"

A "combustible" appeared in another hit comedy with a penchant for incorporating the bizarre, *Picket Fences* (CBS-TV). In its episode "Duty Free Rome" (October 28, 1993), Bill Pugent, affable mayor of Rome, is discovered transformed to an outline of ashes upon his smoldering bed.

Avante garde *Max Headroom* (ABC-TV) in 1987 viewed SHC as the end-result of watching too many time-compressed television commercials in too short a time.

In the future-time series about adventures on space station *Babylon 5* (Fox-TV), the June 8, 1994, episode titled "Mind War" has a fugitive rogue telepath die by SHC, explained as a mind-merge gone awry that disassembled his body's subatomic structure.

The British cult hit *Dr. Who* had its time-lord (played by Tom Baker) face SHC in an episode titled "The Mask of Mandragora." Having ascertained that SHC was caused by a cyclical Alien energy, Dr. Who was asked how to safeguard oneself: "Keep an open mind. *That's* the secret!"

Good advice for any traveler in uncharted territory—

Scriptwriters are not alone in using SHC for its attention-getting power. Strange as it sounds, advertisers have used SHC as a marketing tool too. Perhaps the oddest example—and having one of the briefest runs (I saw it only once)—had to be a 1993 TV commercial by Pepsi Cola. A thirsty customer, facing vending machines for Pepsi and a competing product, put his money in the latter. Big mistake! conveyed the Pepsi ad agency, as the customer spontaneously combusted after swallowing the "wrong" brand of cola.

SHC has played memorable roles on the silver screen, too, though perhaps overlooked by most audiences.

Did you realize SHC is depicted in *The Wizard of Oz* (1939), when Margaret Hamilton as the Wicked Witch of the West melts odor-free into a smoking pile of unscorched garments before the disbelieving eyes of Dorothy?* (Interestingly, this dissolution of evil incarnate by SHC begins with the young heroine tossing water onto the wickedness standing before her... and some historical accounts mention water's flame-*intensifying* affect on SHC.)

Scanners (Avco Embassy Pictures, 1980) dealt with exploding craniums and flamed assassins, via brainfire that director David Cronenberg called "a derangement of the synapses we call telepathy." *Altered States* (Warner, 1980) graphically depicted the flaming transformation of matter into antimatter humans. *Firestarter* (Universal Pictures, 1984) burned up eighteen hundred gallons of propane to achieve its pyrotechnic special effects triggered by mentally induced pyrokinesis. Steven Spielberg's blockbuster *Raiders of the Lost Ark* received an Academy Award for its special effects, notably the powerful climax in which evil Nazis melt down before the unleashed power of the Ark of the Covenant.

Lifeforce (Tri-Star, 1985) employed SHC as a science-fiction horror that disclosed extraterrestrials' human feeding-frenzy; its special effects are stunning and rival *Altered States* for accurate depiction of classic SHC—if not, hopefully, why it happens.

The campy *Repo Man* (Edge City Productions, Universal/MCA, 1984) featured an unearthly '64 Chevy Malibu with the ultimate antitheft device: a trunk with something atomic and alien that, when opened, instantly incinerates any would-be trespasser. In the opening scene, a California motorcycle cop pulls over the Malibu and asks its driver, "What d'ya have in the trunk?" Driver answers, "You don't want to look in there." Cop pops the trunk and is vaporized, leaving behind only a pair of smoking boots on the shoulder of the highway as the Malibu drives off.

"What could've done that? Gasoline? Napalm?" asks an investigating officer.

"It happens sometimes. People just explode. Natural causes," answers a government agent at the scene, who appends on her report of the incident: "Do not tell police." Art imitating life, again...

Young Einstein (Warner, 1988) has fire erupt from the shoes and smoke from the shoulders of star Yahoo Serious, threatening to transform him into a blast of light. When warned that he is absorbing too much energy, young Einstein expresses unconcern, relatively: "It's only electrons!"

---

* A 1995 television commercial parodies this pseudo-SHC scene: the pink Energizer rabbit has time to douse the Wicked Witch after her non-EverReady batteries fail; that is, reliance on other than the sponsor's product can lead to disastrous consequences.

This Is Spinal Tap features rock musician drummers who party too hardy, and then self-combust.

And in *Spontaneous Combustion* (Media/Heron Communications, 1990), the son of a couple of atomic-bomb human guinea pigs grows up to be a sizzling pressure cooker with an arm that throws flames (not very effectively) in this ineffective horror film on which this author served as a (thankfully) uncredited technical consultant.

SHC became the culmination of Iran's *International Guerrillas* (1990). This propaganda film implored Muslim faithful to pray that Allah avenge Iran's clerics and himself for Salman Rushdie's blasphemous assaults in *The Satanic Verses*. Their prayers are answered when a laser beam from Allah ignites a Rushdie stand-in in a pathetic attempt at cinematic SHC.

Science fiction's *The Lawnmower Man* (1992) portrayed what might happen when a scientist uses a low-IQ man to test computer simulation theories of virtual reality, and then must regain control of his creation after he discovers his nonvirtual ability to induce SHC.

The same year saw release of *Wilder Napalm*, an off-the-wall tale of two brothers born with the ability to mentally ignite conflagrations in buildings and people, and how one chose to restrain his wild talent while the other sibling sought to wreak flaming havoc for personal aggrandizement. The finale is an Old West-style showdown of human pyrotechnics.

Lastly is *Like Water For Chocolate*, whose theme is to-eat-is-to-have-passion. It ends with a lady fulfilling her passion by eating matches... and then spontaneously igniting.

FICTION... FANTASY... FACT

Science fiction and fantasy embody, respectively, events outside present human experience yet grounded in realism and fanciful ideas not thought possible. The challenge to forteans—and for anyone who seriously studies the human experience of spontaneous combustion—is this: Where does science fiction and fantasy end and science fact begin?

For instance, science fantasy would include the Human Torch, the 1940s *Marvel Comics* do-gooder who combatted evil with his mental ability to set fires. It would include Charlie McGee, Stephen King's heroine in *Firestarter* (1980), who is normal in every way except that she has the pyrokinetic power to set objects afire with just a glance; a wild talent she is forced to fire against a government determined to control her.

Are these yarns wholly imaginary, though, to be relegated only to fiction-fantasy? "Science fiction very often leads to science fact," remarked *Star Trek*'s

Leonard Nimoy on *The New Explorers* (PBS-TV, January 18, 1995). In the realm of pyrophenomena, it's the other way around: science fact *predates* science fiction.

In the 1880s, the curious and the skeptical repeatedly handed their handkerchiefs to A. W. Underwood of Paw Paw, Michigan. Why? Because this Afro-American was a *real-life* human firebreather! When he placed cloth to his mouth and breathed, it would suddenly burst into flames and be consumed. Dr. L. C. Woodman attempted to prove Underwood a trickster, and submitted him to rigorous examination before and during his kerchief-combusting demonstrations (on which Underwood never sought to capitalize). Dr. Woodman reported in the *Michigan Medical News* (1882) that there was no chicanery. Underwood was, simply, a human firestarter whose wild talent began when he was twelve and continued for at least a dozen years.

How many handkerchiefs were incinerated during that time, Dr. Woodman wouldn't hazard a guess; but he did note that Underwood, when out hunting, would "lie down after collecting dry leaves and by breathing on them start a fire."

Much earlier, at the beginning of the sixteenth century, a German named Hans Odhar made a name for himself in Paris society. His feat? An ability to breath forth fire, according to Warren Smith's book, *The Strange Ones*.

More recently, Lawrence Blair, in his research on cultural traditions around the Pacific Ring of Fire, videotaped an Indonesian who could kindle wadded paper by passing his palm over it. (What will anthropologists say to my suggestion that these men were rekindling a long-dormant, campfire-breathing skill once available to man's earliest ancestors when they needed to warm their caves and cook their mastodons?)

Science fiction-fantasy takes a back seat to science fact when it comes to the human ability to create spontaneous combustion. As human fire-starters live and breathe, so too might they occasionally and unconsciously set *themselves* aflame?

DID A FAMOUS FICTION WRITER DIE BY FANTASTIC FIRE FROM WITHIN?

Fiction and terror have been spun out of SHC by authors for more than a century. Has circumstance ever turned the tables, begetting out of the literary community a real-life self-flaming horror?

The front page of Sunday's *The New York Times* (December 17, 1933) announced that best-selling American mystery writer Louis Joseph Vance had died the preceding morning in a small blaze in his three-room apartment on the thirteenth floor of Town House at 108 East 38th Street, New York City. Vance was found "encircled in flames from a blazing upholstered armchair.... All the upholstery had been burned off the chair, leaving only its wooden frame." Otherwise, fire touched only a coffee table next to the chair and drapes on the nearby window.

The fatal blaze was first attributed to the fifty-four-year-old novelist's known penchant for carelessly dropping his cigarettes; its localized damage credited to quick discovery and extinguishment. Tragic, but not unusual.

On Monday the story, like Vance's own novels, took a twist as baffling as any he concocted in fiction when *The Times* ran this headline: "Vance Fire Death Becomes Mystery." It had been determined that Vance had been unclothed from the waist up, yet one witness said it looked "as if his head and shoulders had been pushed into a blazing furnace, or made into a human torch by dousing them with some combustible fluid."

Assistant medical examiner Dr. Robert C. Fisher quickly ordered an autopsy because of what he termed "the unusual extent of the burns received by Mr. Vance." Assistant medical examiner Milton Helpern, after performing the autopsy, refused to rule Vance's death accidental because he too was puzzled by the extent of burns and the flames' narrow range. A nearly empty can of benzine (a flammable cleaning chemical) in Vance's bathroom raised suspicion in the doctors' minds that it might somehow explain the blaze's ferocity.

Dr. Helpern asked police to investigate. The police could find no evidence of foul play in Vance's apartment. Nor could they link the fire in any way to the benzine. Vance's relatives said it was his habit to economize by cleaning his own clothing, and they saw nothing suspicious about its presence. So how to explain his fiery demise?

Vance's death is attributed to SHC by J. Bryan in Hodge Podge (1986), and by Courtlandt Bryan in his recent book, *Close Encounters of the Fourth Kind: Alien Abduction, UFOs, and the Conference at M.I.T.* (1995). I am not so certain.

It is plausible, though unprovable, that Vance had indeed been cleaning a benzine-soaked garment on his lap and ignited it with a cigarette as he slumped off his chair after suffering a heart attack or simply because, as chief medical examiner Dr. Charles Norris announced following the autopsy, "Mr. Vance was intoxicated to a point which usually produced stupor." In other words, a perfectly nonparanormal accidental death due to mishap. On the other hand, Louis Vance did possess traits found in previous SHC/PC episodes having less uncertainty: he was an easy drunk, and he lived alone as a quasi-widower estranged from his wife.

Unfortunately, history cannot reveal the true nature of this literary artist's death. As Dr. Helpern told *The Times* (December 18, 1933), "It is a very strange case considering the extent of charring and burning of tissue, especially since nothing in the apartment caught fire except the chair in which Mr. Vance supposedly had fallen asleep. The real explanation of what happened may never be known."

PSYCHOANALYZING THE ART OF SHC

We have come full circle in the adaptation of SHC to art. From SHC as a godly force beyond human understanding that punishes Herr Wieland for not keeping his promise to God; to SHC as an equalizer that symbolically burns away social evils and moral injustices in man; to SHC as a holy weapon under the control of human invocation to avenge blasphemy against a submissive God and reestablish righteousness; to SHC as possible fate of a literary artist, the power of SHC in art is extraordinarily rich. As are the theories advanced in literature to explain it.

Literary luminaries Brown, Irving, Marryat, Melville, Zola, and Balzac all intrigued their readers by using SHC as a vehicle for mystery, ignorance, drama, satire, punishment, justice, and symbolism. Interestingly, each of these classical writers described SHC from the perspective of children; perhaps the authors feared they would arouse both public indignation and questions about the credibility of any adult who would present it seriously, whereas the "impossible" can be safely expressed through the naive eyes of young innocence. Only Dickens has SHC seen by adults. Dickens apparently felt secure in having mature eyes view the evidence of human fireballs, but he underestimated the hackles that SHC raises.

Their fiction follows so closely the classic pattern of SHC found in the historical and medical records that one marvels at the accuracy of these writers' words and inspiration.

The literary philosopher Gaston Bachelard did not appreciate this precision, however. In his *Psychoanalysis of Fire* (1964) he rebuked Zola for being "prescientific" when summoning SHC to punish Macquart. "Such a story, entirely a product of the imagination," he said of *Le Docteur Pascal*, "is particularly disturbing when it comes from the pen of a *naturalist* writer [and] leads one to think that Zola built up his image of science on the naive reveries..."

For Bachelard, using SHC even as a metaphor (let alone a real incident) was scandalously grievous. "In the nineteenth century there are virtually no reports of cases of spontaneous [human?] combustion," Bachelard declared in relying on Lewes's and Liebig's pronouncements. *Really?* Spontaneous combustion, Bachelard continued, "gradually became metaphorical and gives way to ready jokes about the red faces of drunkards, about the rubicund nose that a match could set on fire. These jokes are, moreover, immediately understood, a fact which proves that prescientific thought lingers on for a long time in the spoken language. It also lingers on in literature."

Linger it does, but only as metaphor? I wonder how a dozen nineteenth-century French physicians who attested favorably to SHC, some of whom saw its aftermath firsthand, would respond to having their opinions libeled by a twentieth-century literary critic who must never have looked beyond his pen point? They all, I suspect, would miss the joke Bachelard thinks he saw. Dr. E. G. Archer, who recounted a classic case of SHC to the Royal College of Surgeons in 1905, surely would have missed the humor.

Bachelard missed something which, by now, should be apparent: that novelists, as well as writers of nonfiction, have every right to invest their art with spontaneous combustion in humans, with or without using it as metaphor or allegory.

THE EXTRAORDINARY DEATH THAT DR. BOFIN SAW

The doctor entered the tiny, soot-blackened, five-sided room and began looking around. In one corner, he discovered two human feet dressed in stockings with just a few inches of lower shins. Nine feet away from the fire-severed legs sat a television, its plastic frame contorted by heat and melted around the picture tube. A picture frame next to it sat intact, its glass uncracked.

Nothing in between had caught fire in the seventy-square-foot death chamber devoid of the smell of death. Not the legs of the wood-rung chair tipped atop the peak of ashes where once had stood a person... not the linoleum beneath the amorphous ashes, though it had charred just a bit... not the eiderdown comforter covering the bed two feet away... not the mound of bright white salt spilled on the floor from its split-open plastic box next to the pair of feet.

His eyes returned to that mound, once flesh, and peered closer. Aghast, he plunged his hands into the powdery residue. He scooped out only the merest hint of bone. No skull, no spine, no pelvis, no bodily anything the doctor had been taught is most impervious to fire. His mind reeled back to *Bleak House*, this scene did so resemble that of ol' Krook.

Who wrote this fiction? No one. This was no Victorian melodrama or scriptwriter's fantasy. This was real life.

The doctor who looked around the room is Patrick J. Bofin. The

> ...a classic example of fiction becoming fact.
> —Joe Nickell, ridiculing SHC; personal interchange (May 17, 1989)

place is 33 Prussia Street, Dublin, Ireland. The time is 9:30 A.M. on March 28, 1970. The mound of ash *had* been Mrs. Margaret Hogan.

As coroner for the city of Dublin, Dr. Bofin arrived at the scene after being alerted by police to an odd death that they wanted him to look at. He realized immediately the circumstances were unusual: "I didn't see a body. And I had to look with, with care to find the small pieces of body that were, that were there." Even in recall, he stammered to get the words out. "I would have expected greater fire damage in the room. And this I didn't see."

What *did* Dr. Bofin see in the room of this eighty-nine-year-old widow?

He saw on very close examination, in one corner of the very small room, but *very* small damage by fire—except in the victim. Said he: "The only parts remaining of the lady are her two feet. Just her—two feet encased in the stockings there." Mrs. Hogan had literally, wholly, and unmistakably disintegrated right down to her shinbones in a *very* localized fire. She did so sometime since four o'clock the preceding afternoon, after friend Kathleen Rigney had visited to wash her elderly friend's feet and hair, and left her sitting near a fire grate.

"I recognized this then as a case of what we in forensic medicine regard as a—what we call amongst ourselves—a case of spontaneous combustion. Now this we regard as an unusual, rare and worthy of note," he said. The scene did remind him of Dickens, except that this wasn't literature.

Bofin got busy taking note of what lay before his wondering eyes.

He expected to find great heat; he did not. He expected to make a connection between the room's small coal fireplace and the ashening of this widow; he could not. He expected to see a link between the incinerated woman and the melted television; there was nothing to find. He expected to see the picture glass cracked: "That didn't crack. Why that should not occur, I don't know." Though he didn't mention it, the very fact he *failed* to remark on the olfactory offensiveness of investigating this blaze in its extraordinarily small confines is proof that the characteristic, overpowering noxious odor of *normal flesh fires* was not present here.

These, he said, "are characteristics that go to make up what forensic scientists call—and always using inverted commas—'spontaneous combustion'." In a human being, he meant to add.

At the inquest he became apologetic, almost embarrassed for admitting to SHC: "Spontaneous Combustion does not mean that the fires are in fact spontaneous in origin. It's simply a term carried on in forensic literature to describe a set of circumstances in which a person is burned to death without an obvious source of fire..."

A professor of Forensic Medicine, Royal College of Surgeons, Dr. Bofin not only confirmed for his profession the accuracy of the definition of SHC, he attested to the reality of its internal nature by being unable (as were police investigators) to identify any source outside Mrs. Hogan that could explain the facts that so fascinated yet befuddled him.

In his twenty-year medical career he handled about fifty burn fatalities annually, and this was the *first* time he had encountered what he could—and did—describe as spontaneous human combustion. Some of his fellow coroners, he said, had served in the profession much longer than he and had never seen SHC.

On April 3, the inquest court rendered verdict on cause of death. "Unknown."

The case has haunted Dr. Bofin ever since. "Well, it *was* unusual," he admitted in 1985, "and it is unusual still. I haven't seen one since then. It's difficult—it's difficult to explain, yes. But the thing I have to stress is that it's extraordinarily rare."

Rare, yes. But rare does not mean impossible.

Dr. Bofin offered a philosophical viewpoint: "The unfortunate thing about it is that I've seen only one case. I saw it when I was unprepared. The next coroner who sees a case will also see his case unprepared. And we see so few. There is not adequate material for a sense of proper disciplined research..."

Sometimes, the most effective means to educate people about life's anomalies is through art. Art that faithfully, accurately, imitates life. The more the artist knows about the facts of SHC, the more realistic he can be in crafting his incendiary metaphors and the closer his creative inspiration may come to describing what it must be like to see SHC; to experience SHC; even to explain why it may happen as it does.

As Lord Byron's poem about Don Juan says, "truth is stranger than fiction."

20
On the Road to Incredible Incinerations

Strange is it not?...
Not one returns to tell us of the Road, Which to discover we must travel to.

—Edward FitzGerald
The Rubaiyat of Omar Khayyam (1859)

"How *do* you find these cases of spontaneous human combustion?" I am often asked. The ways are many, nearly as numerous as are the detractors of SHC.

Library archives have proven a fruitful source, and judging from the layer of dust atop some of the cartloads of medical tomes perused, I must have been the first person in this century (if not longer) to open their pages and resurrect "new" incidents buried in historical obscurity. Research associates, lecturing and television appearances (notably *That's Incredible!* in the early 1980s), and even supermarket tabloids have yielded useful leads.

Happenstance, synchronicity, library angels, luck: call it by any name, this too has a key to discovering and researching some cases.

And there is no avoiding the one theme that runs through all these routes to the strange: forbearance, dedication, plus just *plain hard work*.

However, study of SHC is not always relegated to tedious thumbing through stacks of antique medical books. The summer of 1980 was one of the many times I spent tracking down SHC. Literally.

An SHC Odyssey Begins

At a conference in July that year, LaVerne Olsen heard me talking about SHC and told me about a possible case in Plymouth, a town near her home in Minnesota. About the same time another tip put me in contact with Officer Tipton, who said he'd be glad to sit down and recollect the Reeser mystery should I ever find myself in North Carolina. Then WLS-TV flew me into town to guest on its top-rated *AM Chicago* and talk about SHC. Dan Weinberg happened to see the show that day, and thoughtfully telephoned me at home to say I should talk to the fire marshal in Bolingbrook, Illinois. I called Fire Marshal Vincent Calcagno, whose Bolingbrook Fire Department had confronted a recent problematic fire. Having just heard about my research, he urged me to come to Bolingbrook to review the case with him. An offer I couldn't refuse—

I packed my motorcycle with provisions and maps of the Midwest and headed down the highway. The high-road to investigative adventure had begun.

The first destination was North Carolina to meet with Officer Tipton, whose comments about the Reeser case you have already read. Then it was on to Bolingbrook. But first, a stop at Nokomis, Illinois.

A news clipping from the Bloomington-Normal *Pentagraph* (June 13, 1975), forwarded by noted fortean writer Brad Steiger, led me to this town in the agricultural belt of central Illinois. Its story said a Richard Johnson had awakened to the moans of his unclothed wife, Sarah Johnson, sixty-five, lying on the bed beside him. She was "badly burned" in a room devoid of fire and died a few hours later. "I'm mystified by the whole thing and have called the crime laboratory in Springfield for help," Nokomis Police Chief James Cohan told the newspaper.

SHC? It surely sounded strange. Worth a look. How strange, even I wasn't prepared for.

I rode into Nokomis on a delightfully sunny morning, passed the grain elevators, stopped at the newspaper office to read about Mrs. Johnson's death in detail, then interviewed officials involved in the case. Yes, Mrs. Johnson *had* been found burned (obviously not to ash) on her unburned bed; yes, Cohan *had* admitted his department was baffled. At first. There were additional facts not contained in the initial newspaper story.

Cohan told me he had discovered incontrovertible evidence linking the woman's injuries to a strange rite her family had engaged in at a cemetery one mile from their house. Her belongings were found there, said Cohen, along with "matches all over the place." And: "There was *definitely* an accelerant there"—the hospital staff had found "a quart of gasoline in her stomach!" Ugh.

I stopped by the police station and met the current police chief, William D.

Haines, and his assistant chief, Steve Zuck. Both said they agreed with Cohan: that the woman had been carried to bed *after* she had burned elsewhere.

Why she had swallowed a quart of gasoline, and how she survived *that* (her burns notwithstanding) for twelve hours, were questions no one could answer. Sarah Johnson died not as a human fireball in circumstances devoid of accelerants, but in circumstances as strange—if anything can be—as SHC. What happened, I don't know, except to conclude Sarah Johnson's death is, while bizarre, not anomalous.

Deputy Zuck confessed that he couldn't believe in SHC, *until* the Johnson case made him realize just how strange human life *can* be. As he and his chief both puzzled over photographs of Dr. Bentley's baffling burning that I handed to them, Chief Haines remarked: "You hear more and more about it all the time."

I was on the verge of hearing a lot more about it. Next stop: Bolingbrook.

THE BEATRICE OCZKI CASE

Sometimes you have to forget what you know to make a breakthrough. Mrs. Beatrice Oczki—victim of a chair fire in Bolingbrook, Illinois. So ruled the Will County coroner's court.

Not everyone in this community twenty-five miles southwest of Chicago agreed. Bolingbrook Fire Marshal Vincent Calcagno was among the staunchest opponents. Why?

Uncertainty began Friday evening, November 23, 1979, when Frank Oczki left home for the weekend, leaving his forty-nine-year-old mother alone. It had been a tough year for him and Beatrice, whose health problems had prompted her to move in with him just a few months earlier. Frank thought things were changing for the better. He left around 7:00 P.M., bidding his mother goodbye. No one knew that life for him and his mother would soon be turned upside down again.

Early the next afternoon, Frank's ex-wife opened the front door of the Oczki suburban split-level house. Standing in the foyer, she realized it was exceedingly dark and overly warm. A coating of baked-on soot stained the light switches next to the door. She glanced downstairs and noticed a pile of charred material where her mother-in-law often sat watching television. She called the police to report a chair fire.

The call was logged in at 12:30 P.M. Bolingbrook police arrived and called the fire department, whose firefighters arrived at 12:40 P.M. Beyond this point normal procedures abruptly ended, to be replaced with confusion, uncertainty, and wonder.

In the center of the den on the lower level they found what appeared to be the aftermath of a simple chair fire, until flashlights focused on it. "That's when it struck everyone—*there was a human form there*! But to look at it in the dark of the house," said Calcagno, "it was a problem to see."

It was the first of *many* problems.

The scene confronting the investigators defied logic. The upholstered, high-back chair was burned almost to disintegration. Its back seemed to have been forced backward, snapping off before dropping to the floor. Shreds of charred fabric hung from the wood frame. The seat cushion was burned almost completely away, revealing its still-coiled springs beneath. Ocher carpeting was blackened in a tight circle beneath the chair. The lower front corner of the chair collapsed, intact, onto the carpeting between the legs of Mrs. Oczki.

Of Mrs. Oczki herself, little remained. "Every drop of blood was consumed," stated a mystified Calcagno. "There were some missing fingers. Total charring! The only thing that was relatively intact was from about mid-calf down to the feet. That was relatively intact. You could tell it belonged to a human. But that was it..."

The third-degree burns on her lower legs stopped abruptly at the top of her white cotton socks, which provided a vivid contrast to her seared skin and the black shoes she wore, themselves as pristine as the socks. She wore a brace on her left leg: its leather straps had burned away, leaving the metal supports still attached to her shoe but without a full leg to support.

The fire—should it be called that—had burned deeply into the sides of her torso and devoured her left arm to the bone. Her watch and its gold metal band now dangled from a charred wrist bone. Her right shoulder, right arm, and right hand no longer existed. Her head, though badly burned, appeared whole.

Mrs. Oczki had surrendered to her chair and the blaze that swirled around them both, as had Mary Reeser done before her. Essentially, Beatrice rivaled a 2500-plus°F cremation interrupted halfway.

Yet the expected evidence of concurrent combustion and extreme fire damage elsewhere in the room was, like the woman's right side, virtually nonexistent.

Investigators could not ignore a green vinyl-covered hassock only *four inches* from Beatrice's right foot, nor the sofa to the immediate left of the hassock. The fire managed to largely overlook both! Vinyl on the hassock melted only where it faced Mrs. Oczki. The sofa, six inches from her left foot, showed *not a trace* of fire or heat; indeed, the entire sofa was sheathed in *a clear plastic cover that showed but a hint of the slightest heat distortion!* On the floor lay a pink crocheted afghan, actually under the toe of her left shoe. The afghan looked as if it had just been laundered!

To the immediate left of Mrs. Oczki sat a hexagonal end table. Heat from the fire *only three inches* away melted one of its plastic decorative panels, now oozing surrealistically toward the floor like stretched taffy. The table corner closest to Mrs. Oczki had not charred. A white Trimline phone gleamed unsullied, its coiled cord *but slightly heat-distended* as it hung in front of the table's melted panel. A table lamp's shade was lightly coated (where closest to Mrs. Oczki) with a six-inch wide band of caramel-brown soot; otherwise, it and its bulb were undamaged. (After ten minutes at 900°F, a light bulb will soften and distort toward the heat source.) A dog collar, a *TV Guide*, and other magazines lying there could have been mistaken for new.

A pillar of heat had risen straight up to peel paint in a three-foot circle on the ceiling, but not hot enough to damage the drywall underneath. Beyond this burned-away area the ceiling was speckled with miniature stalactites of bubbled paint resembling inverted chocolate kisses.

Other factors stood out.

On a shelf along the top of the wall behind Mrs. Oczki, one of three beer bottles had exploded, spraying its contents like abstract art onto the wall and ceiling. Nearby, floor-length drapes sagged from melted drapery rods. At the other end of the room, plastic curtains had melted off their hooks and lay on the window sill, while a television below broadcast the roar of a football game in the otherwise stone-silent room.

Those inside the Oczki house weren't interested in football this Thanksgiving weekend. They faced something far more intimidating than the front line of any collegiate football team. One Bolingbrook officer, hand in pocket, head bowed, stared in profound puzzlement at the incinerated chair in front of him and pondered: How the hell could this happen? How the hell will we explain *this*?

Upstairs, soot had baked itself to a mirror in the bathroom, onto the vanity sink, and onto towels and bottles on shelves above the toilet. Nearby lay a copy of Stephen King's *The Shining*, about a haunted hotel consumed by fire. On the floor lay two pet dogs, suffocated by a lack of oxygen or killed by toxic gases of combustion (most probably carbon monoxide).

This *real-life* fire had started to haunt its investigators. No one could say exactly *what* Mrs. Oczki's death looked like...except that it looked like nothing any of them had seen before!

PROBING DEEPER

Bolingbrook police quickly overhauled the fire scene, released the house back to the family, and wrote the death off to "smoking materials." When Calcagno arrived back in Bolingbrook after a weekend away, he expressed strong disagreement with this determination. *"That is not correct!"* he fired back at the police. "I *can't* buy that! There's too much damage."

Fire Chief Terrence Droogan wasn't satisfied either. "It baffles the hell out of me," he said, noting that extreme temperature was not indicated—except in Mrs. Oczki herself.

Fearing useful clues might have been overlooked in the haste with which the police swept up the evidence, Calcagno insisted a toxicology be performed. No hydrocarbons were found though, proving that no accelerants fueled the blaze. A crematorium then completed what Beatrice Oczki had herself begun.

Nine months later, as I sat with him in his office at the ultra-modern Bolingbrook public safety headquarters, Calcagno confessed: "We *still* have our doubts about what happened."

Bill Ferguson, a lieutenant coroner for Will County assigned to the case, had doubts too. Initially.

"God, this looks like spontaneous human combustion," he thought to himself when he first saw the fire photographs. He checked with experts. Officials at the National Fire Protection Association headquarters told him they never heard of SHC (even though the NFPA knew about our research since 1975). Therefore, Ferguson announced that Mrs. Oczki had dropped a cigarette on herself, igniting a smoldering blaze that slowly burned the cotton batting of the chair plus the body that sat upon it.

Now, how many times have you heard that before? What did Calcagno think about this determination?

"Call me a realist!" he said sardonically, his voice bristling at any insinuation that he was incompetent to identify a normal chair fire. "I've seen photographs of people that have died in overstuffed chairs from smoking materials, and the damage is *no where* near what that is!" he said with rancor, pointing to the photos of Mrs. Oczki. Calcagno and Droogan had no patience for the dropped-cigarette-and-smoldering-chair theory, that was certain.

"It was not a homicide. It's impossible! I mean if somebody took a blow torch, they couldn't do it," Droogan insisted heatedly.

"Well, that's it!" interjected Calcagno. explaining how his men had searched for flame-abetting materials. "We chipped the concrete and tile off the ground, we went *into* the concrete, and nothing was found. How'd it happen?"

Wouldn't you want to know, too?

In the presence of Calcagno, Droogan, and other Bolingbrook firefighters who were fascinated by the case, I probed, examined, and questioned every angle I could think to pursue. The repartee bounded quickly back and forth.

What about smell?

"I walked in there and I didn't smell nothing, *nothing at all!!*" Droogan answered immediately. "I've smelled a lot of smoky places, and they *stink!* This did not smell the same way!"

Was there enough fuel loading to sustain combustion long enough and hot enough to cause this much damage to a human?

Calcagno: "There's *no way* that that chair could have developed such temperatures to do that amount of damage. There wasn't enough fuel loading to even initiate such a sequence of events."

The fire marshal recalled a motorist trapped by a gasoline tank truck fire: "thousands of gallons of gasoline burning like a bitch...as we took him out, he fell apart in pieces. That's the only thing I saw that could be comparable. That's conventional."

Calcagno clearly found the fire in Mrs. Oczki to be unconventional. As did Droogan, who pointed to a photo of a huge industrial plant blaze that he had helped extinguish. There, he said, despite a mix of flammable chemicals the temperature "wasn't much above 1700 degrees!"

What *had* the temperature been in the Oczki household?

"It was 98 degrees when I walked into the building," Droogan replied.

My questions about thermal gradients in the room led to a consensus that the temperature had to have been at least 120°F but not much above 400°F. When this was fully realized, prolonged silence engulfed the fire marshal's office.

More surprises awaited. A line of demarcation ringed the entire room at five feet. "Below that, it was clean! *Nothing!*" remarked Chief Droogan. Smoke rises in a fire, then spreads out and downward, I said for the sake of argument. "But there's got to be some kind of settlement on the furniture," Droogan countered. "And there wasn't anything like that!...You just can't have that kind of heat over that duration and, and not burn the house down."

Obviously one can. The firemen were relieved to learn the same distinct smoky demarcation had been reported in other similar cases. In this instance, every drywall nail and every stud was identified by dark spots and lines where the soot of combustion was attracted (perhaps by ionization) to the iron and wood behind the walls and ceiling. The same situation was found upstairs.

The firemen all figured the house's interior had gradually cooled during the night as the fire died down. They were in for another surprise, as I asked about her wrist watch.

"Here's her left arm, it's completely destroyed," said Calcagno, looking at a photograph. "Yet *that watch is in perfect operating condition. Perfect!* It doesn't even look like it was scorched...you could clearly read the numbers." Some firemen who had gathered around, hearing the case discussed in this depth for the first time, looked deeply puzzled. Calcagno again expressed his own amazement.

For several moments, everyone was in the silence of their own thoughts.

(In the background a radio played Gilbert O'Sullivan's "Alone Again, Naturally"—symbolizing just how each one felt as the case was replayed for my benefit.)

This led to a crucial consideration in any attempt to verify SHC: *the time factor*. The watch had stopped ticking because, presumably, heat had melted some of its delicate components. What time did the watch show? "About 12:15," Calcagno recollected.

Up to this point everyone had assumed the fire that killed Mrs. Oczki began just after midnight, smoldered (or raged) for the next twelve hours, cooled down to 98°F and died out due to lack of oxygen before her daughter-in-law opened the front door about a half hour after noon. In the back of my mind was a more startling possibility.

"The clock upstairs stopped too," added Droogan. He read from the fire report: "'clock on the second floor fell off the wall and stopped at 12:20 A.M.' That makes sense," Droogan reasoned.

"But not necessarily!" interrupted Calcagno, alert as a Navy aviator about to make his first night-time carrier landing.

I already knew what he was going to say, as he exclaimed: "*Not* 12 hours! *It could have happened 20 minutes before!*" Indeed it could have.

I pointed out that the police took the daughter-in-law's call at 12:30 P.M. I looked each fireman in the eye as I gently said: "Actually, it could be *only 10 to 15 minutes* from the time the clocks stopped until—"

A silence settled over the fire marshal's office that must have rivaled that in the Oczki home when officials first arrived there on November 24. I was prepared to expect a body incinerating itself in so short a time; the firefighters, on the other hand, were not!

It meant the impossible could have visited Bolingbrook. It meant that Mrs. Oczki could have burned herself up in no more than ten to fifteen minutes; that her daughter-in-law missed by minutes being a witness to the fire within.

Finally, Fire Marshal Calcagno broke the silence as he spoke on behalf of his firemen, slowly enunciating each word for emphasis: "It's just that simple, *I can't figure it out.*"

IMPORTANT NEW REVELATIONS IN THE OCZKI CASE

It was quitting time. "See you tomorrow," Calcagno said. I could only imagine the discussions he and his men would be having that evening over dinner.

Obviously this was *not* a normal fire fatality. SHC fit the bill, though not in the classic form because here it was *the extremities and sides of the body* that suffered the most severe damage.

There were some things these dedicated professionals could not tell me. I wanted to know more about Mrs. Oczki herself. What kind of woman was she? Were there clues in her personality, or actions just before her untimely death, that could shed understanding that so eluded the fire investigators? I wasn't ready to quit for the day.

Her son, I had been told, still lived in town. I hoped to speak to him.

I found the Oczki home, parked the motorcycle, swallowed hard, and rang the doorbell. A man walked over from across the street. It was Frank Oczki. "No, I don't think I want to talk about it," he said politely when I explained why I was there.

"Did you really come all the way from Pennsylvania?" he then asked with incredulity in his voice, having noticed the license plate on the motorcycle.

"Yes. Specifically to learn more about your mother's passing."

He hesitated a moment. "Come in. I'll help you if I can."

Frank showed me a wedding portrait of his mother. She had been a strikingly beautiful, elegant five-foot-seven-inch lady with brown eyes and dark brown hair. "When she passed away," he said, "I guess she weighed about 160 pounds." Mrs. Oczki didn't meet the criteria of old and obese so often advocated as prerequisites for SHC.

Was there a physical condition that might have contributed to her dramatic final transformation?

"She really couldn't get around too well because she had hurt her ankle," Frank explained, pointing to the leg brace. "Her *right* shoulder had been injured at one time," he added, "and she couldn't extend her arm all the way out."

There was a correlation. Perhaps atrophy in her limbs, or blockages of the bioenergy in these parts of her body, contributed to the fact that it was her right shoulder and left leg which had been most severely burned away.

What about her cigarette smoking? "Very heavy! Very, very heavy. She had smoked for years." However, Frank added, his diabetic mother had cut back on her smoking since moving in with him.

Maybe cigarettes are hazardous to one's health in a new way: their toxins and blockage of oxygen transport playing an unsuspected role in triggering a biological smoking accident.

Early scholars always associated SHC with alcoholism. What about this? "She had a drinking problem too at one time," Frank stated. "And she started drinking again—I'm sure she did. She was an alcoholic at one time."

(To the hard-core skeptic who will, of course, see no need to look further than this to explain away this incident: go back and reread Fire Marshal Calcagno's comments.)

Medicine may want to consider the effects of alcohol and combustion; how subtle biochemical or bioelectrical changes resulting from prolonged imbibing could flare up *in a very unexpected way* years later.

Frank spoke about the recent tragedies to the family: the deaths of his mother's husband, father, and a son. This really piqued my interest. I strongly sensed that extremism in a person's consciousness—be it euphoria or rage or, especially, depression—was a crucial element in triggering many spontaneous combustion outbreaks. Did Beatrice show signs of depression?

"Oh sure," Frank answered without hesitation. Her husband's prolonged illness and death had left her "really distressed." Had Beatrice talked of suicide? "No," he replied. "But she'd say 'I wish I was dead' sometimes. 'I don't want to be a burden to anybody'."

So Mrs. Oczki, a widow, did have a death wish though she did nothing consciously to fulfill it. Ah, but what outcome can be initiated by the unconscious, subconscious, or superconscious mind? The interplay between mind and body is extraordinarily complex and inadequately understood by psychiatry and psychologists. Morose, depressed, sullen, perceiving herself and her life as burdensome, there simmered inside Mrs. Oczki a seething desire to end it all. As romance novelists say of their heroines, "she burned to be free."

Was she prone to fits of red-hot, quick temper?

"Yeah, she had a temper!" Frank answered, as if recalling some of the displays he had witnessed. "She was pretty religious," he added. Perhaps her Catholic background (with its dogma of fire and brimstone) united that morning with another aspect of her consciousness—anger? depression? liberation?—to trigger psychobiophysiologically an internal *flame of transformation* that took Mrs. Oczki to a place where she believed contentment awaited her.

(Much later, I learned about another factor, perhaps significant. "My mother was *definitely* terrified of fire!" her son recalled after reviewing this chapter. "She *always* had that in the back of her mind." The complex nexus of depression, hot-headedness, belief and fear: might Mrs. Oczki have unconsciously demonstrated how impetuous can be its manifestation?)

My focus turned for a moment from Mrs. Oczki herself to the condition of the house. What did Frank find when he returned home from his skiing vacation?

In the room where his mother died—the room in which we now spoke to one another—heat-sensitive video cassettes and eight-track audio tapes were still usable after the fire, he discovered. But he did have to replace a fourteen-square-foot area of the ceiling. As for the floor tile *directly under her chair*, he said it "was buckled a little, but not totally burnt."

More indications that this localized inferno directed little heat either upward or downward, but more so downward than up.

Frank had new information about the soot and odor left behind: "It was really funny. It was sort of oily, greasy. I remember it being blackish. I've smelled dead bodies before, and it's a real putrid smell; a strong smell. The house did not have that smell! The house had an almost sweet—"

"Kinda like hickory, but sweeter!" interjected his friend who had been listening to the conversation. "Like *incense*, almost."

A trademark of SHC and atypical of conventional fires.

Another atypical factor involved the smoke detector mounted in the foyer. *It had failed to go off!* Frank gave it to me. With its original battery still installed, I later tested it. It worked flawlessly.

No one in the fire-fighting services has been able to explain why a functional smoke detector would fail in a fire that incinerated a person, and why the fire that burned her generated a redolent smell likened to hickory incense.

I bid Frank and his friend goodnight, thanking them (then and now) for their cooperation in a difficult situation. The questions I posed to them should be asked by all fire investigators who find themselves facing what their associates in Bolingbrook had to face in late 1979.

DISCUSSIONS OVER BREAKFAST

Next day, I met with Calcagno and his morning crew over breakfast. This time *they* were asking the questions as I passed an assortment of SHC photographs around the table.

Calcagno noticed a recurring theme and asked, "Are most of these people incapacitated in some way?" The fire marshal was sharp; he looked for patterns, saw possible connections, and tried to solve the puzzle.

"Invalidity or ambulatory incapacity often *do* crop up. I think it's an important factor," I told the firemen.

Calcagno returned to the cigarette theory. He had searched for but found no evidence of careless smoking habits by Mrs. Oczki. Proponents of the argument that Beatrice suffered a stroke and dropped a lit cigarette or match on herself will have to argue with Calcagno, who reaffirmed that there was inadequate fuel loading to explain the fire damage to her body.

Calcagno had one last, surprising piece of evidence to share about Mrs. Oczki. "I understand the sutures in the skull were intact. If there had been a rapid build-up of heat," the fire marshal reasoned, "you'd think the skull would have ruptured. But that didn't happen."

Once again, we find a human whose body largely evaporated in a moisture-dehydrating fire with heat so minimal the cranium's sutures do not break up.

Not what most people would consider appetizing breakfast conversation, but I'll listen anytime to a fireman unafraid to question his training when evidence demands that his beliefs be turned upside-down. Calcagno was one such person. He had come up through the ranks as a paramedic, a fireman, an inspector, and now the fire marshal of an affluent community that could afford the best. "It's been a long haul, a lot of fires," the knowledgeable fire marshal said of his career. *"But nothing like this!"*

The case was *still* bothering him nine months later. He took pride in solving the fires he extinguished. Here, he had a blaze that didn't require extinguishing and couldn't be explained conventionally.

"I just don't know," the fire marshal lamented. "We were *never* involved in a situation like that before. The puzzle was *so bizarre...so confusing*. I'd like to come up with something concrete, so I can put this away for ever."

All factors considered, SHC seems the only way to properly close the case of Beatrice Oczki. Calcagno agreed.

Over a handshake, Fire Marshal Calcagno spoke of the challenge presented by Mrs. Oczki and SHC, expressing himself with a sincerity that would warm an unclad body in an artic blizzard. "This could be happening today someplace," he feared, "and somebody's just going to bury it. It'll be lost in a file, and you'll never know about it. Nor will anybody else remember it—unless you get to one of those progressive people who will at least give you the courtesy of a phone call."

NORTH TO MINNESOTA

Calcagno's words echoed in my mind as I repacked the motorcycle and looked at the road maps.* He had shown me a letter sent him by Jan Bowen, who just happened to be a great-aunt of Mary Reeser. She lived in Palatine, Illinois. It was another lead and en route.

The highway to discovery took me to Jan Bowen later that day, and she filled in the story about how her family had dealt with Mary's famous death. Then I rode to Minnesota, to track down LaVerne Olsen's ambiguous lead for another possible SHC.

---

* Years later, I have noticed that Bolingbrook lies on a straight line running from the Rooney farm outside Seneca, Illinois, to downtown Chicago, where Mrs. O'Leary's cow allegedly started the Great Chicago Fire of 1871. This coincidence of unusual fires aligning will be explored later.

A trail of clues eventually proved fruitful, and I identified the case as that of a Mrs. Johnson, who had died earlier in 1980 by fire in Plymouth, Minnesota. To condense four days of follow-up effort into one sentence, once I finally got permission to see the twenty fire-scene photographs generously provided by the Plymouth Police Department I knew immediately that this woman had not died by SHC. Several heat-ruptured butane lighters lying on the floor around her wheelchair had provided necessary accelerants to feed the fire in her badly burned apartment... besides, she herself was only superficially burned.

Unlike Mrs. Oczki, Minnesota's Mrs. Johnson *was* typical of a smoking materials fire fatality. I could at least close the book on this lead.

During my investigation of this fatality, I had made an appointment with the Hennepin County Sheriff's Department to examine their file on Mrs. Johnson. Now already knowing the outcome, I decided as a matter of professional courtesy to visit HCSD anyway. The duty officer in the records division had pulled a slide of a fire death he thought might interest me. I expected to see Mrs. Johnson. Surprise! It wasn't. This fire scene *did* look like SHC.

The casualty, Mrs. Eunice Holstrom, had died by fire on October 20, 1971. Her body had pushed back the chair in which she sat—a characteristic I had just seen with Mrs. Oczki (and other self-incinerated individuals sitting in chairs). Her lower body thus leaned against a wall behind the burned away chair. The more startling aspect, however, was the position of her head. One had to conclude that as her torso toppled backwards, her neck utterly disintegrated and her head fell forward—because it now rested surrealistically balanced upright on the woman's chest.

The report said Eunice was a widow. Hmm. It said her deceased husband had committed suicide. Hmm! I got some names from the report, among them Dolores Higgey. I called Mrs. Higgey. She had overheard a fireman talking about the incident: "I remember them saying it was really strange."

What did she, as a friend of the deceased, know about Mrs. Holstrom's personality? Eunice was mean to her sister, Mrs. Higgey replied, "and she was like that to her husband too—and he was a real quiet guy. She felt a great deal of guilt over her husband's suicide. She was alcoholic." Emotionally schizophrenic, Mrs. Higgey described the victim.

A psychological profile for SHC victims had begun to take shape, in which the person is typically morose, melancholy, isolated or alone (often widowed), dependant, impaired, subject to severe depression or bursts of heated temper, and guilt-ridden. Alcohol was a part of victims' lives at some time, as was an incapacitating physical or emotional injury.

Eunice fit the profile to a T. It looked like I'd be leaving Minnesota with a new (heretofore unsuspected) case of SHC, after all.

Homeward Bound

En route home, I made one last stop. In Pleasantville, Ohio, according to Sanderson's *Investigating the Unexplained* (1972) and later sources, Mrs. Cecil Rogers had "burned to a cinder" with no damage extending beyond her slightly singed bed in March, 1956. I hoped some local people would still remember the incident.

Pleasantville *was* pleasant the afternoon I arrived there. But no one at the volunteer fire department remembered a Mrs. Cecil Rogers, or any fire death that even remotely resembled Sanderson's account; nor did the retired fire chief, who had served the town in the mid-1950s. Unless everyone I spoke to was involved in a massive cover-up of one particularly strange death more than twenty-four years earlier, the alleged SHC of Mrs. Rogers seems never to have happened there or in any place nearby. In the quest for truth, that *is* finding something.

Having done what a good investigator must do—get out of the armchair and into the field where real action and facts lurk—I headed home. During the fourteen-day, twenty-six-hundred-mile journey I had located two sources for new information about Mrs. Reeser, dismissed one suspected SHC incident, debunked two cases erroneously published as SHC, and documented *two new cases* having the signature of SHC.

All around, it had been a very good fortnight indeed out there on the back roads and byways that can lead to SHC.

21

Revelations from the Shadowed Past

Just as in science there is no such thing today as a fixed and final, "found truth," but only working hypotheses that in the next moment may require revision in the light of a newly found fact, so also in the moral sphere [of myth] today all the walls have burst.... adventure is upon us, like a tidal wave.

—Joseph Campbell,
"Mythological Themes in
Creative Literature and Art,"
in *Myths, Dreams, and Religion* (1970)

There is strong scholarly evidence to view myth differently than fireman Kennedy defined it at the beginning of chapter 14.

Having examined the mythology of SHC his way—as an *aesthetic* fable meant to entertain as fiction but offering little promise of meaningful discovery—let us now explore SHC through the other category of myth. This is the *explanatory* myth: the literature and science of premodern people who attempted to explain natural phenomena as they observed and could understand it.

Classicist Edith Hamilton says in her book *Mythology* (1950) that "real myth has nothing to do with religion. It is an explanation of something in nature; how, for instance, any and everything in the universe came into existence."

James Barr concurs. "Mythology is not a peripheral manifestation, not a luxury, but a serious attempt at integration of reality and experience," he says in discussing the meaning of mythology for *Vestus Testamentum* (1959). "It has then its aspects which correspond to science, to logic, and to faith, and it would be wrong to see myth as a distorted substitute for any of these."

And Joseph Campbell, the master of comparative mythology, explained in his PBS-TV series how an important class of myth relates to and attempts to explain nature and the natural world.

Equally strong historical and archaeological evidence proves it is incorrect to see our premodern ancestors as intellectually inferior; incapable of observing and describing their world without resorting to superstition and other fanciful images. Early Egyptians not only knew the Earth is round—the flat-earth theory is actually a recent delusion—but accurately measured its size and built many of their magnificent temples' foundations upon precise celestial astronomy; the Greeks possessed instruments that measured movements of the solar system; neolithic Britons incorporated remarkable astronomical knowledge in their megalithic engineering; twelfth-century Hindu metallurgists out-paced Western science by half a millennium when recognizing the importance of a flame's color for analyzing metals; Chinese techniques of acupuncture, still ridiculed by unbelieving physicians, nonetheless are now being successfully applied at European and American hospitals.

Who can categorically prove that our ancestors weren't just as capable of describing an event that, as it does today, happened rarely in their midst and left them no less baffled? Are myths automatically invalidated if, to use Campbell's term, we dare to "concretize" *some* of them as literal events rather than philosophical symbols? I don't believe so. That *some* myths are not recognized as potentially literal is because those who interpret them sometimes lack the knowledge to recognize the variety of paranormal events that can occur in the natural world.

Could, then, this second type of myth—based on mysterious but real events—reveal useful information and provide real answers to nagging questions about SHC? Like: Is SHC only a modern phenomenon?

GOING BACK, BACK, BACK IN TIME

As you'll recall, the first case described in the medical literature as SHC probably was Polonus Vorstius in the 1400s. Is it possible this curious affliction had been around before, maybe *long before* the fifteenth century? And if so, might its longevity, unsuspected, be a reason belief in SHC has *endured* despite vigorous "scientific" denial?

Answers depend on the type of testimony you are willing to accept, and on its interpretation. To go far back into the events of man's shadowed past, the sources one must look at are those that have survived from those times: the myths of our ancestors.

Like Schliemann who unearthed glorious Troy after following the clues in a blind bard's millennia-old *Iliad* which others had dismissed as fable, let us dig into the lore bequeathed by ancient ancestors to discover whether their myths speak to a reality for SHC long ago. With an eye that sees ancient history as mythological scholars do, rather than as fireman Kennedy does, we may find important revelations and unsuspected answers to the questions raised above.

According to Professor Albert Krantz's *Historia Ecclesiastica de Saxonia* (1482), the army of Godfrey of Bulloigne in A.D. 1099 confronted at Niverna (now Nivernais in central France) an unassailable enemy: "a strange disease made its appearance, where men burnt, being touched with an invisible fire, that pierced into the very bowels and vitals, by occasion of which, the hands of some, and the feet of others, fell off."

Marcellus Donatus asserted that a "similar disease" befell eastern France ten years earlier. In his *De medica historia mirabili libri sex*. (1586), he recorded that in A.D. 1089 "people of a certain locality" in the Lorraine found themselves "burning of an invisible fire in their entrails, and...some cut off a hand or a foot when the burning began, that it should go no further."

What was happening at these two places in medieval France? Mass hallucination? Gastrointestinal indigestion? Leprosy? Collective SHC, of the kind singularly experienced by Susanna Sewell and Jack Angel? Something *else*?

Theodoric the Great (ca. A.D. 454–526) was a powerful Ostrogoth king who became ruler of Italy. Peace and prosperity marked his reign, but an unnerving peculiarity marked his kingly body: when rubbed, it would emit flames. Did Theodoric live on the verge of more serious self-combustion more than fourteen hundred years ago?

Early in the fifth century(?) A.D. a bizarre experience that reads like SHC survival befell Dr. Alexandrinus Megetius. As Brewster reported in his *Natural Magic*: "Ezekiel de Castro mentions the singular case of Alexandrinus Megetius, a physician, from one of whose vertebrae there issued a fire which scorched the eyes of the beholders." One can only guess how the doctor diagnosed his own ailment—

The fourth-century Christian historian Eusebius, in his *Hist. Ecclesiastica* (Lib. IX), stated that Maximinius died in an odd manner: "flesh was taken with a secret fire from Heaven, so that it was burnt, and little by little, turned to ashes. There was no more any shape of man to be seen in him, nothing left but a carcass of bones, all dry." This reads like SHC, though I wonder at what velocity it unfolded.

Pliny the Elder (A.D. 23–79) recounted, in Book II, chapter CXI of his thirty-seven-volume *Natural History*, several astonishing strange fires in the ancient world. Two cases of "sudden fire" involved humans. Pliny wrote that when Servius Tullius was a boy, "a light fire shone out of his head." Further, according to the acclaimed historian, "a flaming fire of the same sort" burned from the head of Lucius Marcius when he was exhorting his soldiers to take revenge following the death of two Scipios. Soldiers would have witnessed Marcius's SHC in 212 B.C. The youthful Tullius survived his flaming face only to be assassinated by his daughter and son-in-law after becoming the legendary sixth king of ancient Rome in the seventh century B.C.

INDICATIONS OF SHC FROM MYTHOLOGY?

The preceding episodes are objective history. History can be found through studying mythology as well, though interpretation is more subjective.

"The Voyage of Bran and His Adventure" is a Welsh chronicle of Bran, a king of Britain at the turn of the first millennium. One adventure involved his warrior Nechtan the son of Collbran, who, seized by home-sickness, beseeched Bran to take him back to Ireland. Bran reluctantly agreed.

They sailed to Srub Brain in West Kerry, Ireland, despite warnings that the decision would lead to sorrow if anyone stepped ashore. There, Nechtan leaps from his wickerweave coracle into the surf, and prophecy is fulfilled. "As soon as he touched the earth of Ireland," says Kuno Meyer's translation of *The Voyage of Bran* (1895–1897), "forthwith he was a heap of ashes, as though he had been in the earth for many hundred years."

No explanation is posited for this extraordinarily strange and straightforward report. Yet what more accurate a description of SHC could history provide than these lines from an ancient Welsh tale? Especially noteworthy is its mention that the incinerated Irishman suffered melancholy and depression, a trait of several modern victims of SHC.

The "Voyage of Bran" can be an example of myth embodying an event so obscure and atypical that it hasn't been recognized for what it could be: not a metaphor or allegory but a true depiction of an historical event. SHC.

If Nechtan had not been the son of Irish nobility and a heroic warrior, would history have recorded the death awaiting him so we might ponder its meaning today? Surely not... which in turn suggests that Nechtan's bizarre fate could have been happening over the centuries to the unnoteworthy peasantry, whose disintegrating misfortunes are forever lost in unrecorded history.

THE SKIES RAIN FIRE... AND HUMANS BURN

Along with this precise description of SHC, Celtic mythology mentions a related item: rains of fire. Three gods of the Tuatha Dé Danann, the proto-ancestors of the Irish, called down "frogs, rain of fire, and streams of blood against the Firbolgs" while a celestial ally of the Firbolgs likewise "promised to rain showers of fire on the foe and to remove from them two-thirds of their might," says Celtic scholar John Macculloch in *The Mythology of All Races* (1918).

To an conventionalist scholar, these deity-generated firestorms are mythological. "Little of this is actual history," says Macculloch. To a fortean, however, these searing showers can be taken literally because all three mythic precipitations—sent by three! gods—have counterparts in the real (however one defines real) modern-day world.

Consider Robert Burch, an electronic mechanic at the Puget Sound Navy Yard. On November 6, 1951, he encountered the legendary wrath of the Tuatha while standing before a mirror in his room at the YMCA in Bremerton, Washington. In the mirror he saw the reflection of a red-orange fireball rushing toward him through the open window. A blinding flash and loud concussion knocked Burch to the floor. Searing pain zapped up his right arm.

His roommate rushed in to discover the aftermath of this fire from heaven: a window sill burned black and too hot to touch, luggage beneath the window charred and smoking, two radios burnt, a wastebasket in the corner with its contents aflame, his spontaneously burnt roommate lying on the floor. As he frantically extinguished the fire, a Bremerton policeman who had witnessed this fire-from-heaven swoop through Burch's window ran in and confirmed the entire incident.

"I don't know what happened," said Burch the next day as he was being treated at the Bremerton Navy Hospital for shock and second-degree burns, "but it sure was peculiar."

Burned by heaven's fire? Burch is not alone. There is historical precedent for atmospheric fire that inflames people and more.

February 1972: two hundred students at Whitchurch were preparing for a pedal-car race when they had to be treated for sudden "mystery face and eye burns." Scientists had nothing to say, except "The cause was never discovered." The enigma was a repeat of an incident that occurred four years earlier in this west border English town, when more than three hundred Whitchurch residents were burned in a comparable event of unexplained origin, reported the *Sun* (February 2, 1972).

During the fall of 1966, a fireball swooped from the sky to singe the hair of Mrs. Louis Matthews in Philadelphia, Pennsylvania. On January 23, 1955, a ball of fire struck William C. Cunningham and burned his hands in the same city.

In August or September 1878, a "meteorite" knocked two people insensible and left a third person at Haraiya "charred and dead" in India. On July 1, 1826, an exploding "fireball" shot from the heavens and killed two woman at Malvern Hills south of Birmingham, England; others were injured. A correspondent to the *Philosophical Magazine* (1811) said that around July 15, 1810, "a very large ball of fire" fell from the clouds and burnt five villages, including crops and some men and woman near Shahabad, India. In 1678, a half-hour "shower of fiery matter" burned the German town of Sachen-Hausen, said London's *Philosophical Magazine* (1814). On March 3, 1557, a "flame" entered the bedroom of François and Diane de Montmorency, burning the Frenchwoman's hair and clothing but nothing else.

Just as fireballs have wrought limited human combustion upon these people in modern times, might those legendary Celtic "showers of fire" have burned some real-life Celts whose names were never recorded in the annals of SHC?

Add to the annals this scene in Euripides' play *Medea* as translated by E. P. Coleridge in *Seven Famous Greek Plays* (1950): A Colchian princess, Medea, used her powers as a sorceress to aid Jason in retrieving the Golden Fleece and bind his love to herself. Jason deserted her, however, and wed the daughter of King Creon. Medea, spurned in love, plotted revenge. She gifted King Creon and his daughter with a cursed robe and chaplet, and soon afterward a messenger arrived to tell Medea that Jason's bride has died. Medea asked how, imploring she be told her rival "perished miserably."

The messenger answered that no sooner did Jason's wife don the gold-embroidered robe and gold crown[1]

> When lo! a scene of awful horror did ensue.... The chaplet of gold about her head was sending forth a wondrous stream of ravening flame...and she starts from her seat in a blaze and...the flame, as she shook her locks, blazed forth the more with double fury. Then to the earth she sinks...

All told, said the messenger to Medea, it was "a fearsome sight to see. And all were afraid to touch the corpse."

Coleridge sees *Medea* as another Greek morality play: "Ultimately, the abortive alliance between Jason and Medea has destroyed them both." So can it be. For this particular scene, though, I offer another interpretation: the long-ago observed death of a royal daughter by SHC, wrongly attributed by the Greeks to magical garments (magical, perhaps, because they ignited fire but themselves did *not* burn).

Euripides' play dates to 431 B.C. If through new eyes we view the Greek playwright recounting an actual SHC, shrouded in the "magic" of a fiery raiment and golden crown but remembered because it involved royalty, then history may be giving evidence for SHC dating back almost twenty-five hundred years!

IN EARLIER TIMES: MACROCOSMIC SHC?

Can SHC be identified even earlier in history? Possibly.

Circa 701 B.C., the Assyrian warrior-king Sennacherib swept into Judah. His massive army encamped around Jerusalem, preparing to attack. Isaiah prophesied that Jerusalem would not fall, assuring King Hezekiah that the Lord would send a mighty "blast" against the enemy. Hezekiah, gazing upon the thousand-score campfires of Sennacherib's soldiers, must have thought Isaiah mad.

But prophecy *was* fulfilled: "Then the angel of the Lord went forth and smote in the camp of the Assyrians one hundred and eighty-five thousand: and when the soldiers arose early in the morning, behold, their comrades were all dead. So Sennacherib departed and went and returned and dwelt at Nineveh" (II Kings 19:7 and Isaiah 37: 36–37).

Sennacherib fled Judah in defeat.

Bubonic plague is the explanation usually offered by historians[2] for the defeat of Sennacherib. Dysentery, mice, a hurricane, even meteor showers have also been suggested for this fantastic *blast*—"the precise nature of which is not specified" says Sean Mewhinney (who advocates the meteor theory) in *Kronos* (1981). Could any of these theories *really* be the "angel of the Lord" that *overnight* defeated the greatest military force then on earth? Each theory fails to resolve one key element in the Talmud's account.

Now why do I mention this ancient event in a book about SHC? The Bible gives you no clue. Other ancient sources do offer tantalizing clues, however. Egyptian texts, according to G. A. Wainwright in *Journal of Egyptian Archaeology* (1932), refer to the Hebrew triumph as "the night of fire for the adversaries." The Talmud and Midrash sources, books of Jewish law and lore, record that overnight this "blast"[3] foreseen by Isaiah burned 185,000 warriors in the army of Sennacherib: "their souls were burnt, though their garments remained intact."

These are very strange statements! Ginzburg, in his *The Legends of the Jews* (1968), even considers the event "miraculous."

How to interpret this?

Dwardu Cardona, after advocating in *Aeon: A Symposium on Myth and Science* (1990) that a hurricane wiped out Sennacherib's army, expresses uncertainty: "If the sources are to be believed, the suddenness of the slaughter as the army lay resting during the night plus the "burned" nature of the victims, *with their garments remaining intact*, do not imply the effects of a hurricane." Indeed!

There is one theory that I think *can* explain Sennacherib's ashed army, one in line with the Hebraic description of a miraculous holy "tempest" and with B. Childs's statement in his *Myth and Reality in the Old Testament* (1960) that "unexplainable events were attributed to the direct intervention of the Gods."

The resemblance of this bizarre long-ago episode in holy scriptures to some modern individual deaths by SHC is indisputable. What I propose is this: that *Sennacherib met defeat through the unprecedented miracle of mass (collective) SHC.*

I know no better explanation to resolve this curious tale of a sleeping army's sudden burning within unburned garments than SHC on a *macrocosmic* scale. SHC can resolve another enigma that has perplexed scholars, changing what some have thought to be the fable of Sennacherib's defeat into a historically accurate portrait of his surrender to a force beyond any king's might to vanquish.

As to the source of the awesome fury that Isaiah foresaw defeating his enemy? It might have involved a macroscopic version of whatever burned Robert Burch in Bremerton... or burned the rebel General Pablo Castilliano and two aides on the eve of a planned attack against the Nicaraguan army in 1907. There, a "flaming mass" from the night sky struck their tent, killing the general instantly and mortally wounding the aides. Said Frank Edwards in *Strange World* (1964), this was "the only known case where a war was brought to an end by direct intervention of a celestial object." Maybe not.

These episodes involved a source of fire external to the victims, to be sure, but one that burned its victims spontaneously upon impact.

Or there could be another factor, which I will suggest later.

If the idea of thousands of humans being reduced to ash overnight by a fire of uncertain origin exceeds belief, the Old Testament offers a less macrocosmic but even earlier account that should be considered.

One day, Moses and Aaron made three burnt offerings to God (Leviticus 9:24). Then Aaron's sons, Nadab and Abihu, stood before the altar and offered "strange fire" in some inappropriate fashion. The consequence was immediate: "And there went out fire before the Lord and devoured them, and they died before the Lord." Relatives of the boys-who-played-with-strange-fire "drew near and carried them in their vestments out of the camp; as Moses had said." Moses ordered the house of Israel to "bewail over the victims, whom the Lord burned" (Leviticus 10:2–6).

A strange (paranormal?) fire of divine (paranormal?) origin that flamed suddenly and quickly and completely, leaving no time to react until the men burned so completely they could be gathered together in robes (either robes of the relatives *or* the unburned vestments of Nadab and Abihu themselves). Parallels to modern-day deaths by the fire within suggest, thanks to Moses' writing in Leviticus, the observance of SHC—*a double SHC*—more than thirty-five hundred years ago!

SHC, singly or doubly or by the thousands, would certainly have been a pressing problem for Moses. (As it would be today). If my interpretation is correct, the Hebrews preserved not a fable but instead documented a historical mystery that confounded its observers then, and continues to confound... a mystery that remains current and convincing thousands of years later.

Catastrophic Conflagrations in Modern Times

If the idea of tens of thousands of Assyrian troops reduced to ashes overnight in the ancient Middle East is disquieting, two recent conflagrations are only slightly less so.

During November and December 1902, a "fantastic array of fire from the skies" plagued many Australian and New Zealand communities with blazes and attendant explosions, reported contemporaneous *Sydney Herald* and *Otago Witness* newspapers. One meteorologist said "There has been nothing like it in the history of the colonies!"

But not unique in the world. A similar firestorm had rained on America's heartland a few decades earlier, with fearsome results.

Sunday, October 8, 1871. At 9:00 P.M.—"by singular coincidence precisely the time at which the Chicago fire commenced," historian Rev. E. J. Goodspeed determined—the heavens "burst into clouds of flame" that swept ferociously through the upper regions of the Midwest, leaving behind mysteries that perplexed the survivors but have ever since been mostly overlooked by historians.

A maelstrom of fire began almost instantaneously across large expanses of Wisconsin, Minnesota, and Michigan, touching down in Illinois and Indiana as well. It was the night Chicago burned, ignited ostensibly by Mrs. O'Leary's bovine (although reporter Michael Ahern would later admit he had invented the story of the lantern-kicking cow, thus creating a modern myth believed by most people to this day). It was the night that saw many people disappearing from the face of the earth in a way that emulated the ancient Talmud's description of the encamped Assyrian army.

Because the northern Midwest had many logging camps, residents knew well the dangers and effects of forest fires. But this night was unlike any ordinary timber fire. This fire played many strange tricks. Blackberries petrified on bushes in the searing heat; sixty dozen axes melted into one mass inside a hardware store; iron on a fire wagon melted, yet the wooden tongue affixed to the iron did not even scorch. Goodspeed said in *History of the Great Fires in Chicago and the West* (1871) that this fire "was able to *so* destroy human bodies that not a trace of them should remain. This *fact* serves to account for the utter loss of many persons known to have been in the vicinity where the fire appeared..."

Most of these human disappearances happened around Peshtigo, Wisconsin. The frightful firestorm swept through the town quickly and left in its wake a number of victims who could be just as quickly swept up.

A nephew of a Peshtigo man was identified only by his penknife and a few relics, his body so thoroughly incinerated that his remains fit in the "palm of the hand." Of Peshtigo's ten-member Doyle family there was "not a trace—vanished from the Earth." Elsewhere more Peshtigo people (estimated around ten in number) had become a "heap of indistinguishable calcined bones and charred flesh...giving no clue to sex or number." A burly Peshtigo lumberjack became "a mere streak of ashes, to fill a thimble" and no more. Of three Peshtigo adults, there was "only enough ashes to fill a two-quart measure." Residents running to escape the flames that descended from the sky into Peshtigo noticed Helga Rockstad seeking her own escape from her flaming hair; she was later found reduced to a "mound of white-gray ash." A pamphlet published by the Peshtigo Historical Museum says that a group of Swedes were so completely consumed that "rescue workers could find nothing to mark the place where they had been except ashes and the metal from their axes and shovels." Other victims of unknown numbers were found reduced to "two-thirds size"—*shrunken* instead of incinerated by the fast-moving fire that snuffed out their shortened lives. In Peshtigo's hotel kitchen, "numbers of people" became mere ashes; one hundred guests perished elsewhere in the hotel, said the *St. Paul Daily Pioneer* (October 12, 1871).

Some who weren't combusted left behind other evidence to the strangeness of this evening. For example, copper coins had melted together in the pockets of a Peshtigo man "*yet the clothing and the body of the man was not even singed.*" One example of this peculiar monetary fusing, repeated several times that night, is on display at the Peshtigo Historical Museum.★

How odd that the atmosphere could be hot enough to melt metal yet not blister skin or kindle fabric. Or was something more esoteric than hot air unleashed that night?

Across Lake Michigan in Sigel Township, Michigan, a Mr. Sarniac was discovered "burned to a crisp" with "only wristbands and an axe to identify him." A case of severe third-degree burns or something more thorough? South in Chicago, where stone buildings "melted" amid blue, red, and green flames, Fire Marshal Williams was frustrated by fires breaking out spontaneously (sometimes below ground). Survivors of that terrible night spoke about the complete "disappearance of people known to be in vicinity of blaze."

★ Silver coinage melts at 1600°F, copper at around 2000°F.

Numerous commentators remarked on the electric-like nature of the fire that night, a fire that descended from the heavens and at the same time seemed to burn up from the ground. Many trees were later found to have their roots burnt out.

Normal forest fire behavior alone cannot explain the phenomena that happened that Sabbath night. Is history once again providing evidence, difficult though it may be to consider, that SHC can happen macrocosmically; that nature rains electrifying precipitation that can fuse metal in the pockets of some while transforming others to streaks of dust?

A New SHC Mythology

Fire has always been a magico-religious power for man, and society views with awe those shamans and ascetics who demonstrate "mastery of fire" through immunity to both flames and freezing weather. Comments Mircea Eliade in *The Forge and the Crucible* (1971): "The mastery over fire and insensibility both to extreme cold and to the temperature of burning coals, translated into ordinary terms, signify that the shaman or yogi have gone beyond the human condition and have achieved the level of spirits."

As you have seen, the ability to inordinately warm the body is not merely mythological; nor is it limited to shamans and yogis and religious visionaries. It is, occasionally, a *real* ability for any person who accesses tumo or Kundalini or some other means to generate superhyperthermia internally...a superhyperthermy that may culminate in SHC.

Applying Barr's mythological premise to these ancient (and a few modern) myths, we can view them not as "a distorted substitute" for science and logic but as precise accounts of observed enigmas. Reality is what the myths themselves sometimes describe happening. In this instance, contrary to fireman Kennedy's lament, it is *good* that self-sustaining myths die hard.

Like the myth of the death of Semele, the mother of Bacchus, who requested of her paramour Jupiter that he visit her in all his great god's glory. When Jupiter (mythological representation of air) did so, Semele was reduced to ashes on the spot. Nineteenth-century physician Edmund Sharkey—himself a proponent of the "disease" of SHC, as he called it in the *Dublin Journal of Medical and Chemical Science* (1833)—saw in the Semele story "another of the many instances in which the ancient mythologists have conveyed an account of the operations of nature under the garb of beautiful allegories."

The line between myth and reality at times seems invisible.

"Every man has the right to his opinion," observed Bernard Baruch, "but no man has a right to be wrong in his facts." One fact is that people die in fires, some *very* strange. My opinion based on that fact, based on a review of history itself, is that SHC has been with mankind for a *very long time*.

22

Earthfire! The Cartography of Combustion

> ...future research may show that the well-being of living things on the surface of our earth is more dependent upon its radiations than has heretofore been imagined.
>
> —John Milne,
> *Nature* (July 6, 1911)

Charles Fort conjectured in his book *Lo!* that cases such as those being chronicled might relate "not to 'spontaneous combustion of human bodies,' but to things, or beings, that, with a flaming process, consume men and women."

Things? Perhaps this avatar of the Unwelcomed was again on the right track.

Things were getting pretty hot in the early 1970s in the busting Midlands city of Birmingham, England. Here, a bevy of blazes baffled firemen, who are trained to determine the origin of fires; coroners, who are employed to ascertain precise cause of death; and Birmingham residents, some of whom did not survive their burning outbursts.

February 17, 1972: Mrs. Edith Thompson, seventy-five, inflamed in her Cheslyn Hay, Staffordshire, home, north of Birmingham. Announced the *Birmingham Evening News* (same day), damage is restricted to her and the chair she was sitting in. It's said she was trying to light a fire *"but had not succeeded in doing so,"* reported Harrison (1976; 1978); police strove to locate the cause. "Death by misadventure" closed the case.

April 8, 1973: John Joseph McRory, sixty-four, was the victim in his Handsworth house of a fire for which "no definite cause" could be found, reported the *Birmingham Evening Mail* (April 20, 1973). Birmingham coroner George Billington did not disappoint: he ruled "accident death."

August 5, 1973: Alan Bricknell, twenty-four, was burglarizing a building in Birmingham when five sudden fires prevented him from escaping the scene of his crime. He confessed to theft, but adamantly denied arson. He was found guilty of arson anyway, reported the *Birmingham Evening Mail* (October 26, 1973). How he managed to set the fires remained unproven. Perhaps he was a bumbling arsonist caught in his own ineptitude, or maybe he got trapped by spontaneous combustion courtesy of cosmic justice.

August 18, 1973: On Coventry Road, Birmingham, a car suddenly burst into flames. A woman was dragged out alive, but an unidentified man died inside the auto, inside a localized firestorm, according to the *Sunday Mercury* (August 19, 1973).

October 20, 1973: William McLeod, octogenarian, was found in a fire-damaged room in Nechells, Birmingham. No mention of how much or how *little* damage was done to the room. *The News* (January 1974) implied a possible SHC.

October 26, 1973: An eighteen-year-old female student became a "flaming torch" at Birmingham's All Saints Hospital and was admitted to the Burns Unit of Birmingham Accident Hospital (BAH) for treatment of serious back and chest burns. Matches are suspected, reported the *Birmingham Evening Mail* (October 26, 1973), but no one went on record to explain how her flame-enshrouded body could but smolder the bed sheets she was lying on.

October 26, 1973: twenty-three-year-old Diane Mold ran six hundred feet across her yard in Coseley, Dudley, in flames. She was taken to BAH Burns Unit, too. It was believed she had fallen upon hot coals, reviving in time to rip off her burning clothes and don a coat before making a dash for help. The *Sunday Mercury* (October 28, 1973) did not state whether Miss Mold agreed with this scenario.

April 14, 1974: Smethwick, Staffordshire (greater Birmingham). In a corner of the Kaur living room, seven-month-old Parvinder Kaur lies in her crib. Both appear to be ablaze. Neighbors threw water on the flaming infant; the fire died. She was rushed to BAH Burns Unit, where concern focused on not letting her die. Fire officials concerned themselves with origin of the fire. They came up with nothing, save that "kids could have been playing with matches." The matches, if there were any, were magical or the target of teleportation: investigators couldn't find them. Cause of fire is undetermined, said the *Birmingham Evening Mail* (April 16, 1974).

Circa April 22, 1974: One week later BAH Burns Unit—busy place!—was treating the badly burned feet and legs of Mark Bradbury, seven, who slept through a fiery destruction of his own body. When he did awake, he screamed. Whether from pain or from surprise, the reporter for the *Birmingham Evening Mail* (April 22, 1974) didn't ask. No one understood how the boy was burned.

Curious fires occur in April, then August, then October, then April again around Birmingham. If the Burns Unit had a clinician interested in cycles, he might have made a mental note to be on the lookout for more baffling blazes during the forthcoming August—

Circa August 26, 1974: Lisa Tipton, six months old, died in a localized fire limited to one room of her parents' house in Highfields, Staffordshire. All electrical appliances in the home worked properly. Officials concocted a complex theory that fell short of the observed facts and remained baffled, said the *Birmingham Evening Mail* (August 26, 1974). No cause was given for the fire, which singled out another innocent infant.

The next seasonal cycle of Birmingham's baffling blazes bursts forth a month ahead of schedule, just in time to herald the beginning of spring.

March 22, 1975: William Cashmore, eighty-two, died in his home at Autumn Close, High Heath, Walsall, just north of Birmingham.[1] Investigators again said all appliances were in working order. What they could *not* say was how the fire began, and why it restricted itself to burning only Cashmore's body, clothing, and the chair on which he sat. To those who claim carelessly handled smoking materials are inevitably to blame in cases of this type, put away your pipe and contemplate this: according to the *Sunday Mercury* (March 23, 1975), "he was a nonsmoker."

June 1, 1978: Miss Andrea Deeley, twenty, of Hunters Road in Birmingham's Handsworth district, asphyxiated by carbon monoxide poisoning from a fire in the small settee she was found stoically sitting on. Her father, bedridden in an adjacent room, was not affected. Firemen and police are mystified, said the *Birmingham Evening Post* (June 2, 1978). Remarked fireman Andrew Hodgkins: "You would have thought that she would have made some attempt to get out." They wondered, too, how the fire began in the first place.

I look for the unconventional. Could these outbreaks of unusual fires around Birmingham in the 1970s be related causally; connected by some common, unsuspected force that united nine people in a personal yet strangely shared holocaust?

Though my two decades of investigating anomalies has found no one theory to explain the many paradoxes inherent with SHC and PC accounts, one surprising factor does *link* quite a few of these baffling burnings. That factor is geography.

EARTH ENERGIES: POWER POINTS AND HINTS OF FLAME

The Earth spews fiery magma from volcanoes at tectonic plate boundaries and points of crustal weakness. At other points it spits forth a different kind of fire, potentially destructive in some very strange ways.

In Britain, death by pyrophenomena has been a recurrent motif since at least the early 1700s. In fact, my research indicates *Great Britain has more cases of SHC per capita than any other nation*. Thus the British Isles became the ideal place to test a radical new idea about SHC, not only because of the large SHC database required by the hypothesis, but also because it is a center for pioneering research into another, perhaps related, anomaly.

That research centers on the fact that many ancient, sacred megalithic sites (e.g., cairns, burial barrows, and stone rings) can be linked by straight lines sometimes scores of miles long, which often pass through modern places having spiritual or esoteric significance as well. Though the nineteenth-century astronomer Sir James Lockyer noted some megalithic sites formed a grand equilateral triangle (with legendary Stonehenge at one apex), it was Alfred Watkins who pursued these alignments with visionary zeal. Astride his horse on June 30, 1921, Watkins scanned the English countryside from the Bredwardine Hills and sensed a gigantic network of laser-straight alignments blanketing the land. Each alignment he called a *ley*.

The complexity, even the concept, of the ley system raises disquieting questions for archeology about how and why allegedly barbaric Neolithic man could organize, survey, and engineer such a massive project. Seven decades of research have followed, and answers remain elusive—though the leys themselves are well-established throughout Western Europe (to a lesser extent, in the Western Hemisphere, too). Insights into their purpose and their phenomena are constantly being gained, *including both subjective and quantitative documentation of luminous and electrical anomalies at traditionally powerful sites*.

Geologically the earth has a long (though generally ignored) history of emitting unexplained energies. Evidence for telluric currents—which today's geologists define as very *weak* electric currents flowing through the planet's crust, and which the Chinese for thousands of years have called *lung mei* or Dragon Lines—is well established. What is less acceptable to modern geophysics is a range of planetary phenomena that seems best explained by aberrant electrical fields in the Earth that produce powerful luminous, even fiery, enigmas underfoot and overhead.

Accounts of phosphorescent displays from the earth date to 373 B.C. in ancient Achaea, Greece. Sometimes it is associated with seismic activity.[2] In A.D. 1257 the Japanese saw bluish flames flare from fissures opened by the

Kamakura earthquake. In the big quake of November 2, 1931, no fewer than 355 Japanese saw in the sky "rays of a bluish search-light pointing upward from below the horizon" (*Literary Digest*, February 4, 1933).

Immediately after the strong 1872 quake at Cerro Gardo, California, men "whose judgment and veracity are beyond question...saw sheets of flame on the rocky sides of the Inyo Mountains," reported the *Inyo Independent*. These flames, lasting only a few minutes, "waved to and fro, apparently clear of the ground, like vast torches." Harry Clawson was sitting three inches from a cast-iron stove at a nearby mill, rubbing his hands together when the earth shook: "As soon as his hand touched his knee he felt a shock, and immediately after, and for a couple of seconds, a stream of fire ran between both knees and the stove." Clawson probably thought he'd be fried alive in this fiery spew of electrical static, or electrical *something*.[3]

> The world's longest natural electrical current has been discovered looping through the heartland of Australia's outback, encompassing almost one-third of the continent; it runs 30 to 125 mile wide, 9 to 19 mile deep for 3,750 miles.
> —*New Scientist* (March 30, 1991)
>
> This Australian anomaly is about a third longer than the North American Central Plains anomaly, which extends from Wyoming to Hudson Bay.

Earthfire of the nonmagma kind need not be associated with earthquakes and volcanoes. Pliny the Elder recounted in *Natural History* (Book II, chapter CXI) several astonishing examples of fiery springs and "funeral pyres" in the earth; of flames that came "out of a rock that is kindled by rain"; of a field near Arezzo where "the earth is set on fire"; of "a sacred rock" from which "flame at once shoots up" when wood was laid on it.

Between nine and ten o'clock on the evening of June 25, 1875, Charles Gape observed "very narrow streaks of a pale blue colour" darting "up from the earth towards the heavens" near his English home. Before midnight of August 28, 1883, William Noble witnessed "a kind of radial illumination upward" from the landscape beyond his observatory in Sussex, England. Knowing the moon wasn't due to rise for another two hours, he suspected this skyward cone of light resulted from a nearby fire—"but there was none."

On March 29, 1905—a year whose early months were replete with fortean events including pyrophenomena—several people around Cardiff, Wales, saw rising from the earth "an appearance like a vertical beam of light, which was not due (they say) to a searchlight or any such cause!"

On the evening of February 6, 1955, residents of Tucson watched the southernmost ridge of the Santa Catalina mountains glow with flames estimated to be between thirty and fifty feet high along a fifteen-hundred-foot-long line of fire. Next day reporter Earl Zarbin hiked seven miles into the rugged mountainscape to find "something out of science fiction," he reported to *The Arizona Daily Star* (February 8, 1955). "There wasn't enough tinder up there to keep a fire going for an hour and yet Tucsonans watched the blaze for three hours and...there was nothing but an irregularly burned out area of 1,000 square yards."

Cacti leaves were crystallized; other cacti had their roots seared off or had been blasted out of the ground and hurled into nearby canyons. The earth was scorched two and three inches below the surface at some points; large rocks were burned black, indicating intense heat; huge yucca plants had been uprooted and partially burned. It looked like someone blasted the area with a giant blow torch (though analysis proved the absence of accelerants), and he concluded: "There was a fire but what caused it is still a mystery."

It was as if the earth itself had detonated and belched forth flame.

Even more unconventional was an *entire plain* ablaze with many thousands of flames, which jetted from the countryside to the north and east of Nandidroog, India, on the night of June 15, 1872. Each flame, at least five feet in length, was visible from ten to fifteen miles away, reported the *Madras Mail*. The newspaper said this field had similarly inflamed in former years and conjectured

> The existence of man upon a thin shell beneath which mammoth forces constantly operate, cannot be over-emphasized nor is it exaggerated in perspective.
> —Michael Persinger and Gyslaine Lafreniere, *Space-Time Transients* (1977)

these lights were "caused by the ignition of some inflammable gas escaping in jets from the surface of the earth."

Or, perhaps, by discharge of telluric electricity or plasma. Like the inexplicable lights and flares that have shocked people standing near megalithic sites such as Abergavenny, Wales...like the "several small fires with no apparent source" that erupted at sunset for a month in 1965 around an amusement concession at the Gateway Shopping Center in Edwardsville, Pennsylvania; fires that miraculously injured no one (though accompanying bombardments of "pebble-like stones" did shatter a nearby greenhouse)...like the twin mile-wide *circles* of baffling fires I discovered in the summer of 1983 with the assistance of Fire Chief Jerry Grimmett and the Gilbert Volunteer Fire Department, who were stumped by a bevy of combusting chaos (e.g., self-igniting mattress, towels, toys, clothing, Christmas decorations inside a sealed cement-block building) that bedeviled the family of Rev. Gene Clemons in Wharncliffe, West Virginia...like the fiery day and night bombardment between Keklujek and Ziaret, two peaks in Mesopotamia's Taurus Mountains, described by respected climatologist Ellsworth Huntington in the *Monthly Weather Review* (July 1900) as "balls of light" that are "sometimes seen to start from one mountain and to go like a flash to the other."

Could some spontaneous combustion happen because a telluric "hot spot" fires off a blazing energy? Could people, for instance, burn up as the result of unknowingly standing, sitting, sleeping, walking, or dancing over the epicenter of an earthly *fire-quake*?

The idea seems outlandish, yet captivating. It suggests a literal fire of Hades underfoot, a geophysical power that occasionally bursts forth to burn whatever is overhead.

Intriguingly, etymology already supports this radical concept. Watkins noticed the abundance of "Brent" place-names along the alignments he discerned and remarked in his magnum opus, *The Old Straight Track* (1925), that this was an old form of the word "burnt." Watkins further noticed unexpected associations with lights in his philological work: among examples he gave, the *New English Dictionary* says an archaic meaning of *leye* is "flame or fire," and the Sanskrit *lelay* means "to sparkle, to shine, to flame."

Ley, therefore, can connote light; shining; fire; flame. Whereas Watkins saw this etymology relevant only to Neolithic bonfires "reflected in water"—that is, a series of lights glimmering off aligned ponds—I sense there is a more fundamental, even geophysical, concept at work here. I submit that some leys generate at certain points an etheric fire, which, under certain conditions, becomes sometimes visible to the human eye and sometimes a source for igniting whatever sympathetically combustible substance is on the site.

Paul Screeton relates in *Quicksilver Heritage* (1982) the curious series of disasters that befell an Irish farmer *after* he had the audacity to deface (disrupt the telluric power of?) an ancient religious site on his property. After his house burned down, he moved away, only to have his new home soon burn down. Both fires went unsolved. Coincidence? Or synchronicity?

Could it be that Celts, for example, when conducting rites at sites with "Brent" place-names for instance, were more attuned to the natural environment than is today's materialistic society and knew about and strove to mollify the powerful energies beneath their feet and homes—thus safeguarding their lives and property from the ravages of untamed subterranean fire?

Could it be that Birmingham in the 1970s was a focus for anomalous and periodic discharges of telluric energy about which its citizens knew nothing, and which led to a series of perplexing human combustion?

For *whatever* reason, the mysteries that haunted Birmingham in the 1970s are not unique; nor is this city alone in playing host to a concentration of bedeviling blazes.

PECKHAM, LONDON: MORE STRANGE COMBUSTION

Leaving Birmingham to cope alone for the moment with its burning mysteries, we travel southeast to England's capital, where, writes Frank Edwards in *Strange World*, strange fire kept Easter-time appointments with the Graham Stringer family in their modest home in the Peckham district of London.

The uninvited fire-starter arrived about three weeks before Easter 1958, when the baby's rocking chair and toy basket began emitting smoke and left the fire brigade facing a mystery. Good Friday, 1958, left the Stringers with a less-than-holy memory as mystery fire resurrected again, spontaneously. This time the flames ate a hole in a pile of their infant's clothing. In 1959, the phantom flames came out of hiding to strike at Mr. Stringer's shoes, just as he was putting them on! He *literally* had a hot foot (feet) that morning.

As the last week of Lent was being observed in 1960, the Stringers observed the rebirth of their phantom fire-starter: a pile of just-sorted laundry ignited, and in less than two minutes, the garments were ashes. A "luminous smoke" swirled through Mr. Stringer's darkroom. The firemen came and went away baffled, while a priest came to perform an exorcism and simply went away.

The fire went away in 1961. In 1962, it revisited with vengeance, as furniture now self-ignited in the Stringer living room. Scotland Yard suspected the obvious—arson!—until learning the Stringers had no fire insurance and couldn't get any, either!

The Carl Berkenpas family endured a similar fiery persecution from Spring 1979 through February 6, 1981. After eighteen uneventful months in their home in Ashton, Iowa, incandescent lights suddenly began to light themselves; buttons fused to the plastic box that contained them; his child "acted as if struck to the floor;" Berkenpas himself experienced "electric shock" and a sensation of "burning" while lying in bed; the bedsheets themselves burned up.

How to explain these outrages to common sense; to conventional knowledge? Is there truth in the ancient belief that Hades rages in underworld recesses, where unknown forces (quite apart from magma) sometimes unleash fire to haunt and burn the lives of men and women above? Is there more reason to believe that such fire underfoot still inflames in ways beyond man's control?

FIRES UNDERFOOT

"A plumber stuck a shovel into the earth to get at a broken water pipe, in Newman, Calif. Earth flamed. He tried again. More fire. Rubbing a clod with his fingers brought more fire. No gas leaks. Sample sent to State Fire Marshal. No report at press time," reported Tiffany Thayer in a 1952 issue of *Doubt*.

Elsewhere, other fires flare up from the earth below.

"I have been wondering for some time about just who makes so many little fires all around on fields and meadows at every season," mused Horst Freidrich of Bavaria, Germany, in a letter to the Society for the Investigation of the Unexplained, "for in my opinion the number of apparently quite recent (max. several days) places where little fires must have burnt and now only ashes etc. remain is astonishingly large."

Ivan Sanderson, founder of SITU, also had noticed numerous isolated patches of burnt vegetation during his fieldwork as a naturalist. "Frankly," he wrote of these hot spots for *Pursuit* (July 1969), "there are just too many of these little fires, month after month, year in and year out, and apparently all over the world, and...the vast majority of them are nowhere near a road from which an idiot could throw a lighted cigarette, none ever seems to cause enough rumpus to prompt a call to the local fire department, none is seen burning, and there is very seldom any record of lightning at the time. But there they are, just small burned-out patches of ashes."

> At Yanartas, Turkey, the legendary "Ever-Burning Rocks of Lycia" inflame on the slopes of Tahtali Mountain. Mentioned in the fifteenth century, the rocks' flames remain impervious to wind and precipitation; try to douse the fire, it will reappear nearby... but you *can* cook a meal over these firestones.
> —*Bugün PS Magazine*

> A thirty square foot area of sod in Zimbabwe's Masekwe Valley suddenly begins to burn—posing a threat to any person and animal which trod upon it.
> —*Harare Herald* (July 20, 1993)

> The Geological Survey Department of Sri Lanka announces discovery of a 300°F ground temperature—highest ever recorded in the country—in a dry river bed exuding smoke. Geologists are baffled; the military cordons off the smoking ground to keep away the sight-seers.
> —*AP* (September 4, 1992)

As fate would have it, the same month Sanderson published his thoughts the earth itself inflamed in Clifton, Tennessee. This time, in late July 1969, people saw flames flicking up out of the ground, flames that proved nearly fatal. "I've never seen anything like it," confessed Perry Davis to the *Sacto Union* (July 28, 1969). "There are burned leaves all over the place," he said of his farm. "Blazes will jump out of the ground for four or five inches. It looks like the earth has been scorched from the ground up."

The phenomenon drew crowds from miles around to stare at this two-acre twilight zone of combustion. Davis posted signs to warn of danger. The foolhardy ignore signs: W. J. Baker and another man walked onto the burning earth, which collapsed under their footsteps and plunged them into a subterranean oven from which they barely escaped alive.

Various theories blamed lightning; spontaneous combustion of underground charcoal from a Civil War pig iron furnace; and trapped gases, which state forest officers said "could be very dangerous." Perry ruled out each theory.

Less than a decade passed by before the next subterranean blaze attacked... this time roots of trees on the property of Fred Bradford, in Central Butte, Saskatchewan. The enigma, accompanied by an awful odor and shockwaves, forced the Canada family to move out and, according to *INFO Journal* (no. 31), left authorities "completely baffled."

Officials, failing to explain why two spots of earth burned hot, recommended letting the fields burn themselves out. The damned shall die of neglect—

As proved to be the case for Fred Bader of San Antonio, Texas. "Attacked By An Electric Flame" headlined the *Shreveport Times* (November 18, 1880), about this eleven-year-old lad. On November 11, 1880, while sitting indoors, he suddenly was enveloped by a "ball of fire" that burned his eyebrows, singed his hair and cap, burned off his shirt, then "passed around the house to an irrigation ditch and was lost." Bader suffered pain from injuries and blindness "produced by electric fire," said his doctors.

Did Bader "spontaneously" burn from sitting upon an earthfire unleashed underneath his chair? Did the Berkenpas home sit upon a telluric hot spot? Did the Stringers live atop another spot in a telluric fire grid, whose fury was timed not by a Christian festival but by a peculiar combination of terrestrial and/or cosmic lines of force combining in just the right way (like saltpeter, sulphur, and charcoal) to explode forth in combustion? Had Mr. Stringer known as much about energies as I believe the ancient Britons knew, he might have chosen to heavily insure his home against fire; in fact, he probably wouldn't have lived in that house at all.

Remember eleven-month-old Peter Seaton who burned in his crib in early January 1939 in Peckham Rye, London? It would be interesting to know the proximity with which the Seatons and Stringers lived, two decades apart.

THE FIRE-LEYNE THEORY: MAPPING THE SITES OF SHC

Line of fire is a phrase usually associated with conflicts on the battlefield. *Fire line* denotes a defense against the advance of a crackling holocaust of lands aflame. *Fire lane* designates a portion of a highway that must be kept open—that is, unblocked—to allow access by fire-fighting equipment. I proposed in 1976 another meaning to these words, one that captures both combustive and destructive aspects in a linear fashion.[4]

This new concept is the *fire-leyne*: a specific type of *ley* with energy that can generate *fire* in objects sited upon its pathway. Fire-leynes may be ground-based (I'll call these *telleynes, tell*uric = earth) or above ground (*aerleynes, aer*ial = air/above). The distinction between a *ley* and a *leyne* is that the former represents a geometric alignment, possibly having an energetic factor; the latter is *primarily* an energy flow indigenous to the earth that can manifest linearly on or near the surface.[5]

I later learned about an ancient tradition called the *Secret Arrow*, a condition of hyper-accelerated telluric energy the Chinese feared. If its flow was too fast and too arrow-straight then things burned out through hyperactivity. I submit that this is not mere allegoric lore but can be, when conditions are right, a *literal inflaming* of physical objects.[6]

Are fire-leynes anything more than an imaginative theory, though? There is nothing like practical experience to drive home a theoretical point. I decided to put the idea to the test.

With several British maps, a good gazetteer, and an index of SHC cases, I created in 1976 a cartographic countryside filled with a maze of dots designating the fifty-one places in my database where strange fires had at various times baffled the townsfolk. Now the straight-edge, quickly. Any two points establish a line. I looked for lines that connect three dots, hence three cases. I find several three-point alignments. Of the fifty-one cases of pyrophenomena known then, a total of forty-three—*84.3 percent*—fall upon at least three-point fire-leynes! (See fig. 22-1.)

But Watkins (like most ley hunters) wanted four-point evidence, below which number coincidence might creep in. Would hours at the drafting table now be reduced to an exercise in futility? No! A four-point fire-leyne is found; then another. Then *five-* and *six*-point alignments are found. Finally, the straight-edge revealed along the North Sea coast an *astonishing alignment spanning more than four hundred miles through towns harboring ten (perhaps more) probable mysterious blazes*. I designated it Fire-Leyne I.

BLAZING THE TRAIL: EVIDENCE FOR FIRE-LEYNE I

It defies coincidence that incredibly bizarre fires spanning more than two hundred years can be joined by arrow-straight trackways. At the very least, the cartography of combustion is a concept sure to fire the imagination.

Let's explore Fire-Leyne I briefly... and the mystery blazes that define it. We begin with its northern terminus (as presently known to be) and move southeast along Britain's eastern coastline.

EARTHFIRE! THE CARTOGRAPHY OF COMBUSTION

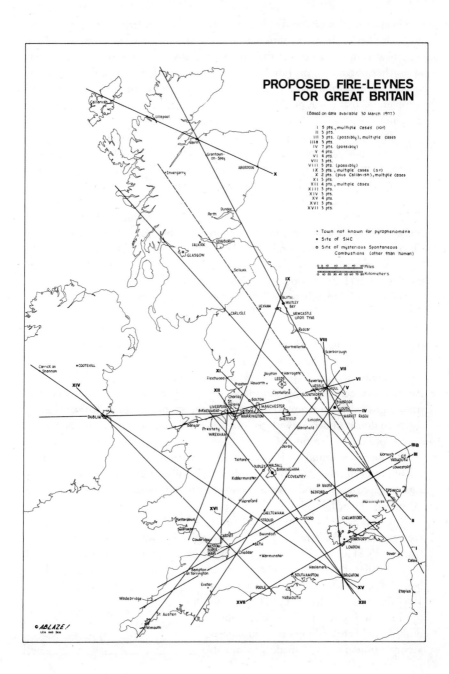

• POINT 1: A weathered carter of wood, John Anderson, fifty, notorious dram-drinker of short and slim stature, rode into a fiery death just beyond the tollgate at Harmuir on the Darnaway–Nairn tollway, seven miles outside Nairn in northern Scotland, on July 29, 1852. He tumbled, already ablaze, from his wagon filled with wood and straw. Upon the ground he quickly burned so completely "that it was doubtful if he could be lifted without falling to pieces," *The Abstainer's Journal* (March 1853) would later report.

Nairn county medico-legal officials requested that Dr. John Grigor inspect the body and report back on this curious incident. Dr. Grigor fulfilled his charge and testified that the charring of combustion was so thorough "the person was only recognised from his horses and carts being known." He subsequently submitted his detailed findings to *The Monthly Journal of Medical Science* (1852), in which he described Anderson's remaining anatomy as "a black, incinerated, and stiffened body." The head and upper body was most horribly burned; the back "not so much destroyed." There is no indication of incinerated limbs or midsection.

Dr. Grigor's interviews uncovered the fact that Anderson had wanted his pipe lit and had asked a sober friend to do it for him. The friend, not thinking this wise to do, "merely put a little fire on the old tobacco ash," and, after one draw on it, Anderson complained the pipe went out. The two men conversed for ten minutes, then the friend turned to walk homeward. Anderson, cold pipe in hand, drove off in his wagon. The friend later emphasized that Anderson seldom lit his own pipe and almost never carried matches.

Moments later, a shepherd saw smoke rising from Anderson's cart; saw Anderson descend from his cart, stand, then stagger and fall. In a few minutes, smoke gushed forth from Anderson, and the shepherd and a woman joined forces to douse Anderson with pails of water. All this, Grigor determined, took "not more than 15 minutes."

The pipe was found with its cap on, below the body, indicating to Grigor it had not been relit. Anderson's clothes "were all consumed, except the lower legs of the trowsers" and a hat and bits of fabric on his back. "There was none of the hay burned."

Not unexpected was the fodder Anderson's demise offered to some political agendas. For example, *The Abstainer's Journal* was quick to imply that Anderson's SHC was divine retribution for his succumbing to "the demon of Intemperance by flaming up as does the vile liquor which permeated his body." Nickell and Fischer contended in 1984 that John Anderson didn't succumb to SHC for any reason because he, obviously, ignited after falling on his lit pipe— a contention not at all made obvious in

> They've fantasized away the facts.
> —Abigail Schearer

contemporary documents such as Wharton and Stillé's *A Treatise on Medical Jurisprudence* (1855); and that he was only superficially burned—again, unsupported by original evidence.

I present that original evidence, which is rather instructive, in Dr. Grigor's personal remarks:

> The case at first sight appeared to me to have arisen from the clothes having by some means caught fire, and the smoke therefrom producing death by asphyxia—the subject being much intoxicated; but second thoughts demonstrated a few points not reconcileable [sic] in my mind with this view, such as the position on the back, etc.—the event taking place in the open air—rigidity of the limbs—no trace of fire—and the rapidity and extent of the combustion, whilst this latter (compared with the accounts of the martyrs, the suttees, and others who have been consumed, and the great quantity of fuel and the time that have been required) and no apparent struggle or attempt having been made to cast off the burning garments, or to quench the flames in the brook running alongside, whilst the man was not at all in a state of insensibility from his potations [alcoholic drink], led me to the belief, that it was no ordinary combustion from the application of fire. I have then been induced to regard it as a case of progressive igneous decomposition, commencing during life without the application or approach of any hot or burning body...

Dr. Grigor had coined a new phrase—progressive igneous decomposition—for what he said other respected physicians called spontaneous combustion, or *Catacausis Ebriosa*, of the human body. Though referring to Bischoff's unsuccessful experiments to burn up alcohol-soaked tissues, Grigor thought alcohol might do the trick, given enough time. Dr. Grigor concluded there was *far too little time* for John Anderson to die this way.

Cause of death for John Anderson? Said Dr. Grigor: "It must be an animal body converted into a combustible body."

Smokin' John Anderson forms the northern terminus of Fire-Leyne I—or to use Watkins's jargon, "the beacon point" on a fire-laden south-south-easterly trackway.

One beacon point shining with the light of a spontaneously combusted body, decreed medically. There are more points, more unsuspecting casualties.

• POINT 2: The trackway enters England and the coastal town of Blyth. This seaside community harbors more than boats: *several* baffling blazes have berthed here, anchored on the fire-leyne that extends from Nairn, 195 miles to the north-northwest.

First, Barbara Bell. On February 23, 1905, smoke issued from beneath the seventy-seven-year-old widow's door. Her daughter groped on her hands and knees through the smoky house until she bumped into her mother, who was "quite dead then." A deputy coroner pronounced the body "fearfully charred."

After superficial examination of the disconcerting facts, the inquest jury ruled "accidental death" because, *The Blyth News & Wansbeck Telegraph* (February 28, 1905) reported, "it was quite clear that the poor woman had accidentally come in contact with the [grate] fire"

This "quite clear" finding contradicted every one of the facts given by witnesses at the scene; yet it was accepted unquestioningly. SHC would have fit all the facts; but SHC is *unacceptable*, unquestionably.

A month plus three years later, a second affront to normalcy appalled the citizens of Blyth. (The event has remained unknown, lost to history, until I came upon it while reading through newspapers at The British Museum.) This time it's Charles Scholl, a German seaman in his mid-fifties for whom Blyth would be his last port-of-call.[7] Scholl had asked his landlady to awaken him the next day. This she tried to do Tuesday morning, March 31, 1908. But to her shock, said *The Blyth News & Wansbeck Telegraph* (April 3, 1908), she discovered his "bed was in flames...whilst upon the burning bed still lay Scholl." According to a doctor called to the scene, the sailor was "badly scorched about the hands, feet and body."

Because the fire was so localized, limited to the bedchamber only, the fire brigade had no worries *except* to explain the origin of the fatal blaze. "The idea of the deceased having been smoking is dispelled," ruled the fire brigade. "How the fire originated is unknown," said the newspaper in a journalistic effort promoting truth.

As rare as perplexing human combustion incidents are, more perplexing is that *two* occurred in the same small town within thirty-eight months! Yet there's more.

• POINT 3: Seven miles farther, the line of combustion zings over Whitley Bay. A few days before tragedy struck Scholl in neighboring Blyth, a similar fire sprang up and forever separated the Dewar spinsters.

On Sunday evening, March 22, 1908, Margaret Dewar entered her home and called out to her sister. No answer. She went upstairs, only to flee in horror. Neighbors responded to her screams and discovered upstairs the charred body of Wilhelmina Dewar—on a bed showing no trace of fire. No sign of fire elsewhere, either.

Margaret testified at the inquest that she found her sister under the circumstances corroborated by some of her neighbors: aflame in the upstairs bed. This did not satisfy the coroner. He called upon a constable, who testified that Margaret had been terribly intoxicated—no such accusation had been made when she stood before the witness stand—and therefore she could know neither what she saw nor said that fateful Sunday evening. The coroner urged Margaret to reconsider her words; she would not. He plead with her to recant; she refused. Margaret was a retired school teacher; she prided her reputation. She would not perjure herself.

The coroner adjourned the inquest until April 1, when Margaret suddenly changed her story, to appease the coroner, it's been said; to appease the townsfolk, who mocked her story; to appease the media, which ridiculed her; and to appease anyone else opposed to her outlandish story. She apologized. She had been mistaken and foolish, she said on this April Fool's Day. She had, she now said, instead discovered her sister burned but alive on the ground floor of their home, and had helped her upstairs and put her into bed, where she then died.

The coroner was satisfied, even though this version failed to address the absence of fire damage on the ground floor; the absence of a trail of cinders falling from Wilhelmina's clothes and body en route to the bedroom; and the absence of soot, soiling, and stains on Margaret's clothing as she supported her shockingly burned sibling up the stairs (*unless* Margaret had murdered her sister and then tidied up both herself and the house before running out to feign alarm in front of her neighbors). The coroner saw no reason to pursue criminal charges. That meant there was no final justice in the fire that killed Wilhelmina, either.

To quote *The Blyth News & Wansbeck Telegraph* (April 3, 1908): "The Coroner, in summing up, said they were in the same position as at the opening of the inquest. They could not tell how the accident happened"—so the jury settled on death from shock and accidental burns.

This solves nothing. *What* was the accident? *How* were the burns generated? *Why* could no one testify to finding burnt debris on the ground floor, thereby sparing poor grieving Margaret from the coroner's badgering?

I'm tired of reading "accidental death" when *so much more* is at stake. "Accidental death" is a confession of ignorance, pure and simple. I think the nature of Wilhelmina's "accident" involved an encounter between herself and a line of fire blazing its way through Whitley Bay.

And like its sister town to the north, Whitley Bay's involvement with enigmatic fires didn't end with just one incident.

I find another. It's a bit different; gives a change of pace.

In that year of mysteries 1938, when things and humans were combusting all over Great Britain, a man wearing a trilby hat walked along Whitley Bay

Promenade at seven o'clock, October 23. A lady shouted "You are on fire." And lo! to his amazement, he raised his hat to find it smoking. A firecracker set off by mischievous lads is to blame, presumed *The Newcastle Journal* (October 24, 1938). No exploding firecrackers were seen or heard, nor were irresponsible boys arrested—British bobbies can't capture what isn't seen or heard.

"Accidentally" the hat ignites. From what? Did a fiery meteor strike the chapeau, which saved its owner from a nasty headache and scorching by fire from heaven?

Maybe clothing *first* ignites inside telluric zones of combustion. SHatC, perhaps? The trilby hat is not alone in the annals of attire spontaneously afire:

• One day in January 1959, an Iranian fire chief battled eighteen separate fires in his own Tehran house, all due to causes which could not be determined. In one particularly annoying outbreak, Fire Chief Hussein Meshauyekh's underwear "caught fire—while he was in them."

• On July 16, 1960, the clothing on Rose Howes ignited and then transformed her into a human fireball in Barre, Massachusetts, according to *Weird Unsolved Mysteries* (1969).

• Next year—October 19, 1961—eight diapers hung out to dry on Mrs. Jack Roll's clothesline in Evansville, Indiana, instead charred mysteriously. Said Edward's *Strange World*: "no fire...no lightning."

• On August 11, 1970, the staff at Fordham Hospital in New York City responded to a different kind of emergency: exploding clothes. Specifically those belonging to Asit K. Ray, thirty-nine. It happened once before he was admitted, then his garments "exploded" twice more inside Fordham. Probable cause is chemical, reported the *Daily News* (August 12, 1970), adding cryptically that the "explosions occurred between skin and clothing."

• On February 5, 1971, Erling Bakke suffered first- and second-degree burns after his coat inflamed for reasons "undetermined" at 10:30 o'clock on this Minnesota morning. The fifty-two-year-old man wasn't even aware of his predicament until someone told him, reported *The Minneapolis Star* (February 5, 1971).

• Adolph Heuer Jr. remarked on the time a wet chore coat burst into flames before his eyes. "It has always puzzled me as to how wet cloth can 'burn'. After all," he wrote to SITU in 1976, "you can use wet cloth to put *some* fires out!" In Whitley Bay a hat burns, without injuring its wearer.

The threads of evidence I call clues to SHC are now found in fabrics, too. Coats and clothing that flame spontaneously; in Whitley Bay, a hat ignites spontaneously. Through Whitley Bay passes a fire-leyne, undiscerned except by the phenomena of its passing.

The alignment of fire, now two hundred miles long, stretches still farther.

• POINT 4: Fire-leyne I intersects *dead-center* with Kingston-upon-Hull, where four fire-leynes intersect. A fire-leyne hot spot, one could say.

Elizabeth Clark could not say *what* it was that severely seared her septuagenarian flesh about 6:00 A.M. on January 6, 1905, at the Trinity House Almshouse on Carr Street, Hull. "She was unable to give an articulate account of the manner in which she received her dreadful injuries," stated the *Hull Daily Mail* (January 6, 1905). Perhaps the impoverished widow could not explain her predicament because it baffled her as much as it did the investigators. She had, after all, felt no pain, and her room and even her bed were unburned. I suspect no one in Hull knew that humans could spontaneously combust, and that this victim appears linked to a similar fiery death fifty-two years earlier in Scotland, three hundred miles away.

(Remember: it was at Hull where G. A. Shepherdson burned mysteriously in his truck on April 7, 1938.)

• POINT 5: Hull was not alone in contending with the problematic in early 1905. The hospital at Louth, thirty miles to the south-southeast in the midst of the Lincolnshire Wolds, was dealing at this time with two oddly burned patients. One was a farm girl from nearby Binbrook; we'll discuss her later.

The other patient was Ashton Clodd, seventy-five, stricken fatally at the end of January 1905 by what *Lloyd's Weekly News* (February 5, 1905) called "mysterious burns." Allegedly he fell onto a fire grate, though no one testified to seeing hot coals therein. Ashton—a name of combustible significance?—is not the *only* person to have burned up in a fireplace devoid of hot embers.

Fire-Leyne I goes right past Louth; precisely, on my map it tracks about one mile west of the town. Perhaps Mr. Clodd 's home was about one mile west of Louth proper and directly atop the fiery trackway in question. I can't say for sure.[8] I do know that I had plotted Fire-Leyne I *before* finding Louth to be in Lincolnshire; before assigning Ashton Clodd to its fifth point. A nice confirmation of theory—

Let's review. We find *eight cases* of unusual, enigmatic fires in towns along a 339-mile straight line from Nairn to Louth. Watkins would surely have viewed the discovery with excitement. Furthermore, three cases cluster in the first two months of 1905, and at least two more are compressed into *one month* (March 1908).

The Nairn-Blyth-Whitley Bay-Hull-Louth alignment may include even *more*.

• POSSIBLE POINT 6: Paging through *The Blyth News & Wansbeck Telegraph* for March 6, 1908, I discovered that three days earlier Mrs. Elizabeth Nicholson, forty, of Whitfield, Northumberland, was found "in flames" in a shed. Unstruck matches lay near her lifeless body. Further details are absent. It may be carelessness; may be suicide; *may* be spontaneous human combustion.

• POSSIBLE POINT 7: Another fire in Northumberland—exact location not given, perhaps near the coast—occurred in a "working man's club... in an industrial town" and baffles two fire experts. The blaze is thought to have two seats (that is, two separate points of origin), and investigators J. Anderson and H. J. Walls find this "rare enough" to write about for *Medicine, Science & The Law* (1961–1962).

Fires broke out simultaneously in two areas of the club, a disconcerting situation especially after the investigators had already ruled out arson. An elaborate hypothesis is then constructed: in essence smoldering cigarettes in one room inflamed, whereupon a column of hot air rose up a partition and caused paint on the other side to "flash-over," whereupon flaming gases and burning particles were convected along the other room's ceiling, down the far wall and into the window curtains, which subsequently ignited and, in turn, burned through a duct cover allowing the fire to spread to the upper floor and roof. Whew!

There's a problem though, which the investigators note with apparent embarrassment: "a rather surprising" circumstance is that the ceiling and exposed insulation show "no signs of intense heating"—in other words, the surprising fact is that what was proposed doesn't fit what was observed.

Two fires, and nothing connects them physically. Great heat, isolated seats of fire, localized initial damage, the combustion of fire-retardant materials, and perplexed authorities: the *very circumstances* associated with the type of pyrophenomena being discussed here.

I propose a different supposition: that the club is sited upon Fire-Leyne I, whose activation this night singled out two objects on the ground-level floor of the establishment through which the telleyne passed and over-energized their atoms, causing atomic thermal heating which culminated in spontaneous combustion.

• POSSIBLE POINT 8: Christine Arnold (no relation), forty-seven; daughter Sarah, seventeen; and Sarah's husband Paul Jupp, twenty-one, moved into Spinney Cottages, Rougham, in June 1989. The Arnold home is said to be "near Bury Saint Edmunds" in the West Suffolk countryside. Unspecified "ghostly happenings" commenced almost immediately, usually around 9:20 P.M. A vicar, invited in by the frightened family, sensed "a bad feeling" around a clothes closet. On November 5, around 9:20 P.M. and just before the family returned from an outing, fire broke out in that very closet, destroying its contents and producing heavy smoke damage. Clairaudient and psychokinetic phenomena ensued, reported the *Bury Free Press* (November 10, 1989), but apparently there were no more fiery outbreaks.

Bury Saint Edmunds is four miles west of Fire-Leyne I. Having uncovered this incident the night before this book is to be delivered to the publisher,

I won't have time to find tiny Spinney Cottages on a map of England. You might want to do that for yourself.

• POSSIBLE POINT 9: Remember Grace Pett, the fishmonger's wife found "consumed by a fire without apparent flame" at Ipswich in 1744? Ipswich is six miles east of Fire-Leyne I. Contemporary accounts indicate that Grace Pett lived in the countryside, not in Ipswich proper. If she lived some six miles west of downtown Ipswich, perhaps along the road to Hadleigh (*ley?*), then her white-ashed body could be assigned as *the eleventh uncanny blaze* on Fire-Leyne I that would now measure at least 430 miles long!

Scientific inquiry becomes possible when a phenomenon exhibits patterns and constancy that can be classified and studied. Uncanny blazes that baffle investigators and are ignored by every specialization within modern science, point to patterns of alignment on the earth upon which people and possessions spontaneously inflame.

Certainly, everything upon these alignments does not burst into flame. If the statistics that underpin the hypothesis of fire-leynes indeed reveal a heretofore unsuspected, unseen external force capable of igniting people and places, there must be a very precise and complex set of circumstances to trigger ignition. Circumstances that fortunately occur infrequently, it would seem.

There is more to consider, cartographically and conceptually, as the mapping of combustion paths disclose additional puzzles and new revelations.

23

TRIANGLES OF FIRE: PUTTING FIRE-LEYNES TO THE TEST

> Establish the triangle and the problem is two-thirds solved.
>
> —Plato

Early 1905: the lonely Yarn Walk farmhouse belonging to Mr. White in the Lincolnshire Wolds, near Binbrook. Uncanny occurrences. Things fly about. Other things, including plates, disappear—for a while. A pan of milk spontaneously overflows when no one is near it; in the pan is found one of the plates that had disappeared. Rev. A. C. Custance of Binbrook Rectory writes to the Society for Psychical Research, saying *things are bursting into flames!*

There is a young girl. Aha, the culprit! This time one must reconsider such an "obvious" perpetrator of the uncanny, because this servant girl at the Yarn Walk "whilst sweeping the kitchen, was badly burnt on the back."

Mr. White gave this account to the *Louth & North Lincolnshire News* (January 28, 1905):

> Our servant girl...was sweeping the kitchen. There was a very small fire in the grate: there was a guard there, so that no one can come within two feet or more of the fire, and she was at the other end of the room, and had not been near [it]. I suddenly came into the kitchen, and there she was, sweeping away, while the back of her dress was afire. She looked around, as I shouted, and, seeing the flames, rushed through the door. She tripped, and I smothered the fire out with wet sacks.

Triangles of Fire: Putting Fire-Leynes to the Test 363

Taken to the Louth hospital, the injured girl was visited by a reporter on January 27. He verified her now-painful critical condition, and remarked that "She adheres to the belief that she was in the middle of the room, when her clothes ignited."

If the youngster was the culprit in the Yarn Walk mysteries, she also seemed guilty of SHC.

The newspapers had a field day with rural Binbrook. The esteemed *Liverpool Echo* (January 26, 1905) chastised the peasantry for their "firm belief in the resuscitation of witchcraft" and acceptance of "extraordinary terpschichorean [sic] performance" by objects that are supposed to lie still where they are placed rather than dance about. To the editorial chagrin of the *Echo*, a reporter sent to debunk these silly tales found them to be unnervingly true.

Weird fires *did* break out. So too did other things, like the spontaneous breaking of chickens' necks on the farm. A ghoulish force was on the loose in farmer White's chicken coop. Out of 250 fowls, 226 had their skin pulled off and their necks twisted and their windpipes drawn out and snapped, despite Mr. White testifying that his chicken-coop had been watched night and day.

The Yarn Walk near Binbrook is only four miles from Fire-Leyne I as it runs its straight-line course from Nairn to Louth—another example of Watkins's principle that ley plotting leads to unanticipated revelations (which is what a successful scientific theory should do). As I will show, that distance is as significant as if the Yarn Walk sat atop the fire-leyne.

More curiosities. Seven miles to the west of Binbrook is Market Rasen, a town situated at the intersection of two proposed fire-leynes. On January 16, 1905, a chicken coop and fifty-seven fowls burned up there. Not noteworthy, you think? I note that no one discovered how the fowl-house caught alight, and that only a few miles away Mr. White lost 90 percent of his chickens in another weird way within the fortnight.

The *Echo* admitted to "some curiosity as to the force which made the pots and pans jig about the kitchen, and we should certainly like to know the truth about the servant's burnt shoulders." So might the doctors at Louth hospital, who in a little more than a week would be treating Ashton Clodd for burns of a similar and equally mysterious nature!

Locally mysterious fires; then localized floods of water. Nature balances one phenomenon by overcompensating with the excesses of another?

Many writers have lumped the Binbrook-Market Rasen episodes into the very broad category of poltergeist outbreaks centered around adolescent females. Consider a different avenue of thought: lines of force, telluric energies, and fire-leynes that channel energies that manufacture poltergeist-type outbreaks.

Nature did seem to go berserk in 1905, as waves of SHC and other paranomalies washed over the shores of this reality. Was this because the Earth's ley grid, over-charged that year, discharged its excess potential at power points or through zones of weakness along the lines of telleyne transmission? And did Lincolnshire find itself in the *midst* of one such anomaly area?

"Evidence is slowly building up that there is indeed a correspondence between types of Fortean phenomena and proximity to Leys," observed Robert Rickard, probably the world's foremost fortean scholar.

Binbrook is located near the center of a triangle formed by three fire-leynes: on the east by Fire-Leyne I (discussed in the preceding chapter); on the west by four-point Fire-Leyne V passing through Lincoln and connecting pyrophenomena in Market Rasen, Coventry, Cheltenham, and Stroud; on the south by Fire-Leyne IV, spanning the heartland of England and perhaps connecting six paranormal combustion sites in Louth, Market Rasen, Sheffield, Warrington, Liverpool, and Birkenhead, plus possibly a seventh episode involving Mary Hogan's SHC in Dublin, Ireland—a total distance of 260 miles.

This fire triangle at Binbrook *is* equilateral, each side 12.5 miles long; its area about 68 square miles. And the Yarn Walk farm is *almost precisely opposite each angle* of the fire-leyne triangle that surrounds it.

As light and sound can be gathered by a receiving cone and concentrated in a powerful beam along its axis, might the energies producing the phenomena at two of the triangle's apexes (Market Rasen and Louth) have concentrated across from those apexes? Therefore, the calamities on White's farm would not be associated *directly* with the Nairn-to-Louth fire-leyne—that is, not being atop the alignment itself—but resulted from the focus of two fire-leynes discharging their power simultaneously into the *center of the fire triangle*, where "X marks the spot" of their crossing.

Let's hypothesize.

Late December 1904: fire-leynes around Lincolnshire become active, as energy begins to flare. A pan jumps off the shelf in Mrs. White's kitchen; pots spin about. Mid-January 1905: the leynes are crackling with power as huge energy potentials begin breaking through the boundary between the unseen and the physical. A chicken coop and fifty-seven hens at Market Rasen spontaneously combust. Late-January 1905: fire-leynes can no longer constrain the tremendous forces flowing through them in the Lincolnshire area. At their angles of intersection vast overloads arc, to unleash titanic bursts of telluric energy toward the point opposite each apex. The fabric of reality is rent asunder around Binbrook, as White's farm weathers the brunt of an invisible, telluric storm as violent as a tornado. Pots terpsichorate; plates teleport; etheric flames spontaneously sear objects of wood and of flesh; floods of water condense out of the ethers; vortices of frightful strength attack chickens and turn them inside-out.

Then the fury abates; balance is being restored. The southeast corner of the Binbrook triangle flickers as its energy flow subsides: Ashton Clodd's frail old body happens to be in the inauspicious spot at that time, and over-energized bioenergy sears him from within as the reordering process continues. He dies in Louth Hospital, never knowing what hit him or that another patient there was struck down by the same kind of firestorm when it passed through farmer White's kitchen.

The fire-leynes dissipate their concentrated power away from the Binbrook Triangle's center now, along a network that stretches for a hundred miles or more. There may be other tottering, fragile souls living along these pathways whose fiery fates remain undiscovered. Late February 1905: the dispersing energy flares at Blyth, and aged Barbara Bell ignites on her kitchen floor.

Then normalcy is restored. The inflammatory holocaust that raged, mostly linearly, through northeastern England ends—for the moment.[1]

THE PLOT BECOMES MORE COMPLEX

Reasoning along the macro-to-microcosm principle of "As above so below," an extension of the Binbrook triangle concept might provide clues to the global horde of anomalous phenomena in 1904 and 1905. That as there may be *leynes of the world* invisible to geophysicists, so too could there be *leynes of the cosmos* undetected by astronomers but nevertheless vital to the maintenance and functioning of the universe. Their angles of intersection form zones of interplanetary (or interstellar or interdimensional) disturbance, through which one can pass and be affected by the energies of a planet or a star or a galaxy, depending on the location and power of the cosmic-leynes encountered.

Whereas Binbrook became the focus for energies requiring one month to peak and another month to dissipate (approximately December 31, 1904 through February 28, 1905), the Earth as a whole needed nearly two *years* to pass through an area of tortured space and time in the cosmos.

The concept of unusual blazes along the bisectors of triangles of fire is revealed elsewhere on the fire-leyne map of Great Britain. For example, Fire-Leyne II with three points of pyrophenomena at Ipswich, Brandon, and Carlisle, intersects Fire-Leyne I. Bisecting the angle formed by these two lines results in a meridian that cuts right through Market Rasen.

Fire-Leyne I also passes very near another site of weird fires—Scunthorpe—where Terry Nelson, thirty-three, would be treated for leg burns after his mattress ignited underneath him at his home on Digby Street on October 18,

1974. Cause is unidentified, reports the *Scunthorpe Evening Telegraph* (October 19, 1974). An SC survivor, Harrison calls him.

Fire-Leyne I passes even closer to Hexham. Hexham? *The Newcastle Journal* reports that a husband found his wife, Isabelle Purvie, in flames on (presumably) October 25, 1938. Her charred body was "a mystery which haunted her survivors" and for which the inquest jury could find no solution. Indications of SHC. The coroner pronounced the official verdict—which you undoubtedly have already anticipated—"Death from shock as a result of burns accidentally received." About as illuminating as a candle in a bucket of water—

If the coroner had considered the *location* of Mrs. Purvie's death, which happened along the meridian bisector of Fire-Leynes I and II, his verdict might have been decidedly progressive regarding a plausible cause of her shocking "accident"—another case of being at the right place at the right time for SHC.

The intersections of fire-leynes may be a focal point, a hot spot, for pyrophenomena. For example, Fire-Leynes XV and XVI converge at Newchurch, a small community along the east border of Wales, where, you'll recall, back in the 1880s a Welshman's body burst into blue flames. I did not know about this case when the map of fire-leynes was plotted in 1975— another confirmation of theory— and I discovered the connection when checking a gazetteer for the whereabouts of Newchurch, only days before delivering the updated manuscript to the publisher.

Should anyone doubt the geometry behind the cartography of combustion, maps and newspaper articles are available in libraries to anyone who wishes to avail himself of the secrets they hold. Go, do the work yourself. Science is awash with examples where a few hours of research have negated decades of defamation dished out by those who reject discoveries without first looking for the facts. The real excitement of science lies in discovering new things, not in defending the status quo.

> St. Neots and Papworth fire crews dash to a house in Graveley, where at about 5:15 A.M. on January 13 a pair of socks caught fire in the bathroom. They are too late—the fire was already out.
> —*Cambridge Evening News* (January 13, 1993)
>
> St. Neots is cradled by the three-point intersection of Fire-Leynes IIIa, XIII, and XI in Cambridgeshire.

A Crescendo of Combustion in Cincinnati

"There is nothing so far removed from us as to be beyond our reach or so hidden that we cannot discover it," affirmed the philosopher-scientist

Descartes. In that spirit, let's reach across fifty-three hundred miles from eastern England to America, where another discovery might lie hidden.

November 16, 1976, 8:00 A.M.; 1711 Harrison Avenue, Fairmount, Ohio. Engine Company 21 arrives at the basement residence of Mrs. May Caplinger. They find the fifty-nine-year-old woman lying on the floor, her clothing and the rug beneath her still aflame. "Nothing else in the apartment was burning," said a firefighter to *The Cincinnati Post* (November 17, 1976). She was covered with second- and third-degree burns over 70 to 80 percent of her body. "She was apparently cooking breakfast... when her clothing caught fire" announced a doctor who tried but failed to save the victim's life.

Firemen spent hours at the scene and found nothing rational, nothing identifiable as to cause; yet a physician, not at the scene, identifies the cause precisely—"apparently." Shame on the fire department for overlooking a hot kitchen stove with overcooked breakfast on it!

November 17, 1976, 12:12 A.M.; 2534 Liddell Street, Fairmount, Ohio. A subterranean inferno rises from the basement of the Preston Sandlin family, routing all eight members into the frigid night air. Fire Marshal James Eversole is mystified: "I'm going to check today to see if we can come up with a reasonable explanation..." His investigation begins with the Sandlin children. Children are almost always suspect when found near fires—they are said to be not only careless but notoriously clever with matches.[2] This time children are absolved of guilt. Twelve hours after the fire's outbreak, the fire marshal still had not found an explanation.

Two weird fires, originating below ground level within a sixteen-hour period in Cincinnati's Fairmount district. Heat is generated when energy of a predescribed pattern reacts with other energy patterns.

Did Mrs. Caplinger react to a peculiar telluric energy underfoot that created within her body a volcano of fire called spontaneous human combustion? Did the same quirky, fiercely heated incendiary energy pattern focus on the Sandlin household—in a house with its own peculiar energies?—and erupt as nonhuman spontaneous combustion?

An over-charged fire-leyne hot spot beneath Fairmount climaxes in a crescendo of combustion and a flash of flames and, once discharged, becomes quiescent again. No more problematic blazes.

What do you think? The Cincinnati fire investigators sounded mystified. But then firemen haven't been trained to consider geology, telluric energy, and fire-leynes when investigating fires. In truth, they generally are not even aware of the weird blazes you have been reading about.

Putting to the Test the Fire-Leyne Theory

I have been taught since elementary school that scientific methodology involves formulating a theory and then testing it. A theory's merit can be judged on (1) repeatability, and (2) ability to predict future events in accordance with the theoretical model. If the fire-leyne concept has merit, it should reveal the existence of unknown data that fits its model.

Following a talk about SHC that I presented to the Rev. Donald Galloway's international symposium at England's Arthur Findley College in 1979, a woman from Hull (where four fire-leynes converge) was moved to confide a very curious story from her childhood. Always on two days each year, she told me, her mother insisted that the front and back doors to the ground-floor hallway (which ran the length of their house) remain open *or their home would catch fire!* Mere superstition? Or a residual memory whose purpose, now forgotten, was to warn of the combustible consequences should the cyclical flow of earth energy be impeded by closed doors where the hallway lay?

> What, to the scientist, constitutes a really satisfactory sort of success for a theory? The answer lies largely in the words generality, elegance, control, and prediction.
> —Warren Weaver, "The Imperfections of Science," *Proceedings of the Amer. Philos. Soc.* (Oct. 17, 1960)

She pointed out the home of her childhood on a city map of Hull. We both discovered that she had lived *very near* (if not atop) one of the four fire-leynes passing through Hull—the city where Elizabeth Clark inexplicably inflamed in 1905—and had been none the worse for the experience. Because, perhaps, her mother knew best?

Thus a fascinating (and unanticipated) oral tradition supported the cartographic evidence for combustion geography, and vice versa.

But the *big* test for the merits of combustion cartography would be to track the fire-leynes themselves and find an unknown case of SHC. As Watkins noted in *The Old Straight Track*, in the hunt for leys "Both indoor map and outdoor field exploration are necessary." It was time to go to work in the field—

Lincolnshire, England, 1979. A research associate and I are traveling through the manicured countryside in search of fortean mysteries; in particular, additional information about SHC cases already in my database. We were on the lookout for new leads, too. After all, the multitude of pyrophenomena along Fire-Leyne I had historically focused on this region.

We stopped first at Binbrook and met Arthur Maultby, a life-long resident of Binbrook, who recalled for us the 1905 commotion at the Yarn Walk farm. Today, however, Binbrook appeared as quiet and serene as an English tea social.

Speaking to other residents produced a new lead about a recent, possibly strange fire at nearby Horncastle.

We went next to Horncastle (about twelve miles south of the Binbrook triangle of fire), where we met Chief Fire Officer Russell Danby. We explained the nature of our research, and through his cooperation the lead quickly proved to be a normal fire.

Chief Danby was intrigued by our pursuit of pyrophenomena. I took out a photograph of Dr. Bentley's death, handing it to him. "Surely this was foul play!" he exclaimed, excited and mystified. No! "There's *nothing like this* I've seen! It's like these photos were *made up!*" Two other Horncastle firemen, looking over their chief's shoulders, agreed.

Here were firefighters who could quickly differentiate between a normal fire scene and one decidedly abnormal. We thanked Danby for his help; in turn, he thanked us for showing him something "the likes of which I've never seen before."

On our last morning in Lincolnshire we visited the Division Fire Brigade headquarters. Fire Brigade Commander John J. Shenton welcomed us into his busy office filled with mementos from many years of fire service. We asked if he knew of any fires that might be SHC.

"Oh! no," Commander Shenton laughed. It was obvious he gave no credence to the idea. Shenton listened politely as I explained the fire-leyne theory that had led us to Lincolnshire. Then I showed him photographs of classic SHC cases, several from his own country. What did he think had happened to these individuals, almost wholly consumed?

The chief fire officer for Lincolnshire pushed himself away from the photos, away from his desk, and leaned back in his executive chair. Was he about to berate us for wasting his time? There was pained, prolonged silence. We waited. His far-away gaze returned to his office, then to us.

"You know," he said pensively, "a few years ago we had a death—"

He hesitated again. His words trailed off into quiet thought. I could *feel* the mental machinations going on inside his head. This must be about *something* which clearly troubled him; that he had almost forgotten or rationalized away. What was it?

He spoke again, recalling a fire call a few years ago that took him to the house of an eccentric recluse.

For the first time in print I introduce the world to the hermit in Kirkly-on-Bain, who lived in a hut Shenton called a veritable fire trap. The earthen floor was piled high everywhere with stacks of newspapers, rags, and old soiled clothes; the walls were lined with dried-out newsprint; oiled paper served as window panes. It seemed straight out of Dickens' *Bleak House*.

The old man had not been seen for two days. When constables checked on him they found remnants of a meal on his table, and in the narrow aisle between stacks of newspapers they found...

"We found a line of ashes," Shenton said. "It was the darndest thing! There were two feet, with stockings, and his upper torso—above the shoulder blades—and head. That was it. His lower torso and down to his ankles, it was all ash. You could see where his legs *had been*—now there were just *two lines of ash* where his legs should be. Half the body was in good condition, *the other half there was nothing*.

"He was lying on the hearth, his head in the firebox but it wasn't much burned. Below the shoulders, nothing, until the two feet. His arms, below the shoulders, were ash too."

The more this Fire Brigade Commander recalled about this astounding fire-death, the closer he came to embracing spontaneous human combustion.

"One thing that has always bothered me, and the more I reflect on this the more bothersome it becomes," he continued, "is *why nothing else in that place burned up in the fire!* I mean, with all that newsprint—even oil paper on the windows. It was a tinderbox!"

The incident happened in the Fall of 1972, and the hermit was about seventy-nine years old. Shenton thought the alarm was called in around 9:00 A.M., with the estimated time of death placed at ten o'clock the preceding evening.

Though uncertain about some details, Shenton recollected the problems posed by the fire-scene itself with crystal clarity. "Quite a bit of smoke damage in the room," he remembered, "but I can't recall any [fire] damage to other objects." A heavy layer of greasy soot or carbon covered the upper area of the hut, "but we were unable to assess the lower level due to the amount of junk in the room." The characteristic smell of burned flesh was absent—further strengthening the case for SHC. And the powdered femur bones? "It takes terrific heat to do that," acknowledged the Lincolnshire fire brigade commander. A terrific heat that *failed to kindle* dry, or oil-soaked newspapers inches from the victim's legs.

What had been the official disposition of the case? "The coroner's court said death was accidental due to infirmity of the person and his mortal condition that caused him to fall over into the fire," Shenton replied. "I'm sure the verdict was accidental death, which it usually is." Naturally.

This *assumes* a fire was in the grate. Yet Shenton had already said "nothing was on fire when we arrived—other than a little mat under the body which was burnt, of course." Of course. And *had* there been a fire in the grate and *had* the old recluse fallen head-first into it, then what would be most severely burned? The hermit's head and shoulders, of course, since those would be in contact with the supposed external source of fire. Yet that's the part of his anatomy that *wasn't* burned! Save for the two stockinged feet, what burned of his body was *farthest* from the alleged source of fire.

No, this grate-fire explanation was problematic and needed rethinking. The fire brigade commander was already engaged in his own reassessing. "I can see what your problem is," acknowledged Shenton.

What of the intensity of the blaze that could ash the recluse's leg bones yet not ignite his tinderbox home? "I wouldn't be able to put an answer to it," Shenton conceded. "You just take the circumstances as they come. You can only imagine what the eleventh hour was... that's all the coroner does."

After further reflection, Shenton spoke again. "It *doesn't* make sense," he said. Under normal circumstances the hermit's body, he now concluded, was "not conducive to burn to the extent that it was." Yet it did.

Shenton rethought the case in our presence, deciding the circumstances were *not* normal at all. *Far from it!* Spontaneous human combustion made eminent sense, the fire brigade commander agreed.

We left Lincolnshire later than planned but elated. A new, heretofore unpublicized case that had all the attributes of classic SHC had been uncovered, found by persistence and *by following the fire-leyne trackways* that had guided us to it. It was a very good morning.

A Welsh Fire-Leyne Finds Henry Thomas—Or Vice Versa

Less than three months after our fieldwork within the Lincolnshire fire-leyne complex and the discovery of an incinerated hermit therein, another astounding event with phenomena no less amazing occurred some 160 miles to the southwest. In Wales. There, at Ebbw Vale, Henry Thomas succumbed to a "rather unusual death by fire" in early 1980.

Thomas was a widower who left behind a pile of burned clothes and disintegrated bones in a smoke-filled room where a plastic settee—only one yard away from his remains—was "completely untouched" by heat. It sounded odd. Coroner Kenneth Treasure blamed ashes in the fireplace grate as the cause of the fatal fire; Dr. J. S. Andrews, Home Office pathologist, blamed death on almost total burning of the deceased. It sounded more odd. When *Fortean Times* editor Robert Rickard told me he got soundly rebuffed by the authorities when asking them for information, the case of Henry Thomas became simply irresistible. I had to know more about it, beginning with the whereabouts of Ebbw Vale itself.

Pulling out the map of British fire-leynes, I ran my gaze along the four alignments that crisscross Wales. And there was Ebbw Vale—*within a mile of a* three-point (perhaps four-point) fire-leyne running 125 miles between sites of pyrophenomena in Cardiff, north of Hoy-on-Rye, west of Wexham, and on to Birkenhead in England!

Could Mr. Thomas be the fourth (or fifth) point on Fire-Leyne XVI?

Cooperation from the Gwent police was not forthcoming; its information officer insisted any interviews be paid for in advance. Then the British magazine *New Scientist* (May 15, 1986) published an article titled "A case of spontaneous human combustion?" The victim was not named but that didn't matter—matching the date given with known circumstances convinced me this was *the* Henry Thomas case. (After all, SHC cases don't happen every day.) The author who asked the title question was Scenes of Crime Officer John Heymer. Not long thereafter, I contacted Mr. Heymer, who in turn sent his invaluably detailed report describing what he was witness to.

Here, then, courtesy of John Heymer and his twenty-five years' experience of gathering forensic evidence at the scenes of serious crimes and sudden death, is the legacy left by Henry Thomas to fire science and to the world.

"On 6 January 1980," Heymer's statement began, "I was called to a council house in Gwent, to the scene of what I was told was a rather unusual death by fire." The weather was bitterly cold, yet upon entering the house he was struck by the pleasant warmth. Heymer noted the absence of central heating, then proceeded to the living room where the uniformed officers who requested his presence told him the fire had occurred.

"I opened the door and stepped into a cooling oven. There was a steamy, sauna-like heat, and the room was bathed in a garish, orange radiance," he noted, the result of a "sticky, orange substance" that coated the window panes and a bare light bulb—the plastic lampshade had melted over the bulb and oozed onto the floor. "The walls were radiating heat. Condensation was running down the window. Heat had cracked one of the window panes." All surfaces, including walls and ceiling, were lacquered with "a greasy black soot."

Across the room was a television set, still on, its knobs distorted by heat. (Neighbors said they had seen Thomas seated in front of the hearth the night before, presumably watching television.) Heymer examined the hearth. "The hearth was tidy; there were no signs of any coals having fallen from the fire." He examined the open grate closely; it contained dead ashes. Just grazing the outer corner of the hearth lay the reason for the fire alarm.

"On the floor, about one metre from the hearth," Heymer wrote, "was a pile of ashes. On the perimeter of the ashes, furthest from the hearth, was a partially burnt armchair. Emerging from the ashes were a pair of human feet clothed in socks. The feet were attached to short lengths of lower leg, encased in trouser leg bottoms. The feet and socks were undamaged. Protruding from what was left of the trousers were calcined leg bones which merged into the ashes. The ashes were the incinerated remains of a man."

The man was hapless Henry Thomas. Heymer looked closer.

"Of the torso and arms nothing remained but ash. Opposite the feet was a blackened skull. Though the rug and carpet below the ashes were charred, the damage did not extend more than a few centimeters beyond the perimeter of the ashes.... Plastic tiles which covered the floor beneath the carpet were undamaged." One edge of a throw rug beneath the victim's lower legs was unburned, like those legs themselves. His pair of slippers sat neatly placed on the floor inches away, unscathed. A small wood chair was the only nonhuman contribution to the fire.

The heat that warmed this chilling scene must have come from superheated ions radiated from the old man's own burning carbon.

Heymer conceded he was astounded by this wholly unexpected paradox: "Reason told me that the scene I was viewing was impossible. Everyone at the scene experienced the same sensation of incredulity: a strong urge to deny the evidence of their senses."

Heymer knew a quarter-century of crime scenes investigation had ill-prepared him for this affront to his professional skills. He took photographs, and then watched the gathering forensic scientists begin to examine the evidence *in situ*. "I soon discovered that scientists, like policemen," he said, "are human, and consequently fallible."

Heymer suggested to forensics that the cause of death was spontaneous human combustion, and to again paraphrase Dickens a century before, none other of all the deaths that can be died.

With that bold proposal, Heymer became the next victim in the room—the target not of fire but psychological pummeling by his fellow professionals. His colleagues, he recalled, "dismissed my proposal with knowing smiles and stated that the fire was entirely explicable." He awaited their wisdom.

Elementary, Mr. Heymer. You see, he was told, the deceased burned in a room with a coal grate, and human bodies can't burn without contacting an external source of fire, *ergo* the coal grate (or the coals therein) was the cause. Besides, look here: burnt fibrous tissue is adhering to the top bar of the grate. It was the ol' human-candle make-sense explanation again. Thomas had simply fallen head-first into the coal fire.

Too elementary, countered Heymer. "Amazingly, the scientists saw nothing wrong in a man falling headfirst into a fire grate, igniting like a wax candle, then somehow picking himself out of the grate and sitting in his armchair to burn himself and most of his armchair to ash. As I said, the grate was tidy. It certainly did not indicate that anyone had fallen into the fire."

Remember, *no one* found any hot embers in the grate. Some scientists would prefer believing—to the exclusion of spontaneous human combustion—that a *cold* fire grate can inflict burns.

Within a fortnight the fibrous tissue taken from the grate had been forensically analyzed. It was, Heymer's superintendent told him reluctantly, "of bovine origin."

> In order to carbonize a human body, you need an enormous quantity of thermal energy over a long period of time.... Human fat is not oil spread all over the body and soaking the human tissues. In no circumstances will fat itself suffice to keep a fire going on a human body.
> —Dr. Notot, director of medical services, Paris Fire Department; quoted by Pezé in *Le Feu Qui Tue* (1986)

Bovine: as in cows, as in steak. Decidedly not human.

Despite this difficulty, there was no further investigation into the cause of the fire. Case closed. But not as neat and tidy as the victim's hearth.

Is transmutation of human skin into bovine flesh to be invoked in order to sanction the experts and avoid the conclusion Heymer himself could not escape about the nature of Thomas's demise? I think not. More likely, Thomas had cooked his dinner in the coal fire at one time and spilled a bit of it over the grate bar... which brings us back to Mr. Thomas's own overcooked body.

"I have never seen a body, even in the fiercest of fire, where the torso burnt away. Even in the hottest fires," Heymer reiterated, "the extremities may burn away but the torso remains. In this case, as in other reported cases of spontaneous human combustion, the opposite had happened."

Heymer's frustration with experts would not end with the undue closing of the police file on Henry Thomas. He anticipated as much in the closing paragraph of his *New Scientist* article: "I realise I shall bring down on my head... the wrath of 'experts'," he said about his defense of SHC. The wrath began descending two weeks later.

D. J. X. Halliday, employed by the fire investigation unit of the Metropolitan Police Forensic Science Laboratory in London, wrote to *New Scientist* (May 29, 1986) and pointed out that Heymer failed to realize that while cremation "is intended to destroy a body in the shortest possible time... a relatively small fire can consume flesh and calcine bone if it is allowed to burn for a long time." He called this "prolonged human combustion"—PHC? "Indeed," said Halliday after saying victims of PHC are always obese, "all cases investigated by this unit have been resolved to the satisfaction of the courts without recourse to the excuse of 'spontaneous' human combustion."

Which says nothing more than that his unit never dealt with a case like Henry Thomas or Dr. Bentley—because *neither* of these men were overweight.

D. D. Drysdale, in the employ of the Unit of Fire Safety Engineering at the

University of Edinburgh, followed up in the June 12 issue of *New Scientist* to berate Heymer for proposing "pseudo-scientific and even supernatural explanations to account for the alleged phenomenon of 'spontaneous human combustion'...."

What Drysdale wanted to know (in his own words) is "how a body can burn completely to ashes without major damage to the surroundings." Don't we all?! (Well, some of us do.) He then implored that "controlled experiments need to be carried out" to confirm the human-candle theory, concluding that "I would be very surprised indeed if such experiments did not lay the ghost of 'spontaneous human combustion' once and for all."

Instead of Drysdale crying out for experimentation, I wonder why he did not use his excellent academic position and resources to do the experiment himself and prove his make-sense rationalization? Or he could have researched Dr. Gee's work done two decades earlier. Why didn't he?

Heymer anticipated such scorn and had issued this challenge to all objectors: "If the incineration of a large human body in an airless room can be explained by reference to the burning of a 'human fat candle' in a forced draught then you can expect to incinerate a bull by putting a match to its tail." I eagerly await the scientific community's grabbing the bull by the tail, so to speak, and proving Heymer wrong.

BBC-TV sought to disprove spontaneous human combustion in the death of Henry Thomas, and deaths generally, in *Newsnight* (January 13, 1986). The producer and his program reporter, Steve Bradshaw, consulted one of Britain's leading crematorium managers, who, according to *Fortean Times* (no. 46), didn't give them the answer they had anticipated. "It looks as if the fire comes from within," he told the *Newsnight* team about deaths like Henry Thomas. "Don't ask me how—I've never heard of spontaneous human combustion. But it wants a lot of explaining..."

That it does.

Meanwhile, the idea of fire-leynes as one more realistic alternative to the human fat candle theory looks better and better.

Oh yes, one last point: where Henry Thomas died, about one and a half miles north of Ebbw Vale, is *very near* where Fire-Leyne XVI blazes past. And some sixteen miles to the southeast, the spinster Annie Webb would inflame similarly less than a month later.

A Point of Fire in Pocklington

In the sparsely populated plains of Yorkshire county and miles northwest of the fire-leyne center of Hull, lies the little town of Pocklington. It may have no association with Fire-Leyne I, which tracks twelve miles to the east, or with Fire-Leyne II, which is some eighteen miles to the west. Then again, being only three miles from the bisector of these two great fire alignments, Pockington may indeed be a fire-point of cartographic note.

What is certain is that, in January 1991, Pockington became undeniably associated with a startling new case of anomalous combustion.

One morning shortly after the Christmas holidays, seventy-year-old bachelor Wilfred Gowthorpe was helping a friend wallpaper her house. At 9:00 o'clock, Sandra left him alone in her home, stating she'd be back in thirty minutes. As she walked out the door, she saw Gowthorpe walking to the sink with a paste bucket and a brush to clean it out.

When she came back a half hour later, the paste bucket lay in the sink and the faucet was on full-force. Gowthorpe was just standing there, as if in a trance, absolutely saturated in sweat. A terrible stench permeated the place.

Sandy thought he had a stroke. She called a doctor and her husband. The doctor arrived first, then her husband. They could not get any sense out of him, just a few grunts. Thinking that he should be sat down, they pulled him away from the unit he was leaning against, when the doctor gasped, *"My god! what have you done to your hand?"*

The little finger and the next finger of the old bachelor's left hand were burnt black. To a crisp. His other two fingers were badly burned. Most of his hand, too. Up his arm as far as his elbow and inside his shirt and jacket—which *weren't* burned.

"That's a replica of the Jack Angel case, isn't it?" chortled John Heymer, as he recounted this case especially for me. Yes, parallels were present. So far.

Gowthorpe was taken to a hospital, where he finally came around a few hours later. "And when he did come around," marveled Heymer, "he was bursting out laughing. He didn't feel any pain. No pain at all!" Another parallel with Jack Angel.

Surgeons immediately amputated the two fingers burned beyond reconstructive repair.

They did skin grafting on the remainder of his hand, and arm. He was in the hospital fifteen months altogether.

"Jack Angel avoided all that mess by having his hand amputated immediately, didn't he?" joked Heymer, recalling my articles about the Georgia man who, once an amputee, recovered rapidly from his hell-fire injuries. "Wilf didn't do that. The skin grafts took to the arm, but never to the other fingers. And they never did heal."

Wilfred Gowthorpe died about eighteen months afterward.

"Now that bloke was in a trance, definite!" exclaimed Heymer, referring to

Gowthorpe's immobility and insensitivity to frightful injury. Once more, evidence for a catatonic condition and neurotransmission decoupling/deactivation during anomalous human combustion presents itself.

Another point I can make. Nerve damage appears to have been less severe on Gowthorpe's arm, where grafted skin rejoined with subcutaneous tissue; far more pronounced in his hand, where tissue and nerve reconnection never succeeded. Not only does this point to his smallest left fingers being the point of origin for whatever crisped them, it suggests origination was *inside* those fingers. And it points *away* from scalding water as a theoretical cause, doesn't it? (As if anyone would make the suggestion, knowing the circumstances surrounding these third- and fourth-degree burns at the sink in Sandy's house.)

A third point I will offer. Did you wonder whatever possessed Gowthorpe to laugh boisterously in the face of such injury? I did. Such an unwonted reaction made me think immediately that, at some point, this unnerving process unleashed an avalanche of pleasure-releasing endorphins in Gowthorpe's brain. As he regained full waken-state consciousness in the hospital, he found himself uncontrollably awash in a cascade of euphoria.

All are clues that point to possibilities of cause, and to what lies behind the outward horror of this phenomenon. It may not be pretty, but it ain't often painful.

Remember this, as you enter the next chapter in the search for understanding the many variations of superhyperthermic carbonization.

24

It's Only a Matter of Time

He who doubts from what he sees,
Will ne'er believe, do what you please.

—William Blake, poet
and mystic (1757–1827)

> WARNING: The author has determined this chapter to be hazardous to those wishing to maintain a habit of debunking SHC. You should skip this chapter.

Fire! The little girl screamed again the dreaded alarm: *FIRE!!!*

No one answered. Her family had just left for Sunday church services, leaving her home alone with her grandmother. And the fire was in her grandmother's room. In her grand...

Well, let's pause. The discovery of this heretofore unpublished case is a story in itself.

During an interview in 1975 with Corporal William Sweet, then president of the Pennsylvania Association of Arson Investigators and an officer of the Pennsylvania State Police Fire Marshal's Office, he asked if I'd like to attend a special course on arson detection at the State Fire School in a few days. I knew that this course was restricted to advance fire-fighting professionals only and that mysterious fires (including spontaneous combustion) were sometimes labeled arson. What a great opportunity to be among scores of firemen attending such a seminar and maybe even glean a few leads. Yes! indeed, I wanted to attend.

Corporal Sweet made the special arrangements. Soon, Howard "Buzz" Triebold, Pennsylvania State University's security supervisor and an esteemed arson instructor, was teaching me how to determine a fire's origin. "Persistence is the name of the game" when investigating a fire, Triebold emphasized. "Know that you have investigated to the *fullest* of your capabilities."

Then came this gem of wisdom for dealing with problematic cases. "Don't be afraid to admit you can't explain a fire's cause," he admonished everyone. "*Always* look for the unusual. And be *persistent!*"

Triebold ended his lecture with a touch of the unusual. He pulled out four photographs of what some people called, he said, *a case of spontaneous human combustion*. The firemen gathered around, fascinated. While waiting to see what I thought would be pictures of Mrs. Reeser, I struck up conversations with a few of the firefighters. "You mean there's *more* cases like this?" one exclaimed wide-eyed. Yes, I replied, as the photos reached me.

Now *I* was wide-eyed. This victim *wasn't* Mrs. Reeser. In fact, it didn't match any case I knew of!

Pointing to the photos, I fired the obvious questions at Triebold: "Who? When? Where?" He didn't know. He only knew that the victim had died somewhere in southeastern Pennsylvania. Great, just the *most populous corner* of the commonwealth! This would be like trying to find one person in a proverbial haystack of five million people, without even knowing in what decade to begin looking. The challenge of challenges.

Less than a week later, I was sitting in Wilton Krogman's office. As he discussed his involvement with Mary Reeser's impossibly complete combustion, he remarked that he had just received a letter describing a recent case much like Mrs. Reeser. "Would you like to see it?" Of course! He kindly handed me the correspondence, in which fire chief H. Newton Walls of Delaware County described a scene exactly like the one seen just days earlier in Triebold's photographs.

It was the first clue. Eight months of diligent detective work followed, culminating in the disclosure of this elusively spectacular needle-in-the-haystack apparently anomalous human combustion. Now, would it stand up to scrutiny?

HELEN CONWAY: HUMAN COMBUSTION *EXTRAORDINAIRE*

Robert C. Meslin, former fire marshall of Upper Darby Township, Delaware County, Pennsylvania, was key to understanding why the little girl screamed. And what a Pandora's Box he unlocked!

Meslin was a volunteer fireman when he responded to the fire alarm one Sunday morning. The blaze was in Drexel Hill, ironically not far from where novelist Charles Brockden Brown placed the fictitious SHC of Herr Wieland almost two centuries before. Meslin remembered the incident vividly, he wrote me, because "in my 33 years in the fire service, *I never saw anything before or since to equal what I saw at that time.*"

November 8, 1964; sunny, seasonably brisk, typical for early fall in Pennsylvania. At 8:45 A.M., the Upper Darby Fire Department received a call for a house fire. "Upon arrival," recounted Meslin, "we found a considerable amount of smoke on the second floor which necessitated crawling on hands and knees. Firemen encountered a closed bedroom door and upon opening the door, heavy smoke and heat was encountered but little flames, save for some fire in a corner of the room to the right of the door (which was handled by a booster [high pressure water] line)."

An assistant chief went in on hands-and-knees, groping with outstretched arms to find the grandmother said to be somewhere in the smoke-blackened room. His hand sank into something greasy that crumbled like burned charcoal at his touch. In another minute he'd realize what it was his hand had penetrated.

A smoke ejector soon cleared the second story of its oppressive, opaque atmosphere. Meslin continued: "Examination of the fire scene revealed that the two windows of the fire room were equipped with storm windows which were closed at the time. The fire damage was limited to an upholstered occasional chair in which the victim was seated and extending only very slightly to two side tables situated on either side of the fire chair."

And what of Mrs. Helen Ann Conway, the fifty-one-year-old grandmother?

The stunned firefighters stood in numbing bewilderment. For many seconds not a word passed a lip, as they all stood aghast at the astonishing scene before them. Only a small wisp of smoke circling from the chair arm disturbed the somber serenity. "My god!" they wondered, "What happened here?!"

None of the firefighters had ever seen anything like this before! Chief Fire Marshal Paul Hagarty, recovering from his shock, yelled at Meslin to get a camera. *"Quickly!"*

Meslin ran home, grabbed his equipment, returned and began documenting the macabre for posterity. "It certainly was amazing," he admitted with awe more than a decade later. "Really something to see!"

What *was* there to see?

The chair cradling Mrs. Conway was badly burned. Her lower legs were reasonably intact, though the epidermis had pulled away and bubbled in some places; subcutaneous tissue of the upper legs had split asunder. The rest of her anatomy was a blackened, almost amorphous mass of homogeneously fused tissue practically unrecognizable as human. As Meslin quipped: "You can discern the rib cage. There was enough for the coroner to remove in a rubber bag, but that was it."

How very similar to Dr. Bentley. Yet unlike Bentley (and Mary Reeser, for that matter), enough of Mrs. Conway did remain to hazard a reconstruction of the tragedy.

Deep heat fracturing lacerated her upper legs. The right leg had split wide open right to the knee; the upper thigh appeared to have disintegrated. Two large pieces of the chest and thorax (sternum) lay charred and curled away from the organs beneath, fused together. Her upper right arm was barely discernible; the forearm, if it existed at all, was obscured by debris. Her left arm, such as it was, lay flexed back onto itself; a metal charm-type bracelet dangled from the ulna—fully exposed—and rested against some charred upper-arm muscle tissue. This flexing of her arm should not be mistaken for the pugilistic syndrome (muscle contraction caused by high-temperature fires). "There was no pugilistic effect here—which I thought strange," said Meslin. Just another quandary competing for attention in a larger mystery—

To the right of this arm and jutting above the sternum, the lower jaw bone was clearly noticeable though partially obscured by whitish smoke *still rising* from the woman's left shoulder or neck. Smoke still swirled from the chair's right arm, which collapsed onto (or into) its occupant.

Beneath the chair but not extending beyond its perimeter, the carpet had been charred from the ashes falling upon it. Wooden baseboard behind and to the left of the chair had blistered or been charred slightly, as had the varnish on the table that sat inches from the death chair. Curiously the white trash can between the table and the chair was unaffected, its unscorched gleaming surface visible through the burned-out chair arm.

The back of the chair had disintegrated, leaving Mrs. Conway's body to fall backwards into the beckoning void. There her "body" lay in blackness, awaiting the assistant chief's futile rescue attempt. It was into Mrs. Conway's midsection, what *had* been her midsection, that the fireman had plunged his outstretched hand. (It would be a long time before he was willing to enter a fire scene again.)

Meslin summed up the view of everyone at the scene when he confided to me: "This case is *far worse* than any other we've experienced!"

CIGARETTES, ANYONE?

Is anyone going to jump up and yell "Cigarette!" as the cause of this catastrophe? Don't jump too hastily. (Oh, go ahead!) Meslin has something to say to you.

"The victim is reported to have been a heavy smoker with careless smoking habits. Cigarette burn marks were evident about her bedroom. However," the fire marshal quickly pointed out, "the extent of burning to the victim and the complete combustion of the upholstered chair in which she was seated in a span of *twenty-one minutes* (including time for photographs) leaves a lot to think about."

Ah, such refreshing candor. Cigarette, anyone?

And what about the annihilation of this widow in twenty-one minutes?

Remember: a small *open-flame* fire in an overstuffed chair can evolve to room flashover (the point where combustibles spontaneously ignite due to heat) in about two minutes. When a lit cigarette is the potential ignition source, however, even though smoldering will begin in two-to-four minutes it will take two-to-three *hours* before smoldering bursts into open flame. "That's typical," says Andrew McGuire, executive director of The Burn Center. Yet here, a fire that ravaged a chair (and its occupant) never approached flashover.

How? Why? Just one more problem—

What do we know about her last minutes of life? She had been a widow of thirteen years. She was infirm to some extent and used a hand bell to summon assistance. She rang it to ask her granddaughter to bring a cigarette from downstairs. This errand the young girl promptly ran. What isn't known is whether the fire began immediately after the girl delivered the cigarette or if she discovered the room afire upon returning with the requested cigarette.

Regardless: Did this raging incineration of a human being require only an incredibly brief twenty-one minutes, as Meslin suggested? No. It happened *in less time!*

How do I know this? The answer lies in commonsense deduction; in the testimony of Meslin himself and of Harry Lott, assistant fire marshal for Upper Darby at the time of this incident.

"The parameter we put around the time was anywhere from 6 to 20 minutes, as near as we could discern it," Lott informed me in a lengthy interview at his home in late 1994. "There was talk about saying half an hour. And it just wasn't a half an hour!"

Six minutes, did he say?

Lott's estimate for the low end of the time available for this fire to unfold— a fire he remembered with precision—coincided with a determination I'd reached with Meslin almost twenty years earlier.

> I've never seen a case like this since, and I never saw one before like it.
> —Harry Lott, a veteran of more than three decades in the fire services; personal communication (December 17, 1994)

"The amazing part of the incident in my opinion is the *time element*," expressed Meslin in 1975. Amazing it is. The granddaughter turned in the alarm as soon as she discovered the blaze, Meslin declared, within "three minutes" after having last spoken to her grandmother. That meant Mrs. Conway was alive at 8:42 A.M. Firemen arrived within three minutes of getting the fire call, to see only the barest hint of flames. Conclusion: the blaze which swept through Mrs. Conway had wreaked its havoc by 8:48 A.M.

Imagine: *a maximum of 360 seconds was all the time this localized inferno needed to consume a human being and chair, and then die out!*

(About the same length of time in which Mrs. Oczki could have similarly been consumed in 1979.)

Guess we can scrap D. J. X. Halliday's theory of PHC (prolonged human combustion) for *this* case. And scrap too the cigarette-dropped-into-the-chair solution... because it takes from one hour to one-and-a-half hours for upholstery to flash into open flame from a smoldering cigarette.

Given these circumstances and what you now know about fire behavior, could a dropped cigarette have so destroyed her body? No. "I don't believe there is any way in this world that that chair could have generated—from a cigarette—enough heat to decompose her body to the point that it was destroyed *in that time frame!*" Lott stressed.

Could massive quantities of highly inflammable accelerant been spilled onto and saturated her chair? No. "We did check the chair for flammable liquids, or anything like that that would burn at a higher temperature and cause that amount of damage to the corpse, the remains of the victim. And yet we found no evidence of any kind of an accelerant being used in the fire," revealed Lott.

Could there be criminal action involved? No. Foul play was absolutely ruled out, Meslin said emphatically. Years later, Lott echoed that conclusion: "We did all of the investigations that were necessary to determine foul play, and there was no evidence of *any* foul play! She apparently came from a very loving family..."

How about the infamous "human-candle" theory? It would take a hyperquantum leap of faith to apply Dr. Gee's smoldering-fat experiment to this situation. Even if Mrs. Conway was corpulent, and it appears she was not, the time interval is *too awesomely brief* to even insinuate the human-candle silliness.

What of the chair? In the Reeser case, authorities had determined the chair and its seat cushions offered inadequate fuel load. The same applies to Conway. "It takes one to two hours before they go [burn] *that* badly," Meslin said of his experience based on combatting numerous chair fires.

"So that's the crux of the dilemma. That's really the heart and soul of the discussion," summed up Lott. "How you could achieve the damage that was done in that short period of time, and then not see in the peripherals the after-effects of those temperatures. Now I don't know how to explain that, or how to understand that. I mean, I have a lot of experience and a lot of background in the fire service, and I just don't know how it can be."

Could cremation science offer any insight? It was not until 1990 that Industrial Equipment and Engineering Company, a leading manufacturer of crematoria, marketed a retort (price, $100,000!) capable of cremating a body in just one hour—which is *ten times longer* than Mrs. Conway had available to herself.

Not surprising, then, is the response given in the 1970s when I asked cremationist John C. Grenoble, whom I consulted about Dr. Bentley, if he had any ideas about Conway. "From my experience, no," he replied. "There was *more* to it."

> The thing which is disturbing to me about the whole thing was that the amount of damage done to her body—and in the time frame that was allowable, between the six and the twenty minutes—it would seem to me you'd probably have to have a temperature of between 1500 and 2000 degrees in order to do that much damage to that body. And that just *wasn't* the case! There was nothing there to substantiate that temperature. The fact that she was still sitting in that chair: the chair wasn't totally destroyed, it wasn't broken and falling apart. There wasn't debris scattered all over. The lady was still sitting—her remains—were still sitting upright in the chair. And that's amazing to me, that all those things like that could happen. And yet the temperatures that we know it would be necessary to do the damage that was done to her physical being, must have been present; otherwise it couldn't burn like it did.
> —Harry Lott, on the Conway fire scene, personal interchange (December 17, 1994)

But *what* more? To quote Lott: "This lady died from extreme heat. Her body was decomposed by extreme heat in a relatively short period of time.... They were extremely high temperatures to do that damage. Where they came from, I don't know. I mean, I have a lot of background in the fire service, and I just don't know how it can be."

Under the circumstances present, no known source of combustion can explain the facts that surround Mrs. Conway as did the flames that enveloped her. Whatever the "more to it" is, it *must* be beyond today's science. As Meslin exclaimed: "Things *just don't burn* this way!" Yet his photographs—one dozen of them—prove otherwise, documenting undeniable evidence for those who *choose* to look.

Opposing Views

There's one more factor to think about: the odor of burnt flesh.

Crawling through that dense black smoke *should* have been a horrendously noxious undertaking for the assistant fire chief. But as I was then beginning to learn, if this was classic SHC there would *not* be an acrid odor. In a follow-up interview with Meslin I asked about this. "I didn't smell any hydrocarbons," he reflected. "In fact, I don't really recall the odor of burning flesh. I can hear the assistant chief, who was first in the room, never mention any smell."

Something burning... heavy opaque smoke... no odor... crawling through the blackness and bumping into a cremated body: how incredibly reminiscent of the English case of Barbara Bell. Reminiscent of scores more cases, as well.

What is left to label Mrs. Conway's death besides human combustion that is *preternatural* (if cigarette had been delivered) or *spontaneous* (if cigarettes had not been delivered)? Within no more than 360 seconds her body had burned right to the bone without causing significant damage to her room; without generating the odor characteristic of burned flesh; without the aid of accelerants or fuel (save her own meager body tissue).

Surely this is *the* case that breaks the PC/SHC opponents' back, so to speak. Yet years after her passing, PC/SHC is still disbelieved; still officially unrecognized; still rejected by most authorities. Why?

Well, to be fair, I chose not to publicize this case until now. But another reason can be found in the comment Triebold made to me after the arson seminar. "Most investigators in SHC cases admit their contact with such events," he confided, "but *are afraid* to attribute them to SHC—they'd be laughed out of the business!"

Regrettably true. Meslin encountered this attitude when he attended a meeting of the International Association of Arson Investigators sometime after Mrs. Conway's death. "I posed a question to a seminar—all 'top-shelf' people—about this case," Meslin reflected, "and *they didn't believe it*. They sort of brushed it off, some 'other' explanation—which I can't believe! Their placebo is 'Undetermined' or 'Electrical,' and they sneak out of it this way. If I don't know, I'll *say* I don't know. And I don't know *what* happened to Mrs. Conway."

Ah, an advocate of a forthright approach to fire science (and science generally). Such refreshing professionalism.

Meslin's words rang vividly in my mind as I ended the interview at his home and drove the short distance to the new, ultramodern headquarters of the Philadelphia Fire Department. I had a list of enigmatic fire deaths in the City of

Brotherly Love—names, addresses, dates—and hoped to uncover additional information from fire service records there.

For example, there was reclusive Mrs. Anna Martin. At 4:45 P.M. on May 18, 1957, her son Samuel—a fireman—found her lying face down in the basement of her 5061 Reno Street home in West Philadelphia, after he smelled smoke wisping around the slightly ajar cellar door.

The sixty-eight-year-old woman's torso had fearsomely burned, but not to ash; her legs *were* ashes, but not her shoes. The coal furnace she lay before was cold; stacks of newspapers and cardboard boxes two feet away were not scorched; floor joists overhead were stained with oily soot but not burned. For reasons you can easily surmise, Chief Medical Examiner Dr. Joseph W. Spelman was baffled; he could not fathom a source for the requisite 1700 to 2000°F he deemed necessary to burn her up so. Though busy city firefighters answered three extra alarms and fought three major fires on this date, the call to Mrs. Martin was unique. Notable. As Chief Inspector John J. Kelly stated simply: "It's one of our most mysterious cases."

And eleven years earlier, at three o'clock on Wednesday afternoon, January 16, 1946, sixty-two-year-old Howard M. Donovan was discovered sitting in a burning chair. Third-degree burns had already seared into the Philadelphia man's right side. Had he fallen asleep or succumbed to a stroke, becoming victim to a careless smoking accident? Had fire begun inside him, only to be quickly extinguished before his self-combustion had time to spread throughout his body? His death before midnight at St. Luke's and Children's Hospital left such questions behind. Questions not answered the next day by stories in both *The Philadelphia Inquirer* and *The Evening Bulletin*.

I wanted more information. At the Philadelphia Fire Department headquarters, a city fire marshal (to whom anonymity is mercifully granted) looked at me as if I were crazy. He teased me about the overwrought claim that people can self-combust. No, he could do nothing to assist my research. I asked, Had the records been filed elsewhere? Purged? Lost?

"What you're asking about doesn't exist! Therefore, we *can't* have any records about your 'cases.' Spontaneous human combustion is about as likely as a piece of paper bursting spontaneously into flame!" he said with sarcasm, pointing to a memo on his desk.

(He'd just unknowingly admitted to SHC, I grinned to myself inwardly, because that too has happened—right here in Philadelphia to a stack of letter paper in a home on Arch Street at 1:00 A.M. on June 25, 1837![1])

"Would you like to see some photos?" I offered.

"Oh sure, *if* you've got any!" He smirked mockingly, thinking he had called my bluff. I lay visual documentation of Bentley, Reeser, *and* Helen

Conway on his desk.

"Oh, I've seen lots of these," he said hastily, nonchalantly, as he picked up a photo of Mrs. Conway.

"You *have*?"

"Sure! Caused by cigarettes!" was his self-assured retort.

Patiently, I pointed out the problems you already know so well: absence of surrounding destruction amid apparent intense heat, lack of pungent odor, incredibly short time-spans, bodies transformed to ashes. The fire marshal's demeanor changed from smug to somber. He called over two colleagues. They looked, shook their heads, and walked away.

> To disintegrate everything is quite odd! That's unexplainable, how something would be burned that much! It's like space junk—it's unexplained.
> —Robert L. West, retired city fireman, examining photographs of Bentley and Conway; personal interchange at the Fireman's Museum of Philadelphia

After a long, pained silence I inquired again: "How *would* you respond if you were called to a fire and found a scene like one of these? Please, help me to understand this. How would you explain it? What would you do?"

The Philadelphia fire marshal looked up from the photos before him, and murmured a classic xenophobic response: "I'd go out, get drunk, and forget about it."

I've no doubt that he would.

AN EXPLOSIVE TWIST ON SHC

Firemen at the Conway blaze were so mystified, Meslin acknowledged, that no one could even *suggest* a cause for what lay before their eyes. Burning was *so* thorough that legal identification of Mrs. Conway was difficult. Insurance investigators, Meslin recalled, actually *sifted* through the remains to look for a $3000 ring she had been wearing.

That ring, which provided the means to identify the person burned, left its own mystery. Recalled Lott, awestruck: "It was in good shape! Now here's a piece of jewelry—which was, I'm sure, 14K gold—which was exposed to that *intense* heat, and yet there wasn't a mark on it! Now that to me, is almost unbelievable! Because gold melts at a fairly low temperature."

And then there's the time interval: that incredible 360 seconds at most!

After all that has preceded this chapter, what could I possibly say but that, unfortunate as it was for her next-of-kin, *the demise of Helen A. Conway proves that preternatural/spontaneous human combustion is indeed a fact.*

Now that I sound terribly nonfortean by making such a dogmatic statement, let me suggest that Mrs. Conway *wasn't* a victim of this rare fate.

Huh?

There is a way to escape the discomfort of finally acceding to the reality of preternatural/spontaneous human combustibility. I suspect the medical profession will like this prospect even less. As I see it, it's probable that Helen Conway burned in much less than 360 seconds by an even *rarer* form of human transition: SHE... *Spontaneous Human Explosion*.

Come now, surely you are not going to propose—even *hint*—that a human can self-explode? laughs the omniscient debunker.

No, I'll let the testimony of Monsieur Morand do that, because he is the one who made the suggestion originally to (of all places) the Royal Academy of Sciences some two centuries ago.

Morand stood to inform his colleagues about a most startling incident he had recently heard about, which occurred in northwestern Italy in 1751. A woman from Bonne Vallie, about thirty-seven years old, was out gathering sticks with four companions in the Montenere forest about fifty miles north of Genoa. All five then began walking home, she in the middle of the group. "As soon as they arrived at a place called Gargan, this woman," said Morand, "suddenly cried out with great vehemence, and immediately fell down with her face towards the ground."

Immediately her friends turned to her aid, only to have their stomachs turn by what they saw. The woman was quite dead. Her clothes, "and even her shoes were cut, or rather torn into slips, and scattered at the distance of five or six feet around the body, so that they were obliged to wrap her up in a cloth, in order to carry her to the village."

There the astounded villagers examined her body, such as it was. Her eyes "appeared fixed and livid," gaping wounds in her head exposed the skull, and "there were also many superficial scratches upon the face in straight lines." Her sacrum had shattered; her internal organs had become either "livid" or had ruptured; the ligaments and muscles "of the right side of the abdomen were destroyed, and had given way to the intestines;" her left abdomen was lacerated with deep parallel incisions plus more superficial parallel scratches; the pubic bone "was laid bare, and fractured, and the flesh was stripped off quite to the hip," and the femur (leg) bone was snapped off and "forced out of the socket, in which it is articulated;" and her buttocks and thighs were "carried away"—to the extent that about six pounds of flesh had vanished.

Notwithstanding this frightful scene, Morand said the villagers faced an even greater affront: "there was not the least drop of blood to be seen upon the spot where the accident happened, nor the least fragment of the flesh that had been torn away."

One can only fantasize how the academicians of the Royal Academy of Sciences in Paris responded to this report by one of their elected members. Morand then recounted the villagers' supposition for this tragedy, a conjecture with which he concurred: that "subterraneous vapours...issued from the ground directly beneath her" to "produce the death of this unhappy person, with all its extraordinary appearances."

Had Morand primitively grasped the idea of fire-leynes, one of which flared through her body in mid-step? For surely, as *Fortean Times* editor Robert Rickard remarked when bringing this grievous incident to my attention, "It was as though she was the focal point for an instantaneous, silent and deadly explosion."

(The same suggestion can be applied to the explosive fate of Italian priest Gio Bertholi in 1776, and to Mrs. Eliza Collier in 1872.)

From this new perspective, let's look closer at the room in which Mrs. Conway died.

Fire damage beyond her body was remarkably minimal...a melted telephone inches away; across the room a television set, heat-distorted (it could have been the twin to the television found in Mrs. Hogan's room six years later); varnish on picture frames near her, liquified; and a ubiquitous sooty discoloration, extending down the walls to within three feet of the floor.

However, debris was scattered about.

Fragments of upholstery fabric and shreds of material (apparently from her clothing) lay strewn over the floor and tables; several long strands draped over a cabinet and hung from a plant more than four feet away from her chair. Black cinders stuck to the wall behind the waste can, and covered the counter top about fifteen feet away where the television sat. On a low table to the right of the victim, objects had been blown away from Mrs. Conway; a glass lay broken, pushed up against the wall. The lamp shade on the end table to her immediate left had disintegrated so that its two wire rings now ringed the lamp's base—yet the light bulb was not distorted in her direction (as a normal fire scene would dictate).

Did a great force of swift combustion cause this disarray, scattering destruction about like shrapnel when a bomb explodes?

Evidence makes it easy to visualize an explosion occurring between her solar plexus and sacrum as she sat in the chair. From this origin the force of the extraordinarily *rapid* combustion—for that is what explosion is—appears to have moved parts of herself to her front, to her rear, and downward simultaneously. The lateral vectors pushed her torso backward into the void where the chair's disintegrating back no longer offered support; her arms flexed at the elbows until her charred fingers nearly touched her shoulder bones. The downward vector of the explosive combustion forced the center of the seat cushion toward the floor, raising its outer edges and pushing the coil springs to the sides or out the bottom of the chair. Her right leg slipped to an unnatural angle as both lower limbs, now severed above the knees, fell back against the front of the concave cushion.

A brass handbell with a black wooden handle lay atop the lady's burned-away right thigh. The bell did not melt, indicating the fatal explosion (or combustion) in her body did not produce a prolonged temperature higher than 1570–1900°F, the melting range for brass. The Trimline telephone's plastic housing sagged slightly and its lucite dialing ring melted, indicating a temperature range lower than 700°F (the highest ignition temperature for these plastics). The wooden frame of the overstuffed chair charred minimally and the wood baseboard and *bell handle on her thigh* didn't burn at all, so the temperature must be still lower (white oak ignites at 410 degrees and white pine at 507 degrees).

Five hundred degrees is sufficient to liquify varnish on the picture frame above the victim, causing it to run down the wall; it is barely capable of heating the *upper half* of her television chassis on the other side of the room twelve feet away. (Sound familiar?) Here, beside the television, sat another outrageous factor in the form of a plastic doll garbed in tulle (mesh) fabric. The doll's head had drooped forward, yet *at the exact elevation where the deformed television cabinet's melting stopped, the doll's highly inflammable tulle was unaffected!* This distinct, sharp demarcation in temperature, paralleled by the Reeser case, is truly astounding.

So: some process that cannot generate sustained heat above 507 degrees destroyed this woman's body in six minutes. *Or less!* For comparison, remember that a crematorium requires several hours with the temperature at 2200 to 2600 degrees in a retort's optimally controlled environment.

What about the *rate* for combustion? Certainly quick! If explosive, energy would have radiated outwardly instantaneously. If *implosive*, it could have been equally quick while minimizing surrounding damage.

The scenario for spontaneous human explosion (implosion) coupled with photographic evidence and the challenging thermal limitations all fit together in the Conway case like biblical "fire and brimstone."

Did Helen Ann Conway die by Spontaneous Human Explosion, an accelerated variation of SHC? Whether you choose to characterize her passing as preternatural or spontaneous, relatively slow (six minutes maximum) or explosively instantaneous... it cannot be denied that something *quite odd* happened to her.

Epilogue to SHE

Besides Helen Conway and the Bonne Vallie woman, there are two other incidents (discounting the tabloid press) suggestive of SHE.

According to London's *Daily Telegraph* (December 28, 1938), Horace Trew Nicholas was walking along Windmill Lane near his home in Hampton Hill, London, when a sudden bang sent the gentleman up like a rocket. He crashed against a chimney and landed on a roof, his clothes aflame, his hair burned off, and his rubber boots melted. The Gas Light & Coke Company checked for gas main leaks and found none. Sewer gas was next suspected. No evidence for this distasteful culprit, either. The coroner ruled "accident death"—what else *but* that all-encompassing catch-all?—and the gentleman's strange fate was sealed.

Might it have been an explosion within his own body that launched Nicholas's sudden upward mobility?

And circa 1949–1950; Denver, Colorado. "Woman blew up on the street in downtown Denver" is how Cliff Johnson told me he remembered hearing the story some forty-five years ago when he lived outside the Mile High City. "Other people were around, on the streets, and saw it. At the time, officials checked to see if she was carrying explosives!" But, he said, they found none; nor an explanation. To confirm this lead, I immediately called the Western History/Geneology Department of the Denver Public Library and asked manager Eleanor M. Gehres to search the *Denver Post* and *Rocky Mountain News*. She and staffer Lyn Spenst graciously and thoroughly did so, but could find no citation. False lead? News censored by wary editors?

By any accounting, SHE is *far more rare* than SHC. Unless, of course, the absence of historical detail has caused researchers to underestimate its magnitude because these decidedly freakish deaths have been consigned conventional labels—like "accidental death"—rather than a more appropriate diagnosis of spontaneous explosion.

What could trigger such bizarre transformations?

From a metaphysical viewpoint, when people speak of "emotions exploding" they may be unwittingly permitting the body to express that emotional outburst in a literal, very physical way. The body's chemical mill might then produce a catalyst that triggers a swift exothermic reaction, or transmute an element into a water-reactive agent that detonates the whole biological factory, just as water-reactive metal powders (aluminum or phosphorus, for example) fiercely oxidize when becoming wet. Kundalini has already provided one scenario by which this might occur, and there could be a dozen more possibilities.[2]

I asked Lott what his last word would be on the Conway fire—blamed officially on a dropped cigarette and case closed—after thirty years' thinking about it. "I'd have to give you the answer that everybody would give you to that case who has any kind of reputation in the fire service. And that would be: I don't know. I have no idea.... There's got to be some kind of a reasonable answer to it. If there is such a thing as SHC, it's got to come out sooner or later. And if it can be proven, fine. That's the way it is!"

That brought to mind Fire Marshal Meslin's parting advice, years before: "One of the prime prerequisites of an investigator is to have an open mind."

25

The Mott Case: All the Symptoms... and New Puzzles

> Our minds are finite, and yet even in these circumstances of finitude we are surrounded by possibilities that are infinite, and the purpose of human life is to grasp as much as we can of infinitude.
>
> —Alfred North Whitehead

In spring of 1986, the telephone rang at ParaScience International. I'll never forget that call.

"This is Tony Morette. I've heard about you. The thing is—"

Morette paused, grappling for words. "I haven't figured it out," he resumed. "The body was *totally* consumed in the bed. It's really something you have to see. That is, it's hard to explain over the phone!"

Tony Morette identified himself as the fire investigator for Essex County in upstate New York. As I got to know him during the next several weeks, I became more and more impressed by his professionalism. He was the consummate investigator, and exacting as a top sergeant. When Morette examined a fire scene, he left no clue unturned in his quest to find its cause. He was accustomed to finding answers.

Morette now had a case that puzzled him. He was eager to know what had happened.

When it comes to SHC, be prepared to *expect the unexpected*. Just when years of research leads me to think I might be close to understanding this mystery of human fire within, along comes... well, along comes a George Irving Mott to throw things askew.

George Mott was a fifty-eight-year-old retired fireman, living alone in the rolling countryside outside Crown Point. His home, a half-mile from his nearest neighbor, was a wood-framed dwelling built around a mobile home. A veteran of the armed forces, he lived by military routine. His actions were predictable, meticulous; his household immaculate, with every utensil and towel and garment neatly in its place. Though divorced, he maintained a close and loving relationship with his children and grandchildren. He met regularly with friends for breakfast at Ticonderoga's Wagon Wheel Restaurant.

Wednesday, March 26, 1986, around 11:30 A.M. The previous day's mail was still in the mailbox at Mott's rural homestead on Bush Road. The mailman noticed nothing else unusual, however, and drove on. Throughout the morning and afternoon, Mott's daughter telephoned her father. No one answered. She finally asked her brother, Kendall, to check on their dad. Kendall reassured her that dad had "seemed to be doing really well" when he visited on Monday at 8:00 P.M., so well that he had foregone his visit on Tuesday. Yes, he would stop by after work.

Around 6:00 P.M., Kendall Mott drove out to Bush Road. His father's truck was parked alongside the house. Dad's home, Kendall reasoned; wonder why he's not answering his phone? The window drapes were darker than when last he visited, Kendall noticed. Dad must have hung new ones, he thought to himself. He went to open the front door and immediately realized more than the drapes had changed inside. Something was wrong. Terribly wrong.

"The door handle was warm to the touch," Kendall Mott noticed. "When I opened the door, it was warm inside. It was all dark. Very dark. I knew the minute I walked in, he was dead."

He groped his way back to his father's bedroom at the rear of the house, and could just make out through the sticky, opaque atmosphere that the television on the chest-of-drawers on the far side of the bed had melted down. "I knew something was *really* wrong! I couldn't see anything else."

Kendall rushed out. He went to get a neighbor, Lester Joiner, to ask him for help. The two men returned to Mott's residence. Kendall, visibly upset, refused to go back in. Joiner entered, a flashlight in hand. A moment later he was calling the police. George Mott was indeed dead.

New York State Police trooper Richard Lavalley was first to respond. Other officers arrived and quickly secured the scene, then called for specialists from the Bureau of Criminal Investigations (BCI). Coroner Jack Harland was notified, who in turn called in Essex County pathologist Dr. C. Frances Varga. Around 8:30 P.M., county fire inspector Morette arrived at the Mott home, followed by Essex County's director of disaster preparedness, Robert Purdy.

Morette immediately realized this was no ordinary fire call, as had Crown Point fire chief Frances Cook and his firemen. There was no fire to fight, but something far more difficult to battle. The impossible.

Morette requested the assistance of ParaScience International in his investigation, and promised full cooperation should we choose to make the 750-mile-roundtrip to upstate New York. It was another offer I could not refuse.

INVESTIGATION OF THE UNEXPECTED BEGINS

Soon a PSI colleague and I were on the road to Crown Point, a hamlet north of Fort Ticonderoga where history had been made during America's Revolutionary War. Now, in the spring of 1986, revolutionary history had again been made here, although no one immediately realized it.

We met Fire Chief Frances Cook and Fire Investigator Tony Morette at the Crown Point Fire Department and listened with rapt fascination as they recounted what, for them, was a unique fire scene.

But all their words could not ease—indeed, they heightened—my desire to visit the place where this incident had happened.

Would Tony escort us to the house? No. He was uncomfortable with the thought of reentering a fire scene that had profoundly bothered him. It had been unlike anything he'd ever encountered in two decades of fire science work, he said, and he wasn't eager to confront it again even though he (and anyone he authorized) had permission to enter the site. Besides, since Mott's family had talked about razing the house, he wasn't sure it still stood. If we wanted to go, alone, he would tell us how to get there. We got directions to Bush Road.

As we wound our way onto Bush Road, we soon recognized the Mott homestead with its redwood-stained siding. The dwelling's appearance was as deceptive as new paint on a second-hand car. Everything looked so normal that we expected Mr. Mott to greet us as we drove up, though we knew that couldn't happen. We examined the outside of the dwelling, having to look very closely to find evidence for the fire we knew had occurred inside. Wisps of dark soot had seeped around window and door seals; one thermoglass pane had cracked. Little could be seen inside, though, because all windows had a baked-on glaze that darkened the house's interior. Gas and electric lines had been cut and sealed. The front door was shut.

A few weeks later, on a second trip to meet again with Morette, we revisited the Mott home. The grass was higher now, but nothing else outside had changed since our first visit... except that the front door stood wide open.

If Don Gosnell's cry "Dr. Bentley's burnt up!" deserves Understatement of the Year for 1966, then the undisputed Irony of the Year for 1986 belongs to the sign posted on the front door of George Mott's home. "No Smoking" it warned. Behind the door this message went unheeded, unimaginably so.

To step beyond the door's threshold was to enter an absurd netherworld that revealed the deficiency of man's knowledge about himself and this magical world. It was an awesome, unearthly feeling to contemplate stepping into a scene so difficult to imagine; so readily deniable outside the realm of science fiction. Knowing what lay before us in the bedroom at the far corner of the house made it all the more eerie. For an investigator of SHC, this was the step of a lifetime; for science, it would be a doorway into the unknown.

Imagine yourself at the front door of George Mott's home, ready to enter a scene few have ever seen... maybe you can make sense of it.

A Walking Tour through the Uncanny

Beyond the door with its "No Smoking" sign, is Mott's living room. It comprises almost half of the dwelling's eight-hundred-square-foot area.

An alcove to the immediate right houses his library, which showed him to be a literate man with eclectic interests ranging from Zane Gray westerns to international Classics. One volume on the bookshelf is Reader's Digest's *Strange Stories, Amazing Facts*, its dust jacket darkened with a translucent, baked-on soot; otherwise, it (with its chapter about spontaneous human combustion) is undamaged.

On another shelf sits a plastic box, now contorted by heat into a shapeless mass barely recognizable as a tissue canister. Next to it, propped against *The Works of Shakespeare*, stands an indoor thermometer that hadn't burst—indicating the mercury never rose beyond the 120°F mark. A model of a clipper ship sits on a wooden coffee table; neither is damaged, save for a uniform coating of soot.

To the left of the entrance, a television and a stereo along the west wall are both blanketed with soot resembling gray dust; except it is baked on. The record player's plastic dust cover is warped by heat and the tuning scale's glass plate is glazed over; otherwise it looks ready to work. Albums and 8-track tapes stored in the open cabinet beneath the stereo are not affected. Behind the television, between it and the stereo, hang two sets of fiberglass drapes from plastic rings, all intact.

An overstuffed chair in the middle of the room has its own light coating of soot on its maroon velvet upholstery. Behind this chair, along the west wall again and beyond the stereo system, stands a small table. On its plywood surface, Mott had arranged papers noting the dosage of each medication he took to ease blood pressure and breathing difficulties. Firemen had removed some pill bottles, leaving behind bright white circles to mark the bottles' former positions on the soot-darkened tabletop. The caps of two plastic bottles are melted into themselves, yet their sides and paper labels remain intact. Next to them sits a glass bottle with a rubber eyedropper cap, undamaged. Inches from the melted plastics, Dixie paper cups and three dozen Q-tips lie scattered atop a magazine; none show the baked-soot glaze that seems otherwise omnipresent on all horizontal surfaces in the room.

Above this table, a third pair of fiberglass drapes had once covered the window with the cracked pane. Unlike their counterparts only a few feet away, these drapes lay in a coagulated puddle of fused fibers on the floor. Plastic drapery rings dripped surrealistically from their curtain rods like hot candy in a taffy-pull. Paint on the window frame is blistered; some of the trim is charred. Wooden shank shingles on the inside wall below the window are charred, as is a three-foot-long section of white baseboard molding.

Without question, there had been a fire. A fire *very* small; so small that one doesn't even notice it when entering the house. Why it burned up fiberglass, melted *selected* plastic items, and *lightly scorched* only those shingles below the window sill and the very edge of the table, yet failed to damage plastic and paper containers upon it plus other easily combustible materials all just a few inches away seems remarkable—all the more remarkable because directly above the burned-away drapes the white ceiling exhibits only minor damage, limited to carbon deposition.

No less problematic is a portable radio-cassette unit sitting four inches from the charred window frame (and next to the stereo). Its plastic chassis remains intact, even though next to it—*farther* away from the drapery fire—the stereo unit's dust cover was heat-distorted! Pencils in front of the cassette player and a wooden floor lamp with a paper lamp shade between the table and the pencils both show no damage—not even blistering of the pencils' yellow paint!

There are even more remarkable things.

A small gas furnace sits in the rear of the living room. Was this the source of the fire?

"The gas stove I tore apart," said Morette. "There were no leaks, from the tank coming in or anywhere else." Above the furnace, an eighteen-by-twenty-four-inch opening cuts through the wall that separates the living room from the bedroom. The window's louvered doors are soot-blackened but show no serious fire damage. The metal handle of a Coleman-type lantern on a shelf directly below this opening has partially melted into its plastic lens.

Across the room, a fly swatter once hung on the wall. Now only its metal handle remains, the plastic paddle melted into nothingness—but not before the pattern of its webbing became imprinted onto the wall's unburned paneling.

In the kitchen, located (as one enters the house) to the right of the living room and behind the library alcove, the horizontal surfaces are darkened with more caramel-hued glaze.

The gas stove looks ready to cook a meal once the soot is scoured off. Morette had inspected the stove for any malfunction. It worked perfectly. The adjacent stainless steel sink and faucets surrendered their sheen to the dull sooty film. Behind the sink the label of a can of Comet cleanser is partly obscured by the pervasive glaze. A white plastic bottle of dish detergent *next* to it is only partly darkened, however. Plasticware, scotch tape, and pans all sit unscathed in the cupboard above, ready for use. An item whose outline-in-soot resembles a breadboard once hung on the cupboard door; now only a five-inch rivulet of plastic remains. By contrast, a plastic dispenser on the wall next to the cupboard is elongated under the weight of the paper cups it held; now its tortuous shape is worthy of a modern art museum.

On the other side of the kitchen (the inside wall to the living room) the counter and cupboard pose their own curiosities.

There, a cream-colored Rubbermaid silverware organizer has horizontal surfaces blackened with sooty glaze; but its molded sides retain bright color. Knives, forks, and spoons, each impeccably arranged by Mott as was everything else in his home, would be ready for use after light scouring. A bright yellow plastic container next to it is undamaged except for a glaze baked onto its ribbed cover. Inside it, cookies look fresh and edible (though we didn't sample any). Next to it, a bowl of sugar has a blackened plastic lid; the sugar inside is not congealed, again indicating very little heat present in the kitchen.

Overhead on the cupboard door, a plastic placard has curled upon itself. Next to it, a holder for paper towels is deformed and pulled away from the screws that held it in place; a foot-long strand of plastic runs down the cupboard door. It still maintains, just barely, its grip on a roll of paper towels. Inside this cupboard are boxes of cereal and sugar, coffee filters, soup bowls, and other kitchenware that could have just been taken off a supermarket shelf, so unmarred are they.

On the counter directly beneath the melted towel holder sits an unopened roll of Bounty towels, upright. Ironically it and its plastic wrapping are *undamaged* except for a glazed film on the top! Someone had moved it a few inches, however, to reveal a soot-free crescent where the roll had been at the time of the fire. In the enclosed space below this counter Mott had stored a Rubbermaid pan, cans of Raid and Endust, and numerous kitchen cleansers: soot baked itself onto the tops of these items but caused no other damage.

Then there is the refrigerator! It poses *special* problems.

Mott had covered the outside of his frig with yellow-checkered contact paper, which effectively camouflaged the caramel-hued glaze now coating it. But there is no camouflaging what investigators discovered inside. "Everything *in* the refrigerator had melted," said Morette. "The butter, and the butter dish too. Yet the refrigerator somehow started functioning again." A more astonishing discovery awaited Dr. Varga, who had noticed an unopened packet of hot dogs encased in plastic packaging. "The hot dogs looked as if they were parboiled," Dr. Varga professed, *"almost as if zapped by a microwave."*

Perhaps the refrigerator's thermocouple-controlled compressor tripped due to momentary overheating, you think, then reset itself to power-up the compressor. (Information on its set-point was not available to me, however.) That might explain the butter's melting. But it is quite another matter to explain why all the plasticware inside had melted and how a packet of franks cooked themselves inside a sealed (and insulated) refrigerator when the ambient heat outside was insufficient to distort the plastic wrapping on a roll of paper towels two feet away!

There is a paper plate on top of the refrigerator. On its glazed surface lie the dehydrated remains of two bananas. Whether their moisture slowly evaporated over time or flash-fried in the heat that had engulfed the *inside* of the refrigerator below, we could not determine. (What we do know, based on Mott's meticulous housekeeping, is the bananas had not dried up due to neglected disuse before his demise.) Next to the bananas is another Rubbermaid container, still sealed. Several oatmeal cookies inside look quite natural despite what had happened to the adjacent bananas and the hot dogs inside the refrigerator below.

Morette certainly was right when he described the whole fire scene as "bizarre!"

The bathroom is near the rear of Mott's small home. Here, again, dichotomy abounds. And confounds.

Behind the door leading into the bathroom hangs a towel rack; several peach and white towels are neatly folded over it, as they have been since the day Mott died. At the time of the fire this door was open, putting some of the towels within four to five inches of a plastic Dixie cup dispenser on the wall. The dispenser was cut in half by heat; now it partly hangs in a twisted mass on the wall and partly lays on the vanity counter amid the scattered paper cups it once held. Revealing the same curious immunity to fire as the cups, *not one loop of terrycloth is singed on the soot-discolored towels* despite being only inches from the burned away cup dispenser!

A plastic bottle of Cepacol on the vanity has partly deformed, too. Next to it, a toothbrush escaped heat damage (unlike those in Mary Reeser's bathroom).

However, the ubiquitous soot has baked itself onto the bristles of the toothbrush and an adjacent shaving brush, as it did onto the vanity's tile counter top but *not* the vertical tiles that form the edge of the vanity. A clock on the vanity stopped at 3:58, either due to winding down of its mainspring or to mark the time the fire froze its mechanism. A veneer of soot blackens the porcelain sink. Above the sink, an eerie opaque glaze uniformly films the wall mirror.

Across from the vanity are the toilet and the bathtub. A ring around the tub makes the inside of the white tub look as though it had been filled with three inches of jet-black water, now evaporated. The soft-plastic padded toilet seat is blackened with soot but not melted. The toilet tank and lid have a layer of heavy soot, darkest where moisture condensed. Kleenex tissues on top of the toilet tank, however, have only the barest hint of sooting.

Above the toilet hangs a rattan rack, showing no ill-effects from fire or heat. Soot discolors the tops of two stacks of washcloths there; also the label on a Lysol aerosol can between them. The pressurized can did not explode nor did its plastic cap melt, another indication that the fire in the bedroom a few yards away generated very little heat here. Yet *literally* an inch above the intact aerosol can, a transistor radio's polystyrene chassis melted and dripped onto the upraised toilet lid below!

Something else is curious. Here, for the first time, is significant sooting on some vertical surfaces: on the toilet's water tank, on the vanity's sink and the mirror above it, in the bathtub, on one set of towels hanging next to the sink. What could these items have in common? Wet surfaces. It seems that moisture attracted the pervasively baked soot like a magnet draws iron filings.

Beyond the bathroom and its puzzlements is a small utility room. The paint on its seven-and-a-half-foot ceiling is lightly blistered. Otherwise the room offers little of interest—except for one insanely perplexing discovery. A desk-top adding machine's hard plastic casing is grotesquely melted next to a brown plastic laundry basket that looks brand new except for a light glaze of soot on its rim—all the more problematic because several soft plastics elsewhere in the house were melted.

To the left of the utility room and behind the living room is one last room. The bedroom of George Mott.

His bedroom is quite small, approximately ten feet by ten feet. Into it Mott had crammed a bed, centered lengthwise in the room with its headboard abutting the living room wall; a rung-back wooden chair; a pine dresser along one wall; a pine nightstand against the far wall beyond the bed; and a television atop a chest of drawers in a small alcove at the foot of the bed. Underneath the bed Mott had placed an ocher Colonial braided rug. The walls were covered with white-and-turquoise patterned wallpaper.

There had been a portable oxygen-enrichment machine, now removed, immediately inside the doorway. "A glorified humidifier," Morette described it; designed to increase by two to three percent the concentration of oxygen in the room so that Mott could breathe easier. "The thing that puzzles me is the oxygen machine was still running," marvelled Morette; "It was *on* when I got there." That meant any fire in the rear of the house should have burned more fiercely due to a heightened oxygen concentration. That didn't happen... except in one very localized area only.

Furthermore, Morette had noticed that the soft plastic hose and breathing mask laying atop the machine did not melt. Also atop it sat a canister of long wooden matches. "*The matches never ignited!*" said Morette, even though they sat only inches from the edge of the burned-up bed. "I still don't understand the matches myself."

> I was there first-hand. I've never seen anything like this! Ever! And I doubt I ever will again.
> It was just, just amazing!
> You're trying to guess what the hell happened here. And when. And how. Why didn't it do this? Why didn't the house burn down? Why didn't the matches ignite on the stand next to the victim?...
> —Robert Purdy, on the Mott fire scene; personal interchange (December 6, 1994)

Ignition-proof matches weren't the only difficult thing to understand.

The wood chair is not burnt. A mirror above the pine dresser is darkened by more baked caramel-hued glaze, but there is no heat cracking. The white milkglass of a hurricane lamp is soot-darkened; its globe set aside by someone who needed the bare bulb—glazed but not deformed—to shed as much light as possible on the scene they too had struggled to comprehend. Clothes lay neatly folded in drawers, looking as though they had been freshly laundered. On the floor beside the dresser sits a wicker wastebasket with a white plastic liner inside, neither is touched by heat or flame.

At the foot of the bed is a walk-in closet, where soot baked onto the eight-inch-wide floorboards outlines light-colored rectangles where storage boxes once sat. On top of the chest of drawers are three items: a plaster statue, intact; a telephone, partially melted; and a thirteen-inch portable television set that collapsed into itself, its chassis a mass of melted plastic that pulled away from the picture tube. (This is the TV set glimpsed by Kendall Mott, and its appearance is reminiscent of the melted televisions in the Bofin and Conway cases.) The picture tube's glass screen was coated with (yes!) more caramel-hued glaze. Behind the partly melted phone and wholly melted television hangs an orange-colored fiberglass curtain, intact.

On the far wall is the room's only window, eighteen inches from the chest of drawers and twenty-four inches from the footboard of the bed. The window frame's white paint is blistered, mostly around the top sash opaqued by amber glaze; blistered due to the burning of a fiberglass drape now fused into a heap alongside the white baseboard. The baseboard, however, remains white, and the wallpaper around the window reveals no fire damage either. Why this drape burned, when its counterpart a few feet away did not, mirrored the selective burning of the living room's drapes and was just as perplexing.

Further along this wall an electrical outlet still grips a cord running to a digital clock and a second hurricane lamp on the nightstand next to the bed. Closer examination of the outlet, less than two feet from the bed, reveals that it is intact; the cord is intact; the wallpaper behind the cord is intact. The nearby nightstand is intact too and retains its varnished luster. On the nightstand, itself a few inches from the headboard that had charred in the fire that killed Mott, the digital clock had (yes!) melted into an almost unrecognizable puddle of plastic. Below is a Kleenex box, its surface soot-dusted but its tissues usable.

In the middle of the room is the bed where Mott died, burned in such a way that the outline of his body can still be seen. (Reminiscent of Mrs. Pococke, whose burnt body left its own outline in the eighteenth-century Irish almshouse.) In a room remarkable for the fire-immunity of most of its furnishings, this bed is the one piece of furniture that didn't escape.

Its frame is charred. The varnish along the top of the headboard is seared; a silver-dollar-sized hole has burned through one of three trim boards. Behind the headboard a new pattern adorns the wallpaper: a faintly dark outline of the *entire* headboard—emulating the shadow images burned onto walls at Hiroshima by the A-bomb blast. Behind the headboard is the room's third electrical outlet which, after careful inspection, shows *no trace of any flame damage.*

The footboard is more burned than the headboard, yet charring still penetrated less than one-quarter inch into the wood. Varnish over large areas of the footboard didn't even bubble.

The mattress is largely nonexistent, most (but not all) of its material having burned away. Lengthwise from headboard to footboard there was a deep V—as though Mott's body had been *strongly forced down* upon it. "The springs are intact all the way through," Morette said, but underneath the body they had lost their compression strength. "The strange thing was it would take 1500 degrees for these springs to collapse," he affirmed.

"His slippers were still intact," Morette continued. Were they on the floor? "No, on the bed where he took them off!"

It was all *so* weird. Along with his slippers, part of a cotton blanket still lays on the fire-cooked bed springs. *Too* weird!

This was the scene Kendall Mott had unknowingly stood next to when he saw the melted-down TV set beyond his father's bed.

A layer of dark sooting caresses the upper four feet of all the bedroom's walls. Concentrations of dark soot defined locations of wall studs and nails throughout the bedroom. Shades (so to speak) of the Reeser and Oczki homes, again—

Five feet above the bed and directly over the point of most intense combustion, no fire has damaged the ceiling's white wallpaper. No burning, no scorching, no searing; no peeling of paper. Repeat: *the low ceiling directly above the burned bed is pristine, undamaged in any way save for a microscopically thin film of soot!*

Contrary to everything common sense says about superheated air rising pillarlike from a blaze, *this* fire burned as a one-way column *down* through the bed and the floorboards and into the crawl space below. As Morette said, "*It burned straight through!*" Downward.

The braided rug on the bedroom floor confirms Morette's statement. The portion directly under Mott had disintegrated, but beyond the oblong hole defined by the burned body the rug is only superficially singed for a few inches; the rest of its braiding and ocher color remains.

No wonder the officials who first responded to this fire call kept asking for more assistance.

THE RESIDENT FIREMAN

There is one more thing to describe about the bedroom: George Mott himself. There isn't much to say, because the fire left so little of him to describe. Morette efficiently expressed it in one sentence: "There was three and a half pounds left of him, and he was a 180-pound man!"

I urged him to elaborate.

Only two items remained recognizable as human, explained Morette. One was a right leg. "Just below the knee, that's all that's left," exclaimed Morette. Its bronzed skin contrasted dramatically against the blackened springs which now supported it. And the left leg? "Non-existent! Nothing is existing, even underneath! We crawled under there and tried to dig up bones, and found some very small pieces. Most of the stuff was dust!... Even the rib cage, there were only very small pieces."

What happened when these tiny fragments of bone were touched? "It crystallized right in your hand—just like picking up sand," replied Morette. "It was pulverized. It just fell apart. Nothing!"

Chief Cook agreed. He said the bone—neither he nor Morette *ever* referred to a skeleton—was "a powdery form. There was nothing!"

Whatever burned Mott, it did so more thoroughly than a crematorium.

"It was just strange!" exclaimed Ann Cartwright, who assisted the fire investigators on that unforgettable night. "I was just in total awe."

The other item recovered was the head of George Mott. It provided two invaluable clues.

First, it guaranteed the victim's identity. Though some investigators first doubted that *anyone* had died in this blaze, close examination uncovered a horseshoe dental plate that offered positive ID. "George Mott was the only person whose gag reflex was so sensitive he could not wear a full upper denture," said Dr. J. William Brennan, Mott's dentist, when I interviewed him. "I made George a partial upper plate." It was this plate made of merthacolate plastic (which remained *unmelted*) that enabled Morette and the BCI to say the ashes were truly George Mott. "Besides," said Morette, "there's nobody else missing in this community." No other identifying features existed.

Second, the head provided perhaps the greatest enigma in this intricately enigmatic case. The head, Morette volunteered, "was like a tennis ball, had shrunken right up."

It was a depiction as hard to ignore as a clap of thunder. Momentarily taken aback, I immediately remembered the Reeser case. Mr. Mott's head shrank?

"It shrank," repeated Morette.

Tennis-ball sized?

"Well, maybe a little larger—grapefruit-size, maybe. It wasn't a normal skull!" he insisted. "We put what was there in a small box, because everything was pulverized."

The sutures of the skull remained intact? Yet there was no tissue inside?

"Nothing. It should have exploded apart, you know. But somehow it didn't. It just seemed to shrink up. Just bizarre!...It looked like a shrunken head."

A normal fire death, this? No way.

This revelation by Morette can offer important substantiation for the most controversial aspect of the Reeser SHC. (Remember that Mary's head, too, had been described as grapefruit-sized.) It raises *paramount and profound questions* about what occurs during some anomalous human fire fatalities.

Was Morette certain about this observation? Yes, he was sure, he affirmed. Shrunken by how much? "By about 15 percent," he answered.

A human being reduced in weight by more than 98 percent; by a fire that consumed his body and his *very* immediate surroundings to the exclusion of adjacent items more easily combustible; in circumstances that defy the way fire is supposed to behave. Was George Mott another addition to the long (and long-disputed) history of SHC?

I met Joseph Regan at the Wilcox–Regan Funeral Home, which handled Mott's funeral. "What happened to George Mott?" I asked.

"We took him to a crematorium," Regan answered matter of fact.
"He was pretty well cremated already, wasn't he?"
"Yes he was," Regan agreed.
"Have you ever seen a body burned this badly outside of a crematorium?"
"No, not that I can say," Regan replied.

What incredible irony! Firemen know you can burn whatever's left that didn't burn in a fire, but you can't burn again what has already burned. On that basis, the Adirondack-Burlington Crematorium's retort needed little time to finish the task that Mott himself had so effectively begun!

The twice-cremated ashes of George Mott were scattered at sea, in accordance with family wishes.

ANALYZING THE FATE OF GEORGE MOTT

The New York State Police labeled Mott's death accidental and closed the book. That wasn't good enough for Morette. He was bothered that the case was being closed so quickly, just as his own investigation for the county was beginning. "*How* he died was the question that bothered me," Morette said when I met him at Crown Point's fire department.

George Mott was an intelligent man; methodical, orderly, and "very organized," said Morette. Everything in Mott's house confirmed that. He had been a fireman with Ticonderoga's Defiance Fire Department for about fifteen years, and Morette had served alongside him. "He was one hell of a good fireman," testified Morette, and he was fanatical about eliminating anything that might be a fire hazard. "Mott was concerned about fire. He was nervous as hell about fires. He knew his business!"

Newton Brown, captain of Defiance and a friend who also had fought fires alongside Mott, agreed with Morette's assessment. He also remembered that Mott at one time smoked and drank heavily. But in the 1970s Mott changed, said Brown; he began to take care of himself, carefully doing everything his doctor prescribed, and enjoying the companionship of his beloved grandchildren.

These traits painted a portrait for *the least likely victim* of a carelessly ignited blaze. "So as far as him not being knowledgeable in fire science or being careless, I would say, no," affirmed Morette. Unfortunately accidents can happen, even to the most conscientious person. Did one happen to George Mott?

Morette invested three hundred hours trying to answer that question. He checked the gas lines; he checked the gas appliances; he checked the electrical wiring inside the house and outside at the pole. He looked for incendiary devices and accelerants in the house that could have been used accidentally or intentionally by Mott—or by anyone—to ignite the confined blaze(s). He came up empty-handed, as did the BCI laboratory. Avowed Morette unhesitantly: "There was *nothing in there* that attributed to the fire as far as carelessness or anything else."

Did Mott commit suicide? "I ruled that out. He had just refilled his prescriptions," revealed Morette, who had counted the pills and found them "within reason based on the consumption according to the directions." His respiratory problems and high blood pressure had responded well to treatment, said his doctors. And though Mott had a recent bout with pneumonia, his daughter Kimberly told me, it never developed fully. She thought her father was in good spirits.

Except: as I just recently learned from his son, on that fateful Monday evening George Mott *was* "somewhat depressed. He said he was sick of being sick. He didn't like being in hospitals."

This notwithstanding, no one who knew George Mott believed suicide was the answer. "Besides," demanded Morette, "find me the *way* he did it!"

The fire investigator again emphasized the *absence* of any accelerant; the *absence* of massive fuel loading needed to feed the fire once ignited; the *absence* of damage to surrounding combustibles that a conventional fire would have left behind; and the *presence* of an oxygen-rich environment that should have abetted the blaze but did not.

These problems had been immediately apparent to Coroner Harland when he stood before Mott's ashes. He phoned Dr. C. Frances Varga, pathologist for Essex County, and alerted him to the perplexing situation.

Dr. Varga, who had handled about a half-dozen fire cases in his career, decided to make the long drive south from his home in Lake Placid because, he told me, the case sounded "so bizarre. It just seemed *strange* to me."

He looked around Mott's home and at Mott—what there was of him—and concluded that the retired fireman died "peacefully and accidentally." He confirmed Morette's assessment that no accelerant had fueled the fire and a few days later signed Mott's death certificate.

The certificate of death placed Mott's death on March 28, 1986. This is wrong, since Mott was found dead at 6:00 P.M. on March 26. "Cremation" was entered as the means for disposition of the body. Although cremation technology was indeed utilized, in truth Mott's body had cremated *itself* more thoroughly than a retort does. Official determination of death, Dr. Varga certified, was "accidental" due to these factors:

IMMEDIATE CAUSE OF DEATH: "Asphyxiation" in "minutes"
DUE TO: "Smoke inhalation" in "minutes"
DUE TO: "Fire with advanced charring" in "hours"

Reminiscent, certainly, of the death certificate filed on behalf of Dr. Bentley's remarkably similar incineration. Unlike the claim made for Bentley, however, George Mott was *not* autopsied. As Dr. Varga quipped wryly to me, "There was nothing to autopsy! You can't do an autopsy on ashes."

To imply that this is a normal fire with attendant advanced charring is to flagrantly understate the situation—*unless* the natural end of "advanced charring" is always disintegration of a human skeleton. As you have learned by now, that *isn't* the case unless...

In fact, the idea of SHC *did* occur to Dr. Varga when he first heard about this fire fatality. But as the certificate of death shows, he backed away from SHC. "It's most implausible," he later told Ticonderoga's *Press Republican* (July 1, 1986) about SHC. "It doesn't sound rational."

> I think it's too sensational.
> —Dr. C. Frances Varga, on SHC;
> personal interchange
> (December 9, 1994)

But then Mott's death itself was most implausible; not at all rational. "There's nothing there!" exclaimed a coroner's assistant when helping Morette search for Mott amid the ashes.

Coroner Harland reflected studiously about how *truly little* remained of George Mott. "We could fit everything in a shoe box" but the leg, Harland told me as I sat with him on the veranda of his home. He reinforced Morette's statement that the skull was hollow, all its tissue burned out. Harland also confirmed that Mott's head had shrunk in size—because a normal-sized adult head could not fit inside a normal-sized shoe box.

Let's review the findings that characterized the Mott fire scene, one atypical of normal fire but typical of SHC:

• *Localized heating.* Temperature in the bed springs had to reach 1500°F—that's the temperature houses generate when they burn to the ground. Yet Mott's wooden house obviously remained standing.

• *Selectivity of the fire.* Conventional wisdom would predict widespread damage to items with low kindling temperatures, certainly in the bedroom if nowhere else. Yet electrical cords and paper tissues and even barnburner-type matches within inches of the 1500°F-damaged springs failed to ignite, as had his slippers and a portion of the cotton blanket on the mattress...all escaped the apparently torrid heat! Soft and hard plastics selectively and nonsensically melted throughout the house, no example more incongruous than the plastic merthacolate which escaped melting inside Mott's mouth! The burning away of two sets of fiberglass drapes to the exclusion of matched drapes nearby made no sense, either. But it happened.

• *Medical anomalies.* Standard medical knowledge says that at 1500°F the human skeleton will be largely impervious to destruction; instead, here it was utterly destroyed. To a skull, however, 1500°F is an elevated temperature and it should have fragmented; instead, here the fire hollowed out and uniformly contracted the skull. As Morette avowed repeatedly, *the head was smaller than could be accounted for by mere heat-contraction of skin around a skull.*

• *The odor of burning.* Entering the small house should have been odiferously unpleasant. Instead, fire chief Cook, fire investigator Morette, coroner Harland, and pathologist Dr. Varga all told me the insufferably pungent smell of burned human flesh was *not* present. "It seemed strange," remarked Morette. "It was kinda a sweetish smell, basically. I attributed it to the pine pitch on the headboard." In any way typical of a burned body? "Geez, no!" shot back the fire investigator.

(Had not, also, the smell in Dr. Bentley's house been called "sweetish"... and likened to "hickory incense" in the home of Mrs. Oczki?)

Three months after the fire, only a slight musty odor likened to damp charred wood was discernible in the Mott home. One would never know olfactorily that anyone had died in a fire there.

• *Heat pattern.* If all this wasn't damnable enough, everyone accepts the simple fact that heat and flames rise in a blaze. Here the bedroom ceiling—*only five feet above the cremating body of George Mott*—showed no heat damage whatsoever!

This fire was everything *but* simple. If this wasn't spontaneous human combustion, *what* was it?

"I thought this is crazy!" was Morette's first reaction when hearing about SHC in a fire seminar years earlier. Now, having faced the crazy nature of the fire he had thoroughly investigated, he told me he had to change his viewpoint. "From reading and talking to you," he declared on June 21, 1986, "I pretty well lean to spontaneous human combustion."

Four days later, ParaScience International issued its preliminary report on the cause of George Irving Mott's unfortunate demise: *Accidental death by spontaneous human combustion.*

Not everyone involved in the Mott case was willing to accept this. One BCI official, John Beck, told me he had identified the fire's real cause—a conventional cause. An electrical arc had shot out of the outlet in Mott's bedroom and ignited the man's clothing, Beck explained. End of mystery!

Frankly, I couldn't believe I was hearing this from a professional criminologist. I knew Morette had already explored this possibility and found it insupportable.

I also had carefully examined all three outlets in Mott's bedroom: not one—repeat—*not one* of them showed any arc scorching on its receptacle or cover plate.* What the BCI investigator was looking at, I don't know, unless it was for a way to avoid looking at SHC.

> It sure wasn't electrical arcing!...
> I took those outlets apart in my hand. There was no arcing. There was *none*! That was to 'close the book' for me.
> —Robert Purdy, personal interchange (January 27, 1995)

Harry Dish, a member of the Media Resource Service that nationally disseminates (supposed) scientific information to the public, cavalierly solved the enigma over the telephone when he told Ticonderoga's *Press Republican* (July 1, 1986) that "some seeming spontaneous-combustion cases"—George Mott included—"have been explained as the result of small, undetected gas leaks."

Dish exemplifies the shoddiness that gets passed off as science, because as you know, no leak was found in Mott's home. Outraged that someone was claiming to have solved Mott's death over the telephone without conducting an investigation, a livid Morette called me to say: *"As far as I'm concerned the case is closed. He died of spontaneous combustion."*

The director of Essex County Disaster Preparedness, Robert Purdy, also immediately realized that he had strode into a very special situation when entering Mott's house that fateful Wednesday evening. As a USAF airman 1st-Class during the Korean War, he had seen many horrible fire deaths caused by airplane-fuel explosions; and he had been to many more fatal fires in his career as a no-nonsense public servant in emergency operations. About the fire inside the wood house out on Bush Road in late March 1986, he admitted frankly that "I've *never* seen anything like this!"

Purdy had absolute confidence in Morette's ability to conduct a proper investigation, even into a death so unusual as this. He fully backed his fire investigator as well as ParaScience International in their joint determination that George Mott died by SHC. "It makes me feel a little more at ease that this is one possibility," he told the *Press Republican* (July 1, 1986). "I feel good that they've pursued it this far and come up with what they feel is a logical conclusion."

In 1986, Trooper Lavalley told me he, too, accepted SHC as the correct verdict for Mott's death, adding: "It gives me an eerie feeling."

More than a year later, Essex County's fire investigator had found no reason to change his conviction about SHC. "Nobody's been able to blow any holes

---

*Copper wires begin melting at 1981°F, which is well below the temperature of an electrical arc. Wiring behind the unblackened outlet plates in Mott's bedroom was unmelted, further ruling against electrical arcing as an external ignition source.

in it," Morette affirmed. Then adding, "I *still* have a hard time accepting it! You say to yourself, *Holy Christ, how can that happen?*"

Time has not diminished Bob Purdy's assessment of the amazing scene he witnessed firsthand and would never forget. "There was two pieces of bone! The rest was powder!" he said emphatically to me in early 1995.

"I saw it! I saw it *with my own eyes!*" Purdy continued. "He burned from the inside out, not the outside in. That's spontaneous human combustion in *my* estimation. Or somebody's gotta tell me what *else* did it. And *I* don't know what else!"

ALL THE SYMPTOMS... MAYBE AN ANSWER... NEW PUZZLES

Tony Morette had conducted among the finest, most aggressive investigations of a fire scene that I had encountered in a dozen years of researching anomalous fire phenomena. If not for his dogged determination to make sense of it all, the significance of Mr. Mott's passing would have likely been forever lost. Morette's three-hundred-hour investigation ruled out every suspected cause in this fire fatality until only one was left, the one least anticipated.

That last choice offered the best solution.

The characteristics of classic SHC are certainly present. As Kendall Mott acknowledged: "I was kinda flabbergasted. He definitely died of a *weird* cause! There wasn't anything left of him."

The facts, as known, seem to allow no other explanation: *George Mott died by abnormal internal thermogenesis.*

Yet there are new puzzles. How to explain the selective melting of plastics? How to explain three separate burn sites: Mott and his bed, one bedroom drapery, another pair of drapes in the living room? How to explain the frustratingly bizarre burning of two sets of draperies, when identical drapes only a few feet away hung with only a coating of soot?

I have no satisfactory answers yet. Do you?

There are new revelations, too.

I sense that Mott bathed Monday evening after his son left, then walked into his living room to take his medications. What happened after that can only be guesswork.

I can, however, speak confidently about the time of Mr. Mott's death, which the death certificate dated at least two days too late. And time is always an important factor when investigating any fire episode, no more so than in the type that is the focus of this book.

Unquestionably, the fire that killed George Mott occurred within no more than forty-six hours—between eight o'clock on Monday night, March 24, and

Wednesday evening at six o'clock. Some investigators argued for a slow smoldering fire throughout this period, but several factors argue *against* this.

First, a smoldering fire is incapable of sustaining the 1500°F temperature that caused compression failure of the mattress springs. Second, to my knowledge, a smoldering bedding fire cannot transform a human skeleton to dust. Third, a smoldering mattress will generate a lot of *warm* moisture that loosens and peels wallpaper from walls and (especially) ceilings; I have seen this happen at other fires, but it did not happen anywhere in the Mott home. Fourth, the house was not hot when investigators arrived, indicating the fire had likely been out for some time.

A fifth factor also argues against a lingering blaze. Everyone who knew Mott agreed that he lived life based on predictable routine and attention to detail; it would be uncharacteristic for him not to retrieve his mail each day, for example. Mott didn't pick up Tuesday's mail, delivered around 11:30 A.M., which indicates to me that he almost certainly had passed away by then.

Therefore, time of death is *not later* than 11:30 A.M. on Tuesday, March 25; duration of fire, an *absolute* maximum of fifteen and a half hours. Of course, it might have occurred in much *less* time.

Another point I can make with certainty. Similarities in the Mott case to Bradshaw, Thomas, Conway, Oczki, Reeser, Bentley—especially the last two—are too striking to ignore. Repetition in upstate New York in 1986 of the same characteristics found in scores of other anomalous fire deaths throughout the world, reinforces the conclusion that George Irving Mott succumbed to the accidental death of spontaneous human combustion or (if you prefer) superhyperthermic carbonization.

> My secretary asked me if I *really* thought SHC is what killed Mott. "What else *could* it be?" I said.
> —Robert Purdy, personal interchange (January 27, 1995)

In death, George Irving Mott of Crown Point had rewritten history...his enduring legacy will be a testament to *truly* anomalous human combustion.

26

More Clues to Human Combustion Conundrums

> Progress depends on the raising of questions. The answers will almost always prove disquieting to the old order of science, and utterly destructive of its central dogmas.
>
> —Trevor James Constable,
> *The Cosmic Pulse of Life* (1976)

For many people the three hardest words to utter are *I don't know.* Since the days of von Liebig, scientists have found it easier to scoff at the prospect of pyrophoric people than to take the subject of SHC seriously. Instinctively relegating it to the category of folklore, they need not know anything about it. Nor be bothered by it.

As history has shown, naysaysers sweep away the mystery of SHC/PC in one of two ways. They deny the evidence, sometimes purposefully refusing even to look at it. Not very scientific. Or they invoke the farcical human-candle theory, and argue that the more fuel (fat) is available to an external flame the easier and more thorough the combustion. Unfortunately for this rationale, cremationists say excessive fat renders the human body *more* difficult than it normally is to burn.

Besides, when was the last time you were at a restaurant or backyard BBQ, and an apologetic chef had to announce "Sorry! your well-done steak just burnt to ashes on my grill"?

In the *Journal of Criminal Law, Criminology and Police Science* (March–April 1952), Dr. Lester Adelson, a graduate of Harvard Medical School, repudiated PC and SHC as "monuments to bygone days." A generation later, on January

14, 1986, this forensic pathologist and deputy medical examiner for Cuyahoga County, Ohio, sat in his office and looked at fire-scene photographs of Dr. Bentley and Helen Conway I had loaned to Hana Gartner, host of CBC-TV's top-rated *the fifth estate* which was doing a segment titled Great Balls of Fire.

"These are not natural phenomenon [sic]!" he declared. These two persons had been murdered and the photographs were "hoaxes," he insisted. On what evidence? Because, said the Harvard-trained forensics specialist, "I do not for a single second buy the overheated concept of spontaneous human combustion."

Never mind that these images had been taken by photographers whom I met and found no reason to doubt their authenticity (photographer or picture). Never mind that these photos had been taken decades before images could be easily subjected to all sorts of indignities and manipulation by software-controlled electrons inside a desktop computer. Never mind that scores of humans had paranormally burned in the interval between his two statements. Never mind that he had not investigated any of those cases in his thirty-five-year career (a fact Dr. Adelson affably acknowledged when I spoke to him in December 1994).

No, the good doctor would stand by the conclusion of his "tired, old paper," as he termed his *Journal* article. His opinion still is that all similarly ashed, disintegrated bodies described over the centuries as SHC *must* be fakes or hoaxes!

Would Dr. Adelson prefer belief in a murderous cabal? Whose stealth is so perfected they can target victims randomly engaged in almost every human activity and callously commit perfect crimes without leaving one clue to their presence? Whose legacy has consistently challenged what medical science says must not be?

Surely Occam himself would choose SHC over a cult of clandestine incendiarists who for centuries have possessed a lap-top cremation technology which surpasses today's large retorts engineered by modern science.

Nor can I believe, contrary to Dr. Adelson, that these deaths are hoaxes perpetuated by liars. Recalling Charles Fort's words in *Wild Talents*: "when I come upon the unconventional repeating, in times and places far apart, I feel—even though I have no absolute standards to judge by—that I am outside the field of ordinary liars."

> If, occasionally, historical evidence does not square with formulated laws, it should be remembered that a law is but a deduction from experience and experiment, and therefore laws must conform with historical facts, not facts with laws.
> —Immanuel Velikovsky, Worlds in Collision (1950)

SHC is the antidote for the toxic mental shroud which blinds one's perception of reality and possibility. That medical scholars have (albeit infrequently) acknowledged the occurrence of such cases testifies to high strangeness, not hoax. These deaths were, and are, different; special; outstanding; noteworthy. And memorable. Unless the scores of fire marshals interviewed for this book are just plain stupid, these deaths are also unexplained.

Not a cinder of evidence substantiates Dr. Adelson's slander. If he, or you, can demonstrate that *every* case in this book is a hoax—or that I have everywhere erred in interpreting history's witness to SHC—then I will accede and scrape egg off my face.

I'm pretty confident my face is, and will remain, egg-free.

SHC has been a problem apparently for thousands of years, yet has not chewed its way into the mainstream. And a problem it will be for some time to come.

Denying it will not rid the world of it. Psychopathically invoking the human-candle inanity, even to ignoring the evidence of bodies incinerated inexplicably, will do nothing to safeguard lives in the future. What is needed is more of the attitude promised me by Marshall Considine, director of the Chicago Police Crime Laboratory: "Be assured of our cooperation should we receive a case where a death is attributed to SHC."

Not all cases alleged to be SHC are, of course. The death of Countess von Görlitz in 1847 was initially ruled SHC, you will recall, but evidence vigorously pursued compelled a manservant named Stauff to confess that he had murdered then torched her. Diligent investigation has likewise conventionally explained other cases initially thought to be SHC.

Still, cases not facilely resolvable remain.*

Since anomalous fires cannot be repeated under controlled conditions whenever desired, they cannot be classified truly as scientific. But collected as they have been in this book, they can be studied. If but *one* of the above cases is genuinely paranormal, then the axiom that SHC does not happen is compellingly dismissed forever and a new world is revealed as fantastic as our present world would seem to those reluctant observers who peered at the wonders disclosed by Galileo's telescope.

---

*Confidential sources tell me about still more cases: (1) a newspaper reporter says witnesses "won't talk" about a woman walking past a diner in the 1970s who "suddenly caught fire" along U.S. Route 30, between Chambersburg and Gettysburg, Pennsylvania; (2) a forensics engineer asks me in 1991 to testify about SHC as a defense in a Minnesota arson-murder trial, but I hear no more from his firm; (3) a convict strapped to the electric chair in America beats the executioner via self-incineration. Each is intriguing; each unverified so far.

A Manual to Define the Scene of SHC

Bridging the gap between the known and the unknown can pose very real dilemmas. And dilemmas abound with SHC, as this study has shown.

Study without reflection is a waste of time, however, and reflection without study is dangerous. Combining the two is, on the other hand, worthwhile, and holds out hope of narrowing the gap between what is known and what remains to be known about these baffling blazes. To quote John Barracato, former Deputy Chief Fire Marshal of New York City and internationally respected authority in fire investigation: "Few people realize, including some fire fighters and investigators, that fire doesn't destroy evidence; it creates it."

Mercurial in its outbreak and aftermath, unpredictable yet unforgettable once encountered, SHC should be immediately considered by any fire department and law enforcement agent whenever a fire scene presents as evidence these conditions:

> Things are sometimes defined best by that which surrounds them.
> —Charles Fort

• an intensely localized blaze that significantly destroys a human, even to disintegrating the skeleton—especially when accelerants are absent. This localization can include an immunity of clothing and furniturefabrics (such as bed sheets) to the fire that burns the person who is in contact with those inflammables.

• dehydration of tissue and vaporizing of blood. "In any normal-type fire there will always be blood left in the body," instructs Fire Marshal Klages. By that criteria, then, many of the fatalities previously listed are *automatically* abnormal: eg. Dr. Bentley, Mrs. Kazmirczak, Beatrice Oczki, George Mott.

• nearly total incineration of the torso with extremities largely (or wholly) intact—especially if the limbs appear to have been seared through scalpel-like. This signature of classic SHC is occasionally inverted, as the fire leaves extremities internally burned to the exclusion of the torso.

• absence of noxious odor customarily associated with burned flesh. A sweet redolent perfume-like aroma pervading the fire scene heightens the probability for SHC.

• baked-on glaze of caramel or grey color throughout the premises where there is no other indication of prolonged, intense heating.

• selectively paradoxical damage to surrounding combustibles. The melting of soft plastics and television sets to the notable exclusion of other damage—to newspaper, wood, fabrics, and rubber tips on a walker for instance—would define the standard SHC scene.

When a burn suddenly appears on the epidermis in the absence of external heat or fire, consideration of SHC is mandatory.

To paraphrase Forrest Gump: "SHC is as SHC does."

Should these criteria be discovered at a fire scene, special procedures need to be implemented.

First, I recommend that a scintillator be brought in posthaste to measure for low-level radiation. Should these human Hiroshimas be caused by a pyrotron (or other factors that unleash subatomic disruption and radiation releases), a radiation counter would alert emergency personnel to the immediate need to safeguard themselves from radioactive injury, and at the same time, provide important evidence to understanding the physics of the phenomenon.

Second, I urge that emergency personnel go beyond the requirements of their basic (and immediate) duties at a fire and aggressively document the scene and retrieve evidence for later analysis. I have long pondered why so many SHC-style fire scenes, clearly perplexing, seem to be closed so casually. And quickly. Harry Lott suggested why this is often the case:

> The people who do the investigation would love to know the reasons and the causes and all of the things behind a fire death, but the primary thing they're interested in is whether or not there's been a crime committed. And if there has been no crime committed, the interest of the public official usually wanes at that point. So that may explain, to some extent, why you don't have more notification of this type of thing. It may be a lot more prevalent...

I believe Lott is right on both counts. Until more data can be acquired by, and from, officials at the fire scene, the sought-for reasons and causes will remain elusive. Air sampling is one simple procedure that would preserve airborne postcombustion particulates, as evidence for analysis, which could shed insight into the combustion chemistry of these odd fires.

Identifying those causes will be further thwarted until officials permit themselves to be less intimidated by the unexpected than I have found many to be. I can sympathize with the anxiety that can result from a profound affront to one's years of training and expertise. But going out to get drunk will solve nothing, scientifically; nor will refusal to confront the problematical protect a potential victim in the future.

In the case of Mrs. Conway, for example, assistant fire marshal Lott was standing behind his fire chief as the troubling photographs of her fate were passed among firefighters gathered in the chief's office. Marveling over the degree of destruction, fire chief Haggarty said: "She was a smoker. So that's all we're going to put down as cause of death. Accidental, dropped cigarette." Lott objected.

"Paul, I'd agree with that," Lott interjected, "except for one thing. The time element! The time element doesn't permit that to have happened."

The chief turned to him, and conceded: "Yes, but if we don't say dropped cigarette, we'll look strange."

No one will never know how many fire deaths have already been misrepresented, written off as accidental use of smoking materials, dropped cigarette, case closed. I hope there will be fewer mischaracterizations after *ABLAZE!*

Finally, a new category needs to be established to define these situations. To the International Association of Coroners and Medical Examiners, and to the National Fire Protection Association, I propose that superhyperthermic carbonization injuries—exemplified by Dr. Bentley, Henry Thomas, George Mott et al.—be classified as *5th-degree burn* injury. Because these cases fall *well* beyond the criteria for the present four categories of burn description.

> **What procedure would you pursue, if called to an SHC-style fire?**
> That's an interesting question. You'd have to think about it. You've lost the body pretty much for the forensic types, from what I understand. And I think that would give you one heck of a dilemma.
> You'd have to assume—make a jump in reasoning, anyway—that there has to be something wrong. You'd have to look for signs of forced entry.... You'd scan for hydrocarbons and accelerants and so forth. I'd want to eliminate all those.
> And going with your premise that that's what it is [SHC], those things shouldn't be evident. That's one of the things that would make me start thinking your way.
> The next thing I'd have to do is, I'd want to do some research. I'd probably re-read your article—I still have it.
> And what else would I do? I guess I'd have to look at the physical history of the decedent... do a psychological profile, at least a physical profile on him. And try to go from there.
> —Jack Lotwick, fifteen-year veteran as a Pennsylvania State Police Fire Marshal; personal interchange (December 9, 1994)

UPDATING THE CHARACTERISTICS OF SHC

The criteria that Lair established in the early nineteenth century for SHC, criteria which have remained largely unchanged since, are only partly correct at best. It appears several are decidedly wrong.

Based on the many cases which have occurred since his time, plus those I have uncovered that eluded Lair, I construct this new set of criteria:

• SHC appears most often, most likely to be an electric-type fire; not the oxidation combustion that fires a match head.

• SHC has a low thermal coefficient; or if temperature *does* reach the bone-ashing 3000°F range of a crematorium, it is a radiant-heat combustion

affording no chance to auto-ignite nearby objects having low ignition temperatures.
- SHC seems to be painless at the time it is happening.
- SHC has been witnessed. Color of combustion ranges from orange-red to bluish green to silver to white, though electric-blue is without doubt the bioluminescent hue most often seen. This speaks against the oxidation combustion heretofore popularly implicated, suggesting instead a cool chemiluminescence of excited molecules or an electric or plasma energy. *If* an oxidation-type combustion is involved, blue would indicate a *very hot* (2550°F) fire.
- SHC has been observed to consume a body slowly over several hours (as the next chapter will confirm) or transpire in an instant.
- SHC is not an ailment of nationality, nor it is limited to France (as once the Germans had declared) and to Germany (as the French had affirmed). Rather, SHC has occurred on every continent save Antarctica. SHC is a universal potential.
- SHC occurs outdoors as well as indoors, despite Dr. Thurston in 1961 having relegated its venue to the latter.
- SHC is not restricted to winter months. It can occur in any season. As Rev. Ferguson remarked about the 1808 death of Mrs. A.B. in Ireland, this is a phenomenon everyone must be prepared to face...and face at any time.
- SHC can apparently happen to people engaged in almost every common human endeavor, save one. Did you notice that, too? Sleeping, sitting (especially near a television?), walking, dining, driving, even boating comprise situations that have proved fatally fiery. The only type of human activity conspicuously absent in the annals of SHC is, shall I delicately say, amorous. Should you desire to set your love life on fire, SHC seems not to be an issue. At least there's something to be thankful for, regarding this strange malady—
- SHC need not be fatal. (Something else to be thankful for.) Commander Edward Sullivan is—as is each SHC survivor I've interviewed—living testimony to the spirit in man to triumph over outrageously strange misfortune; to escape an energy that can inflame from within, thus allowing (sometimes) life to continue nonetheless.
- Finally, SHC is an *extraordinarily rare* phenomenon. (Something to be very thankful for.) If this were otherwise, researchers would have long ago been begging for federal grants to study it instead of rejecting it freely.

Unlike heart disease, which afflicts one in four Americans and claimed nearly a million lives in 1990 (more than all other causes of death combined), SHC is a killer easy to ignore *because* it is so rare. Some three hundred cases culled from the last four centuries means the odds approach hundreds-of-millions-to-one *against* any one person dying of SHC.

As the chance of experiencing SHC is infinitesimal, just as certainly the odds are inversely great that *more cases of SHC have happened* than anyone presently knows. I believe that hundreds, if not thousands, of cases have been lost (some permanently) to history because their true nature was not realized... or their mystery *was* recognized, but for fear of ridicule, the facts were misrepresented or quietly buried along with the victims' ashes.

PROFILING THE VICTIMS OF SPONTANEOUS COMBUSTION

Do the odds for SHC favor you? The answer depends on the cause underlying any particular SHC episode. Knowing that English author Lady Mary Montagu said "General notions are generally wrong," some general observations at least can be ventured.

• First and least comforting, it seems everyone is susceptible.

Having said that, what does our research reveal about the profile of a person who is predisposed to what generally is rapid and painless burning?

• Lair said that SHC "took place only in women." In 1967 Gaddis wrote, "I would roughly estimate that there must be forty or fifty female victims for every male." On the basis of many more cases than these gentleman accessed, neither claim is supported. Nor is the finding of Randles and Hough, who concluded from their one hundred cases that

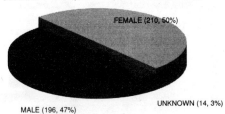

SHC: GENDER STATUS

FEMALE (210, 50%)
MALE (196, 47%)
UNKNOWN (14, 3%)

"Death rates for males were slightly higher than for females." PARASCIENCE INTERNATIONAL's database indicates gender is inconsequential to classic SHC, with females statistically just *slightly* more predisposed to SHC.

• Age affords no particular immunity when the range apparently spans six weeks to a hundred and fourteen years old. However, the elderly, not unexpectedly, account for most of the fatalities.

• Widows and widowers *do* appear regularly throughout the history of SHC. Perhaps loneliness or longing prepares the way to a sometimes dramatic destiny.

• Alcohol cannot be overlooked. Although youngsters and teetotalers have experienced paranormal burning, the overwhelming number of victims who were regular imbibers (whether or not intemperately) does warn—as did the Temperance Movement—that alcohol and SHC are associated. At least statistically.

- The psychological profile of a person predisposed to SHC has been sought diligently, with difficulty. Data suggest that persons who explode with quick-to-rage tempers or who are chronically depressed, morose, even contemplating suicide, have a greater risk of sudden autonomic self-immolation. Being physically *and* psychologically moribund is a red flag for internal fire.

> Consider yourself forewarned: do not fear the [exterior] fires that rage out of control. For they can be extinguished. Beware of that fire that burns within—that quiet fire. For it cannot be so quickly put out.
> —Letitia Adair, letter;
> *Patriot-News* (May 7, 1992)

- Invalids and persons with ambulatory problems *do* seem prone to SHC. Perhaps harboring inner rage at being handicapped turns to raging combustion; perhaps being helpless infuriates them into flame; perhaps impaired limbs restrict the flow of bioelectricity until it bursts forth, capacitor-like, in fiery discharge.

- Victims are sedentary, said early SHC advocates. Many burned in bed. While there are exceptions, a state of physical and mental relaxation does seem important, as does a slothful dispirited attitude. Hence, languor and relaxation are traits encountered in victims too frequently to ignore.

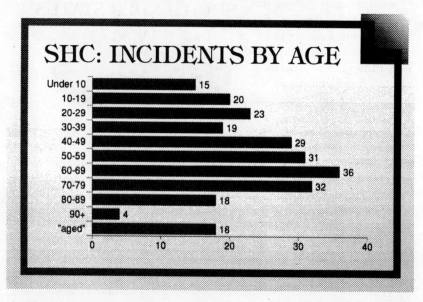

- Early researchers assigned membership in society's lower class as a mandatory qualification for SHC. I revoke this criterion. Professionals as well as paupers can become victims of inner fire, as well as witnesses to it. Physicians and businessmen who have survived SHC are as perplexed by the behavior of their bodies as have been the less educated who likewise had no explanation for their flaming fate.

• Whether nutrition is an important factor, I cannot say. Aside from Dr. Bentley's regimen of coffee and shredded wheat and the little girl who ate too many chestnuts, little is known about the victims' diet and what role, if any, it might play in SHC.

Two proposals have linked diet and digestion to SHC. "The freak eruptions known as spontaneous human combustion can be caused by eating eggs, followed by alcohol-based medicine such as a laxative, says British researcher Jenny Randles"—so says the tabloid *Star* (September 15, 1992). The second proposal is, well, stranger. It comes from a Sidney Alford, identified as a Wiltshire "explosives engineering consultant" in his rebuttal to Heymer's pro-SHC article about Henry Thomas in *New Scientist*. Alford offered to explain why the elderly are so often victims of alleged SHC. It's because, he says, they consume "liquid paraffin" as a laxative, whereupon this inflammable chemical reacts with digestive gases such as methane in their digestive tract to create—this is his term—the "dreaded phosphinic fart."

O Horror! may the solution to SHC be anything *but* this.

SCIENCES' ROLES IN RESOLVING HUMAN SELF-CREMATION

Einstein observed that "The most beautiful thing we can experience is the mysterious. It is the source of all true art and science." Whether or not SHC is beautiful, it is *certainly* mysterious.

As a source for discoveries scientific, what can be learned by studying it further? The burning question, the core of the conundrum of these very personal conflagrations, is: *How?*

If there is one place in America where one could expect to find this question answered, it would be the United States Fire Academy in Emmittsburg, Maryland. The USFA, part of the Federal Emergency Management Agency, is the nation's premier fire training school. Here, the best firemen in the country receive cutting-edge instruction in advanced fire science. A good place to inquire about how SHC might happen.

I discovered that the impressive collections of the USFA library contained—as of May 27, 1994—just nine citations that reference SHC. Of those nine, one was missing the day of my visit; one was pro-SHC; the remaining seven dismissed the phenomenon as impossible. (The collection contained not one of the many SHC articles published under my name. Interesting.)

I next showed, or tried to show, photographs of two classic spontaneous/preternatural human combustibility cases to members of the USFA curriculum staff.

One senior fire instructor said he had never seen, or heard of others who had seen, fire scenes like these. He expressed shrug-of-the-shoulders disinterest and walked away. Another instructor outright refused even to look, recoiling as though he was being handed the Bubonic plague rather than a pair of innocent photographs.

Then: USFA fire studies specialist Ken Kuntz, who happened to glance at the photographs as he walked by, stopped dead in his tracks. Expressing an enthusiastic desire to know more, he invited me into his office and eagerly posed questions. He studiously examined the photograph of Dr. Bentley's ashed remains, finally saying: "This certainly fits the definition of spontaneous human combustion. This is the kind of stuff that leaves us scratching our heads. Not a clue..."

Which is what I was there for. Clues. Clues to explain (not explain away) SHC.

Searching for clues to any phenomenon demands, as scientists must admit, not ambivalence but a certain disrespect for accepted, conventional explanations. Unless discovery is solely by happenstance, broader awareness is a prerequisite to comprehending an unconventional, more complex universe. Every scientific breakthrough inevitably means advancing into

> Nature always observes laws and rules which involve eternal necessity and truth, although they may not all be known to us...
> —Baruch Spinoza, seventeenth-century philosopher

unknown territory where habits of thought must be replaced by bold initiative and imagination. One no longer knows what will be found. Or where. Nor precisely quite how.

At present, I know only two ways to replicate SHC artificially outside of a laboratory, excluding an A-bomb test. One method is by High-Temperature Accelerant (HTA) fires. When first identified in 1984 after a steel-roof truss vaporized and concrete turned to glass in an extraordinary Seattle warehouse blaze, firefighters knew they had a problem. And a mystery. "It was like science fiction," Seattle Fire Department investigator Richard Gehlhausen told *The Wall Street Journal* (October 7, 1993). HTA arsons have a unique signature. They generate temperatures in excess of 5000°F; can render a building "a total loss" in 2.25 minutes; can be made even worse by applying water (extreme heat splits it into explosive hydrogen and flame-feeding oxygen). The accelerant is thought to be a variant of rocket fuel. Its exact composition, and identity of the arsonist(s), remain unknown.

The second method was demonstrated in 1995 by George Goble, a participant in the Purdue University engineering department's annual hamburger speed-cooking competition. Goble lit upon perhaps the ultimate process for the technophobe barbequer: using what I'll call a Low-Temperature Accelerant (LTA)—super-cold liquid oxygen. Dial up Goble's World Wide Web address (http://ghg.ecn.purdue.edu/) to see how three gallons of liquid oxygen dumped onto sixty pounds of charcoal, touched off with a match, yielded a 10,000°F. fireball and charcoal ready for cooking in a world-record three seconds! "Basically, the grill vaporized," said Goble. Presumably the same can be said for his overly well-done hamburgers.

Is HTA to blame for SHC? Only by the *remotest* superstretch of imagination, and only for a few *recent* cases. SHC that occurred decades ago, definitely requires other hypotheses. And LTA? Not even by the remotest stretch of imagination—

Hundreds of people who exhibited paranormal thermal phenomena have demonstrated that titanic forces exist in nature—inside human beings—which are not understood. These hidden, rarely acknowledged, and infrequently experienced forces lurk unobtrusively until the moment when potentiality flares into a firestorm of blazing reality. How?

Arson educator Fred Klages instructs firemen that "If you don't find the origin, you'll never find the cause." In many if not all the human fires I've chronicled, the origin of combustion has been found: it's *inside* the body.

When you have a situation where fact conflicts with theory, the only honest resolution is to change the theory; redesign the model. So if I hear one more, excuse the slang, *flaming idiot* (you know what I'd *like* to say) churlishly fire off this damned human-candle whitewash of an explanation for what moments earlier was a live human and now has plumes of smoke exuding from flesh en route to becoming a mound of dehydrated dust, I'm gonna strike a match and light a stogie and stuff both down his shirt! Then I will step back and wait. I'll probably get sued for battery, but it will be my satisfaction to having proven—by the protagonist's very ability to walk away and take offense—that his explain-away is as valueless as a eunuch in a brothel.

Okay. Now, to look for a realistic cause for SHC. Or causes.

In truth, I have no single theory to offer for SHC. I do not believe there is only one explanation for the phenomenon. Conditions and circumstances appear to be too varied; too capricious; too individualized between the hundreds of these fire fatalities to even hope for a singular, never-can-it-be-anything-but-this solution. That's why I have proposed several diverse theories already; avenues awaiting additional exploration.

If one factor *can* explain the fate of all SHC victims, I confess inability to see its omnipresence. (Nor, if I thought I did see it, could I prove its singular presence in SHC.) Maybe it is so obvious that I am blinded to it. And once revealed, in retrospect it will seem amazing that neither I nor anyone else noticed it sooner. Perhaps you can see what I do not.

At the same time, one must not err in thinking—as did von Liebig—that a single cause must be at the heart of singularly identical outcomes.

"Why?" is usually asked linearly; yet sometimes it is necessary to go beyond the linear, to a multiplicity of considerations, of whys. There can be many routes to the same destination. For example, on November 13, 1987, Joe Nickell and I sat together in the Toronto studios of CFTO-TV's *Lifetime* to argue SHC; a viewer would see two men together, but not the behind-the-scenes diversity of different philosophies, different scholarship, and different travel routes that converged at that moment on one soundstage.

So it seems with SHC. In truth, the vagaries of the phenomenon speak to a multitude of underlying factors, none common to all cases save that a human is involved. Behind the varied conditions reported for SHC and PC are factors that, now as in the nineteenth century, cry out for broad interdisciplinary consideration. Investigating the paranormal is not yet a mature science; it would be unwise to restrict one's considerations where limits are unknown.

After two decades of research, I am less sure than ever about an "ultimate" theory for SHC. I will therefore present a number of ideas. You can decide which (if any) make sense to you. As we delve deeper into some cubicles of science for the possible cause(s) of SHC, I ask two things: (1) that you recall the hundreds of cases you've just read about, and (2) that you remember these words about modern medicine written by Dr. Albert Paul Krueger in *Immunology and Allergy Practice* (July-August 1982): it is "the modern trend which recognizes that living forms react to seemingly minor environmental inputs, chemical and physical..."

> There are great ideas undiscovered... places to go beyond belief.
> —Neil Armstrong, on the twenty-fifth anniversary of his setting foot on the moon

MEDICINE. No field of specialization is impacted more by SHC than the medical sciences. And no other offers a greater diversity of possible answers. In the quest for those answers I will categorize this complex discipline, though in medicine nothing can truly be separated from the whole.

BIOCHEMISTRY AND NEUROSCIENCE. Medical researchers and pharmacologists should find speculating about SHC more than an entertaining pastime because the possibilities are many.

Anatomically, the hypothalamus is a prime contender in the hunt to explain SHC. The hypothalamus is the body's thermostat. The hypothalamus is the seat of emotions, such as anger and fear. It acts with the reticular system in the medulla oblongata to keep the brain awake and alert. It also regulates thirst.

Recall that Jack Angel and Dr. Sullivan both said they felt extreme thirst after SHC; that many victims are relaxed or asleep for prolonged periods (up to four days in Angel's case!); that some victims are known to have been hyper-emotional and quick-tempered. The hypothalamus, and factors that cause dysfunction in it, warrant serious consideration.

By recent count some thirty biochemicals control the transmission of neuron signals across the one hundred trillion synapses between the cells in a human body. Brain neurotransmitters fluctuate as their nutrients vary due to food, thought, and emotions. Then there are the human trait variables, determined by one hundred thousand genes containing three billion base pairs in the chromosome structure. Obviously the human body is not a simple system to quantify.

Therefore, any hormone or pharmacological drug that acts as a thermogenic agent or catalyst at the molecular level to dramatically increase oxygen consumption during metabolism can be suspect. As heart surgeon Dr. Lawrence Shaffer instructed me: "There are so many metabolic processes in the body that something could get out of whack, and suddenly—poof!"

Here are some possibilities for runaway thermal aberrations and linkages to SHC traits:

• Inositol (vitamin B_{10}) is a phosphagen essential to maintaining normal metabolism by substituting itself for temporary glucose deficiency due to oxygenation. *Applied Trophology* (December 1957) said this about inositol: "a compound like nitroglycerine, of endothermic formation. It is no doubt so highly developed in certain sedentary persons as to make their body actually combustible, subject to ignition, burning like wet gunpowder in some circumstances." As you know, many SHC victims were said to be sedentary.

• Endogenous pyrogen (EP) is a hormone particularly suspect, because, once released by the immune system, it acts in conjunction with the hypothalamus to turn up the body's thermostat and wage thermal war against viruses or other diseases. EP's beneficial action can backfire, though, in what Dr. Donohue has called "the body conflagration." In 1991 Dr. Redford Williams, a researcher in behavioral medicine at Durham's Duke University Medical Center, showed a link between outbursts of hostility and rise in cholesterol level. Hysteria — as in red-hot rage?—also gives rise to overproduction of endogenous pyrogen. Excessive EP then overstimulates the hypothalamus, causing very high fevers, delirium, convulsions, tissue wasting. And death.

When the EP-hypothalamus interaction becomes hyperactive, can anyone say *how* high one's temperature might soar in a fatal effort by the body to heal itself? A hothead's temper floods the hypothalamus with too much EP...the body's thermostat fires off frenzied commands to increase temperature...heat builds up in cholesterol-saturated fatty tissues...the person feels himself burning up, then...poof?

• The parathyroid offers another hormonal route for superhyperthermia. This endocrine gland regulates magnesium in the body and affects adenosine triphosphate (ATP), an isomeric ester that is an intermediate in the release of energy for muscular (and perhaps all types of cellular) work. One parathyroid hormone thought to be affected by alcohol can rapidly shut down the conversion of T4 cells to T3, thereby crippling the body's immune system and disrupting normal metabolism.

Hyperthyroidism embodies a constellation of symptoms called thyrotoxicosis, whose sufferers say they are "burning up inside."

Interestingly, an overactive parathyroid can dissolve bones. And a large portion of the body's endocrine system is governed by, yes, the hypothalamus.

Might, for example, a weakened immune system and accelerated metastasis lead to a magnesium imbalance, which sets off a destructive exothermic reaction assisted by an alcohol-agitated parathyroid hormone, which then disintegrates the skeleton as ATP flares off in electric-blue radiance?

• A strange toxic malady called Strep-A necrotizing fasiatis may have unexpected side-effects beyond its flu-like symptoms and 102°F fevers. Its eerie hallmark symptom is to make flesh resemble "burned roast beef," says one specialist, who adds that Strep-A fasiatis appears to run in cycles "about every thirty-five years." That's rather close to the apparent thirty-three-year SHC cycle. It's worth thinking about.

• SHC may be a *new* form of etio-cholanone fever—a recurrent fever associated with chemical degeneration of plasma that circumvents the usual safeguards against runaway thermogenesis. Taking the drug sulphazine will raise body temperature. Production by the body of elevated levels of phosphorus and sulfur, combining to make the blood pH alkaline then uniting with potassium to break down the immune system, might offer another thermokinetic route to SHC.

• Abnormal functioning of the lymph nodes may be a factor. These glands help cleanse wastes from the body's cells; furthermore, metaphysical educator Connie Newton tells me the lympathic system "deals with fire energy in the body." Perhaps this complex recycling system occasionally kicks into overdrive and burns up more than the body's accumulated toxins. That the elderly seem more prone to anomalous incineration, may be due to them having more years to store up poisonous substances; an aging mechanism less effective at eliminating toxins; a poor diet that further handicaps the lymph system...until its fire energy overcompensates and burns up the whole body from within. That the lymph nodes, which are most plentiful in the neck, armpits, groin, roots of the lungs, and along the large veins of the chest and abdomen, coincide with the area of the body where most SHCs originate—the torso and abdomen—may only be coincidence. Then, too, there's that New Jersey incident involving Mary, whose neck explosively burned—

• Neuropeptides are chemical messengers that link two centers I think significant to SHC: the brain and the solar plexus. Because neuropeptides are also involved in the regulation of moods and emotions, these complex amino acids may be a catalyst for an aberrant neurobiochemical breakdown that terminates in SHC.

• Any chemical that stimulates relaxation, sleep, or depression is a potential culprit. Serotonin makes the body drowsy; it also affects the way nerve cells communicate electrically, carrying chemical messages that excite or inhibit neighboring cells. A lack of B_3 (niacin) and B_6 leads to depression and emotional instability. The neurotransmitters dopamine and norepinephrine produce lethargy and apathy.

In 1989, Anna Inez Matus developed tremendous internal heat coincident with taking 250 mgms. of Niacin. "I have never experienced anything like this before," the health care professional told me. "The intensity was from the waist up. It felt like these sparks wanted to come out. I felt panic. I had to really keep calm. I think I came as close to spontaneous combustion as I would ever wish to—"

In 1990, an unusual variation of the dopamine D2 receptor gene was proven to be an important chemical link to alcoholism. Oxygen-containing organic alcohols burn with a clear blue flame; ethyl (grain) alcohol, with a flash point of only 54°F and an auto-ignition temperature of 793°F, burns with a pale blue sootless flame. Since dopamine produces lethargy and apathy, several *key* factors of many SHC incidents appear linked through this one receptor gene.

• The stomach is a chemical factory of enzymes and acids that break down food into energy the body can assimilate. Might extraordinary irregularity in diet and digestion produce not mere heartburn but, rarely, *body*burn?

Biochemically, elderly people are especially susceptible to depression if they survive on a diet of tea and toast, says Dr. Todd Estroff, a neuropsychiatric evaluator at New Jersey's Fair Oaks Hospital. Would Dr. Bentley's diet of coffee and shredded wheat also qualify, triggering a complex chain of events that can culminate in inflammability? Fatally?

• Breaking molecules apart typically generates heat and light—what you would expect, and what history shows does happen, in SHC episodes.

About 70 to 75 percent of the average human body is water, a molecule comprised of one oxygen and two hydrogen atoms. Water cools the heat of combustion (which is why firemen use so much of it). Water gradually decomposes into hydrogen and oxygen under great heat, to recombine as water again when the temperature is lowered. This process of physical chemistry, called dissociation, happens every time you boil water.

Total absence of liquid in several bodies examined—Mr. A. M. and Mrs. Oczki, to cite two—points to a process, chemical or otherwise, that can boil off body moisture from **within**. Water then becomes the *fuel* that feeds the fire-like reaction, rather than the means to squelch it.

> Hydrogen gas, in the presence of oxygen, burns with a blue flame and can burn hot enough to cut steel.

If, for instance, the water molecule's covalent bonds between H_2 (a highly flammable atom) and O_2 (which feeds fire) were broken, H^+ and OH^- free radicals would be released. The body would then be a natural fuel dump of explosive potential, because hydroxyl (OH^-) free radicals are highly reactive chemically and easily ignitable. Therefore, I propose a variation on the traditional fire rectangle: a Fire Pentagon comprised of oxygen, heat, fuel, molecular chain reaction...and now *free radical electrochemistry* as a fifth avenue for sustainable combustion in water-containing organisms.

In fact, biophotolysis—biological dissociation of water into hydrogen and oxygen—has already been accomplished at the chemical technology division of Oak Ridge National Laboratory. *Industrial Research & Development* (September 1980) reported that ORNL scientist Elias Greenbaum demonstrated that a biological mixture containing spinach, light, and the special enzyme hydrogenase produces self-sustained "true water splitting." Whether a similar process could unfold in man remains, so far, open to speculation.

Taking this possibility a step further, Powell and Finklestein reported in *American Scientist* (1970) that chemical interactions need *not* produce heat during collision reactions. The energy liberated by break-up or decay can be absorbed and stored in metastable states, then transferred to another molecule, producing light and radiation as byproducts. The relevance to classic SHC with its burst of light and minimal (or self-containing) heat is inescapable.

• The cornucopia of diseases already recognized by the medical profession may fire off inflammations of the charring kind under special circumstances not yet recognized.

One candidate might be fibromyalgia syndrome (FMS), a musculoskeletal pain and fatigue disorder for which the cause is still unknown. Most sufferers of FMS say they ache all over, and at times their throbbing muscles twitch and "at other times they burn." Memory lapses, numbness, dizziness, dry mouth, and impaired coordination are other traits, according to studies by the Fibromyalgia Network. Known aggravating factors for FMS are hormonal fluctuations, anxiety, and depression. These features seem emblematic to many sufferers of SHC/PC.

Wanda Cotton, a sufferer of FMS for twenty years, called recently to suggest her ailment might relate to SHC because of the intensified burning she's experienced inside herself. One theory being explored to cure her disease, she said, is that FMS is a genetic problem in which the body does not slough off phosphate. Phosphates are chemical compounds that contain phosphorus and oxygen in the phosphate radical, PO_4. Phosphorus can spontaneously burn in air. When it burns in a good supply of air, dense white vapors of phosphorus pentoxide form (which is used as a smokescreen agent) and immediately absorb water.

Several survivors of SHC have reported whitish smoke exuding from themselves, you will recall. Maybe a few of them were sufferers of FMS (a recently defined malady), and one day experienced a very rare aberrant variation of it. It would be useful for firemen to take fresh scrapings of the oft-encountered caramel-colored soot deposited throughout a room that hosted classic SHC and have it tested for the presence of aqueous phosphorus pentoxide. One never knows where a clue to resolve the unknown might be found—

Undoubtedly, there are other pathways yet unsuspected. As David Stone of the Genetics Institute told *The Wall Street Journal* (November 13, 1989): molecular biologists and neurologists confront in the cell "a vast array of products, most of which probably haven't been discovered yet..."

BIOELECTRICITY. The concept of SHC is electrifying. As may be the cause of the phenomenon.

Each living cell in every human being has electrical energy. This bioelectric potential can be discharged through the cell membrane or internalized to destroy the cell. (In the three seconds it may take you to read this sentence, you will produce about twelve billion electrical impulses in your brain alone.) Could the body's electrical circuitry malfunction in the way that short-circuited wiring can start fires that burn houses to the ground?

One crucial element could be the hippocampus, the most electrically unstable part of the brain.

A cubic inch of human muscle cells can theoretically generate 400,000 volts of very low amperage, Dr. Mayne R. Coe estimated in *Fate* (July 1959). If amperage were to significantly increase, self-electrocution would be unavoidable. Might, then, the scores of billions of cells in the human body (the brain alone has some fifteen billion cells) suddenly and simultaneously discharge their electrical potential to unleash a massive natural burst that fuses out the body like a current overload melts thin wire? Might such a burst dissociate, by flash electrolysis, the body's water molecules into highly combustible H_2 gas and OH^- radicals, to be ignited by a bioelectrical spark?

As modern as this reasoning sounds, it was addressed as early as 1817 by a precocious German scientist named von Nasse, who explored in *Archiv fur medizinische Ergahrung* (August 1817) the role of "electrical phenomena" in self-combustion (although he considered only electrical energy at the surface of rather than inside the human body).

In the 1940s, French bioenergy researcher George Lakhovsky determined that each cell is an electrical resonator surrounded by a modifiable electromagnetic field with chromosomes functioning as regulators of the cell's energy. In other words, the body is a complex electrical circuit subject to the vagaries that plague any electrical device. Lakhovsky's oncological research focused on one part of a cell, the organelle *lysosome*. He discovered that a reduction in oxygen to the cell plus the introduction of various toxins—such as from cigarette smoking?—could increase the number of lysosomes, which, in turn, rupture individual cells to set off a lethal chain reaction of electrical short circuits racing through the organism.

Might biochemical or bioelectrical imbalances exceed a crucial threshold whereupon rapidly produced lysosomes or lymphotoxin proteins unleash body-wide cellular suicide with byproducts of heat, toxins, electrical arcing, and gases? Gases that burn blue, such as those inside Lucie Bettelsen—whose abdomen suddenly emitted blue flame, which burned the surgeon operating on her at New York's Walter Mandich Hospital in 1975.

What could trigger a runaway lysosome chain reaction is unclear, though medical researchers suspect carbon monoxide—a byproduct of smoking and a hazard faced by Billy Peterson in his car. Interestingly, lysosomes—which, by the way, microbiologists call "suicide sacks"—will destroy bone.

• Another runaway condition to consider is Rasmussen's Encephalitis, an extraordinarily rare malady which causes abnormal triggering of brain electricity. According to Dr. Benjamin Carson, chief of pediatric neurosurgery at Johns Hopkins School of Medicine, it can produce eighty to one hundred seizures daily. Such massive overloading of the brain's fourteen billions cells and connections (Dr. Carson's estimate), could have dire consequences to the temperature regulatory mechanism inside the head of the sufferer. Strokes could be another outcome. And a stroke messes up the body's thermostat. Might some victims of SHC suffered a stroke beforehand, with undiagnosed Rasmussen's Encephalitis compounding the assault on one's hypothalamus? Perhaps further aggravated by thyrotoxicosis, all of which no fire investigator or coroner thought to look for?

• If a radical theory by Dr. Björn Nordenström of Stockholm's Karolinska Institute is correct, then *everyone* is a living, breathing, biologically closed electric circuit (a concept similar to Lakhovsky's). Dr. Nordenström told me that he has identified all the elements of an electric circuit in the human body, where blood vessels serve as cables; white blood cells as negative-charge conductors; enzymes in capillary walls as the electrodes; necrosis in tumors as an AC power source; and dead and dying cells as a battery drive for the entire system. He has applied his unconventional theory to effect dramatic electrical healing of cancer by attacking tumors with electrodes.

Dr. Nordenström could offer me no suggestion about how an antithetical phenomenon like SHC might unfold. However, in explaining that bioelectric circuits become less stable with increasing age and infirmity, he does posit an association with SHC and its elderly and incapacitated victims. Systems breakdown in Nordenström's bioelectric model of man is a contender in the race to explain whole-body electrical arcing that itself models SHC.

Perhaps intuitive awareness of such a breakdown is what prompted John Reno, a registered architect in Virginia, to telephone within minutes of seeing me on NBC-TV's *The Other Side* (February 27, 1995). As he explained:

> My interest in SHC began in earnest last year after the sensation that I was a potential candidate for SHC. The discovery that the human electrical system enters into a state of "chaos" and this resultant chaos of electricity within the body triggers hydrogen as a fuel that consumes the body, is an observation that I am certain is a fact. I am also certain that a chemical (unknown to me) within the body triggers the electrical chaos.

Reno had no suggestions for the trigger, however, but he does think a person's emotional state and consciousness are relevant to unbinding a quantum chaos in the body. "There are periods when I'd get very excited," he related. "I'd have this incredible sensation of a quantum leap in my body, like going from 100 MPH to 10,000 MPH!" This feeling of being accelerated at every atom in his body got him thinking about a friend's research in ultra high-temperature electrical engineering, where 15,000–18,000°F can be generated in a hydrogen-fed fire. "If chaos freed up the hydrogen in my body and somehow ignited, it would be like using a drop of water to put out a house fire!" he postulated.

• Not everyone responds to electricity, to stress, to stimuli in the same way physiologically. Some people are supersensitive to electrical fields; others, so insensitive as to appear almost immune to electricity.

"About 15 percent of the population suffer from allergies, and of those, about two in every three develop a sensitivity to electricity," says Dr. Jean Monro, a leading allergy specialist in London. Dr. William Rea, director of the

Environmental Health Center in Dallas, Texas, has found that symptoms of electric hypersensitivity include depression, skin rashes, and profuse sweating. The inability of these people to tolerate subtle electric fields—fields generated either inside or external to the body—may have a highly inflammable side-effect, heretofore unsuspected. SHC.

Conversely, those with higher tolerance to electricity may have another problem. "I had 240 volts pass through me twice within 30 minutes," medical technologist Anna Inez Matus revealed. "Since then, I have been thirsty. Constantly. People who survived being struck by lightning also have a constant thirst desire." I recall that Jack Angel complained of extreme thirst after waking up. Although lightning was ruled out in his case, dehydration from a four-day sleep certainly cannot be discounted. Neither can partial, internal human electrocution be discounted—given that physicians likened his injuries to those produced by high-energy electricity. A high-amperage *bio*electricity? His attending physicians *did* say the burning was "internal in origin"—

ELECTROSTATIC FORCES. Human dynamos exist. A Zulu boy named Kinkon packed such a charge at age six-and-a-half years that, when touched, "what is understood to be an electrical shock is experienced" whose intensity varied with the weather, said *The Medical Press & Circular*.

Remember Susanna Sewell? Fodor's *Encyclopaedia of Psychic Science* mentions a baby born in 1869 at Saint-Urban, France, who shocked all who touched him; who emitted luminous rays from his fingers; who glowed for several minutes after dying at nine months. In 1895, Jennie Morgan of Sedalia, Missouri, suddenly began emitting sparks from her teenage body, shocking herself and others. In 1952, Englishmen Charles Bockett and Brian Williams, two human generators, lit lamps with a touch of their fingers.

You will recall that during 1982–1983, ten-year-old Benedetto Supino displayed electromagnetic effects that ruined equipment, ignited objects, and burnt himself; yet one scientist called him "normal." As normal as Mrs. Jacqueline Priestman, twenty-two, who in January 1985, burned out thirty vacuum cleaners, five irons, two toasters, and sent sparks flying from outlets in her home near Manchester, England. Dr. Michael Shallis of Oxford University told the *Sunday Express* (January 20, 1985) that this housewife from Stockport, Cheshire, was the worst of six high-voltage people he knew—in that she generated "*more than 10 times* the usual amount of static electricity in her body." Dr. Shallis also studied Manchester housewife Pauline "Old Sparky" Shaw, who blew up three hundred light bulbs, twelve televisions plus other appliances because 100,000 volts of electricity surged through her body. Scores more examples could be cited.

Mrs. Priestman discovered that increasing her intake of green vegetables ameliorated the severe electromagnetic effects she had been broadcasting for four years. The new diet enabled her to live a stable, nonshocking life.

Some human capacitors live with their affliction for years, as did Englishwoman Mandy Boardman from 1983 to 1986. Others manifest it autonomically, unpredictably, instantaneously and perhaps with greater intensity than their long-term suffering counterparts.

Electrostatic cooling is an industrial technique used to lower temperatures of hot steel. Might the inverse occur in people? Could electron vibrations increase in the body until, like a transformer blasted by static oversurge, the unwary person suddenly explodes electrostatically from within?

And what does one say about the "electric woman" of Monnaie, France? In 1756 her electrostatic propensity heightened whenever she stood on mounds of earth, said SHC investigator Dr. Pierquin. Or about Frank McKinstry of Joplin, Missouri? In the 1880s, his extraordinary static(?) charge was so intense that on cold mornings his feet stuck to earth "as though he were treading on fly paper" and, if he stood still, he had to ask passersby to pry his glued-to-the-ground feet from their strange bondage! Or Canadian Caroline Clare? As documented by the Ontario Medical Association, in 1877 at age seventeen she became a "human magnet" for metallic objects. Or Louis Hamburger? This teenage magnet's knack for picking up and suspending a five-pound jar of iron-filings with three fingertips was studied at the Maryland College of Pharmacy in 1890. Or the thirty-four convicts who exhibited undue "magnetism" toward paper as a consequence of botulinus poisoning at New York's Clinton Prison? Their highly charged condition varied with the severity of poisoning and vanished once recovery was complete, chief prison physician Dr. J. B. Ransom told the *Electrical Experimenter* (June 1920).

Perhaps it was this kind of puzzlement that Apollo 14 astronaut and consciousness researcher Edgar D. Mitchell had in mind when he said, "There are no unnatural or supernatural phenomena, only very large gaps in our knowledge of what is natural"...large gaps that allow for the mysteries of bioelectromagnetism and SHC.

GEOMAGNETISM, ASTROPHYSICS, AND CYCLES. Speaking of magnetism, magnetic fields influence human physiological processes.

Broken bones will mend faster in a modulated magnetic field, for example. Dr. Jerry Phillips described to ABC-TV's *Good Morning America* (November 11, 1989) his research that showed electromagnetic fields can increase cancer cell growth rates and diminish white blood cell counts.

V. M. Mikhailovsky at the Ukrainian Academy of Sciences has demonstrated that even a weak 1000-gamma magnetic field (about one percent of the Earth's normal surface field) applied at frequencies between 0.1 Hz and 12 Hz would produce "measurable reactions" in three out of ten people, including increased pulse rates and sweating outbreaks.

America's pioneer researcher in biomagnetism, Dr. Robert O. Becker, wrote in the *New York State Journal of Medicine* that "it is not difficult to conceive of a galvanomagnetic effect between the fluctuations of the magnetic environment and this solid-state electronic system in the brain stem producing the subtle behavioral alterations of biological cycles."

Professor Livingston Gearhart, of the State University of New York (Buffalo), pointed out in *Pursuit* (1975) that "the onset of magnetic storms coincides with much human-related forteana." SHC is one fortean event Gearhart linked to Earth's magnetospheric storms. And University of Florida (Jacksonville) entomologist Philips S. Callahan, Ph.D., has been finding empirical support for his idea of "biological forms as radiators and antennae for electromagnetic bio-information," he told The Center for Frontier Sciences on November 16, 1990.

> In May 1992, Joseph L. Kirschvink, a geobiologist at the California Institute of Technology, announced proof of "little biological bar magnets" made of crystals and magnetite inside the human brain:
> "This really is an exciting discovery.... It is the first new material that has been found in humans since our ancestors found blood, guts, and bones."
> Besides offering a possible link between cancer and electromagnetic fields, he proved that unidentified materials can still lurk inside the human body.
> —*Patriot-News* (May 12, 1992)

I invested a few days at the National Oceanic and Atmospheric Administration's library in Bethesda, pouring over log books that contained records of global geomagnetic activity. Each day is quantified on a scale from "0" (a calm, stable day magnetically) to "2" (the geomagnetic equivalent of a killer hurricane). Was there a correlation with SHC episodes to suggest magnetism may have a malefic aspect?

You already know that 60 percent of the ten cases I classified as auto-SHC coincide with elevated geomagnetic activity, including Olga Worth Stephens who burned up in her undamaged car on a one-day geomagnetic spike rated 2.0.

Among the nonauto autocremations, Elizabeth Clark burned up on her unscorched bed on a day rated 1.8 (a very strong global magnetic storm day); Lois Chapman, in a stormy cycle that peaked at 2.0; Madge Knight, on a 1.9 day; and Billy Peterson, near the 1.7 peak of a six-day bubble in a very calm geomagnetic period. Approximately one-sixth of the SHC database, up to the 1980s when NOAA changed its magnetic storms reporting format, coincide with cycles of peak disturbances in the Earth's magnetospheric or with bell-curve rises amid otherwise very calm periods.

Another cycle that everyone experiences, on a more familiar level, is body temperature itself. And most people live quite comfortably with their thermal fluctuations and within the Earth's magnetic fluctuations, which envelop everyone.

But as some individuals can be allergic to ragweed, or some radios receive short wave and others not, might a select number of people uniquely respond to a narrow-band electromagnetic frequency? A reception that, once "tuned in" to by cellular circuitry, turns up the heat on their temperature cycle to a beat they can't live to? A beat that so vigorously vibrates the body that its water boils (or dissociates by spontaneous molecular hydrolysis) until little remains but dehydrated carbon ash and calcium bone dust—

Several human-related fortean spontaneous blazes coincide with dates of severe magnetosphere storms—which, in turn, are associated with solar flare activity, itself cyclical. Hence, astrophysics becomes another potential avenue for multidisciplinary discoveries about how, and *when*, a human body might unleash its neuro-cellular electrical potential of millions of volts during a magnetic tempest born of solar flares (or something else). Bioastrogeomagnetics will doubtless prove instrumental in researching SHC and other bodily phenomena.

Other cycles and associations may be at work, too.

Biorhythm theory is a possibility to consider. This concept—in which a person's physical, emotional and intellectual aspects are said to cycle through periods of twenty-three, twenty-eight, and thirty-three days respectively—postulates that traumatic (possibly sudden) physical events can occur to a person when high emotional and intellectual energies coincide with a physical critical day (when the twemty-three-day curve crosses the baseline that bisects its amplitude). Such was the case for Mrs. Reeser's biorhythm at the beginning of July 1951.

I do not claim that biorhythm alone will cause SHC—the fact that thousands of people aren't igniting daily offers convincing proof—but do ask whether one's biorhythm, in conjunction with *other* inflammatory factors being proposed, might be the "flashpoint" that sets off the human body in a bone-melting, tissue-dissolving cremation from which there is no escape. Except to the grave.

SHC, despite its infrequency, seems to cluster in cycles itself. Significant cases of pyrophenomena have occurred in, or clustered around, these years: 1749, 1771, 1780–1782, 1804–1808, 1836–1837, 1871–1872, 1904–1905, and 1938. These dates reveal a heretofore hidden periodicity for combustion phenomena:

    1749 to 1771 = twenty-two years
    1749 to 1780–82 ≈ thirty-two years
    1771 to 1780–1782 ≈ eleven years
    1780–1782 to 1804–1808 ≈ twenty-four years
    1771 to 1804–1808 ≈ thirty-three years
    1804–1808 to 1836–1837 ≈ thirty-two years
    1836–1837 to 1871–1872 = thirty-five years
    1871–1872 to 1904–1905 = thirty-three years
    1904–1905 to 1938 ≈ thirty-three years

From whence this periodicity of approximately thirty-three (or multiples of eleven) years? First thought: extraterrestrial. The Sun exhibits sunspot activity peaks on a twenty-two-year cycle, with a smaller interim peak every eleven years. Humans and their property react inflammably to an unidentified solar (cosmic?) phenomenon occurring every half solar cycle, then peaking about every one-and-a-half solar cycles (that is, about once every generation, or third of a century).

As stars occasionally nova in the macrocosm, might not the human body in its microcosm infrequently become its own supernova due to a mystery force in deep space?

In the end, this periodicity may be a false artifact of too little data. The next predicted thirty-three-year peak for SHC was 1971. Our data reveal nine examples of anomalous fire phenomena for this year. Significant? Conversely, twice that number of people might have mysteriously inflamed that year and no one had the courage to publicize "the impossible." I wait to see if hidden data from 1970–1972 will now come to light. (I do note a cluster of anomalous fires for 1980–1982, a forty-two to forty-four year interval from 1938; that is, about a double solar cycle.) The next thirty-three-year cycle will occur in the next millennium, around A.D. 2004. One can but wait to see if 2004 will host the next rampage of paranormal fires, including SHC.

One more area having its own cycles may provide insight into the timing for SHC.

With no small sense of trepidation I introduce it, because it's certainly not a factor firemen and coroners consider when investigating a fire fatality. But then, to begin with, we are beyond the scope of run-of-the-mill considerations. This is the cycles of cosmobiology.

The cosmic patterns in which one is born are unique for each individual, based on time and place of birth. Astrophysicists acknowledge this. The symbolic language of those patterns is found, and interpreted, through astrology. A person's birthchart or natal horoscope, then, contains an individual message that, when knowledgeably studied, can reveal significant characteristics of personality while serving as a celestial roadmap to pinpoint when important potential events in one's life are likely to reach a point of explosion.

Might clues to SHC be disclosed through examining the natal charts of its victims? Specifically, does a birthchart identify a basic flaw in personality or a crucial time in life when the outcome will be incendiary? Literally?

I cast birthcharts for thirteen SHC victims,* five of them fatal. Exact birthtime was available only for Marilyn O—; others were retrocalculated from progressed charts. Hence the analysis that follows is not as precise as I'd like, but given the difficulty in obtaining such information it is a historic initiative.

Several astrologers helped to interpret this data, notably Ruth Brown; Roxan O'Brien; and Canadian researcher Shelagh Kendal (now living in the United States).

Mrs. Brown determined that, cosmobiologically, most victims had been "under mental stress or confusion;" that "they were all powerful individuals whose basic consciousness and self-expression were 'one' with the life force;" that "they all had afflictions to the 6th [health and physical resources] and 12th [interferences] Houses, showing that the energy of the solar constitution was out of harmony with itself and therefore could be used to its own detriment." She discovered that crucial transits/aspects of Saturn (evolutionary moment of truth), Uranus (explosive turning point), Neptune (breaking up of old patterns in the search for perfection), and Pluto (urge for self-transformation and rebirth) "are important" relative to the birth energies in each of these individuals when their internal fire broke out.

Miss Kendal concurred. "My conclusion," she stated about the planetary placements she found in these thirteen charts, "is that, in each case, there was a pyre of misused energy waiting to be torched in all these people, and the positions of the planets lit the match."

Specifically: the planet Mercury symbolically transmits solar energy to Earth, and its movements are linked to changes in electromagnetic fields. Mars, universally identified with fire, represents physical energy and survival; if Mars's energy is not tamed, it may eventually implode in a ramage of heat, anger, and rage.

"All the people in this study had serious blocks to either Mercury or Mars, or both," Miss Kendal discovered. "As well, at the time of the combustion, the victims' progressed and transit charts showed that internalized forces were building to an explosion."

*Melvin and Naomi Anderson; Jack Angel; John Bentley; Helen Conway; Barbara Green; Harry Halfpenny; Elizabeth Holstrom; Elizabeth Norris; Marilyn O—; Beatrice Oczki; Mary Reeser; and Edward Sullivan.

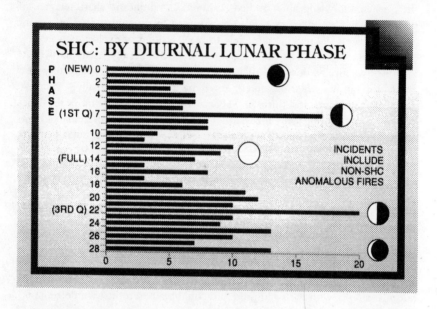

More specifically: the triggering for these explosions. An astronomical phenomenon called retrograde makes it appear that planets periodically move backward. This happens three times a year for Mercury, and symbolizes a time when electromagnetic confusion is greatest. "At the time of each SHC incident, in every case," Kendal discovered, "the previous period of Mercury retrograde had focused on a natal planet." Furthermore, she found in all cases that "the previous solar eclipse and Mercury retrograde had, in some way, coincided with the natal charts." What differentiated fatalities from survivors, she said, was that the latter's charts depicted a means of ameliorating or otherwise releasing this pent-up energy.

More specifically: for Mars, retrograde happens about every other year, and symbolizes a time when Mars energy turns inward—causing extra strain and frustration for people who do not direct their physical energy outward (that is, persons already depressed or morose).

Four of these five people who died from SHC, says Kendal, "had direct correlations of the position of Mars when it went retrograde before their deaths." She suspects the fifth fatality had a similar correlation, though data wasn't sufficiently precise to express confidence.

Uranus and Neptune, representing *sudden* transformative change and stimulating the search for *perfection*, are both prominently aspected in these charts.

Astrology can't predict the physiological underpinnings of SHC... but it may be useful in determining who is susceptible; and when; and how to avoid it. Anyone experiencing unexplained localized tissue scarring or smoke emission—as presursors to classic SHC—would be advised, Miss Kendal suggested, to reconsider his or her behavior patterns; especially in terms of communication (Mercury) and use of physical energy (Mars).

GEOPHYSICS AND GEOBIOLOGY. This planet still controls its best secrets, and the cartography of combustion only *begins* to hint at the power and wonder that lie below our feet. That more than 80 percent of the anomalous burnings identified in Britain by 1976 can be connected geographically surely merits the interest of geophysicists and biologists as they probe the frontiers of the human-planetary relationship.

Anne C. Arnold Silk, a member of the Royal Society of Medicine, researches the effects of piezoelectricity produced by certain rocks under stress and knows that seismic movements at fault zones can also generate telluric radio and microwave energy. She wondered whether these emissions from geologic activity might have biological effects on humans. Examining forty-seven British cases identified as SHC by Randles and Hough, Silk found that an astonishing thirty-one cases correlated with areas known for seismic activity.

Randles and Hough find this intriguing, but are properly cautious: "The figure, almost 75%, must however be weighed against the low sample, and "the fact that many parts of Britain are heavily faulted anyway." I agree with their caution.

Ms. Silk's interests in medicine and geology led her to an imaginative theory, nonetheless. Knowing that certain types of radiation can ignite oxygen, she offered this multidisciplinary speculation to Randles and Hough for their book *Spontaneous Human Combustion* (1992): "Supposing that either a lightning strike, or a seismic energy emission enters the body, it could ionize the gas [oxygen] in the lungs and any flatus in the stomach. This latter would be very common in the sedentary elderly, the status of many SHC victims. The result would be a conflagration."

> Ca. 9:00 A.M., July 1, 1848. The parish of Hinstock, near Market Drayton, Shropshire, England: "an explosion of the electric fluid burst forth out of the earth with a horrid noise, so as to be hear many miles.... tearing up several trees by the roots... stripping several cottages &c of their thatch. Two young men... were suddenly whirled into the air, and carried into the adjoining fields." Devastation spanned several miles, yet in a strip only 300 feet wide.
> —*The Oswestry Advertizer* (Aug. 17, 1898); credit Richard Holland, *Fortean Times* no. 62

Not to be overlooked in geology, and one day to be explained, are numerous examples of electrode-like displays of the earth *aglow*. Only a decade ago, U.S. Department of Mines scientist Brian Brady demonstrated that quartz-bearing rock sufficiently compressed (150,000 psi) will produce a "luminous arc or jet" in the form of "small round glowing balls."

Might piezotelluric electricity—on a much *grander* scale—be what terrorized the Abbé Girolamo Leoni de Ceneda in 1731 when, near Venice, a "flame taller than himself" suddenly burst forth with a roar from the ground where he had just stood? Perhaps the abbot survived instant incineration/explosion by one fortuitous side-step that day, unlike the unfortunate Italian woman from Bonne Vallie twenty years later.

And how will geology explain what a Filipino scuba diver retrieved from the sea off Bataan in 1992? He brought to the surface several fist-sized red rocks, which heated up and smoked once exposed to the air. One child was *severely burned* after handling them, reported the *Bangkok Nation* (August 15, 1992). Scientists of the Philippines government rubbed the rocks together, whereupon the latter burst into flames and burned to powder. So much for scientifically preserving evidence—

Much earlier in history, in A.D. 363, laborers rebuilding the Temple at Jerusalem fared less well than did that Filipino child. As the *Ammianus Marcellinus* recorded, "fearful balls of fire burst forth with continual interruptions close to the foundations, burning several of the workmen and making the spot altogether inaccessible."

Earthfire is not a phenomenon to be ignored.[1]

Standing in the right place at the wrong time might just prove combustibly hazardous to one's health.

METEOROLOGY. Atmospheric phenomena not yet understood by weather forecasters may play a role in SHC.

Many types of airborne whirls of fire are no less perplexing than the *authentic* anomalous "crop circle" agrigrams that, in the past decade, have spontaneously decorated fields worldwide with intricate, geometrical patterns impressed upon uncrushed crop stalks. A few of these enigmatic agrigrams have been the sites for sightings of anomalous globes of light that have been seen (and video-recorded) moving along the pathways of depressed grain.

Here are some candidates, culled from meteorological records, to consider in the context of this study:

• BALL LIGHTNING. This phenomenon is defined by James Barry in *Ball Lightning and Bead Lightning* (1980) as "a single, self-contained entity that is highly luminous, mobile, globular in form, and appears to behave independently of any external force." History has given ball lightning a dozen names, yet, until Barry's definitive work, many scientists rejected ball lightning despite reliable sightings dating back to at least the sixty century A.D.

Barry noted that while heat emission is rarely associated with ball lightning, there are exceptions when humans are struck. You already know about Robert Burch and William Cunningham. Barry adds to the list these additional quasi-SHC casualties of fire from heaven: a man burnt and rendered unconscious circa 1890; a female whose foot was burnt circa 1886; a Wiltshire man burnt on his shoulders on June 20, 1772; the Russian Benjamin Franklin—Dr. Georg William Richmann—struck on the head and killed by fist-sized blue-and-white ball lighting (or something resembling it) that entered his St. Petersburg laboratory on August 6, 1753.

Barry pointed out that ball lightning often detonates in an implosive rather than explosive manner, sometimes leaving behind a tar or sooty residue. Was this what suddenly appeared above the kitchen stove of a young housewife in Smethwick (Warley), England, in August 1975? The "bright blue-to-purple" light, about four inches in diameter, moved across the kitchen and toward her. It hit her below the waist, burning a hole in her polyester skirt, heating her wedding ring, burning her finger, and causing her hand to redden and swell. She brushed it aside, only to hear it detonate with a loud bang! M. Sternhoff estimated for *Nature* (1976) that this particular fireball carried a temperature of at least 480°F in its one-inch core, while Barry calculated that ball lightning just 50 percent bigger could pack a wallop of one megaJoules—which is two-thirds the energy density of TNT or approximately a million watts if released in one second!

Orbs of plasma up to forty megaJoules were artificially created in the air by the United States Navy in 1947. Short-lived fireballs have been manufactured at Yeshiva University by the use of electromagnetic fields of radio frequency. Perhaps nature can equal or better that power,[2] and do so, naturally, in the vicinity of humans... and animals.

Randles and Hough disclose two incidents possibly relevant here. On the evening of February 24, 1975, a postal worker witnessed a greenish-yellow pumpkin-sized buzzing mist that "attacked" a man and his dog on the beach at Sizewell, Suffolk, northeast of London. The ethereal form soon resembled a TV set's picture tube, said the postman, and emitted an electrostatic field. The dog cowered; then, showing the wiser part of valor, fled. Its owner, who maintained his position beneath this object, soon suffered prolonged ill health and perhaps debilitating injury (the nature of which is not given).

And on February 9, 1988, a long-time military man was walking along a road near Oswestry, Shropshire, in west-central England. Ahead, he saw a woman and her spaniel dog by the Shropshire roadside... and in the field beside them a misty cloud of yellow vapor "glowing with angry ferocity." He noticed, he told Randles and Hough, that this vaporous spheroid seemed to vibrate the grass under it and created a sulphurous odor (which may have been ozone created by an electrical field ionizing oxygen in the air). The spaniel ran toward this vapor and vanished into it! Moments later, the radiant cloud itself vanished. Left behind was the spaniel, comatose, panting heavily, and mysteriously afflicted: "Its eyes were red and fiery and its skin was soaking wet and almost scalding hot to the touch."

Left behind was another mystery, of course: the origin and nature of the thing that sucked up and parboiled a hapless pooch. The dog's owner, interviewed by Randles and Hough, corroborated this story but refused to cooperate further; her dog, which seemed to recover fully after three days, passed away a few weeks later.

Here's a horrid hypothesis: that the biosphere creates powerful plasmoids at coordinates of converging energies sometimes just as a hapless human (or pet) strolls into the zone of emergent fire, never to walk away. Did the Smethwick housewife narrowly miss standing at such a manifestation point, to escape with only partial rather than whole-body SHC?

• CONVENTIONAL LIGHTNING BOLTS. These spectacular atmospheric displays can discharge in one bolt about one hundred million volts of electricity, heating the air in its path to more than 60,000°F. Lightning can do some strange things.

On July 26, 1969, for example, Harold Deal was hit with enough electricity to light a small city. The lightning bolt blew the rivets out of his shoes and melted $1.50 in change in his pocket; adding injury to insult, the force of the strike crushed discs and vertebrae in his spine. "I'm two and a quarter inches shorter than I was before I was hit," he told Lightning Strike and Electric Shock Victims International at its fifth annual convention near the Gettysburg Battlefield in 1995.

In June 1984, lightning melted a baseball cap to the head of Canadian Harold Blankert and sheared off his undershorts. "Like somebody had taken a pair of scissors to them," said the Alberta farmer who qualifies for membership in LSESVI. In late July 1994, Veronica Ronn became another unlucky winner in lightning's cruel game of change: a bolt of lightning blasted the Swedish teenager right out of her shinguards and boots at a soccor match in Gavie; the lass's footwear was ripped to shreds and set afire, her heart stopped for four minutes, and upon regaining consciousness her first words were, according to the *Oxford Mail* (August 10, 1994), "How did the match go?" And on September 14, 1994, lightning momentarily handicapped golfer Nancy Wilde on an English countryside, knocking her unconscious and badly burning her hair; when a nearby medical trainee rushed to her aid and bent down to administer mouth-to-mouth resuscitation, smoke spewed from Ms. Wilde's opened mouth. The *Daily Telegraph* (September 16, 1994) reported that she was expected to recover fully from the shocking experience that had effectively rendered her dead for some ten minutes.

And then there are the human lightning rods. Among them is Brad Willard, an Orlando insurance agent hit by three separate bolts within seconds while playing golf in south Florida. And Major R. Summerford, who was also thrice struck, first in 1918, again in 1924, again in 1930, to become paralyzed and wheelchair-bound; yet not even in death was he safe, for two years after his burial a bolt flashed down upon the graveyard to strike only one thing—Major Summerford's tombstone! And one cannot forget Roy Sullivan, a National Park ranger in Virginia and record-holding human lightning rod, who survived *seven* lightning strikes, including one bolt that set his hair on fire.[3]

However many times a person might be struck by a bolt from the blue, I have never found a medical reference to lightning *cremating* a person.

• ANOMALOUS DISCHARGES. Meteorological literature does, however, suggest there are less well-known types of atmospheric discharge that might.

One such mystery has occurred in the Hessdalen valley of central Norway. Here, between 1981 and 1985, hundreds of strange lights appeared in the air. Sometimes they would hover; other times, they sped through this small valley at an awesome 18,600 MPH! A research station set up to probe them netted perplexing data. The glowing orbs produced no spectral lines on film; a spectrum-analyzer registered occasional electromagnetic signals with 80 MHz harmonics; a magnetograph detected changes in the local magnetic field when the lights popped up; laser beams directed at the glowing spheres caused behavior changes in them; and, provocatively, sometimes radar revealed the presence of some additional invisible thing moving in ways comparable to the lights when seen, according to the *Journal of Scientific Exploration* (1994).

Though I have no report of anyone in this valley succumbing to SHC, I can't help but wonder what might ensue should a Norwegian be standing in the trajectory of one of these unseen magnetoelectroluminosities rushing toward him at thousands of miles an hour. Also to consider is whether The Hessdalen Phenomenon, as it's been named, is isolated to this one valley, or can be hurtling invisibly toward others elsewhere around the world.

Another consideration dates back to 1971, when orbiting satellites detected random and brief cone-shaped bursts of light described as "like amorphous glowing octopi" shooting into the stratosphere at 80,000 MPH. They did not coincide with normal lightning. Announced *Nature* (August 20, 1971), "No acceptable mechanism has yet been found to explain these anomalous data spikes." Some twenty years later, scientists at the University of Alaska succeeded in filming these huge anomalies, now called "red sprites" and "blue jets." In 1995, they remain unexplained.

Could one of these aberrant flashes blast downward and transform someone to dehydrated dust? Since the mechanism that births them remains a mystery, no one can say.

Here's a frightful possibility. Tomlinson described to the *Philosophical Magazine* (1888) a phenomenon I'll call *white* (flashless) *lightning*, which killed several people and animals, yet wasn't seen by nearby observers even though it left unambiguous evidence of high-current discharge. Invisible lightning strikes down unsuspecting bystanders and incinerates them to ash, instantly—including Mrs. Wilber Finer, fifty-five, of Dubuque, Iowa? She died two days after being found outside in her yard on September 10, 1969, with all her hair and clothing burnt off; no trace of scorching anywhere beyond a few blades of grass where she collapsed.

• MICROWAVES. Microwaves are radio waves of very short wavelength—between 100 micrometers and tens of centimeters. Besides figuring prominently in a modern kitchen appliance and electromagnetic pollution generally, they are yet another possibility for SHC—particularly since medical personnel have characterized odd human burnings as microwave-like in nature.

Microwaves generate electric and magnetic fields in objects they touch. Microwaves can heat water-bearing substances, as most American cooks know. Microwaves can also burn. One California radar operator suffered a quarter-sized hole which microwaved into his chest as he sat at his console. In Massachusetts, Cape Cod hang-gliders have been burned when they unwittingly flew in front of powerful U.S. Government Pave-Paw radar units there.

Science News reporter Dietrick Thomsen even discussed, in 1983, the capability of high power microwaves to tear apart atoms and molecules under the title, "Zap! You're Disintegrated." He was concerned about the integrity of computer circuitry exposed to microwave power of one megawatt per square centimeter, achievable (ostensibly) only by complex and expensive technology. My concern is, also, about ripped apart atoms inside radar operators and hang-gliders.

Question: Could short microwave bursts be generated *naturally* with similar body-scorching results? G. I. Babat recounted for the *IEEE Journal* (1947) his experiments that showed electrical flames could be formed by the microwave breakdown of air. Cade and Davis reviewed in *The Taming of the Thunderbolts* (1969) several theories postulating that ball lightning could produce short radio waves, then proposed a disquieting consequence similar to mine for the Smethwick housewife: "it is possible for victims to be burned to death... within their skin...just by the action of the intense radio-frequency field which, in the absence of their body, would have formed a lightning ball at that place."

Hence a variety of atmospheric discharges not associated with thunder-and-lightning might mark their short-lived, unexplained existence with SHC.

• SKY FIRE. The skies unleash a myriad of odd deluges with fiery ramifications. During the afternoon of January 5, 1909, reported *The Los Angeles Times* (January 7, 1909), the air electrified over Santa Cruz and then precipitated a "molten rain" of "little red hot metal globules" around several boys, one of whom "carried a burn on his finger" after being struck. On July 4, 1945, what the Spanish government called a "rain of fire" precipitated out of clear skies and spontaneously ignited clothing and plaster walls—especially *white* ones—in the town of Almeria; a series of thermal anomalies so strange the government documented it with a book-length report. And how will meteorologists explain, wondered Frank Edwards in *Strange World*, the fireball storm over Lapland, which "cremated" a man in Parajaevarra in late July 1938?

> On his daily stroll along Ringwood Lake in Victoria, Fred Porter is nearly bashed on his noggin by a heavy, red-hot lump of metal plummeting from the Australian sky—missing him by *millimeters*. He picks it up, and thinks (hopes!) it's a piece of "space junk."
>
> —*Ringwood Mail*
> (December. 13, 1989)

It is both a charm and a curse of forteana that one mystery can almost always be invoked to explain another. That, perhaps, fires and molten raindrops from cloudless skies can drench men and materials with flaming consequences—

CHI, KUNDALINI, TUMO, AND BIOPLASMA. Tradition has long said these energies (1) are in every person; (2) can generate intense heat often likened to an inner conflagration; (3) can affect the neuroelectrical system; (4) can lock up the muscular system to thwart movement and escape; (5) can produce by means unknown an anomalous redolent smell; (6) can create ignis lambens (luminous phenomena) to give the appearance of a body wrapped in flames; and (7) can manifest either by directed will or spontaneously, without warning.

Modern (hard) science has begun to substantiate the soft (subjective) traditions for the presence of these energies within the human body. Energies that have a thermal coefficient.

Being internal and biological, chi and Kundalini can resolve the "burning" problem of the fire's isolation: why humans burn while other more readily combustible materials so often escape unscathed. The propensity of Kundalini to uncoil and activate as a component of consciousness, particularly during relaxation or sleep, ties in with the observation that many people appear to have been in this state of mind just prior to immolation. The number of full-blown Kundalini experiences is quite small, estimates Gene Kieffer; so, too, is SHC.

Itzhak Bentov proposed to me, shortly before his untimely death, that Kundalini could be activated externally due to intense environmental vibration acting upon the body's cells. He favored electromagnetic energy as the stimulus that could excite Kundalini, and other biological functions, to the point of creating hyper-hyperthermia in a person. Naturally, such a process would be invisible to any onlooker, as well as to the subjectee. Only the latter would realize anything was amiss, unless the interaction is *so swift* there's no time for the brain to react in pain or fright.

Plus, I have firsthand testimony from the famed Kundalini practitioner Gopi Krishna, who said this transformative energy is quite capable of reducing a human to ashes in an evolutionary process that backfires occasionally. I take literally Gopi Krishna's statement in DILIP (September–October 1975) that "Kundalini provides the only solution to the most burning problems of the day."

Not that Kundalini is the *only* answer for SHC; but, as a probability, it is a very promising one!

A human being extends beyond the immediately visible bounds of his flesh. Traditions for auric energies in and around humans now have been partly quantified by frontier scientists.

The photographic technique pioneered by Soviet Semyon Kirlian has shown dramatic changes in intensity and pattern of otherwise unseen energies around people, energies modifiable by changes in consciousness; energies whose change in color can forewarn of latent disease. Tradition says the color of chi is blue. Dr. Joseph Pizzo of Lamar University has made a 25,000-volt 7.7kHz Kirlian photograph of his face, to reveal that acupuncture points emit a bright blue light. Comparable to the brilliant electric-blue "fire" reported for most witnessed SHCs.

In 1979, I participated in a remarkable experiment designed by British electronics engineer Trevor Stockhill. Using modified video equipment designed to record ultraviolet and infrared energies but not visible light frequencies, he focused his camera in the pitch-black room on a woman being treated by famed Brazilian healer Alberto Aguas. Playback of the videotape revealed stunning rainbow-hued coruscations of light flowing around and through her, light invisible to the eye even in a well-lit room.

And in research that spanned decades, the Yale University School of Medicine's Dr. Harold Saxon Burr documented in *The Fields of Life* (1972) how electrodynamic "force fields in organisms change in strength and polarity in response to internal (biological) and external (cosmological) events"—events he called Field Profiles.

When these complex interplays of normally unseen human energies and field profiles become catastrophically disturbed, might SHC result?

VETERINARY MEDICINE. Opponents of SHC argue that if spontaneous combustion occurred in humans, it would also have to occur in animals; since there are no reports of the latter, ergo the former doesn't happen. The logic is specious. Besides: paranormal combustion is *not* restricted to human beings, as a few cases of SAC have already proven. Here are others.

• In November 1986, at the Anmer Lodge in London, a "terrific bang" and flash of "blue flames" inside Peppi the catnapping feline hurled him "several feet" into the air—a "build-up of static on the cat's fur" it was said. (Oh really?) During the frightful poltergeist fires that afflicted the Robert Dawson farm on tiny Thorah Island in Lake Simco, north of Toronto, Canada, during November and December 1891, the Dawson's kitten kindled spontaneously; as did a dress being worn by Jennie Bramwell, Dawson's adopted daughter. (An SHC near-miss?)

• Naturalist Ivan Sanderson once discovered in the remote West Virginia woods a large opossum—its back half "literally burned to a crisp, so that all that remained of the flesh was a black cinder through which the bones showed," he wrote in *Investigating the Unexplained* (1972). Carrington told of a kitten dead for an hour that left "a distinct outline" of itself burned on a rug... and leaving behind a case, perhaps, for postmortem fire in felines—

• In my files of European cases, is an instance of exploding turtles—the outcome of an improper diet which produced an overabundance of digestive gases. And from Hungary, comes this new threat to man: *exploding birds*. Some fifty crows "gorged themselves on rice thrown at a pair of newlyweds in Cegled. Some of the crows then exploded," said *The Lawyers Weekly*, "apparently soiling the wedding guests' clothing with their avian entrails." Too late for this wedding party, zoologist Lazlo Urganslai warned against throwing rice because it expands rapidly inside birds and "not only... is a threat to wildlife, but the birds explode so violently, their flying body parts can be dangerous to humans." Ugh—

• And may diners not forget the exploding escargot that blew up in the face of Karen J. Prouty as she celebrated her thirty-sixth birthday on February 26, 1988: the sizzling snails scalded her face, blinding her for ninety minutes. The owner of three-star Pascale's Wine Bar & Restaurant in Syracuse, New York, feared ruin; instead, reported Albany's *Times Union* (Feb. 28, 1988), he discovered "this definitely added to our national reputation." As for Mrs. Prouty, she admitted liking the taste of escargot but would be "hesitant to order it again."

Can SAC strike larger animals? Yes, if you recall the golden retriever with the flame-gushing belly from Jarrow, England. Yes, if you accept a story told to Randles and Hough by Raymond Reed, a member of the Royal Welsh Fusiliers during WW II. Reed spoke of the night he was on patrol along the Weymouth coastline, when a bright fire suddenly erupted about three hundred feet away in an open field. His three-man patrol approached cautiously, to discover a sheep ablaze, blue flames spurting from its stomach area. "We were absolutely astounded. The sheep was a large animal," said Reed, "in no way decomposed, in fact quite fresh looking. I think it was dead, but cannot be certain. We extinguished the fire by throwing earth and clods onto it."

The ways that life can turn on itself seem endless, indeed. As are potential clues to these baffling burnings, and the cases themselves.

NEW CHEMISTRY, PARTICLE PHYSICS, COSMOLOGY, ETHERIC PHYSICS.

The wonders of SHC—and SAC too—require considerations beyond the conventional. Perhaps *far* beyond! In a universe that modern physics finds ever more uncertain, the possibilities are boundless.

(At this very moment, science is announcing a universe it says is too young, moving in the wrong direction, and crisscrossed with heretofore unsuspected intergalactic currents sweeping worlds to who knows where.)

Unexpected discoveries await.

A laboratory accident at the Institute of Physio-Chemical Problems at Byelorussian State University (Minsk) led recently to the unanticipated discovery of a state of combustion heretofore thought impossible: *liquid* fire. Project scientist A. I Lesnikovich described it as a dark red spheroid that pulsated as it burned. "This is definitely a new type of combustion," Jan Puszynski of the State University of New York (Buffalo) told *Industrial Chemist* (January 1987) about this fire that is (almost) all wet.

Might nature create its own liquid fire, perhaps along unseen aerleynes in the sky overhead, whereupon it rains down to earth, and occasionally, onto an unsuspecting human?

The energy state that research scientist Wilhelm Reich called "orgone" can (if one accepts his experimental work) accumulate in and around a body, to discharge with *bright blue* luminescence. Another clue, perhaps, to innovative interplays that end in SHC?

Recent research using supercomputer simulations has spawned what the technology magazine *Research & Development* (October 1992) called "New Chemistry." Studies by Uzi Landman and his associates at Georgia Institute of Technology's School of Physics demonstrate that during nanometer-scale (really tiny!) collisions of atomic clusters with solid surfaces can yield vast pressures and temperatures as high as 7000°F. "We have very little knowledge about chemistry under these extreme conditions," Landman stated. "Compounds that may not form under normal conditions may be produced under these circumstances."

I think back to the theories of quantum physics and hot chemistry I had first explored in the mid-1970s. Landman explained what happens in his microscopic collision experiments with this metaphor: "It's like a multi-car collision on the freeway where a car stops abruptly and cars behind it bump into it and each other. In such a shock-wave pileup, the atoms in front not only stop, but also recoil back to compress the cluster into a very small volume in a matter of picoseconds [trillions of a second]." The tremendous energy released by this ultrafast compression is sufficient to shatter chemical bonds and alter other atomic properties of the material struck. "Most of our analytical theories and knowledge of equilibrium thermodynamics are invalid under those conditions," Landman admits.

His admission suggests to me a wide-open subatomic/chemical theoretical doorway through which, say, a neutrino can speed to trigger within a person or, say, a sheep along the Weymouth coast a chain reaction of instantaneous superheated ions disintegrating in a flash of vaporizing blue light and fire. Is there any proof this mechanism has *not* been around since the birth of this universe, awaiting Landman and his colleagues to discover it in the 1990s?

The farthest horizon in the study of SHC would be to discover a new force in nature that penetrates everything, unmeasurable by current technology. Classic SHC and quantum theory combine to reveal such a force: a rogue subatom of awesome power capable of instantaneously disassembling a body to leave in its wake dehydrated ashes.

This incomprehensibly small pyrotron dovetails nicely with the fourth/fifth sides—molecular chain reaction, and free-radical electrolysis—of the Fire Pentagon. And should physicists accept the evidence for the pyrotron's presence, they will face an upheaval more fundamental than the revolution inspired by

> Incomprehensible? But because you cannot understand a thing, does not mean it ceases to exist.
> —Blaise Pascal, seventeenth-century mathematician/philosopher

the neutrino. Admitting to the existence of the pyrotron will require that current ideas about the universe's evolution be radically rethought, if not scrapped outright. And the energetic, ethereal little pyrotron may help solve one of cosmogony's biggest riddles: the make-up, and whereabouts, of the universe's missing (dark) matter.

(Now *that's* a profound impact!)

Recurring reports of intact but shrunken skulls in the wake of low-heat SHC hint at forces not yet integrated into science's Grand Unification

Theory... forces that could hyper-accelerate atomic motion and actually *condense* matter.

Some theorists have been devising a new world view of Reality beyond entropy (beyond disentropy too) based on soft electrons, scalar waves and orthorotated hyperfield energy, chemical ethers, cosmic electronics, orgone energy—which radiates as *blue light* to those who can see it, researcher Ron Mangravite and others tell me—and other frontier formulations that threaten present paradigms. For example, Dr. Michael Baran Anteski has theorized in *Aura Paradigm* (1988) the existence of a cosmic ether/aura which pervades all matter; its "mass-directed resonance of emitted subatomic particles can explain... SHC as a sub-atomic process involving an acceleration of a natural form of energy." He extrapolates—I believe correctly for some instances of anomalous spontaneous combustion—that "atomic mass is itself being consumed in SHC."

> I am afraid of this word Reality.
> —Sir Arthur Eddington, astronomer and physicist (1882–1944)

Black holes as time warps; hyperspace and cosmic super-strings; faster-than-light tachyons that "interact with ordinary particles," muses physicist Gerald Feinberg; hidden-variable theories; single-mind and multiple-mind theories; supervenience; parallel- and multiple- and holographic-universe theories; realization that the "microcosm" and "macrocosm" are, increasingly, the same subject. We're not in Kansas anymore, Toto. Or are we?

(One thing is certain: we are now quite beyond the career training of firefighters, police officers and medical professionals—the three livelihoods most likely to come face-to-face with a case of SHC.)

In this realm of extreme possibilities, virtual particles arise continuously out of the quantum void of a subatomic universe; electrons jump energy levels within their atoms without actually crossing the distances between orbits. Photon pairs are linked acausally, as proven in experiments by John Bell and Alain Aspect; that is, altering the polarization of one photon can change its companion half a universe away. Chaos theory allows for a minute perturbation to become hugely magnified and randomized, like a tiny noise can launch flocks of screeching birds from quiet roosts.

No thing is isolated; nothing is separate anymore.

Where might spontaneous combustion appear in this cosmic web of superphysics? In property? Animals? Humans? Stars? The polarity shift of a photon in the Pleiades could, theoretically, trigger a chaotic vibration in the gut of a reveler on the streets of Mardi Gras or in the forearm of a sleeping salesman. The karma of one's parallel life may overlap into another life expression, so that one's persecuting *by* fire in one lifeframe counterbalances with a persecution *of* fire in another lifeframe.

And no fireman and no physicist, short of possessing omniscience, could ever know this would be the answer to "How?" in any specific episode of anomalous combustion.

Maybe Professor William A. Tiller, Ph.D. chairman of the Department of Material Science and Engineering at Stanford University, was on to something when he likened the human body to "a multidimensional antenna array" with "the endocrine-chakra pair as transducers of energy between the physical and other dimensions of the universe." More inter-relationships.

Lynn Volpe tells about a childhood experience of a friend, Lydia Stalmaker: "As a child, she reported seeing a column of blue fire in the middle of the living room. It wasn't hot. And there was no damage." Did Lydia see a telluric discharge in her home; a tube of light like the one that immobilized Betty Connell and her friend in 1971; something else, maybe a portal to another dimension that would have whisked her away in a burst of worm-hole energy, never to be seen again, had she been standing closer to it? One cannot say, except to wonder. And ponder. If the New Physics is (for the moment) right, the unpredictability of a quantum universe offers little security. And great uncertainty.

> Real life isn't a series of interconnected events occurring one after the other like beads strung on a necklace. Life is actually a series of encounters in which one event may change those which follow in a wholly unpredictable, even devastating way.
> —Ian Malcolm, the Chaos scientist of *Jurassic Park*

Last, one should not overlook the other-dimensionality that seems to underlie pyrophenomena. This includes so-called fire elementals, fire sprites, and the host of other names for the Spirit of Fire that esoteric traditions for millenia say undergrids the bright flames which kindle our campfires, heat our dwellings, and warm—sometimes incinerate—the human body. I agree with SHC paragnost and survivor Peter Calhoun when he states: "I feel that what happens here is that the physical fire is, at its lowest level, a manifestation of an elemental force found on every level. The ancients had more knowledge of this than we do today. I feel we are on the verge of rediscovering this knowledge—and we better be *ready* for it."

In time, answers to SHC may well be found here as well, though science as I know it to be today awaits astronomic advances in technology—and belief system—before this presence of "the fire behind the fire," as Calhoun calls it, can be quantified.

Establishing the relationship between such leading-edge thinking and the force(s) underlying SHC is beyond the scope of this book. But as one frontier scientist, Trevor James Constable, observes in his book *The Cosmic Pulse of Life* (1976): "Each human is the bearer, at the subconscious level, of a variety of powers that transcend current scientific knowledge." Someday, someone will find in those powers the key that can unlock the mysteries of SHC.

INSURANCE AND HUMAN SERVICES. The American Insurance Association Engineering and Safety Service writes that the phenomenon of spontaneous ignition "is a complex one affected by a variety of mechanisms and factors." AIAESS didn't have SHC in mind, but its statement is applicable nonetheless.

How would you, if a health care provider, treat the physical and psychological trauma of a patient afflicted once or repeatedly by flames for which there is yet no proven mechanism? Would you even know what to look for? Or would you, like Jack Angel's initial care providers, stand around aghast, pondering *what* had happened instead of initiating a proper healing regimen?

"I think it's extremely interesting!" said Dr. Donald (Michael) Slate, when I asked him about spontaneous human combustion in January 1995. "As far as I know, no one has ever mentioned it in school. Never in *medical* schools." Did he, a resident pathologist at Buffalo General Hospital and graduate of the University of Buffalo Medical School, think it would be a good idea for medical schools to begin teaching physicians about SHC? "Because they try to squeeze so much into four year, probably not," he answered. "Probably in specialized training, like forensic medicine, it would be better. But I haven't heard of it being taught there, either."

Dr. Slate wanted to know more about SHC. I would hope the American Medical Association will respond similarly now, because two decades ago the AMA ignored my request for any information about anomalous burn injuries.

Would you, if an insurer, issue a health policy for Peter Jones who twice self-ignited? Would you write property insurance for Blyth, England, or San Gottardo, Italy—the latter under siege by mysterious fires that began February 14, 1990, and continued for months until newspapers being read, automobiles, a wheelchair at its owner's feet, ski boots, an electric razor in the hands of the town mayor, furniture and appliances all ignited enigmatically for more than a third of the town's five hundred residents (some of whom also suffered sudden inflammation of their skin)? As a policy

> I have seen a book on a high shelf turn to charcoal in a few seconds... and lighters explode like small bombs.
> —Capt. Giorgio Rossini, San Gottardo Fire Department; in *Fortean Times* (Fall 1990)

holder, could you afford the premiums if these uncommon fires were actuaried into insurers' risk assessments?*

Interesting is the response (or lack of) to a questionnaire about anomalous fires that I sent to every state fire marshal's office in America, and to major casualty insurance companies. Only three fire marshals replied. Two dismissed all spontaneous combustion other than the wet-hay oily-rag type. James McMullen, California State Fire Marshal and (at the time) president of the National Association of State Fire Marshals, kindly sent back a reprint of Allen Eckert's *True* article, adding: "I do not personally know of any documented cases of the type of burn you are addressing." Fair enough.

No insurance company replied... even though one—The Travelers—had settled a claim for twenty-eight mystery fires which, over fifteen hours, spontaneously burned out a book's inside and a wall calender, et cetera at the Hackler home in Odin, Indiana, in 1941; then advertised the incident (and the firm's prompt settlement for damages) in *Collier's* magazine!†

Developers and zoning officials might wish to know more about SHC, too.

Finally, firemen—those brave professionals who safeguard society from the peril of conventional fires—would certainly wish to know more about the unconventional side of their calling. Injury, death, and destruction by fire does not differentiate between the normal and the paranormal.

---

* Between 1984 and 1986: more than thirty spontaneous fires in the Rome offices of INAIL baffle Italian police; flames erupt within desks, wardrobes, photocopiers, and locked safes. When employees return to their offices after repairs are made, the freakish fires begin again. "We have no idea what is happening," concedes a spokesman to *Sunday Express* (April 20, 1986). "We are a department for accident claims for industry, and we are getting more than our share."

June 1988: for one month sudden fires and rains of stone mystify everyone in Didiga, Chinna Hydrerabad, and other afflicted villages of Andhra Pradesh in southeast India. Police superintendent Venkatiah attests to the phenomena, which includes wet clothes catching fire, but is "not able to know how these incidents took place."

April 1994: mystery fire terrorizes villagers of the Himalayan foothills in western Nepal. Ten houses are reduced to ash; sometimes, one blaze goes out in full view only to shift to a nearby roof. "Nobody knows who or what causes them," relates the Nepalese news agency RSS on April 30, "but the fires continue to erupt on house-tops at Manaparipur in Bardiya district."

† In 1988, The Travelers became involved in another fiery mystery. Between March and August, fire officials in Orland Hills, a suburb south of Chicago, documented twenty-six separate eerie events at one hosue: smokey mists... suphur smells... a one-inch flame "under pressure" roaring out of an electrical outlet to strike a mattress almost two feet distant. The family was advised to move out, and the engineering firm hired by The Travelers never found a solution.

CONSCIOUSNESS AND EVOLUTION. Emotions, attitudes, and consciousness have repeatedly played a role in the generation of inflammatory bioenergy. Let me simply concur with Yale University psychiatry professor Dr. Morton F. Reiser, whose *Mind, Brain, Body* (1984) documented that distinctions between mind and body are simply "meaningless and out of date..."

Human firestarters have shown one way that consciousness influences one's environment, by igniting fires outside themselves. Might there be instances when humans who invoke fire have their invocations backfire, thereby burning up themselves?

"I know about this heat!" exclaimed Lynn Volpe, a meditator and researcher of biophysical phenomena, commenting about the thermal potential of tumo. "It can be caused by emotional telepathy transmitted through the heart chakra." Transmitted internally, or, perhaps, externalized toward another.

> While the traditional model of psychiatry and psychoanalysis is strictly personalistic and biographical, modern consciousness research has added new levels, realms, and dimensions and shows the human psyche as being essentially commensurate with the whole universe and all of existence.
> —Stanislav Grof, *Beyond the Brain* (1985)

Human firestarters who invoke fire...

Professor Beverly Rubik, Ph.D., director of Temple University's Center for Frontier Science, shared with me in 1991 a graphic example of how thought-beliefs can affect physical reality. A group of people, their eyes tightly shut, were instructed to walk across a bed of hot coals. Actually they strode forward, unknowingly, onto a bed of cold granola cereal. "Some got blistered feet—*on granola!*" said Prof. Rubik, bemused yet impressed by the consequences of a wholly imagined reality construct. Illusionary for her, real in the mental image of those walking. One outcome of the exercise? Spontaneous human combustion. Witnessed. And survived.

Human firestarters who invoke fire...

Kretzer's *Lightning Record* (1895) recorded that the previous year an Arkansas fire-and-brimstone preacher at Walnut Ridge had beseeched heaven to send a sign his congregation would remember: in mid-sentence a lightning bolt exploded a nearby tree and knocked many in his audience senseless. Nature has its near misses.

Citizens of Wiltshire, England, erected a monument to commemorate the day a Devizes woman on trial for petty crimes invoked heaven to strike her down in the town square if she were guilty. Moments later, the heavens rendered apparent justice as a lightning bolt executed her on the spot. On July 12, 1912, Elijah Howard scoffed at his family's fear of a thunderstorm approaching

McKean County, Pennsylvania: "The only thing that scares me about lightning is, I'm scared it won't strike me." Shortly afterward, skyfire struck his wagon and killed him instantly, though no other family member riding on the wagon was hurt. And according to Hoose and Naifeh's book, *True & Tacky: More Weird Stories from the World's Newswires* (1992), Louisiana attorney N. Graves Thomas stood up in a boat and defiantly shouted "Here I am!" at a thunderstorm cracking around him. Moments later, lightning instantly fried him to death in front of four shocked companions.[3]

Nature has its direct hits. Then, of course, there is Joe Nuzum and Peter Calhoun—

Incredible coincidences? Or can one's thoughts invoke an unsuspected, yet requested zap with horrific results? Like whole-body incineration.

For perhaps the ultimate theory about SHC and consciousness, see the next chapter.

SAFEGUARDING AGAINST THE FIRE WITHIN

The untapped thermal potential of the body is awesome—comparable perhaps to the human brain whose power Einstein estimated to be 85 to 90 percent untapped. Must you therefore live in constant fear of SHC?

Assuredly not. To repeat: *the odds overwhelmingly favor that you—that anyone—will* never *be singled out for this flaming fate.*

Still, a few people will always beat the odds.

In the words of the 1975–1976 Fire Prevention Week motto, then, can one "Learn Not To Burn" oneself? Can a clinical ecology minimize the already slim chance for SHC? Yes, I think so.

While it would seem impossible to safeguard against a pyrotron's hypersonic impact, its ultra-minuscule size already gives you incredible protection against an internal subatomic chain reaction. To minimize the minuscule risk inherent in other theories, I offer these guidelines.

To safeguard from earthfire, and possible fire-leynes, consult local historians, folklorists, and fire departments to learn if the area where you live (or plan to move to) might have a legendary propensity for anomalous incendiary phenomena, particularly human combustibility.[4] Be prepared for a teasing, even testy, response. But the findings could reward your patience and peace of mind.

For uncoiling Kundalini that threatens to fry from the inside out, Gopi Krishna gave this technique to the All India Institute of Medical Sciences in 1975: "the remedy prescribed is to lay on a coat of wet clay over the body or to immerse it in a pool of water up to the neck. For the heat experienced at the crown of the head or the space between the eyebrows, the rubbing of sandal-paste is recommended."

This procedure presumes Kundalini moves slowly, giving one time to apply these countermeasures. That isn't always the case.

Another approach is more reliable. As Gene Kieffer wrote to me, "The best safeguard against dangers associated with Kundalini is, of course, a clear conscience. Yogis sometimes refer to 'purification of the nadis,' but this is simply a phrase to describe the necessity of right living, thinking, etc."

Here, then, is a key to protecting oneself against the unlikely experience of SHC. It sounds supremely preferable to a treatment mentioned in the nineteenth century by Dr. Edmund Sharkey, who passed along—without endorsement—this SHC cure adopted by a Dr. Swediaur: "no other than the administration of recently voided human urine." (M'thinks not.)

"Right living" would include nutritionally balanced meals and maintaining a strong healthy body, factors Gopi Krishna attributed to his own vanquishing of inner fire. Whereas three years of subsisting on Dr. Bentley's daily menu should be enough to upset anybody's body! Or you might want to try what a contemporary of Pierre-Aime Lair prescribed as an alternate preventative for self-combustion: consuming copious quantities of milk. (Beats urine, any day!)

Right thinking would encompass a healthy emotional stance, one devoid of profound emotional blockages, hot-headed seething anger, brooding depression, or suicidal death wishes which seem common to the psychological profile of many victims of SHC. When one speaks of having "fiery fury," of fuming with "red-hot temper," of being "so mad *I could just burn up!*"—might these factors, these internal ragings, psychobiologically set off self-cremation?

I sense that they can. For example, Dr. Edward Sullivan told me he was a cauldron of wrath, despondency, and suicidal desire the night he lay down to sleep, a sleep that included an intricate "dream" (too complex to elaborate here) that mirrored a classic near-death experience. Did a confluence of these complex psychological circumstances combine with the metaphysical that fateful night to trigger in his body the interior fire that so often kills but instead spared him for tasks still to be accomplished?

"When fires consume old women, is it because there are thoughts of death and hellfire?" pondered Damon Knight in his biography of Charles Fort. "If so, when we are old, let us take care to think only nice, clean, benevolent thoughts."

Right thinking does seem to assure one form of immunity to fire. For years, Vernon "Komar" Craig's thinking enabled him to hold the world's record for walking uninjured on hot coals at 1431°F (that record currently belongs to Steve Bisyak's fire-walk at 1547°F). More than 150,000 people, including the author, have likewise defied the "impossible" by walking on or handling fire with impunity. To do so is exhilarating! Just as it has been exhilarating for me to document its antithesis—catastrophic inner burning.

Yes, one *can* learn not to burn! History, in fact, offers just as many cases of miraculous human immunity to fire as it offers for spectacular human susceptivity to combustion.

There is one circumstance, however, in which prevention of this catastrophe might *not* be desired...

27

Phoenix Fire: Wrapped in the Flames of Ascension

> The aborigines assert that the day will come when the earth returns to the dreamtime.... If this is true, the evolutionary fires that are beginning to flicker and dance through our collective psyche may be our wake-up call, the trumpet note informing us that our true home is elsewhere and we can return there if we wish.
>
> —Michael Talbot,
> *The Holographic Universe* (1991)

Though the ultimate meaning of life eludes intellectual grasp, man still assumes he can—and someday will—understand the laws of nature and how the universe works. So far, however, every space probe, every magnification in microscopic probing, every instance of SHC unveils more mysteries than answers. The world seems destined to be always hidden from ourselves, by ourselves.

Hidden until a divine spark fires the mind to broader, transcendent consciousness.

Dr. Richard Maurice Burcke had such an experience of consciousness inflamed. He described it in his classic book, *Cosmic Consciousness: A Study in the Evolution of the Human Mind* (1900):

> All at once, without warning of any kind, I found myself wrapped in a flame-coloured cloud. For an instant I thought of fire, an immense conflagration... the next, I know that the fire was within

myself. Directly afterward there came upon me a sense of exultation and immense joyousness accompanied, or immediately followed, by an intellectual illumination impossible to describe.

At this seminal time in human and planetary evolution, more and more people will likely (if not already) be experiencing an inner fire and its potential for accelerated movement into more holistic consciousness. As bestowed upon Burcke, its promise is to be welcomed rather than feared; its gift a blessing instead of horror. Indeed, fear could be *the* difference between upliftment and fatality.

Yet the process of divine fire within may not even end here. Ageless traditions teach of overcoming the wheel of life on Earth; of perfecting one's spirit and being rewarded with a lighter, brighter, noncorporeal body.

Consider fifteenth-century Hindu saint, Radha, whose final moments researcher John White chronicled in "Divine Fire: A Little-Known Psychic Power" for *Venture Inward* (1990). To the astonishment of those gathered around her, Radha one day humbly announced her death. She then walked a short distance and, to their endless amazement, burst into a fireball of light, leaving behind a small heap of ash upon the ground where she stood.

Consider the twentieth-century holy man, Sadhu Singh. The *San Francisco Chronicle* (December 17, 1970) reported the news that this 114-year-old venerated Hindu had been sitting in the lotus posture on a straw mat at Ganjudwara, India. A young student approached his teacher, to discover him ablaze in a blue-green "candlelike flame" that grew larger and larger as it radiated from the area of Singh's solar plexus. (A flame similar in color to those seen lapping at the body of Mrs. Ginette Kazmirczak in 1977?) Other awestruck students gathered to witness this phenomenon. Soon policemen joined the other onlookers marveling at this strange occurrence. No one tried to extinguish him—probably because Singh sat in serene stillness, his peace-filled face radiant with light as his glowing body produced a "pleasant aroma" which wafted through the courtyard.

For six hours Sadhu Singh burned. Burned himself. From the inside out. When it was over, according to the *Chronicle*, even his organs were reduced "to ashes."

You can make of this what you choose. From the moment I heard about this incident I felt that Sadhu Singh, said to be very evolved spiritually, had chosen this pure and simple (not a pun) mode to exit a

> The spiritual aspects [of SHC] are probably more interesting than any other.
> —Harry Lott, assistant fire marshal; personal interchange (December 17, 1994)

body his consciousness no longer wished to (or could) reside in. Not wanting to burden his followers with burying his body, he *willfully and in full consciousness* utilized the energy within himself to achieve self-cremation.

Support for this interpretation of physical transition by SHC has since been found in several diverse sources.

First: in his book *Living with the Himalayan Master*, yoga Swami Rama spoke about ancient techniques of *mahāsamādhi*, which means literally to "cast off one's body." That is, death can be induced at will and by conscious design. These techniques are still practiced in the high Himalayas today, Swami Rama learned while there, and can lead to death that is methodical, painless, and conscious. It is those deficient in *mahāsamādhi* who accept normal methods of dying. Swami Rama then described one "very rare way" to cast away one's body.

"By meditating on the solar plexus the actual internal flame of fire burns the body in a fraction of a second," he explained, "and everything is reduced to ashes." Sound familiar? He said this technique had been imparted long ago by Yama, the king of death, to his beloved disciple Nachiketa in the *Kathopanishad*.

(I wonder whether Nachiketa is among the SHC deaths in history's long-ago shadowed past.)

"All over the world," Swami Rama continued, "instances of spontaneous combustion are heard about, and people wonder about such occurrences. But the ancient scriptures such as *Mahakala Nidhi* explain this method systematically."

The power behind SHC is thus both consciousness and energy, he stated. Knowing how to move that energy to affect thermogenesis can warm oneself (tumo) or, if desired, incinerate oneself (SHC).

Second: in his *The Forge and the Crucible*, scholar Mircea Eliade explored the alchemy of fire and those persons said mythologically able to control fire and consciously generate internal combustion. "In theoretical terms," he remarked, "to produce fire in one's own body...signifies the attainment of a state superior to the human condition." I submit such "mastery" is not merely theoretical, but can be actual; practical. That humans have already, and repeatedly, demonstrated an ability not only to generate heightened body heat (tumo) but to kindle *real* fire outside and inside themselves.

Third: in conversation with Ram Dass, revered advocate of human potential and spirituality in the '70s and '80s, I was told about an ancient Tantric concept called *bardo*. At the moment of death, explained Ram Dass, masculine and feminine energies separate in the body; the former moving to the head, the latter to the genitalia. Then they rush to recombine, he said, crashing "in a cataclysmic flash like nuclear energy—and you get the White Light." Might this moment of *bardo* be the brilliant flashes of light associated with some SHC reports?

Fourth: in *The Urantia Book*—a massive, psychically channeled, controversial tome that constructs a detailed revisionist history for life on and beyond the Earth—one reads about the "translation flash" that occurs when advanced evolutionary humans transcend the natural world. This "fusion" with higher divinity obliterates the physical body in a "blazing glory of consuming fire," which, explains *Urantia*, is isolated by "intervening celestial personalities" so that nothing else is destroyed by this "life flash" which can be "well-nigh instantaneous." SHC is not mentioned per se, yet *Urantia* is describing to perfection classic SHC. It may be describing with equal precision why surrounding combustibles—even to clothing and sheets a person lies upon—escape destruction by the energy of internalized bodily cremation.

Fifth: in Christianity's *Bible*, Hebrews 11:5 recounts the final moment of life for the Lord's servant, Enoch. "By faith Enoch was translated that he should not see death; and was not found because God had translated him; for before his translation he had this testimony, that he pleased God." The text differentiates between death and translation; the processes are not the same, though clearly Enoch is no longer in the body. Enoch has pleased his God; that is, he has attained a superior spiritual state. For this mastery, he is specially rewarded. How? Is "translation" a euphemism for the process described by Swami Rama; by Eliade; by *Urantia*?

By these accounts, then, SHC would not be an event to fear; to abhor; to avoid with pathological zeal; but testament that the person has achieved ascended spirituality. SHC would be a means—*one* of the doorways—through which man is liberated from the physical to enter another of the universe's "many mansions." The holy person, through holy fire within, becomes holier.

> In my Father's house are many mansions; if it were not so, I would have told you.
> —Jesus the Nazarene; John 14:2

Is this what Sadhu Singh had achieved at age 114? Had he so perfected his spirit that it could no longer be constrained by physical matter, and did he, with intent, will his body to self-cremate so it would not be a burden for his students to dispose of once he left it behind?

(To qualify for this form of ascension, one need not be an ascetic atop a distant Himalaya peak. In the realm of the spirit one cannot judge a person's status by external appearances; the unassuming vagabond, the poor servant girl, the retired fireman who worked honorably and loved his family can be just as worthy of phoenix fire as is the acclaimed spiritual guru, perhaps more so.)

If this theory of the *phoenix fire* has merit, then SHC speaks to spirituality as well as to science through the challenges its poses. It may indicate that the person consumed has become so evolved in consciousness that he (she) autonomically *must* become the phoenix...to arise from the ashes of a natural evolutionary fire born of transmuted consciousness and realize the ancient alchemic dream of *homo faber*, what Eliade calls "collaboration in the perfecting of matter while at the same time securing perfection for himself."

It is a goal many would rejoice to attain.

A Relic of Perfected SHC Preserved?

This last consideration for SHC raises one fascinating aside: the Shroud of Turin.

This linen shroud, historically traceable to the mid-fourteenth century and preserved since 1578 in the Cathedral of St. John the Baptist at Turin, Italy, is a 3'-7"×14'-3" cloth that features a reverse-negative image of a crucified man. Its very existence causes consternation and controversy among theologians and scientists.

Some say it is a medieval painted forgery (despite its anatomically correct white-for-black reversed rendering). Joe Nickell advocates a "rubbing technique" by an unknown artisan using iron-oxide as an acceptable, non-supernatural explanation for the Shroud. Others argue on behalf of aromatic embalming fluid fumes. Nicholas Allen, a South African photographer, claimed in 1994 that he had solved the Shroud enigma by employing a tenth-century Arab technique that could imprint an image upon linen sensitized in a solution of silver nitrate, using sunlight focused through a quartz lens. Still others vigorously defend scientific evidence that indicates the Shroud dates to the first century A.D.

Many sindonologists—the technologists who study the Shroud—have arrived at a stunning consensus that is quite different. Their analyses suggest the image is best explained by thermoluminescence—what Shroud scholar Frank C. Tribbe calls a "blast of light" likened to a heatless nuclear flash that transposed the body's three-dimensional contours "*through space*, not by contact" onto the fabric.

"What we see on the Shroud is a radiation-form image," optics expert Kevin Moran has testified. "It is *impossible* to duplicate today, let alone five hundred years ago!" When asked how it *had* been formed, Moran spoke of a process beyond our accustomed dimension of space and time: "a radiation particle caused the disappearance of this body into N-space."

On the other hand, Samuel F. Pllicori, member of The Shroud of Turin Research Project, concedes: "We cannot exclude natural processes as causing the image on the Shroud."

Yet the challenge of this curiosity remains. As Shroud scientist Dr. John Heller has admitted, "We are left with the question 'How is the image produced?' And for that, the answer is a mystery."

I have proposed an answer, a radical solution to this vexing problem confounding the sindonologists, which is both natural *and* paranormal: that the person wrapped inside the Shroud experienced with absolute thoroughness what hundreds of people have experienced in diverse ways during the last few hundred years. SHC.

With a difference. In this particular instance, SHC instantaneously transformed *100 percent of the body* into a higher energy state. Every atom became transmuted in an $E=Mc^2$ empyrean radiance—the pure fire (light) of the highest heaven in ancient cosmology—which left behind not one extremity, not one bone, not one gram of ash; only a negative-image imprinted by flash photolysis onto unsinged linen and a residuum of low level radiation that would throw off by more than a millennium the Carbon-14 dating technique that would be applied to slivers of the Shroud in the late 1980s.

With this explanation arises (so to speak) the issue of *why* something comparable to but more complete than Mrs. Satlow's spontaneously flamed corpse might have happened inside the Shroud of Turin. Was it a freak accident of nature? Or the fulfillment of mankind's hope for ultimate human ascendancy, with absolute SHC being a natural consequence of the transmutation of all atoms in the body of a perfected human being?

Is *this* what SHC might ultimately be about? "I want to go through this myself," said Moran about the process that produced the Shroud's imagine, "to see what the resurrection is like." Might he be wishing himself, someday, into SHC?

COMPLETING THE CIRCLE

I am the first to admit that SHC defies common sense and smacks of the unknowable. I don't have all the answers to it; I may have none of the answers. And certainly I don't have all the pieces to this jigsaw of enigmas.

What I can say with confidence is this: spontaneous (as well as preternatural) human combustibility happens, though it has remained well hidden. When Deputy Sheriff Pat Berry investigated Mrs. Oczki's death, he discovered that many fire officials in the Chicago area *did believe in SHC* but wouldn't admit it publicly. Explained Berry: "They don't want to look foolish."

So it has been, for far too long. But more than the fear of looking foolish has hindered the investigation of SHC.

Spontaneous human combustion raises deep psychological xenophobia. How else to explain disinterest and outright rejection of a phenomenon whose very nature intrigues and inspires engrossing possibilities?

> The search for truth is in one way hard and in another easy. For it is evident that no one can master it fully or miss it wholly. But each adds a little to our knowledge of nature, and from all the facts assembled there arises a certain grandeur.
> — Aristotle

Whereas the wide-eyed innocence of elementary school students leads them to ask thoughtful questions about SHC, men with Ph.D.s have often felt compelled to pontificate (usually outside their scholarly expertise) against spontaneous human combustion with other forms of SHC—silly histrionic claims and scurrilous hysterical countercharges. Sadly, even fire science and medical professionals express attitudes about SHC that usually range from benign neglect to brazen denial.

One surgeon quickly turned down the opportunity to look at my photographs of SHC, exclaiming "It's too scary to think about!"

Two leading Soviet physicians, when I asked them in 1989 about SHC and if they knew of occurrences in the U.S.S.R., had diametrically opposite reactions.

Dr. Mikhail Paltsev, Dean for eight thousand students at First Moscow Medical School, showed complete disinterest and said curtly, "I don't know anything about it."

By contrast, his colleague, Surgeon General Eugene Maloman of the Moldavian Ministry of Public Health, responded quite differently. "I know nothing about it," Dr. Maloman admitted. Then, instead of walking away, he asked to see the photographs. He studied photograph after photograph that I handed to him, excitedly and animatedly asking questions in Russian about what he was looking at. "You know," he finally said through his translator, "there are phenomena science knows nothing about. Things science can't explain. One day perhaps this will be known. I wish you luck!"

It may take luck to solve the riddle wrapped in the mystery inside the enigma of SHC. An inquiring attitude, such as Dr. Maloman demonstrated, will be mandatory.

Fortunately, opinions and attitudes are changing about spontaneous human combustion.

"I believe it does exist—there's *something*, even if I don't know what it is," acknowledged Fire Marshal Robert Meslin about SHC. "*Weird* to say the least."

"We have to believe that it *does* happen," voiced Assistant Fire Marshal Edmund G. Knight III, public information officer for the Lancaster (Pennsylvania) Bureau of Fire. Knight had just shared the stage with me on WGAL-TV's *Live!* (November 13, 1990), which that day featured Dr. Edward

Sullivan's own (and only) televised statement about his personal SHC. "It used to be that a fire department would write it off as electricity," said Knight, "and that was that. But we've grown up in the past ten years. If we don't know what causes a fire after investigating it, we will say so."

SHC, discussed in an eighteenth-century edition of the *Encyclopaedia Britannica* but subsequently banned from its pages, warrants resurrection and recognition by science and medicine as the newest (or oldest unrecognized) malady to afflict the human body. SHC deserves listing in *Cumulated Index Medicus*; in *Glaister's Medical Jurisprudence and Toxicology*; and in the National Fire Protection Association's *Fire Almanac* as a cause of death. It warrants inclusion in medical school curriculums, especially in forensics and pathology, and in the course work at the U.S. Fire Academy.

True, SHC severely challenges credibility. This I have always acknowledged. However, that doesn't mean SHC can't happen. Who can prove the limits of human potential? Years of investigating pyrophenomena have prepared me to expect the marvelous; to respect the diverse manifestations of energy within *Homo sapiens* and in the cosmos. Despite remarkable advances in medical science, our bodies remain quite capable of magic and mystery for which explanations still await.

I've offered numerous theories, theories based on the largest database of SHC-type fires yet published, in an attempt to resolve these fantastic occurrences. If you don't like one of the theories, throw it out; if you like none of them, trash them all.

What you cannot do, unless you know more about fire science than scores of professional firefighters who refuse to explain away their firsthand encounters with fierce localized (even witnessed) human burnings, is to dismiss the cases themselves.

The cases stand on their own. Some may not be SHC; some can never be determined with certainty either way. Throw out one, a dozen, or a hundred cases. Accept just one, and SHC becomes a real phenomenon in search of a solution.

Unless you are prepared to call Peter Jones and his wife liars, and to write off Robert Meslin and Harry Lott as inept observers ignorant of fire behavior, you cannot discard all the cases. And you are obligated to develop your own alternative explanation(s), if you find displeasing the ones submitted for your consideration.

In the century past, SHC was hotly debated. In this century, as transportation advanced from the horse carriage to supersonic airflight, the debate has been lukewarm at best; mostly, it has been frozen under the icy repression of officialdom's denial. So long as SHC remains relegated to tabloid sensationalism, and elicits smirks from scientists and from officials whose duty it is to protect the public from the hazards of fire, SHC will remain deniable—though it will continue to happen and to mystify those who confront it honestly.

"I don't understand the whole damned thing," Jack Angel said as he waved the stump of his right arm in front of my face. "I don't know anything about human combustion theories. But something has to cause it, doesn't it?"

The longer we investigate, the less we seem to know. The more answers we think we have, the more questions develop. Will SHC remain a subject of never-ending, enigmatic affronts to common sense? I suspect there

> The truth is out there.—*The X-Files*

are questions we may never know how to form, answers we will never be able to get to. But one can never know that for sure, if one does not first ask the questions.

Somewhere in the asking, I sense that some thing—perhaps many things—can be revealed that can intercede to *prevent* some SHCs if, in the first place, effort is made to understand what precautions would prove most effective.

Therefore, the search for the fire within will continue. Someday that search, with cooperation from alert officials and the public, may find answers that Jack Angel and others have sought about their enigmatic inflamings.

Many mysteries await us, the small accretions of subatomic particles that we truly are, in all of the vast universe. It's so intriguing. Can't imagine being on any path, but a searching one. What a waste it would be not to be enthralled by the unknown.

So many questions remain unaddressed. So much is yet to be discovered—about the human body, indeed about energy in general—through the study of spontaneous human combustion...a pursuit that will someday, hopefully, arrive at knowing how people can literally burn themselves to ash. And how to safeguard oneself from that fate, rare though it is.

Unraveling the mysteries and meaning of SHC has only just begun.

"Every phenomenon is unexpected, and most unlikely until it has been discovered. And some of them remain unreasonable for a long time after they have been discovered."

—Eugene Wigner,
twentieth-century Nobel laureate physicist

Epilogue

Some who claim to investigate SHC-style combustion, after years of doing so, fail to find one new addition to the annals of mystery fires.

In the last months of preparing this book for its final journey to editor and then to you, I came upon three episodes worthy of nomination. Two, I learned about in one morning just by judiciously making a few phones calls. Of course, these candidates' final determination for inclusion into the SHC archives awaits sufficient time to research them properly. However, I thought you might like a hint about what is still out there to be explored:

• On the fifth floor of a prestigious hotel in a major American city, a tenant's foot was found standing upright next to a pile of ashes that had been the rest of his body. The fire, if one wishes to call it that, was limited to a small-radius circle. It became an "X-file"; as in crossed-out, nonexistent. A fire chief, retired from the jurisdiction where this occurred, nonetheless vouches for this case from the late 1940s.

• Near the waterfront of a modern U.S. city, an elderly female burned through the center of her bed, through the floor of her second-floor bedroom, and vanished; twenty to thirty firemen tromp throughout her house, searching vainly for the occupant neighbors adamantly insisted was at home when the fire call was sent in. They *finally* found her—consumed to "clear white ash" and residue "the size of a ten-pound ham"—amid stacks of newspapers she fell between when dropping through the ceiling. Says my informant in his quest for the victim: "I personally stepped on her several times." This episode dates to the mid-1980s.

• Another elderly female—an invalid who used a walker!—burned through the floor; lower leg consumed, light smoke and heat damage. "This one was brought up immediately when I asked," said my source, a senior fire investigator whom I had asked to make inquiries about SHC. "Maybe because it was so recent." How recent? "November 1994."

As poet e. e. cummings urged, "Listen, there's a hell of a universe next door: let's go!"

NOTES

CHAPTER 2
1. Harrison (1976; 17). Cf. Harrison (1978; 3).
2. See Arnold (1976).
3. This report appears as "Observation CXVIII—Sur l'enbrassement spontane d'une vieille femme, par Mat. Jocobaeus [sic]."
4. Small pieces of charred or burned wood.
5. The *Transactions for 1744–1749*, along with Lair (1812; 165), Pierquin (1829) and Sanderson (1972), misspell her name as "Pitt."
6. Harrison (1976; 40. 1978; 25) declares that, outside of one newspaper report, "the first mention of this case is in Sir David Brewster's *Letters on Natural Magic* (1823)." To the contrary, besides the *Transactions* (1744–1745), there are references to Pett in *Hamburg Magazin Bd*. (1748), *Annual Register* (1763), Thornton's *Medical Extracts* (1796), Lair (1812), and LeCat (1813).
7. Credit is due Peter Christie of Sussex, whose diligent research recently uncovered this heretofore overlooked information about Pett in its original archaic English.
8. Hergt (1837) dates the episode on 22 February 1749, but this is too late.
9. Harrison (1978; 142) spells the priest's name "Boineau."

CHAPTER 3
1. Harrison (1978; 17) names her "Clewes"; Godfrey (1870) names her "Clies." Her age varies among accounts, with Lair (1812; 163 and 173) assigning *both* figures in different parts of his essay.

CHAPTER 4
1. Beck (1851; 98) wrongly dates it on January 15.
2. Joly (1837), Hergt (1837) and other sources do not give Bernard's surname. Hergt does mentions a Bernard Larivière, age 75, who fatally burned in Surville on 6 October 1836. I believe the two episodes are, in fact, one episode, which early sources—A. S. Taylor (1865; 507–508) and *Medico-Chirurgical Review*, (XXX: 500)—date for September.
3. Sanderson (1972; 274) dates this case on February 25, 1851.

CHAPTER 5
1. Russell (1955) erroneously assigns the deaths of Pegler, Wright, Ridge, and Hardrill to December 28, 1938. Harrison (1976. 1978) perpetuates the error.

CHAPTER 6
1. Foam rubber *can* combust spontaneously, as I discuss elsewhere, but only when it's repeatedly washed and stored damp—something not likely to be done with an overstuffed chair.
2. Two seconal tablets is a "considerable amount"?
3. A slightly expanded version of Krogman's original article on "The Cinder Woman" appeared in *The General Magazine and History Chronicle of the University of Pennsylvania*; all citations are to this version of Krogman's article.
4. The smoke and smell (what little there was) produced by the fire needed time—a few hours—to dissipate, because by eight o'clock the telegraph messenger "noticed nothing amiss" when he walked up to the apartment building and Mrs. Reeser's door.
"Smoke smelled at 5 A.M. was thought to be coming from an over-heated water pump," reported Jerry Blizin in the *St. Petersburg Times* (9 August 1951). "Mrs. Carpenter turned off the pump and went back to bed." The landlady never confirmed her suspicions. More likely in her drowsiness, she smelled an odor that seeped from Mrs. Reeser's apartment and attrib-

uted it to the pump—having no reason to suspect another source.

The electric clock stopped at 4:20, probably due to excessive heat or electromagnetic phenomena associated with the fire's peak energy that temporarily froze its motor. This evidence points to 4:20 A.M. on 2 July 1951 as the most probably time of death for Mrs. Reeser.

5. I have been unable to identify *who* first characterized the head as shrunken, nor the *basis* for this determination. Dr. Richard Reeser Jr. disagreed with the popular description of this alleged artifact: "There were no identifiable facial features present, but only a portion of the skull bone was found."

Regrettably the FBI report, which presumably detailed the skull's condition, is unavailable to us; nor does the one photo I have of her remains allow a firm conclusion. So this much publicized and disconcerting mystery *may* be explained at last. Then again, maybe not.

6. Under the original FOIA (1966), certain categories were exempt from disclosure, among these the "investigatory files compiled for law-enforcement purposes."

However, the revised FOIA (February 1975) "stipulated that such files as those kept by the FBI would have to be furnished on request unless disclosure would interfere with a pending proceeding, violate an individual's privacy or compromise a confidential source or investigatory technique." Since the Reeser case is public domain and none of the other restrictions apply, the FBI file on Reeser qualifies for release without preconditions.

Nevertheless: "Concerning autopsy reports and/or Laboratory services performed for our Department by another agency," says Sgt. R. G. Kerlikowske of the St. Petersburg Police Department's Internal Affairs Unit, "these are not released without authorization from the Medical Examiner and/or the laboratory which performed these services."

7. That information in 1975 consisted of one newspaper reprint and a woefully inaccurate one-page article by Morrison Colladay titled "They Called It Spontaneous Combustion" from *Blue Book* (March 1951). Colladay says that the victim of SHC in a Charles Dickens novel is "Crook"—it is *Krook*; he says a Dr. Frank published in Berlin information about SHC in "1943"—it was in 1843; he says "victims were brandy-drinkers in a state of intoxication"—history clearly refute this. "There have been no cases of death from spontaneous combustion for nearly a hundred years," he concludes.

8. Jerry Blizin, reporter for the *St. Petersburg Times*, followed the Cinder Woman investigation closely: "While people may discuss the Reeser case at still greater length than has already been done, we must not overlook the tremendous amount of work poured into the baffling case by investigating officers. Police Chief Jake Reichert, Detective Lieut. Cass Burgess, Detectives R. H. Lee and Ross Boyd and others all did back-breaking work. Whether the verdict convinces arm-chair sleuths or not, City police were not lax in their pursuit of answers." Lax, no. But not solving it, either, since Reichert would not close the case.

CHAPTER 7

1. *Super* = "above; beyond; to an especially high degree."
Hyperthermic = "abnormally high fever; application of heat."
Carbonization = "to char; the formation of carbon from organic material."

2. Raymond Davis Jr. of the Brookhaven National Laboratory, attempted to capture evidence of neutrinos with an ingenious detector installed in a deep mine shaft at Lead, SD. Of the trillions of neutrons passing through his apparatus every second over two-month periods, he would find evidence for *only 30 collisions* caused by neutrino impacts (*Discover*, March 1981).

3. A light-year is the distance light travels at

its own speed in one year: six million million miles.
4. If this is so, it means the neutrino—heretofore believed to be massless—must have a tiny bit of mass, or weight. This possibility, which disturbs some physicists, has profound implications for cosmogony as well as particle physics. To quote Reines, speaking to the APS: *"If this is true, the universe is not the way we thought it was."*
5. This amount of currency could never be printed, by the way, and is about the only comparison that makes the federal deficit look inconsequential.
6. Georgi says that a distance of only 10^{-29} centimeters can be sufficient to trigger a reaction. So a shift of the energy orbits within the hydrogen atom (and its leptons and hadrons) need not be very big at all.
7. $H_2O \rightarrow H^+ + OH^-$ or $H + O_2 \rightarrow O + OH$.
8. The pyrotron-X fundamental particle can resolve a significant challenge to physics today: the problem of the universe's missing mass. It also offers an explanation for the apparent predominance of matter over antimatter.
9. The Superconducting SuperCollider (SSC) approved in 1989 for construction in the Texas desert at a cost of $5 billion was designed to produce only 4×10^4 GeV (or 40×10^{12} electron-volts).

CHAPTER 9
1. Sanderson (1972; 274) and Moffitt (1956) spell the name "Peacock."
2. The case is also found in Hugh Gaine's *New York Gazette; and the Weekly Mercury* (7 January 1771).
3. Harrison (1978; 80) misnames the coroner Dr. Floyd Clemens, incorrectly places Mr. Rooney "sitting dead in a chair" and wrongly gives Mrs. Rooney's weight as "over 190 pounds."

CHAPTER 10
1. As early as the thirteenth century, Zen schools developed techniques that emphasized the role of meditation (*dhyana*) in producing biological phenomena esoterically. Ko Ch'ang Keng, the principal proponent of Taoist-Zen alchemy, taught that meditation furnished the necessary fluid and the sparks of intelligence the necessary fire to transform the body through this contemplative alchemy. Keng, also known as Po Yu Chuan, told Waley that "By this method a gestation which normally requires ten months can be achieved in the twinkling of an eye."
   Since the body can gradually warm itself for five-score years, *might not it by the mental inducement of a too rapid thermogenesis sear itself in the twinkling of an eye?*
2. The Carmelite nun St. Maria Magdalena de' Pazzi (1566–1607) lived an austere life said to be "one long ecstasy." Along with her ecstacy came agony. Like St. Philip, she often loosened or shredded her garments because of "the excess and abundance of this celestial flame which consumed her," recorded the *Oratorian Translation* (pp. 235–237). Even during harsh Italian winters she'd run to Pazzi's well and douse herself with its water, "so great was the flame that burned in her breast that even externally she seemed to consume." Often she cried out "I can no longer bear so much love," according to her biographer Father Cepari. Also like St. Philip, her face at times glowed and her eyes "sparkled like stars" during the ecstatic incendium amoris she endured, recounts Thurston (1952; 210).
   St. Peter of Alcántara (d. 1644) echoed the curious wintertime habit of these Italian saints, for the same reason. A Spaniard, he too was often seen uncovering his chest when outdoors in winter, "so torrid were the fires that burned and broiled in his breast."
   The English religious leader George Fox (1624–1691) often walked barefoot in midwinter through the snow-covered streets of Litchfield. "The love of God" gave peace and warmed his body, he said, adding that it inspired him to found the Society of

Friends, says White (1990; 12). Superhyperthermia, tumo, incendium amoris, near-SHC: whatever its name and cause, it *can* have a constructive impact on individuals and society.

St. Benedict Joseph Labre (1748–1783) demonstrated both fire-immunity (when in ecstasy) and internal fire during his brief life. While staying at Santa Maria, Italy, the young priest was observed by colleagues to tear open "his ragged coat at the breast to assuage the fiery flame within."

St. Gemma Galgani of Lucca (1878–1903) lived a brief life of divine ecstasies and raptures that included clock-like insensitivity to pins and open fire on her skin every Thursday evening and Friday afternoon. "There generally was a forewarning," said Summers (p. 110); "a burning flame of love towards God and Our Lady [that] seemed to blaze more fiercely in her heart and consume her." Consume her it did, though not to ashes. Only twenty-five years old, she died of premature burn-out, possibly a direct consequence of her incendium amoris.

The Venerable Orsola Benincasa (1547–1618) experienced almost constant trance-ecstasies since childhood. Onlookers would pinch, prick, even cut the girl with lancets; according to Summers' *Physical Phenomena of Mysticism* (p.106), they "even went so far as to burn her with a naked flame, but all these injuries affected her not in the slightest, although when she returned to herself she keenly felt the result of such ill-judged, and indeed cruel, maltreatment." This masochism (and Orsola's willingness to receive it) reveals a state of transcendent consciousness that is immune to applied pain and to *external* fire as well.

CHAPTER 11

1. Consensus lists seven chakras, though the count varies (depending on source) from six to as many as thirteen. Six chakras lie along the cerebrospinal column: above the anus, near the genitals, behind the navel at the base of the rib cage, around the heart and the throat, and at the pineal "third eye" behind the forehead. The seventh "crown" or *sahasrara* chakra near the top of the head is variously ascribed to the pituitary, the anterior fontanelle or, according to Gopi Krishna, the entire brain itself. It is the center most associated with self-realized illumination—the culmination of Kundalini's ascent. Refer to Leadbeater's *The Chakras* (1927) for classic esoteric tradition and to Bruyere's *Wheels of Light* (1989) for modern scientific corroboration of these energy centers.

2. Down through the ages healers and spiritual teachers have produced a body of remarkably consistent testimony for the existence of an invisible, extraphysical galaxy of energies in the human body. Now that frontier-science has begun to verify this "bioenergy philosophy" instrumentally, no longer can it be scoffed at and cavalierly dismissed as primitive belief and superstition. Rather its transcultural framework *demands* that these traditions be seriously explored by science, indeed examined as credible until demonstrably proven otherwise.

3. Refer to Gopi Krishna's *Kundalini* (1971), *The Biological Basis of Religion and Genius* (1972), and *The Awakening of Kundalini* (1975a); W. Thomas Wolfe's *And the Sun Is Up* (1978; 149–173); and publications by the Kundalini Research Foundation, 10 East 39th Street, New York, NY 10016.

CHAPTER 12

1. Harrison (1976) states incorrectly that the victim "nearly died."
2. Gearhart (1975) lists an Olga Stephens dying in a mysterious fire on October 24, 1963. Being unable to corroborate this incident leads me to conclude it's a misrepresentation of Olga Worth Stephens' fate nearly a year later.
3. Citing the *Sheffield Independent*, Harrison (1976) identifies the victim as "G. A.

Sanderson"... in Harrison (1978) the name changes to Shepherdson without explanation.

CHAPTER 13
1. Norman's *Weird Unsolved Mysteries* (p. 105) dates the episode *much* earlier: 30 July 1937. Sanderson (1972; 275), however, dates it later: 30 July 1938.
2. Harrison (1978; 266) dates it 27 December 1938, referencing Russell (1955).

CHAPTER 14
1. Eg., Beck (1851; 100); Gaddis (1967; 224–225); Harrison (1978; 69–71); Sanderson (1972; 242–243); Reader's Digest (1982; 83–84).

CHAPTER 15
1. Based on medical documents in my possession, the date has to be Friday, November 15. Dating of the Angel case has always been a challenge; sources give conflicting dates, and one document even contradicts itself several times. I now believe that the span of "November 11–15" represents the correct dating.

CHAPTER 16
1. During this period a friend of Dr. Sullivan's deceased wife had a dream, which she later told to him. Sullivan recounts it: "During the 24 hours in which whatever it was happened, she had had a dream in which she and I and a third, unidentified individual were on a small boat at the edge of a mist-covered stream. I insisted on being taken across. The other refused. She tried to talk me off the boat, and eventually succeeded."

Sullivan (1982) thinks the woman in this dream was the dreamer herself and that the anonymous figure was Charon, the ferryman in Greek mythology who guided souls across the river Styx to the land of the dead.

I propose another viewpoint that in no way negates his interpretation. From a metaphysical vantage this dream accurately depicts a near death experience (NDE) by Dr. Sullivan, who argued at some level of consciousness to be released from a world he found distasteful but was instructed in the classic NDE sense that his "time for transition" from the physical was not yet to be. Being "denied" death, he returned to his strangely injured body lying on the hotel bed.

Whether his injuries occurred before this "dream" transpired or coincided with it and his reincarnating, whether the third dream individual was the mythical Charon or instead his deceased wife telling her husband *his* "time" was not yet, and why his wife's friend four hundred miles away became involved in the Sullivans' personal drama will remain unsolvable questions surrounding this case.
2. Researchers of poltergeist-fire phenomena have also noted an emotional aspect with these (usually) nonhuman spontaneous combustions, as I point out elsewhere in *ABLAZE!* and in works pending.

CHAPTER 17
1. Date based on Keyhole (1973); Stringfield (1978) places the incident on March 19, 1968.
2. The year 1966 seemed to concentrate a variety of fortean phenomena, as did 1938 and late 1904–early 1905; phenomena included several SHC episodes.
3. The U.S. Army, for example, has admitted administering LSD to unwitting American citizens in public places during the 1950s and as recently as 1964.
4. Aerial Phenomena Research Organization, of Tucson, AZ.
5. Episodes involving beams directed from UFOs at people are too numerous to catalog here. *Official UFO* (July 1976) details a variety of these incidents, as does Jacques Vallee (1990) in his important book, *Confrontations*.

I detail only one incident here:

- A recent UFO sighting, indeed dozens of them collectively referred to as the Gulf Breeze Incident, occurred to Ed and Frances Walters beginning the night of November 11, 1987. Not only did Ed, a respected contractor in the western panhandle of Florida, and his wife repeatedly see one, two or three UFOs near their home not far from Eglin Air Force Base, it turned out many of their townsfolk were seeing the same things. Ed took dozens of color polaroid photographs of the UFO(s) under varied circumstances. The case has been widely publicized—Fox-TV's *Current Affair* (March 16, 1990), *PM Magazine* (March 27, 1990), *Hard Copy* (April 27, 1990)—and described in the Walters' 1990 book, *The Gulf Breeze Sightings*.

Of special interest to this discussion is the fact that twice, on 2 December 1987 and 8 February 1988, Ed Walters *photographed* what so many UFO witnesses had been reporting for years: *a bright blue beam emanating from a UFO*. On one occasion, the beam seemed aimed directly at him. Only an eave above a doorway and his instantaneous reaction to jump aside probably prevented him from being struck and perhaps burned like so many others had been before that night.

CHAPTER 18

1. In what looks like an typesetter's error, Ogston (1870; 187) dates Vatim on February 22, 1821.
2. Ring (1984) surveyed only a *minute* percentage of the 25,000 to 50,000 suicides annually reported in the United States at that time—making for a very limited sampling base.

CHAPTER 21

1. The emphasis Euripides places on the gold raiment and crown in association with death is not inconsequential, I submit. Alchemically and mythologically, gold is noble; it is the bearer of a higher spiritual symbolism; as Mircea Eliade (1971) points out in *The Forge and the Crucible*, gold is immortality. Gold holds the power for transformation, and ultimately, represents nature in perfection. For King Creon's daughter, symbolically crowned and gowned with gold, what is intended to be punishment meted by the jilted Medea becomes noble transformation through the purification of fire. The innocence of King Creon's daughter is rewarded with immortality, whereas, as E. P. Coleridge remarks, "the abortive alliance between Jason and Medea has destroyed them both."
2. The Hebrew word is *ruach*, meaning "wind," "tempest," and/or "whirlwind," according to J. Strong's *Hebrew and Chaldee Dictionary* (1980; 107). In the context of this incident in the *Talmud*, I suggest "tempest" is the appropriate translation.
3. See, for instance, the comments of A. D. Godley, *Herodotus* (Leob Classical Library, 1921), footnote to ii:141.

CHAPTER 22

1. One marvels at (or is amused by) the irony of fate: the beginning of spring, symbolizing rebirth, brings death (via combustion) to Autumn Close.
2. Tabulation of earthquake-luminosity relationships through March 1910 by Dr. Ignazio Galli in *Bollettina della Societa Sismologica Italiana* (XIV–6) runs 184 pages! Yet not until 1964, when earthquake lights were undeniably photographed in the Orient, did most geophysicists give credence to these reports.

3. *New Scientist* (1978, 78:896) reports that Professor Gold, director of the Center for Radio Physics and Space Research at Cornell University, contends in a 1978 lecture at London's Imperial College that spark ignition of "outgassing" methane explains all episodes of atmospheric luminosity when "flames shot out of the ground." I suggest this explanation is not universally applicable.

4. This concept was formulated in 1976; published in 1977 as a three-part article for *Fortean Times* (nos. 22, 23, 24); and subsequently condensed for *Frontiers of Science* (January–February 1982).

5. Mircea Eliade, the esteemed modern religious historian, sees *lelay* as conveying the notion of fire, light, or spirit, and derived from the root *lila*, meaning cosmic creation. I see *leyne* as the most recent derivation upon this ancient root, encompassing all aspects of its root meaning into the straight tracks it defines.

6. The association of light, fire, and burning to straight Dragon Lines (*lung mei*), and legends of fire-breathing dragons slain by St. George, and the preponderance of ley place-names to St. George, suggests a deeper, hidden level to our mythologies—one that embraces physics and physical phenomena as well as philosophy and the subconscious.

7. I discovered this case in 1977 while scanning local newspapers at the Colindale Branch of the British Museum for firsthand information on the Dewar sisters SHC controversy. It appears on the *same page* with the Dewar story, yet has been overlooked by other SHC researchers, including Charles Fort. In a sense, the fire-leyne led me to it.

8. I visited the Louth town hall in 1979, seeking land deeds or other records that would indicate exactly where Clodd lived. Despite kind help from the Louth staff, that information seems no longer to exist.

CHAPTER 23

1. Elsewhere on the planet, people watched an earlier inflammatory discharge from the Earth that echoes the one proposed for Binbrook.

    This exchange (in 1900?) ensued between Keklujek and Ziaret, two peaks in the Taurus Mountains south of Harpoot, Mesopotamia. "The weapons were balls of light," the respected climatologist Ellsworth Huntington was told by resident natives. "A ball of fire is sometimes seen to start from one mountain and to go like a flash to the other." This fiery bombardment occurred day or night, but only when the sky was clear. Wrote Huntington in *Monthly Weather Review* (July 1900): "I became thoroughly convinced of its truth."

2. E.g., the blazes in the home of Adam Colwell in Brooklyn, New York, in 1895. In front of scores of witnesses, fires spontaneously erupt in furniture and wallpaper; the house eventually burns to the ground. "I do not want to be quoted as a believer in the supernatural, but I have no explanation to offer, as to the cause of the fires," the fire marshall tells the *New York Herald* (January 6, 1895). Rhoda Colwell, the family's young daughter, is above suspicion—but only for two days. Then, firemen and police are said to have been "artfully tricked" by Rhoda's alleged antics that enabled her to toss matches onto and through walls to ignite seemingly incombustible objects. There shall be *no* unsolved cases!

CHAPTER 24

1. Beginning January 6, 1922, the Alexander Macdonald farm at Caledonia Mills, Nova Scotia, Canada, likewise played host to an uninvited guest: thirty-one fires spontaneously ignited in six and a half hours, a total of thirty-eight fires by the time the visitation ended fifteen hours later. Flames that were "pale blue...not hot" erupted in wood, at the ceiling, in a *chair* that smol-

dered by itself—and in *wet paper that burned of its own volition!*

2. Small fusion microexplosions are possible by using ablation (compression) techniques similar to that used in laser fusion experiments. Yields of 0.002 kilotons were achieved in 1976, resulting in emission of superthermal x-rays, fast ions, and temperatures characteristic of "thermonuclear burn." Perhaps Mrs. Conway fell victim to a less efficient fast-ion (pyrotron?) collision than did Mrs. Reeser; that is, the lower resultant energy was inadequate to burn-up and disintegrate her as thoroughly as the St. Petersburg woman. But she came close.

CHAPTER 26

1. I investigated a series of spontaneous combustions that centered on the family of Rev. Gene Clemons in Wharncliffe, West Virginia, between May 27 and early July 1983. No human spontaneously combusted, but just about everything else did: mattresses, bedding, sofa cushions, wall outlets, trash cans, a towel rack, dolls, shoe skates, carpeting, even an artificial Christmas tree inside a cardboard box inside a locked cement-block building!

   Most people with rudimentary familiarity about such odd combustions would immediately have labeled these as "poltergeist fires." But with fine cooperation from the fire chief and Deputy Fire Chief Kendell Simpson, both of whom were mystified (and properly so) by dozens of spontaneous fires, I learned about *other* enigmatic blazes occurring in their county at the same time that also could not be readily explained.

   Plotting the locations of these baffling blazes on a map, revealed that *most of the fires fell on the perimeters of two circles*. This discovery is published here for the first time.

2. On October 5, 1989 Antony Fraser-Smith, monitoring very low frequency radio waves around Corralitos, California, noticed a sudden twenty- to thirty-fold jump in background airwaves below one Hertz; his instruments showed a gradual decline from the anomalous high spike until the afternoon of October 17, when a low frequency *airquake* (my term) two hundred times background jolted his equipment. "We *never* see signals like that," said Fraser-Smith, himself jolted by the enormity of the signal leap.

   Three hours later San Francisco was rocked by a large quake, its epicenter four miles from Corralitos. Fraser-Smith sees an association. "I look at these low frequencies as a window into Earth that hasn't been considered," he told *Discover.*

   I know he is right. And I believe that discoveries like his will give new insights to both seismic quakes and fire-quakes.

3. Roy Sullivan was said to be nature's favorite high-voltage target: Hit 1—lightning passed through his body from head to foot in 1942; Hit 2—July 1969, lightning burns off his eyebrows; Hit 3—1970, lightning left his shoulders raw and burned; Hit 4—16 April 1972, a strike sets his hair aflame; Hit 5—7 August 1973, lightning blasts a hole the size of a 50-cent piece in his hat and then, doctors confirmed, "scorched him inside and out;" Hit 6—5 June 1976, struck while cranking his car window shut, his head catches fire; Hit 7—25 June 1977, lightning interferes with his fishing by burning his chest and stomach.

4. The corollary to fires associated with sites is that certain *shapes* might also create energy patterns capable of igniting fires within a body or a house. I would have spontaneous combustion as a factor of architecture as well as of geography. Someday fire science may choose to consider further this possibility.